Major Evolutionary Radiations

Proceedings of a Symposium held in Durham, UK,
September 1989

The Systematics Association
Special Volume No. 42

Major Evolutionary Radiations

Edited by

P.D. TAYLOR
Department of Palaeontology,
The Natural History Museum,
London

and

G.P. LARWOOD
Department of Geological Sciences,
University of Durham,
Durham City

Published for the SYSTEMATICS ASSOCIATION by
CLARENDON PRESS · OXFORD
1990

Oxford University Press, Walton Street, Oxford OX2 6DP

Oxford New York Toronto
Delhi Bombay Calcutta Madras Karachi
Petaling Jaya Singapore Hong Kong Tokyo
Nairobi Dar es Salaam Cape Town
Melbourne Auckland

and associated companies in
Berlin Ibadan

Oxford is a trade mark of Oxford University Press

Published in the United States
by Oxford University Press, New York

British Library Cataloguing in Publication Data
Major evolutionary radiations.
1. Earth. Evolution
I. Taylor, P. D. (Paul David) 1953- II. Larwood, G. P. (Gilbert Powell) 1930-
III. Systematics Association IV. Series 551
ISBN 0-19-857718-4

Library of Congress Cataloging in Publication Data
(Data available)

Typset by Colset Pte Ltd, Singapore
Printed in Great Britain by
Courier International Ltd
Tiptree, Essex

Preface

The fossil record is a rich repository of data on patterns of biological origination and extinction through long periods of geological time. With due regard for the limitations and constraints imposed by an imperfect data base, palaeobiologists are now actively studying macroevolutionary patterns and the processes which may have driven them. Extinction has undoubtedly formed the main focus of this research, following the impetus given by the discovery of geochemical evidence supporting an extraterrestrial cause for the famous mass extinction at the end of the Cretaceous. In response to this interest, the Systematics Association convened a symposium on extinctions which was held at the University of Durham in September 1986. The results were published as *Extinction and survival in the fossil record* (Systematics Association Special Volume No. 34, edited by G.P. Larwood). During the extinction symposium it became clear that a complementary symposium would be welcomed dealing with 'the reverse side of the coin', that is, major evolutionary radiations. Accordingly, the symposium of which this volume is the proceedings was held at the University of Durham on 7 and 8 September 1989.

A total of 22 contributions were invited with the intention of providing as comprehensive a treatment of major radiations as possible within the limitations imposed by the format of the symposium and its proceedings. Three of the published papers deal with general aspects of radiations; the remainder focus on specific taxonomic groups which are generally well-represented in the fossil record. A late withdrawal explains the absence of a paper on mammals, regarded by many palaeobiologists as the classic example of an adaptive radiation. Additional contributions on the Cambrian 'explosion' (S. Conway Morris), corals (B.R. Rosen and J. Pandolfi), and gastropods (J.D. Taylor) were presented at this symposium.

Contributors were asked to address a series of questions relating to major evolutionary radiations. (The term 'adaptive radiation' was avoided for teleological reasons.) These questions were:

1. How important were major radiations in shaping the evolutionary history of particular taxonomic groups?

2. Do major radiations bear any relationship to times of mass extinction, or to periods of diversification in other groups?
3. How does an improved knowledge of phylogenetic relationships influence the perception of major evolutionary radiations?
4. Do major radiations show any particular palaeogeographical or facies-related patterns?
5. Which biological or geological factors may have been important in stimulating evolutionary radiations?
6. Do concepts of adaptive zones and thresholds have any relevance?

As expected, not all these issues were discussed, and those that were addressed were tackled from different perspectives, using a variety of approaches. For example, many contributors felt that adequate description of evolutionary patterns demanded cladistic analysis of phylogenetic relationships. Others did not go beyond enumerating taxonomic diversity changes through time.

Summarizing the numerous conclusions reached by the contributors to this volume is impossible in the space available. Nevertheless, it is worth noting that 'major radiations' appear to have played a role of disproportionate importance (relative to their duration) in the evolution of many biological groups. Short-lived, highly 'creative' evolutionary phases of rapid diversity increase and morphological divergence characterize the evolutionary histories of many groups.

Perhaps surprisingly, few contributors were willing to implicate changes in the physical environment (e.g. climate, sea-level) as causal factors for major radiations. Instead, evolutionary innovations in the groups concerned were more often quoted in this context. An interesting contrast therefore exists between current ideas on the causes of extinction, normally attributed to changes in the physical environment, and of major radiation.

Among several important themes raised in more than one paper are: (1) origination of higher taxa in marginal, stressed environments (e.g. nearshore marine environments, Polar latitudes); (2) macroevolutionary lags (i.e. delayed diversification following origination); and (3) possible decoupling between taxonomic diversification (numbers of new taxa) and morphological radiation (divergence in form). Future studies can be expected further to develop these themes.

We wish to thank all of the participants for helping to make this a successful symposium, and the Department of Geological Sciences and Hatfield College for providing facilities. Several referees generously gave of their time and expertise to review manuscripts. Financial assistance from the Royal Society towards the holding of the symposium is gratefully acknowledged.

London P.D.T.
Durham G.P.L.
February 1990

Contents

Contributors

M.J. BENTON
Department of Geology, University of Bristol, Wills Memorial Building, Queen's Road, Bristol BS8 1RJ, UK.

DAVID J. BOTTJER
Department of Geological Sciences, University of Southern California, Los Angeles, California 90089-0740, USA.

D.E.G. BRIGGS
Department of Geology, University of Bristol, Wills Memorial Building, Queen's Road, Bristol BS8 1RJ, UK.

E.N.K. CLARKSON
Grant Institute of Geology, University of Edinburgh, West Mains Road, Edinburgh EH9 3JW, UK.

J.A. CRAME
British Antarctic Survey, High Cross, Madingley Road, Cambridge CB3 0ET, UK.

PETER R. CRANE
Department of Geology, Field Museum of Natural History, Roosevelt Road at Lake Shore Drive, Chicago, Illinois 60605, USA.

M.S. DAVIES
School of Pure and Applied Biology, University of Wales College of Cardiff, Cardiff CF1 3YE, UK.

D. EDWARDS
Department of Geology, University of Wales College of Cardiff, Cardiff CF1 3YE, UK.

T.S. ENGESER
Geologisch-Paläontologisches Institut, Universität Hamburg, Bundesstraße 55, D2000 Hamburg 13, Germany.

R.A. FORTEY
Department of Palaeontology, Natural History Museum, Cromwell Road, London SW7 5BD, UK.

B.G. GARDINER
Biosphere Sciences, King's College London, Campden Hill Road, London W8 7AH, UK.

E.M. HARPER
Department of Earth Sciences, The Open University, Walton Hall, Milton Keynes MK7 6AA, UK.

M.B. HART
Department of Geological Sciences, Polytechnic South West, Drake Circus, Plymouth PL4 8AA, UK.

DAVID JABLONSKI
Department of the Geophysical Sciences, The University of Chicago, 5734 South Ellis Avenue, Chicago, Illinois 60637, USA.

G.P. LARWOOD
Department of Geological Sciences, University of Durham, South Road, Durham DH1 3LE, UK.

SCOTT LIDGARD
Department of Geology, Field Museum of Natural History, Roosevelt Road at Lake Shore Drive, Chicago, IL 60605, USA.

ANDREW R. MILNER
Department of Biology, Birkbeck College, Malet Street, London WC1E 7HX, UK.

CHARLES E. MITCHELL
Department of Geology, State University of New York at Buffalo, 4240 Ridge Lea Road, Buffalo, New York 14260, USA.

N.J. MORRIS
Department of Palaeontology, Natural History Museum, Cromwell Road, London SW7 5BD, UK.

R.M. OWENS
Department of Geology, National Museum of Wales, Cardiff CF1 3NP, UK.

MICHAEL J. SIMMS
Department of Geology, Trinity College, Dublin 2, Ireland.

P.W. SKELTON
Department of Earth Sciences, The Open University, Walton Hall, Milton Keynes MK7 6AA, UK.

ANDREW B. SMITH
Department of Palaeontology, Natural History Museum, Cromwell Road, London SW7 5BD, UK.

S.M. STANLEY
Department of Geological Sciences, Case Western Reserve University, Cleveland, Ohio 44106, USA.

P.D. TAYLOR
Department of Palaeontology, Natural History Museum, Cromwell Road, London SW7 5BD, UK.

ROBIN J. WOOTTON
Department of Biological Sciences, University of Exeter, Hatherly Laboratories, Prince of Wales Road, Exeter EX4 4PS, UK.

1. Adaptive radiation and macroevolution

S.M. STANLEY
Department of Geological Sciences, Case Western Reserve University, Cleveland, USA

Abstract

Many adaptive radiations have produced substantial evolutionary changes during relatively brief intervals of time. There is evidence that punctuational speciation events have played an important role in many radiations. Diversification early in adaptive radiation usually approximates exponential increase and can be shown often to have occurred because probability of speciation exceeded probability of extinction. There is a general correlation among taxa in characteristic rate of speciation and rate of extinction during adaptive radiation. This can be attributed to control of both rates by the same causal factors. A few 'supertaxa' are characterized by relatively high rates of speciation and low rates of extinction.

Most adaptive radiations can be attributed to the provision of new ecological opportunities by the operation of extrinsic agents or by the evolution of key adaptive traits. In cascading radiations, diversification of one group has promoted expansion of another and this, in turn, has fostered radiation in a third group. The late Cainozoic radiation of grasses and herbaceous plants, for example, has promoted spectacular radiations of rats and mice and songbirds that have contributed to the dramatic diversification of colubrid snakes. Some radiations may have resulted not from the appearance of new ecological opportunities but from the acquisition of traits that simply produce a propensity to speciate.

Introduction

Adaptive radiations have produced most taxa above the genus level. In the process, they have accounted for most adaptive breakthroughs in the

Major Evolutionary Radiations (ed. P. D. Taylor and G. P. Larwood), Systematics Association Special Volume No. 42, pp. 1–15, Clarendon Press, Oxford, 1990.

history of life. For this reason, it is of paramount importance that we understand the patterns of evolution that have typified adaptive radiation and the factors that have produced these patterns. Also warranting interpretation is the fact that some adaptive radiations have proceeded more rapidly than others. More fundamentally, we must explain why certain taxa have characteristically radiated more rapidly than others.

Adaptive radiation and the punctuational model of evolution

One of many ways in which Charles Darwin proclaimed himself a gradualist was in declaring his puzzlement over the sudden appearances of diverse taxa of invertebrates in the Cambrian System and of diverse taxa of angiosperms in the Cretaceous System (Darwin 1859). The fossil record reveals no long prior history for either group; for years, palaeontologists have searched in vain for an extended earlier record of expansion for both groups. In fact, we now recognize that the angiosperms underwent a spectacular diversification during the Cretaceous (Doyle and Hickey 1976; Lidgard and Crane 1988) and that marine invertebrates expanded even more dramatically in latest Precambrian and Cambrian time (review by Glaessner 1984).

Even as late as 1978, Durham concluded that a vast interval of Precambrian evolution was required to produce distinctive Cambrian echinoderm taxa because an average Cainozoic echinoid species survived for about 6 Ma and similar durations should have characterized early species of the same phylum. Durham's reasoning was that in order to reconstruct the early phylogeny of the group one would have to align end-to-end large numbers of species, each changing gradually into the next, to produce a distinctive class of Cambrian echinoderms from some morphologically distant ancestor. This appraisal unwittingly amounts to what I have termed the test of adaptive radiation (Stanley 1975*a*, 1979). This example of the test opposes the gradualistic model of evolution. It is now evident that the Proterozoic fossil record is of high enough quality to reveal that there is no possibility that echinoderm evolution spanned hundreds of millions of years of Precambrian time. We must invoke episodes of rapid evolutionary change to produce distinctive Cambrian taxa.

Closer to the present day, data for mammals also constitute a failure for gradualism. During the Palaeocene and early Eocene adaptive radiation of mammals, an average species in North America survived for between 1 and 2 Ma without evolving enough to deserve a new name. Nonetheless, hundreds of new genera and many new families evolved during this interval. There appears to be no recourse but to rely on episodes of sudden evolutionary change to account for the diverse

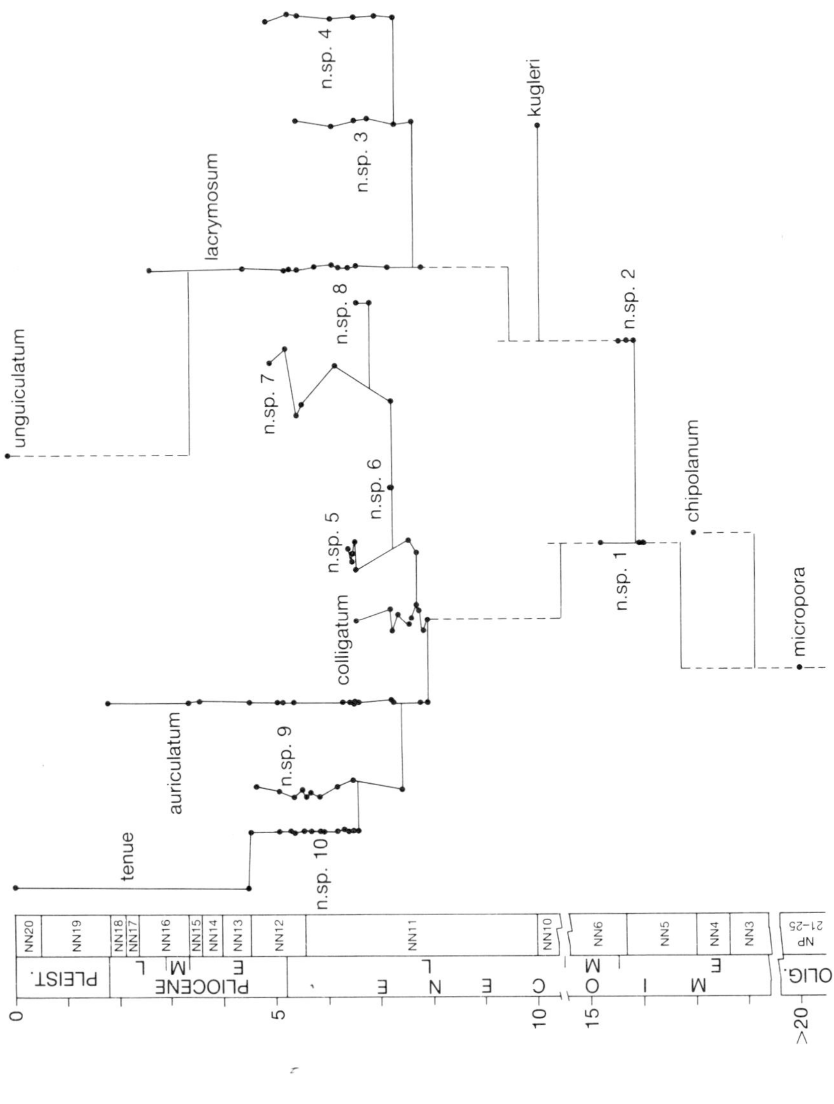

Fig. 1.1. Stratophenetic tree for the Caribbean bryozoan genus *Metrarhabdotos* (Cheetham 1986). Morphological distances (horizontal axes) between ancestral and descendant species are Euclidean distances based on discriminate analysis for 46 morphological characters. Fluctuations of morphology within species are scaled to distances between species. Note that most new species must have arisen by rapidly divergent speciation because, without branching, an ancestral species can be transformed into only one descendant species.

group of mammalian taxa that existed in early Eocene time.

Of course, even when approximate stasis is the norm, sudden change may be concentrated in rapid phyletic shifts without speciation: a staircase pattern may prevail. Where the fossil record of phylogeny is exceptionally complete, adaptive radiation offers a way of testing the staircase pattern against the phylogenetic pattern in which rapidly divergent speciation prevails. The critical fact here is that a species that appears early in adaptive radiation can be transformed rapidly, without speciation, into only one younger species. Other distinctive younger species can have arisen only by speciation. The elegant analysis by Cheetham (1986) of evolution within the bryozoan genus *Metrarhabdotos* in the Dominican Republic illustrates this argument. Here, individual species, assessed by comprehensive multivariate morphometrics, are remarkably stable (Fig. 1.1). New species appear in the record abruptly and undergo no statistically significant morphologic change. The fact that the clade was diversifying during the interval that Cheetham considered means that most of the distinctive new species could only have arisen by way of rapidly divergent speciation.

The mathematics of radiation

Rates of adaptive radiation in which the units are species or higher taxa constitute taxonomic frequency rates. These rates are not properly assessed in terms of arithmetic increase (Simpson 1944, 1953). Instead, when evaluated as instantaneous rates, they represent exponential increase, which is to say fractional change per unit time (Yule 1924; Stanley 1975*a*, 1979). Even measured over millions of years, these rates will approximate exponential change if adaptive radiation is not strongly braked by intrinsic or extrinsic factors. It turns out, for example, that a generally linear trend emerges when one plots Recent species richness for currently radiating clades of bivalve and gastropod molluscs on a logarithmic scale against interval of radiation. There is a rather well-defined maximum exponential rate of adaptive radiation for these groups. In comparison, balanoid barnacles have radiated much more rapidly (Fig. 1.2). This kind of contrast illustrates the utility of approximating rates of increase as exponential phenomena. Some intrinsic biologic difference must explain the discrepancy. The rapid rate of adaptive radiation for balanoid barnacles probably reflects the fact that internal fertilization characterizes this group, promoting the reproductive isolation of populations.

One might question this conclusion, claiming that such differences could be stochastic accidents. Raup *et al.* (1973) suggested that many large radiations documented by fossil data might have resulted from

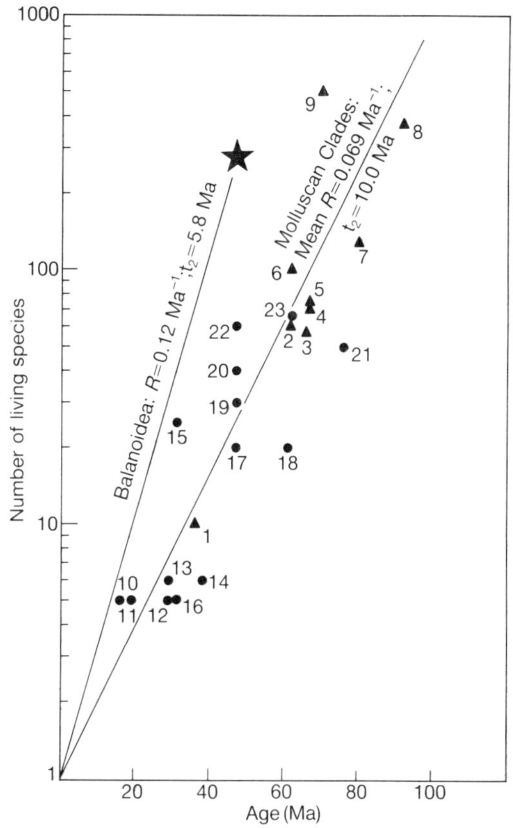

Fig. 1.2. Net rates of diversification for rapidly radiating clades of extant molluscs (circles), compared to the rate for the extant barnacle family Balanoidea(*). The molluscan taxa yield a mean rate of radiation (R) of 0.069 Ma^{-1}. The balanoid barnacles have been radiating more rapidly than any large molluscan clade ($R = 0.12\ Ma^{-1}$). From Stanley and Newman (1980).

chance fluctuations in rates of branching and extinction. When species diversity and rates of speciation and extinction are scaled realistically, however, it can be shown that important radiations of the past could not have been accidental (Stanley *et al.* 1981). Stochastic simulations for post-Cambrian orders of brachiopods, for example, reveal that chance fluctuations in realistically scaled rates of speciation and extinction would have produced only minor changes in species diversity.

Chance factors have only weak potential to account for diversification even on a much smaller scale. For example, analytic calculations reveal that there is less than a 1 per cent probability that a bivalve family nucleated from one of an estimated 10^5 speciation events for Early Cretaceous marine bivalves would by chance have produced a clade containing as many as 56 species today (Stanley *et al.* 1981). 'By chance' here means that probability of speciation equalled probability of extinction.

The Veneridae, which contain about 500 living species, and the Tellinidae, which contain about 350 living species, are families of siphonate burrowing bivalves that originated during Early Cretaceous time. Clearly, in both of these families probability of speciation has exceeded probability of extinction. Mantle fusion appears to have been the key to success for both of these taxa (Stanley 1968, 1986*b*). Interestingly, the siphonate families have actually experienced lower rates of speciation than have non-siphonate clades of burrowing Bivalvia (Fig. 1.3); the success of the siphonate clades has resulted from extremely low rates of extinction, which apparently have resulted from avoidance of predation by rapid burrowing and maintenance of deep life positions in the sediment. Siphonate species, on average, have larger

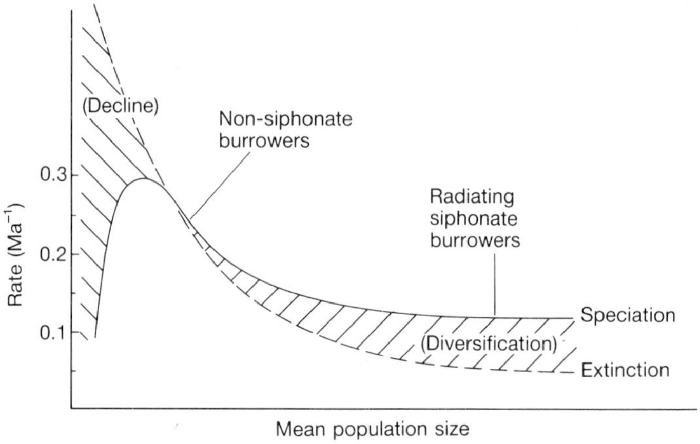

Fig. 1.3. Graph depicting differences in rates of speciation and extinction for extant taxa of siphonate and non-siphonate burrowing bivalves. For radiating clades of siphonate burrowers, typified by populous species, rate of speciation (S) is relatively low, but rate of extinction (E) is even lower so that there is a high rate of adaptive radiation ($R = S - E = 0.069\ \mathrm{Ma}^{-1}$). A typical clade of non-siphonate burrowers, on the other hand, experiences little net diversification or decline because the typically small size and instability of the species' populations yield high rates of extinction but also high rates of speciation. The taxon in which mean population size becomes diminished to a very low level suffers a decline in diversity because rate of extinction continues to rise and new species rarely become established; speciation is favoured by the frequent fragmentation of populations. From Stanley (1986*b*).

populations than non-siphonate species (Stanley 1986*b*). This example illustrates an important principle. Adaptive radiation need not result from unusually high rates of speciation for a higher taxon; it can also result from abnormally low rates of extinction.

Controls of rates

One way to identify biological variables that influence rates of adaptive radiation is to establish correlations between these variables and measured rates. Testimony also comes from another general correlation: the fossil record reveals that taxa that experience relatively high rates of speciation during adaptive radiation also suffer relatively high rates of extinction (Stanley 1979, p. 231; in press). This relationship implies causal linkage, which constrains our assessment of controls for rates of speciation. Factors that elevate rates of speciation must also tend to elevate rates of extinction. (Note that rates of extinction are not dictating rates of speciation here, by making way for new species, because rate of speciation exceeds rate of extinction during adaptive radiation.)

I have concluded that rates of extinction and rates of speciation for radiating taxa are fortuitously linked by multiple causal factors. Both rates tend to increase with levels of stereotypic behaviour, for example. Clearly change in complex behaviour represents an important isolating mechanism. On the other hand, complex stereotypic behaviour in general constitutes a narrow dimension of niche breadth that renders a species vulnerable to environmental change and at greater risk of extinction than a less specialized species. Similarly, rate of extinction tends to increase with a decrease in overall niche breadth, but so does rate of speciation, apparently because ecological specialists tend to occupy patchy, unstable habitats (Slobodkin and Sanders 1969). Rates of both speciation and extinction also tend to be inversely related to the size and stability of populations that constitute species (Stanley 1986*b*). It appears that the effect of dispersal ability is more complex. For species whose populations are patchy and unstable, effective dispersal favours speciation by generating spatial isolates, but, because of their instability, such species also exhibit relatively high rates of extinction. On the other hand, for species whose populations are stable and continuously distributed, effective dispersal suppresses probabilities of both speciation and extinction. It produces large, widespread populations that can survive regional crises, and yet, at the same time, it reduces the probability of isolate formation.

By definition, species selection strains the linkage I have described, tending to elevate rates of speciation and depress rates of extinction. A few taxa stand out as exceptional, in the sense that their very high rates of speciation during adaptive radiation are not matched by exceptionally high rates of extinction. The resulting net rate of radiation is markedly high for such groups, which I have labelled supertaxa (Stanley 1979, p. 281). The murid rodents represent an example. Their extraordinarily high rate of speciation during late Cainozoic time appears to result from traits associated with their small body size. Their weak dispersal promotes speciation, yet their enormous populations offset the effects of narrow distribution so that their rates of extinction are relatively low. Cheilostome Bryozoa with short-lived larvae may also qualify as supertaxa. Their patchy distribution promotes speciation, yet the effective dispersal of their colonies by rafting makes their populations large, widespread, and resistant to extinction (Taylor 1988). The key point here is that dispersal by rafting, while in the long run effective in establishing broad distribution, is too haphazard to maintain strong gene flow among all populations of a species.

Triggers of radiation

Simpson (1953, p. 207) noted that, in order to enter a new adaptive zone and experience adaptive radiation, a taxon must have physical, evolutionary, and ecological access to the zone. Physical access may arise through migration to an otherwise available habitat or through the origin of a new habitat within the geographical range of the taxon. Because islands and lakes are habitats with discrete boundaries, their origins have yielded the most dramatic experiments of radiation within newly formed habitats. Among such radiations are those of cichlid fishes in extant African lakes (Fryer and Iles 1972; Greenwood 1974), and those of more primitive fishes in rift valley lakes represented by strata in the lower Mesozoic of eastern North America that were deposited during the early stages of the rifting that formed the modern Atlantic Ocean (McCune *et al.* 1984). Certainly the most famous insular radiations are those of the Galapagos Islands that contributed prominently to Charles Darwin's conception of organic evolution.

Simpson emphasized the role of biotic competition in obstructing ecological access to a new adaptive zone, but intense predation can impede access as well. Almost certainly both factors contributed to the dinosaurs' suppression of the mammals during the Mesozoic Era.

Simpson also emphasized the key role of adaptive breakthroughs in providing evolutionary access to new adaptive zones. The identification of such breakthroughs represents one of the most valuable applications of functional morphology to the fossil record. Explanations in this area run the risk of appearing to be *ad hoc* and untestable. They are most convincing if we can test their predictions in the modern world. Thus, I feel secure in the main conclusion of my first article in evolutionary palaeontology—the contention that the adaptive radiation of burrowing bivalves from late Palaeozoic time to the Recent has resulted from the evolution of mantle fusion and siphon formation (Stanley 1968). As noted earlier, observations on living species demonstrate that these traits of siphonate taxa provide for more rapid and deeper burrowing. Furthermore, predation is the most important limiting factor for populations of marine bivalves in the modern world, and we can easily see that siphonate taxa are disproportionately well-represented among the most abundant species in modern seas.

Similarly, ecological patterns in the Recent bolster the contention that the evolution of a tubiferous wall structure triggered the adaptive radiation of balanoid barnacles since early Eocene time (Fig. 1.2; Stanley and Newman 1980). By virtue of this skeletal trait, balanoids competitively displaced chthamaloid barnacles, which represent a more ancient clade, by growing more rapidly and attaching more firmly to hard substrata. Balanoids characteristically overgrow and undercut chthamaloids on hard substrata (Connell 1961).

There is also no doubt that evolutionary events in one taxon have sometimes triggered adaptive radiation in another. The predicted pattern here is an extended early evolutionary history of non-radiation or weak radiation for a group followed by an interval of rapid radiation when the beneficial group evolved or diversified. Lucinid bivalves may exemplify this effect. They evolved during mid-Palaeozoic time, but then existed inconspicuously at low diversity for about 300 Ma. Today lucinids are quite diverse in shallow tropical seas, where they thrive especially in substrata stabilized by submarine grass. Probably, modern lucinids owe their success in stable grassy habitats to the fact that, uniquely among bivalves, they and other members of the Lucinacae feed with the use of a fragile mucous tube. Notably, the diversification of the lucinids began in the Cretaceous Period, when marine grasses evolved (Voigt 1981). I doubt that the temporal coincidence was accidental.

Synergistic and cascading radiations

Ecological opportunities, adaptive breakthroughs, and beneficial biotic interactions are all important triggers of adaptive radiation, but there are also more complex patterns of interaction that have fostered radiation on a larger scale. These include synergistic interactions, in which two taxa promote each other's diversification, and cascading radiations, in which the radiation of one group sets in motion a chain reaction of radiations of benefiting taxa.

Perhaps the most clear-cut example of synergistic radiation is the simultaneous diversification since mid-Cretaceous time of angiosperms and their pollinators (chiefly insects, but also such taxa as hummingbirds and fruit bats). Species-specific pollinators provide a special mechanism for plant speciation because pollinator populations that exhibit aberrant behaviour provide for reproductive isolation of plant populations. Diversification of pollinated plant species, in turn, provides additional niches for pollinators.

What I propose to label cascading radiation represents a different mechanism of large-scale biotic diversification. Here, the pattern is a linear chain reaction involving at least three steps: the radiation of one group promotes the radiation of another and this in turn promotes the radiation of a third group. Illustrating this phenomenon is the late Cainozoic history of terrestrial vegetation, seed-eating herbivores, and snakes. For purposes of explanation, it is useful to begin a discussion of this example high in the food web.

Until the Miocene Epoch, snakes were a conservative reptilian group, represented by species allied to modern boas and pythons. Beginning about 13 Ma ago, however, the family Colubridae began an adaptive radiation that has produced most species of snakes in the modern world—about 1400. It seems evident that the colubrids' spectacular success has stemmed from their special ability to enter the holes of rodents, to traverse narrow limbs in search of birds' nests, and to penetrate aquatic habitats (Stanley 1979, p. 230). Rats and mice and songbirds have themselves been radiating spectacularly during the snakes' radiation, and it appears that the snakes' history has been tracking their expansion (Fig. 1.4). Adaptive radiation of ranid frogs may also have contributed to the snakes' success, but the fossil record of frogs is too poor to document when the great expansion of ranids began. There is no unequivocal record of the Ranidae prior to the middle Eocene (Carroll 1988).

The success of the rodents and songbirds themselves has resulted in large part from the adaptive radiation of seed-producing grasses and herbaceous plants since Eocene time (Fig. 1.4). The plants' success brings us to the abiotic cause of the chain of evolutionary events (Stanley 1986*a*, p. 564). Climates throughout the world have become cooler and drier since the Eocene, promoting the spatial and taxonomic expansion of grasses and herbs as forests have shrunk. Certainly beneficial biotic interactions other than trophic relationships have the potential to cause cascading radiations, but the plant–herbivore–carnivore pattern that I have described has perhaps been the most likely kind of cascading radiation to develop in the history of life.

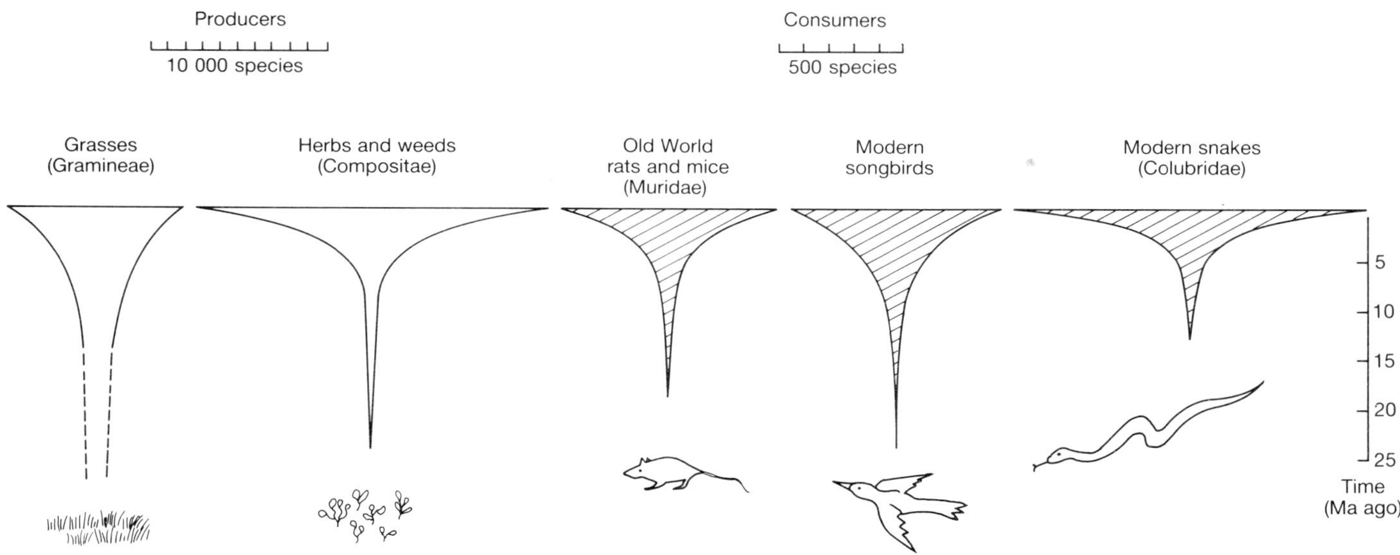

Fig. 1.4. Cascading radiation for terrestrial taxa of the Neogene Period. This example illustrates propagation of radiation upward from the base of the food web. Radiations of grasses and herbs, fostered by cooling and drying of climates, have supported radiations of murid rodents and songbirds, and these, in turn, have led to the radiation of the dominant modern family of predatory snakes (Colubridae). Note that none of these highly diverse living groups except for the grasses originated appreciably before the beginning of the Neogene.

Radiation without adaptive superiority

Only recently has it become apparent that some radiations have resulted not from beneficial ecological traits but simply from the evolution of a propensity to speciate (Stanley 1975*b*, 1979, p. 205; Vrba 1980). The evolution of non-planktonic development in volutid gastropods (Hansen 1978) may have been such an event, and Taylor (1988) has developed a strong case that the evolution of a non-planktotrophic larva triggered the adaptive radiation of cheilostome bryozoans in mid-Cretaceous time by promoting allopatric speciation without providing ecological benefits. Presumably, non-planktotrophic larvae at least sometimes also confer adaptive advantages at the level of the individual, but this complex issue remains controversial (Strathmann 1980). The evolution of animal pollination for angiosperm plants (Stanley 1979, p. 281) presumably accelerated the angiosperms' rate of adaptive radiation, although angiosperms appear to have adaptive advantages as well (Doyle 1977).

The most pervasive biological trait promoting diversification is sexual reproduction (Stanley 1975*b*). Asexual taxa diversify slowly because they lack genetic recombination equivalent to that of sexual taxa. Because sexual recombination plays a major role in divergent speciation, sexual clades expand in total niche breadth more rapidly than clones and are therefore less vulnerable to extinction (Fig. 1.5). In other words, asexual clones are feeble at radiating, and this may be the

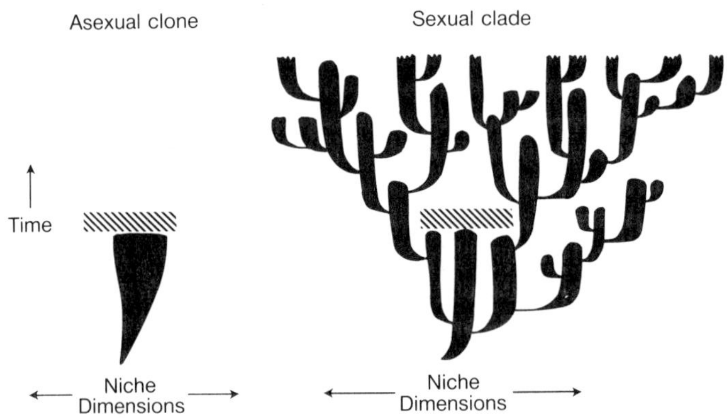

Fig. 1.5. Hypothetical phylogenies depicting the early extinction of an asexual clone compared to the survival and continued diversification of a sexual clade. The clade persists because it is composed of numerous species, each of which has diverged rapidly from its ancestral species. An extinction event (hatched bar) eliminates populations only within a narrow range of niche dimensions; yet the entire clone suffers extinction. From Stanley (1975*b*).

primary reason that, while they frequently crop up in phylogeny (Stebbins 1950, p. 417), they typically display narrow adaptive zones and are geologically short-lived.

Pattern of radiation

Owing to the imperfection of the fossil record, we cannot reconstruct the detailed pattern of most adaptive radiations. Of interest, however, are changes in rates of both speciation and extinction. We can predict that mean longevity of species within a clade (the reciprocal of mean rate of extinction) should often increase with time. This is one component of species selection. If longevity is a partially heritable trait for species, influenced by ecological characteristics, then long-lived species should tend to accumulate (Slatkin 1981). Viewing the situation from the opposite perspective, we can expect to see a progressive dwindling of types of species that evolve in restricted environmental contexts but ultimately prove to be poorly adapted: they should die out quickly and therefore leave relatively few descendant species or none at all. This pattern amounts to a kind of evolutionary 'experimentation'. We appear to see such a pattern in the successive adaptive radiations of Cambrian trilobites, although the data assembled to date are incomplete (Stitt 1977).

The typical configuration of a plot of species diversity against time for adaptive radiation is also unresolved. I have approximated radiation as an exponential phenomenon, recognizing that this represents a simplification—but one that allows us to identify important contrasts between taxa (see Fig. 1.2). Walker (1985) investigated other approximations, including a logistic curve and a circumferential model, in which species are added at the margin of a growing adaptive zone in multidimensional niche space. I am dubious that such particularized models will have general utility. Ecologists have employed the logistic equation to describe population growth because it is mathematically simple, not because it necessarily describes reality. The basic assumption is that the impact of crowding within a population increases linearly in proportion to species richness. Overlooked in 'crowding' models are three facts. First, intensity of interspecific competition varies enormously among taxa; it is relatively weak for a group such as the bivalve molluscs, upon which predation is intense and among which behavioural interactions are virtually absent and niche overlaps enormous (Stanley 1973, 1977, 1986). Second, the course of radiation in one taxon must often be strongly affected by competition and predation from other taxa; both of these factors can be expected to vary through time in unpredictable ways. Third, not all species are sympatric; allopatric species can occupy identical niches regardless of the intensity of interspecific competition between sympatric species; species richness within a higher taxon should vary with total geographic range and total geographic range may fluctuate greatly through time.

Not only ecological interactions but also morphogenetic constraints brake rates of adaptive radiation. Buss (1988) has shown that total Recent species diversity within metazoan phyla is strongly correlated with the ontogenetic timing of germ-line determination. Early determination appears to restrict diversification. This and other forms of morphogenetic canalization must often stifle adaptive radiation, but the details of their operation remain to be explored.

References

Buss, L. W. (1988). Diversification and germ-line determination. *Paleobiology*, **14**, 313–21.

Carroll, R. L. (1988). *Vertebrate paleontology and evolution*. W. H. Freeman, New York.

Cheetham, A. H. (1986). Tempo of evolution in a Neogene bryozoan: rates of evolution within and across species boundaries. *Paleobiology*, **12**, 190–202.

Connell, J. (1961). The influence of interspecific competition and other factors on the distribution of the barnacle *Chthamalus stellatus*. *Ecology*, **42**, 710–23.

Darwin, C. R. (1859). *On the origin of species*. John Murray, London.

Doyle, J. A. (1977). Patterns of evolution in early angiosperms. In *Patterns of evolution as illustrated by the fossil record* (ed. A. Hallam), pp. 501–46. Elsevier, Amsterdam.

Doyle, J. A. and Hickey, L. J. (1976). Pollen and leaves from the mid-Cretaceous Potomac Group and their bearing on early angiosperm evolution. In *Origin and early evolution of angiosperms* (ed. C. G. Beck), pp. 139–206. Columbia University Press, New York.

Durham, J. W. (1978). The probable metazoan biota of the Precambrian as indicated by the subsequent record. *Annual Reviews of Earth and Planetary Sciences*, **6**, 21–42.

Fryer, G. and Iles, T. D. (1972). *The cichlid fishes of the great lakes of Africa*. Oliver and Boyd, Edinburgh.

Glaessner, M. F. (1984). *The dawn of animal life: a biohistorical study*. Cambridge University Press, Cambridge.

Greenwood, P. H. (1974). Cichlid fishes of Lake Victoria, East Africa: the biology and evolution of a species flock. *Bulletin of the British Museum (Natural History), Zoology Series*, **12**, 315–57.

Hansen, T. A. (1978). Larval dispersal and species longevity in lower Tertiary gastropods. *Science*, **199**, 885–7.

Lidgard, S. and Crane, P. R. (1988). Quantitative analyses of the early angiosperm radiation. *Nature*, **331**, 344–6.

McCune, A.R., Thomson, K.S. and Olsen, P.E. (1984). Semionotid fishes from the Mesozoic great lakes of North America. In *Evolution of fish species flocks* (ed. A.A. Echelle and I. Kornfield), pp. 27-44. University of Maine, Orono.
Raup, D.M., Gould, S.J., Schopf, T.J.M. and Simberloff, D.A. (1973). Stochastic models of phylogeny and the evolution of diversity. *Journal of Geology*, **81**, 525-42.
Simpson, G.G. (1944). *Tempo and mode in evolution*. Columbia University Press, New York.
Simpson, G.G. (1953). *The major features of evolution*. Columbia University Press, New York.
Slatkin, M. (1981). A diffusion model of species selection. *Paleobiology*, **7**, 421-5.
Slobodkin, L.G. and Sanders, H.L. (1969). On the contribution of environmental predictability to species diversity. *Brookhaven Symposia in Biology*, **22**, 82-95.
Stanley, S.M. (1968). Post-Paleozoic adaptive radiation of infaunal bivalve molluscs—a consequence of mantle fusion and siphon formation. *Journal of Paleontology*, **42**, 214-29.
Stanley, S.M. (1973). Effects of competition on rates of evolution, with special reference to bivalve mollusks and mammals. *Systematic Zoology*, **22**, 486-506.
Stanley, S.M. (1975*a*). A theory of evolution above the species level. *Proceedings of the National Academy of Sciences, USA*, **72**, 646-50.
Stanley, S.M. (1975*b*). Clades versus clones in evolution: why we have sex. *Science*, **190**, 381-3.
Stanley, S.M. (1977). Trends, rates, and patterns of evolution in the Bivalvia. In *Patterns of evolution as illustrated by the fossil record* (ed. A. Hallam), pp. 209-50. Elsevier, Amsterdam.
Stanley, S.M. (1979). *Macroevolution: pattern and process*. W.H. Freeman, San Francisco.
Stanley, S.M. (1986*a*). *Earth and life through time*. W.H. Freeman, New York.
Stanley, S.M. (1986*b*). Population size, extinction, and speciation: the fission effect in Neogene Bivalvia. *Paleobiology*, **12**, 89-110.
Stanley, S.M. (in press). The general correlation between rate of speciation and rate of extinction: fortuitous causal linkages. In *Causes of evolution: a paleontological perspective* (ed. R.M. Ross and W.D. Allmon). University of Chicago Press, Chicago.
Stanley, S.M. and Newman, W.A. (1980). Competitive exclusion in evolutionary time: The case of the acorn barnacles. *Paleobiology*, **6**, 173-83.
Stanley, S.M., Signor, P.W., Lidgard, S. and Karr, A.F. (1981). Natural clades differ from random clades: Simulations and analyses. *Paleobiology*, **7**, 115-27.
Stebbins, G.L. (1950). *Variation and evolution in plants*. Columbia University Press, New York.
Stitt, J.H. (1977). Late Cambrian and earliest Ordovician trilobites, Wichita Mountains area, Oklahoma. *Oklahoma Geological Survey Bulletin*, **124**, 1-79.

Strathmann, R. R. (1980). Why does a larva swim so long? *Paleobiology*, **6**, 373–6.

Taylor, P. D. (1988). Major radiation of cheilostome bryozoans: triggered by the evolution of a new larval type? *Historical Biology*, **1**, 45–64.

Voigt, E. (1981). Upper Cretaceous bryozoan–seagrass association in the Maastrichtian of The Netherlands. In *Recent and fossil Bryozoa* (ed. G. P. Larwood and C. Nielson), pp. 281–98. Olsen and Olsen, Fredensborg, Denmark.

Vrba, E. S. (1980). Evolution, species and fossils: how does life evolve? *South African Journal of Science*, **76**, 61–84.

Walker, T. D. (1985). Diversification functions and the rate of taxonomic evolution. In *Phanerozoic diversity patterns* (ed. J. W. Valentine), pp. 311–34. Princeton University Press, Princeton, New Jersey.

Yule, G. U. (1924). A mathematical theory of evolution, based on the conclusions of Dr. J. C. Willis, F. R. S. *Philosophical Transactions of the Royal Society of London*, **B 213**, 21–87.

2. The origin and diversification of major groups: environmental patterns and macroevolutionary lags

DAVID JABLONSKI
Department of the Geophysical Sciences, University of Chicago, Chicago, USA
DAVID J. BOTTJER
Department of Geological Sciences, University of Southern California, Los Angeles, USA

Abstract

The fossil record contains a wealth of information on the origin and radiation of higher taxa, and on the evolutionary dynamics of lower taxa within them, but these have been relatively neglected from an ecological perspective. We have analysed the environment of first appearance for 41 major groups of benthic marine invertebrates that originated in post-Palaeozoic times (including all currently recognized orders plus the echinoid superorder Microstomata). Of the 25 taxa with relatively good preservation potential, 18 first appear in onshore environments (i.e. settings above mean storm wave base). This pattern contrasts significantly with the more even environmental distribution of first occurrences for the 16 post-Palaeozoic orders with poor preservation potential (e.g. stoloniferan cnidarians, echinothurioid echinoids, and the asteroids). This difference between the two preservation categories suggests that the preferential onshore pattern of first occurrences seen for the novelties marking the well-preserved orders is not an artefact of sampling or preservation. In contrast to ordinal origination, genera and families within the well-preserved groups show no strong onshore bias, but instead appear to be constrained by the bathymetric diversity gradient of their clade: environments richest in species, genera, and families are generally the sites of more origination at those levels. A number of

Major Evolutionary Radiations (ed. P. D. Taylor and G. P. Larwood), Systematics Association Special Volume No. 42, pp. 17–57, Clarendon Press, Oxford, 1990.

different kinds of mechanisms might produce the observed patterns: the patterns could be: (1) strictly artefact; (2) ecologically constrained (e.g. echinoids in early Mesozoic); (3) externally determined (e.g. by mass extinctions); (4) internally determined by an origination bias or an extinction bias.

In contrast to many early Palaeozoic orders, these post-Palaeozoic groups rarely show rapid radiation immediately upon origination, although episodes of sudden diversification may occur later in their histories. Such *macroevolutionary lags*, in which radiation is preceded by a long interval at low diversity, may be the result of several very different processes that range from release following extinction of a pre-emptive competitor, to accelerated speciation rates owing not to adaptive breakthroughs but to changes in rates of gene flow. These delayed radiations post-date, rather than coincide with, expansions across the continental shelf for at least some groups. Ecological patterns of diversification thus reflect a complex interplay of processes at a number of hierarchical levels.

Introduction

The origin and radiation of a higher taxon cannot be divorced from the ecology of its constituents. The first species bearing the group-defining novelty (or novelties) must persist over ecological time-scales, subject to the vagaries of its physical and biotic environment, before any diversification can unfold over geological time. The later persistence and multiplication of species, the quantitative or demographic aspect of a radiation, will be influenced by a host of organismic, populational, and species-level traits, and here too diversification should be viewed within an ecological context. Ecological questions become even more pressing if we consider the qualitative or functional component of radiations, i.e. divergence, morphological or otherwise, from the ancestral design.

Here we survey environmental patterns of origination and diversification in the skeletonized marine invertebrate orders that have appeared since the beginning of the Mesozoic. This new analysis corroborates and extends our earlier studies (on more limited and preliminary data sets) by showing: (1) a general tendency for major groups to appear first in onshore settings; and (2) that this environmental bias is evident in groups with good preservation potential but not in those with poor preservation potential. The contrast in first occurrences between the preservation classes suggests that the environmental pattern among the well-preserved groups has biological significance, and that the absence of pervasive environmental patterns at lower taxonomic levels within those groups is an evolutionary discordance that deserves further investigation. Finally, we note another kind of evolutionary discordance, between the environmental expansion of a group and its taxonomic diversification, and discuss this in the context of *macroevolutionary lags*, or delays between first appearances and diversification of higher taxa.

Environmental histories of higher taxa

In order to assess the environment of first appearance and later ecological histories of major groups, we have developed the following protocols. First, we need well-defined higher taxa. These are not always readily available, and we have relied on the published literature in combination with the advice of a number of experts on the groups involved. We have attempted to use strictly monophyletic groups in most instances (e.g. the bryozoan order Cheilostomata) but, because we are also interested in functional or ecological units, we have retained certain paraphyletic groups that seem useful. For example, we are interested in the first occurrence and fate of the stalked crinoid order Isocrinida, even though it is rendered paraphyletic by separation of the stalkless Comatulida (see Bambach 1985 for a survey and review of the adaptive significance of higher taxa). We do not imply that higher taxa are ecological units in any strict sense (see Eldredge and Salthe 1984; Eldredge 1989 on the distinction between genealogical and ecological hierarchies). But we do suggest that the traits that delimit higher taxa can have functional significance and can tell us something about the life habits of members of that taxon taken as a group or viewed as a genealogical unit. As Eldredge and Salthe (1984, p. 201) argue, 'In general, the genealogical hierarchy affects the ecological hierarchy by supplying the interacting players of the ecological game.'

The work of Kier (1974) and Smith (1981, 1984, 1988*a*) on echinoid phylogeny and functional morphology permits us to examine this class more directly in terms of functionally significant novelties and the monophyletic groups they define, and we note the relationship between the appearance of these novelties and the traditional ordinal taxonomy. We include in our analysis the first occurrence of the irregular echinoid superorder Microstomata (a monophyletic clade whose oldest representative combines several traits associated with the shift towards an infaunal deposit-feeding habit) as well as the first occurrences of the traditional orders within that clade.

A second prerequisite to analysis of the ecological history of major groups or adaptations is an independent environmental framework. We use a simple five-partition model for marine environments, ranging from nearshore through inner, middle, and outer shelf to slope/deep basin, all recognized on the basis of physical sedimentary and stratigraphic criteria (see Bottjer and Jablonski 1988; Jablonski and Bottjer

1990*a*). This palaeoenvironmental scheme is based on physical energy levels and not on the geographical or topographical position of environments, so that a barrier bar is classified as nearshore and a lagoon as inner shelf. These subdivisions correspond closely to Sepkoski's (1988) categories, with our inner shelf including both his 'protected subtidal' and 'offshore wave-agitated' categories. Our nearshore and inner shelf categories can be taken as the suite of onshore, more frequently disturbed habitats, as distinguished from our three more offshore, less frequently disturbed settings.

We have also developed a method to assess negative data, i.e. absences of a particular group from a given set of time–environment coordinates. For this purpose we have designated *taphonomic control taxa* (Bottjer and Jablonski 1988) that approximate the preservational tendencies of a given focal group. For example, cyclostome bryozoans serve as the control for cheilostome bryozoans, and the absence of cheilostomes is only taken as meaningful for those localities yielding cyclostomes.

Taphonomic controls alone do not provide adequate confidence limits for comparisons of first occurrences over the broad range of skeletal types considered here. The flimsy, cap-like internal aragonitic shells of the gastropod order Anaspidea could not have produced as reliable a fossil record as the robust calcitic tests of spatangoid echinoids, or even one comparable to the sturdier aragonitic shells of most neogastropods. In order to take these differences into account, we have grouped the post-Palaeozoic orders into two preservation categories, according to our assessment of the potential for recovery of identifiable skeletons or skeletal elements. We have been conservative in assigning orders to the poor preservation category, thereby minimizing the number of taxa whose first appearances in the fossil record are taken to be controlled more by preservational factors than by evolutionary ones. Additional preservational categories might be added to our binary scheme, for example by factoring in intensity of study: the hydrozoan order Stylasterina builds a robust skeleton but has enjoyed little palaeontological attention and thus may have a less reliable *known* fossil record than the more intensively studied anthozoan coral order Scleractinia. However, such elaborations become increasingly subjective and further subdivide an already limited sample; for the present analysis we use only two preservation classes.

Ordinal patterns

Based on our environmental framework, taphonomic control taxa, and assignment of groups to preservation-potential classes, we have addressed the question: are there ecological patterns in the origins of major groups in the post-Palaeozoic? As documented in the Appendix to this chapter, 40 skeletonized benthic marine orders first appear in the post-Palaeozoic (plus the superorder Microstomata). These data supersede our earlier, preliminary work on echinoderm orders (Jablonski and Bottjer 1988).

The crinoid clade Bourgueticrinida is of uncertain affinity. As Simms (1988; Chapter 13, this volume) shows, this stalked group may have originated by paedomorphosis from the stalkless comatulid crinoids, thereby representing another significant shift in function, and thus should be included in our analysis of major novelties. On the other hand, if they were derived directly from the stalked Isocrinida they may be less appropriate for our study. In light of these alternatives we have analysed the data with and without bourgueticrinids.

The onshore bias in first occurrences, argued from more limited data by Bottjer and Jablonski (1988), is seen only for the orders that fall into the good preservation category (Fig. 2.1). The 16 orders consigned to the poor preservation category show no environmental pattern, so that their first occurrences are distributed almost evenly across our environments (Fig. 2.1). The numbers are small, but the two distributions are significantly different using a chi-square test in which the onshore and offshore environments are treated as alternative cells in a 2 × 2

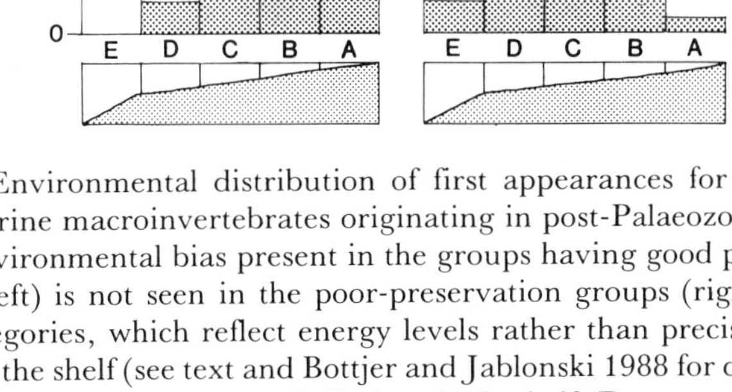

Fig. 2.1. Environmental distribution of first appearances for all orders of benthic marine macroinvertebrates originating in post-Palaeozoic times. The onshore environmental bias present in the groups having good preservational potential (left) is not seen in the poor-preservation groups (right). Environmental categories, which reflect energy levels rather than precise geographic position on the shelf (see text and Bottjer and Jablonski 1988 for details), are as follows: A, nearshore; B, inner shelf; C, middle shelf; D, outer shelf; E, slope and deep basin. The open box in environment D (left) represents the bourgueticrinids, of uncertain rank (see text). See the appendix to this chapter for documentation.

contingency table ($p < 0.001$). The general contrast between the preservational groups is encouraging, because it suggests that the onshore peak in originations on the left-hand side of Fig. 2.1 is not simply an artefact of such factors as preservation or available outcrop. If this skewed environmental distribution was generated solely by taphonomic processes, we would expect the poor-preservation group to be especially sensitive to such biases, and thus to show an onshore peak as well. The more even distribution on the right-hand side of Fig. 2.1 suggests that something other than taphonomy is creating the striking environmental pattern of the well-preserved orders.

There are many other processes that could give rise to such an environmental pattern of origination (see Bottjer and Jablonski 1988; Jablonski and Bottjer 1990*a*, who also review Palaeozoic patterns). One of the more straightforward biological hypotheses, which might be termed extrapolational, would be that the orders are simply originating in the environments expected to impose the most diverse and heterogeneous *selection* pressures on individual organisms. Such heterogeneity, it could be argued, might elicit greater divergence and more frequent production of novelties onshore than offshore. However, if this were true, we would expect an onshore concentration of new *genera* and new *families* as this greater tendency for divergence manifested itself across the hierarchy. Jablonski and Bottjer (1990*a*) found no such onshore bias in origination for isocrinid and bourgueticrind crinoid genera, or for novelties within the bryozoan order Cheilostomata (Fig. 2.2).

Jablonski and Bottjer (1990*a*) also documented first occurrences of the 11 families of the bivalve superfamily Tellinacea, which is one of the major post-Palaeozoic groups of infaunal bivalves, comprising more than 200 genera and subgenera (see also Skelton *et al.*, Chapter 5, this volume). We found that tellinacean families (and almost certainly genera, though these are not yet fully assessed) *do* show preferential origination in nearshore and inner shelf deposits, unlike the other within-group novelties discussed here. This is not surprising given that today and throughout their history the Tellinacea have been most diverse and abundant onshore. Jablonski and Bottjer (1990*a*) argue that this correspondence between bathymetric diversity patterns and origination at low taxonomic levels is a general phenomenon: the more genera present, the more likely a family will originate in that environment. Thus, for families and genera origination is itself seemingly diversity-dependent, whereas the origin of orders and major novelties tends to be environment-dependent, showing a significant peak in onshore settings regardless of the later history of the group.

Having undermined a simple extrapolational hypothesis, many possible explanations remain for the preferential origin of higher taxa

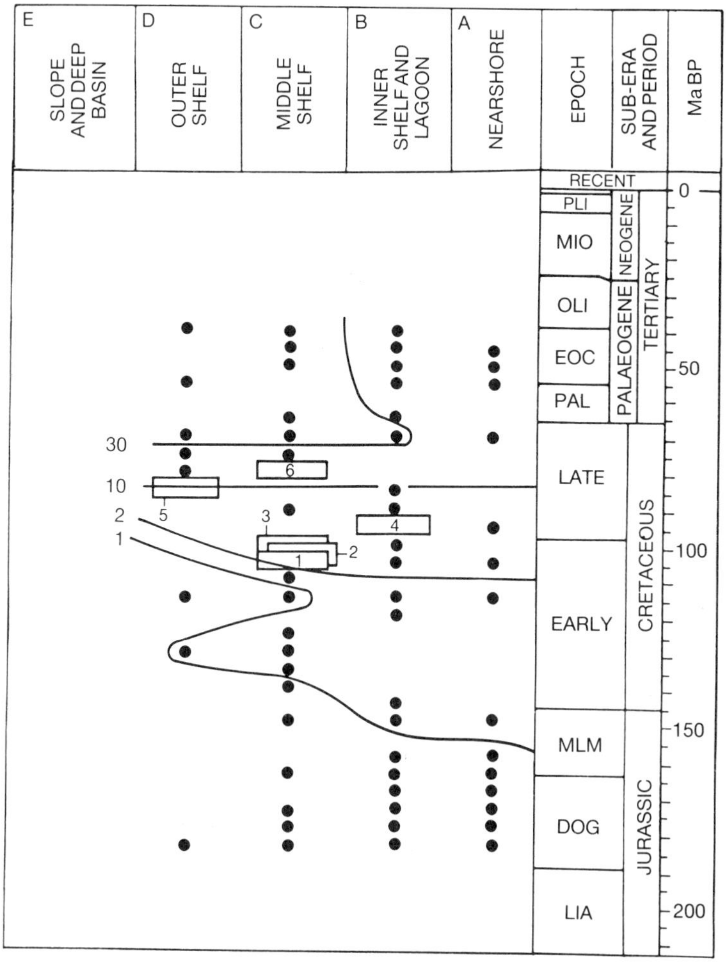

Fig. 2.2. Environmental history of the bryozoan order Cheilostomata, contoured by within-habitat generic diversity. Data points below the one-genus contour are taphonomic control points, where cyclostomes but not cheilostomes are recorded. Numbered boxes represent the first appearance of six novelties within the clade; note that these show no environmental bias. Note also that the environmental expansion of the group precedes their within-habitat (and global) diversification. Evolutionary novelties: 1, ovicells; 2, erect colony form; 3, cribrimorphs; 4, articulated colonies; 5, ascophorans; 6, lunulitiforms. From Jablonski and Bottjer (1990*a*), who provide documentation.

onshore and the apparent discordance with origination patterns of the taxa within them. These explanations do not require ordinal rank to be strictly comparable across phyla, or that orders originate abruptly with their full complement of derived characters in place. They do imply that the first members of a clade possess a novelty that is unusually conducive to further ecological and morphological change. Recent hummingbirds and geese resemble theropod dinosaurs less than does *Archaeopteryx*, but *Archaeopteryx* (and its immediate precursors) exhibits novelties that laid the foundation for the later diversification of the avian clade. Thus, for marine organisms, it appears that some aspect of the onshore environment fosters origination or enhances survivorship of more derived morphologies, of concatenations of derived morphologies, or of species with sufficient long-term evolutionary malleability ('evolutionary potential'), so that a disproportionate number of groups assigned high taxonomic rank can ultimately be traced to onshore species. We can rule out a number of additional alternatives, but, as yet, a satisfying, well-corroborated general hypothesis is elusive.

The distribution of taphonomic control records around the times of first occurrences (Jablonski and Bottjer, in preparation), as well as our results regarding groups judged to have poor preservation potential, argues against a wholly artefactual explanation. Similarly, if the onshore-origination pattern were simply a function of temporal variations in rock exposure or Lagerstätten occurrences, the poor-preservation groups should be most sensitive to these biases. Contrary to this expectation, they show an even distribution among environments. Nor are the poor-preservation taxa constrained to live in deeper, quieter water by less robust construction, with differences in the preservation categories attributable to a primary biological separation: at least two-thirds of the extant poor-preservation taxa occur at inner shelf depths today (Parker 1982). There is room for much additional work, including improved quantification, but the available data suggest other processes are at work.

A related possibility might be termed ecological constraint. If the phylogenetic sources of major novelties or new orders were taxa that lived exclusively in onshore settings, there would have been no opportunity for originations offshore. This problem is particularly notable because the onshore peak in ordinal originations is dominated by members of a single class, the Echinoidea. However, our preliminary survey of the data suggests that environmental restriction of higher taxa is detectable mainly for the Triassic (although this may be an artefact of facies available for sampling and not a true ecological restriction). More work is needed, but the onshore origination of orders later in the Phanerozoic, not only in the late Jurassic (cheilostome bryozoans) but in the early and mid-Cretaceous (holasteroid echinoids, helioporacean corals), and even the early Tertiary (clypeasteroid echinoids), argues against such constraint as the full explanation. The strong bias towards northern European first occurrences also makes us uneasy. Again, because both non-European records (e.g. Morocco, west Africa, United Arab Emirates, Japan)—and ongoing discoveries or improvements in phylogenetic information for European taxa—tend to be consistent with the patterns documented here, we are willing to treat those patterns as a working hypothesis.

A temporal bias may also be present: 83 per cent of the well-preserved taxa but only 50 per cent of the poor-preservation taxa appear before the Albian Stage of the mid-Cretaceous (a significant difference using a binomial test). By analogy to our earlier arguments, this difference between preservation categories suggests that the general decrease in originations of major groups since the early Mesozoic (reviewed by Jablonski and Bottjer 1990*b*) is not simply a preservational artefact. The failure of poor-preservation taxa to conform to the trend is probably taphonomic rather than evolutionary, however: these forms are on average more vulnerable to destruction by diagenetic processes and thus are at increasing risk of removal from the fossil record with increasing age. The frequent occurrence of Lazarus effects (*sensu* Jablonski 1986) among poor-preservation taxa—including Anaspidea, all three echinoid orders, several asteroid orders, and so on—further attests to the unreliable stratigraphic record of the poor-preservation groups.

For well-preserved taxa, the timing of first occurrences is wrong for onshore origination to be fostered by externally imposed physical events such as mass extinctions, sea-level fluctuations, or oceanic anoxic events (Bottjer and Jablonski 1988; Jablonski and Bottjer 1990*a*). This leaves intrinsic biological explanations, which can be divided into those involving either a true origination bias or an extinction bias. By origination bias we mean that onshore environments in some way favour the assembly of derived characters into what authors will later regard as higher taxa. This effect is presumably related to the greater disturbance and heterogeneity of onshore settings. Heterochrony may be an important vehicle for macroevolutionary change, but is unlikely in itself to have imposed environmental bias: too many non-progenetic orders arise onshore, and too many progenetic ones appear offshore to apply Gould's (1977; see also McKinney 1986) suggestion regarding modes of heterochrony relative to ecological factors. Jablonski and Bottjer (1983) speculated on possible population-genetic mechanisms, but more data are needed before these and other alternatives can be subjected to rigorous tests.

Preferential onshore origination might be driven by differential extinction if onshore species are geologically long-lived relative to offshore ones, as seems to be the case for at least some groups (review by Jablonski and Bottjer 1990*a*; but see Sepkoski 1987 for contrary patterns among Palaeozoic genera). Groups that happen to originate in more extinction-resistant onshore species might be more likely to survive beyond the low-diversity phase of early diversification than would similar groups arising in more extinction-prone offshore species. Such subtle differentials would be more important if these origination events were relatively rare, whereas they would be swamped out by the flood of new genera dictated by the environmental location of the group's diversity peak; this might explain the discordance across taxonomic levels. Again, more empirical evidence and quantitative modelling are required to test this class of hypotheses.

Even if higher taxa are recognized by hindsight (the old argument that *Archaeopteryx* would just be another aberrant dinosaur were it not for subsequent avian radiations; see Runnegar 1987 for a modern statement of this view), then those groups that begin in extinction-resistant species are the ones most likely to survive and continue to capture the derived traits that prompt us to delineate higher taxa. This in no way implies that higher taxa are meaningless, but it may help us to rephrase our question in terms of character sets rather than taxonomic names. On the other hand, perhaps the appropriate approach to extinction bias is to ask whether offshore environments somehow dampen the origin of higher taxa but fail to suppress the production of genera or families.

Macroevolutionary lags

In contrast to many early Palaeozoic orders, post-Palaeozoic clades rarely show rapid radiation immediately upon origination, although episodes of sudden diversification may occur later in their histories (e.g. Gilinsky and Bambach 1987; Gould *et al.* 1987). We term a prolonged low-diversity phase prior to diversification a *macroevolutionary lag*. (We use the term macroevolutionary to suggest temporal and phylogenetic scale rather than any particular mechanism, and as a contrast to previous, smaller-scale use of the term evolutionary lag, e.g. Stenseth and Maynard Smith (1984).) In at least some instances, such delayed evolutionary bursts tend to post-date, rather than coincide with, expansions across the continental shelf (the situation is less clear-cut in Miller's (1988) Palaeozoic bivalves). Cheilostome bryozoans typify this pattern (Fig. 2.2), achieving their full environmental distribution while still at very low within-habitat and global diversities. Disasteroid echinoids may be another example, beginning onshore at the beginning of the mid-Jurassic (Aalenian) but reaching bathyal environments within about 10 Ma (e.g. the Bajocian *Cyclolampas killiani* Lambert; see Jablonski and Bottjer 1988), at an early stage in a diversification that continued into the Middle Jurassic (Bathonian–Callovian) and into the Late Jurassic (especially Oxfordian) (see Mintz 1968 and Solovjev 1971 in Appendix references).

Macroevolutionary lags may be rather common in the fossil record, especially among taxa that post-date the early Palaeozoic radiations. As with the onshore bias in origination of higher taxa, a host of alternative explanations must be considered, and taxa may differ in the mechanisms that underlie the general pattern. Beginning with two null hypotheses, one geological and one biological, apparent macro-evolutionary lags might represent:

1. *Sampling biases*. If a group originates in a time that provides few or poor samples, then a long interval of apparent scarcity and low diversity might be followed by sudden radiation if preservation or sampling quality improves. Such an artefact might occur in the Early Triassic; and Smith (1988*b*) argues that the stepped nature of the Cambro-Ordovician echinoderm radiation is shaped by a dearth of Late Cambrian assemblages. This geological null hypothesis can be tested, as Smith did, using Lazarus taxa (Jablonski 1986) and by using taphonomic controls.

2. *Exponential diversification*. As Stanley (Chapter 1, this volume) points out, diversification is a multiplicative rather than an additive process, so that a clade radiating from a single ancestor will inevitably have an early lag phase of low diversity, with total origination rates mounting rapidly as standing diversity increases. This biological null hypothesis can be tested mathematically by fitting an exponential curve to the relevant diversity trend, as Sepkoski (1978) did for early Palaeozoic metazoan orders (although using regression statistics to test for significance in a time series is problematical). He found that seemingly explosive origination was underlain by a constant per-taxon origination rate from the Early Vendian into the Cambrian; on the other hand, a long metazoan history in the Precambrian at low diversity (Conway Morris 1989) might suggest a macroevolutionary lag here as well. Even if diversification is roughly exponential, interpretations should be made cautiously. Miller and Sepkoski (1988) argued that the prolonged exponential increase in bivalve generic diversity could not be attributed to unencumbered radiation, because brief rebounds from mass extinctions occurred at a much greater rate.

3. *Pre-emptive competitors*. Initially dampened diversification followed by release after the extinction of a pre-emptive competitor is a plausible

explanation for many macroevolutionary lags (Benton 1987; Hallam 1987). The classic example is the prolonged coexistence of mammals and dinosaurs during the Mesozoic, followed by exuberant mammalian radiation upon the dinosaurs' demise. The general hypothesis is, however, difficult to test beyond temporal correlations (see Benton 1987 for discussion), and potential competitors are rarely obvious for specific invertebrate orders. Decapod crustaceans appear in the Late Devonian but fail to diversify until the Mesozoic and extinction of competitors may have been important in this macroevolutionary lag (Briggs and Clarkson, Chapter 8, this volume).

4. *Key innovations*. An evolutionary burst might be triggered by an adaptive breakthrough or key innovation. Although this has been perhaps the favourite hypothesis for most evolutionary explosions, it is difficult to test and few potential examples have been analysed critically (see discussions by Benton 1987 and Jablonski and Bottjer 1990*b*). Not only is the recognition of a key novelty fraught with circularity, but cause and effect are difficult to separate: any group is more likely to capture innovations while undergoing rapid diversification than while at low, relatively stable diversity. Bona fide examples probably do exist; for example, Skelton (1985) rejects a series of alternatives in favour of the hypothesis that changes in ligament growth patterns in rudist bivalves provided the springboard for diversification of the group in the early Cretaceous. The macroevolutionary lag in the history of lucinacean bivalves, which originated in the early Palaeozoic but did not radiate until the late Mesozoic (Stanley, Chapter 1, this volume), might be another example but more data are needed.

5. *By-product of other evolutionary changes*. Diversification after a long lag might be triggered by a new trait that does not in itself impart adaptive superiority or open new habitats, but instead accelerates diversification rates as a side-effect. For example, Taylor (1988; Taylor and Larwood, Chapter 10, this volume) has argued that the explosive phase of cheilostome diversification was initiated by the acquisition of a low-dispersal, brooded larval stage, which accelerated speciation rates through reduced gene flow (see also Stanley, Chapter 1, this volume). Similar arguments have been made for angiosperms, where rapid diversification (possibly after a long low-diversity phase) may be due to the evolution of reproductive characters that heightened speciation rates rather than to any newly-acquired competitive superiority (see Crane and Lidgard, Chapter 17, this volume). These indirect effects, in which a seemingly modest evolutionary change has far-reaching consequences across hierarchical levels, are difficult to test due to the inherent

complexity of hierarchical systems, but should constitute a rival hypothesis when mechanisms are sought for evolutionary radiations.

Summary and conclusions

When the 40 post-Palaeozoic marine benthic invertebrate groups (41 orders plus a superorder that includes six of those orders) are traced to their environment of first occurrence, we find that those with good preservation potential tend to originate onshore. While the origin of higher taxa thus appears to be at least partially environment-dependent, the origin of species, genera, and families is evidently diversity-dependent. Thus, in contrast to the ordinal pattern, at lower taxonomic levels diversity begets diversity, with evolutionary patterns dictated by the bathymetric diversity gradient of the clade. This discordance between the locus of diversification at different taxonomic levels is intriguing but still not fully understood, though we have been able to eliminate some of the possible explanations.

In at least some clades, environmental expansion precedes global and within-habitat diversification. This is another unexpected discordance, in this instance between environmental distribution and global diversity. This dynamic ecological pattern that occurs despite the macroevolutionary lag (i.e. the interval between origin of a group and its diversification) can be used to test hypotheses on the factors that drive the radiation and ecological expansion of particular groups. For example, the pattern suggests that expansion offshore is not fuelled by the pressure of mounting species richness onshore. Other aspects of environmental distribution clearly deserve study; for example, Palaeozoic bivalves differ in their bathymetric distributions, relative abundances, and diversities in carbonate and terrigenous sediments (Miller 1989). Overall, the data presented here indicate that, when viewed ecologically, evolutionary radiations involve a complex interplay of processes at a number of hierarchical levels.

Acknowledgements

We thank the following for their generous assistance and advice: F. M. Bayer, D. B. Blake, H. J. Campbell, K. Chinzei, B. Clavel, S. K. Donovan, M. L. Droser, H. Hagdorn, W. Hinsch, H. A. Kollman, S. Lidgard, N. J. Morris, R. Mouterde, A. Oravecz-Scheffer, J. A. Schneider, J. J. Sepkoski, Jr., M. J. Simms, A. B. Smith, G. D. Stanley, S. J. Suter, P. D. Taylor, and E. Voigt. Susan M. Kidwell and Andrew B. Smith gave valuable reviews of the manuscript; Jay A. Schneider assisted in data collection and management. This research could not have been performed without assistance from the following libraries: Crerar Library, University of Chicago (especially Kathleen

Zar); Hancock Library, University of Southern California (especially Suzanne Henderson); Earth Sciences Library, University of California, Los Angeles (especially Michael Noga); and the Palaeontology Library, British Museum (Natural History) (especially Ann Lum). For support of this research, we thank the Donors of the Petroleum Research Fund, administrated by the American Chemical Society, and the National Science Foundation (grants EAR 84-17011 and INT 86-2045 to DJ and EAR 85-19941 to DJB).

References

Bambach, R. K. (1985). Classes and adaptive variety: The ecology of diversification in marine faunas through the Phanerozoic. In *Phanerozoic diversity patterns* (ed. J.W. Valentine), pp. 191–253. Princeton University Press, Princeton, New Jersey.

Benton, M.J. (1987). Progress and competition in macroevolution. *Biological Reviews*, **62**, 305–38.

Bottjer, D.J. and Jablonski, D. (1988). Paleoenvironmental patterns in the evolution of post-Paleozoic benthic marine invertebrates. *Palaios*, **3**, 540–60.

Conway Morris, S. (1989). Burgess Shale faunas and the Cambrian explosion. *Science*, **246**, 339–46.

Eldredge, N. (1989). *Macroevolutionary dynamics*. McGraw-Hill, New York.

Eldredge, N. and Salthe, S. N. (1984). Hierarchy and evolution. *Oxford Surveys in Evolutionary Biology*, **1**, 182–206.

Gilinsky, N. L. and Bambach, R. K. (1987). Asymmetrical patterns of origination and extinction in higher taxa. *Paleobiology*, **13**, 427–45.

Gould, S.J. (1977). *Ontogeny and phylogeny*. Belknap Press, Cambridge, Massachusetts.

Gould, S.J., Gilinsky, N.L., and German, R.Z. (1987). Asymmetry of lineages and the direction of evolutionary time. *Science*, **236**, 1437–41.

Hallam, A. (1987). Radiations and extinctions in relation to environmental change in the marine Lower Jurassic of northwest Europe. *Paleobiology*, **13**, 152–68.

Jablonski, D. (1986). Causes and consequences of mass extinctions: A comparative approach. In *Dynamics of extinction* (ed. D.K. Elliott), pp. 183–229. Wiley, New York.

Jablonski, D. and Bottjer, D.J. (1983). Soft-bottom epifuanal suspension-feeding assemblages in the Late Cretaceous: Implications for the evolution of benthic paleocommunities. In *Biotic interactions in Recent and fossil benthic communities* (ed. M.J.S. Tevesz and P.L. McCall), pp. 747–812. Plenum Press, New York.

Jablonski, D. and Bottjer, D.J. (1988). Onshore–offshore evolutionary patterns in post-Paleozoic echinoderms: A preliminary analysis. In *Echinoderm Biology—Proceedings of the 6th International Echinoderm Conference* (ed. R.D. Burke, P.V. Mladenov, P. Lambert, and R.L. Parsley), pp. 81–90. Balkema, Rotterdam.

Jablonski, D. and Bottjer, D.J. (1990*a*). Onshore–offshore trends in marine invertebrate evolution. In *Causes of evolution: a paleontological perspective* (ed. R.M. Ross and W.D. Allmon), pp. 21–75. University of Chicago Press, Chicago.

Jablonski, D. and Bottjer, D.J. (1990*b*). The ecology of evolutionary innovation: the fossil record. In *Evolutionary innovations* (ed. M.H. Nitecki), pp. 253–88. University of Chicago Press, Chicago.

Kier, P.M. (1974). Evolutionary trends and their functional significance in the post-Paleozoic echinoids. *Paleontological Society Memoir* 5. (*Journal of Paleontology*, **48** (Supplement to No. 3).)

McKinney, M.L. (1986). Ecological causation of heterochrony: a test and implications for evolutionary theory. *Paleobiology*, **12**, 282–9.

Miller, A.I. (1988). Spatio-temporal transitions in Paleozoic Bivalvia: an analysis of North American fossil assemblages. *Historical Biology*, **1**, 251–73.

Miller, A.I. (1989). Spatio-temporal transitions in Paleozoic Bivalvia: a field comparison of Upper Ordovician and Upper Paleozoic bivalve-dominated fossil assemblages. *Historical Biology*, **2**, 227–60.

Miller, A.I. and Sepkoski, J.J. Jr (1988). Modeling bivalve diversification: the effect of interaction on a macroevolutionary system. *Paleobiology*, **14**, 364–9.

Parker, S.P. (ed.) (1982). *Synopsis and classification of living organisms*, Vols 1 and 2. McGraw-Hill, New York.

Runnegar, B. (1987). Rates and modes of evolution in the Mollusca. In *Rates of evolution* (ed. K.S.W. Campbell and M.F. Day), pp. 39–60. Allen and Unwin, London.

Sepkoski, J.J. Jr (1978). A kinetic model of Phanerozoic taxonomic diversity. I. Analysis of marine orders. *Paleobiology*, **4**, 223–51.

Sepkoski, J.J. Jr. (1987). Environmental trends in extinction during the Paleozoic. *Science*, **235**, 64–6.

Sepkoski, J.J. Jr. (1988). Alpha, beta, or gamma: Where does all the diversity go? *Paleobiology*, **14**, 221–34.

Simms, M.J. (1988). The phylogeny of post-Palaeozoic crinoids. In *Echinoderm phylogeny and evolutionary biology* (ed. C.R.C. Paul and A.B. Smith), pp. 269–84. Oxford University Press, Oxford.

Skelton, P.W. 1985. Preadaptation and evolutionary innovation in rudist bivalves. *Special Papers in Palaeontology*, **33**, 159–73.

Smith, A.B. (1981). Implications of lantern morphology for the phylogeny of post-Palaeozoic echinoids. *Palaeontology*, **24**, 779–801.

Smith, A.B. (1984). *Echinoid palaeobiology*. George Allen and Unwin, London.

Smith, A.B. (1988*a*). Phylogenetic relationship, divergence times, and rates of molecular evolution for camarodont sea urchins. *Molecular Biology and Evolution*, **5**, 345–65.

Smith, A.B. (1988*b*). Patterns of diversification and extinction in early Palaeozoic echinoderms. *Palaeontology*, **31**, 799–828.

Stenseth, N.C. and Maynard Smith, J. (1984). Coevolution in ecosystems: Red Queen evolution or stasis? *Evolution*, **38**, 870–80.

Taylor, P.D. (1988). Major radiation of cheilostome bryozoans: triggered by the evolution of a new larval type? *Historical Biology*, **1**, 45–64.

Appendix

Occurrence of earliest known members of all benthic macroinvertebrate orders listed by Sepkoski (1982) as originating since the beginning of the Mesozoic. For echinoids, innovations associated with the origin of other major, monophyletic groups are given and related to traditional taxonomy, following Smith (1981, 1984). Asteroid orders after Blake (1987, 1989). * Poor preservation potential.

Porifera

Class Hexactinellida

Order Hexactinosida: Range extended to the Late Devonian (Frasnian) of the Holy Cross Mountains, Poland (Rigby *et al.* 1981).

Order Lychniscosida

Taxon: *Triadocoelia magyara* Vinassa de Regny.
Age and environment: Late Carnian, inner shelf.
Unit: Veszprem Formation, Veszprem (Bakony), Hungary.
References: Galacz *et al.* (1985); Oravecz-Scheffer (1987).

Class Sclerospongiae

Order Muranida: Subsumed within order Axinellida by Wood (1987).

Coelenterata

Class Hydrozoa

Order Spongiomorphida: Recorded from the Late Permian of South China by Fan *et al.* (1982) and Sheng *et al.* (1985).

Order Milleporina: Recorded from the Late Permian of South China by Fan *et al.* (1982) and Sheng *et al.* (1985).

Order Stylasterina

Taxon: unspecified stylasterines.
Age and environment: Maastrichtian, outer shelf?
Unit: White Chalk, Denmark.
Reference: Floris (1979).
Notes: The oldest named taxa comprise a diverse fauna from the Danian Coral Limestone, Fakse Quarry (Floris 1979, 1980; Cairns 1983).

Class Anthozoa

*Order Stolonifera

Taxon: *Scyphopodium ingolfi* (Madsen).
Age and environment: Late Pleistocene, slope and deep basin.

Unit: Dredged from 2216 m, attached to coral with radiocarbon date of 18 200 ± 200 years.
Reference: Bayer (1981).
Notes: Oldest records in the *Treatise* are for *Epiphaxum*, which is placed by Bayer (1979) in Helioporacea.

Order Helioporacea (= Coenothecalia)

Taxon: *Proheliopora kysylkumensis* (Kusmicheva).
Age and environment: Albian, inner shelf.
Unit: Unnamed, Central Kyzylkum', Dzhengel'dy State Farm, Uzbekistan, USSR.
Reference: Kusmicheva (1975).
Notes: 'Inhabited the rocks and shingle of the littoral zone of the Albian Sea' (Kusmicheva 1975, p. 280).

Taxon: *Pseudopolytremacis hanagaensis* Kusmicheva.
Age and environment: Barremian–Aptian, inner shelf.
Unit: Unnamed, Armenia, USSR; unnamed, Marmarosh zone, Carpathians, USSR.
References: Nalivkin (1973); Kusmicheva (1975); Borsuk and Sholpo (1983).

Order Pennatulacea: Range extended to Silurian, with *Protovirgularia* M'Coy.

Order Scleractinia

Taxon: *Montlivaltia chonocalyx* Weissermel, *Pamiroseris silesiaca* (Beyrich).
Age and environment: Early Anisian, nearshore (and inner shelf?).
Unit: Gorazdze Beds, Lower Muschelkalk, Upper Silesia, Poland.
References: Zawidzka (1975); Morycowa and Roniewicz (1986); Morycowa (1988).
Notes: Scleractinians occur in similar environments at a number of localities in the Late Anisian (ranging from south China to Poland to the Dolomites and the Vincentinian Alps), but the Polish Early Anisian records are still the oldest known (Stanley 1988; Morycowa 1988).

Mollusca

Class Gastropoda

Order Neogastropoda

Taxon: '*Fusus (Chrysodomus)*' *grodischtansus* Ascher.
Age and environment: Early Hauterivian, outer shelf?

Unit: Hradiste Beds (= Grodischter Schichten = Grodziszcze Beds), Tesin–Hradiste Formation, Silesian Nappe, West Carpathians, Czechoslovakia — transported material?
References: Hanzlikova and Roth (1963); Mahel' *et al.* (1968); Ksiazkiewicz (1975); Malik and Olszewska (1984); H.A. Kollman (*in litt.* 18 October 1988).

Taxon: *'Buccinum' incertum* d'Orbigny [includes *Fusus minutus* of Trautschold; Glazunova 1973].
Age and environment: Late Hauterivian, outer shelf.
Unit: unnamed claystone sequence, Ulianovskoi Pavalsh, Volga Region, USSR.
References: Glazunova (1973).

Taxon: *'Fusus' brunvicensis* Wollemann [placed in *Cantharulus* Meek by Kollmann 1982].
Age and environment: Late Barremian, outer shelf.
Unit: Upper clay sequence of Mutterlose and Harding (1987), Moorhütte bei Braunschweig, NW Germany.
References: Kollmann (1982); Mutterlose (1984); Meyn and Vespermann (1987); Mutterlose and Harding (1987).
Notes: The identity of the oldest known neogastropod is problematic, because diagnostic features are often lacking in these small, poorly preserved, and uncommon fossils; confusion is augmented by the presence of two extinct families (Colombellinidae and Purpurinidae) that may be sister-groups to the Neogastropoda or even early members of the order (Taylor and Morris 1988). The possible early neogastropods cited above may constitute a single lineage close to the origin of the order, recorded from offshore deposits; however, Beizel (1977, 1983) considers at least two of these species to be members of the Colombellinidae genus *Khetella*. Alternative candidates include *Fusus neocomiensis* d'Orbigny, *Fusus valangiensis* Pictet and Campiche, and *Fusus ornatus* d'Orbigny.

*Order Sacoglossa (= Ascoglossa)

Taxon: *Berthelinia (Berthelinia) elegans* Crosse, *Hemiplicatula solida* Deshayes, *Gougerotia orthodonta* Le Renard.
Age and environment: Late Ypresian, inner shelf.
Unit: Sables de Cuise, Facies de Pierrefonds, Pierrefonds, France.
References: Mégnien (1980); Le Renard (1983).

*Order Anaspidea (= Aplysiomorpha)

Taxon: *Akera tanganyicensis* Cox.
Age and environment: Mid-Late Callovian, inner shelf.

Unit: Upper part of Mandaea-Mahokondo Series, along Mbaru stream, 1 mile north-west of Mbinga, south-eastern Tanzania.
References: Aitken (1961); Cox (1965); Bate (1975).

Taxon: *Akera mediojurensis* Cossmann.
Age and environment: Callovian-Oxfordian; Callovian environment unknown.
Unit: Callovian locality unspecified, 'Sarthe', France.
Reference: Cossmann (1895-96).
Notes: (1) Slightly older but of uncertain taxonomic affinity is the poorly preserved '*Acera?*' *primaeva* (Deslongchamps) of Cossmann (1895-96), known from two specimens (including the missing type) from the Late Bathonian Calcaire de Ranville, Ranville (Calvados), France; this unit was deposited in a nearshore environment (Mégnien 1980).
(2) The Akeridae have sometimes been placed in the Cephalaspidea (Bullomorpha) (e.g. Marcus 1972; Thompson 1976), but recent work (Morton 1972; Thompson and Seaward 1989) supports anaspidian affinity. If Akeridae are bullomorphs, the oldest anaspidean is *Dolabella aldrichi* Dall, also from inner shelf environments, in the Late Burdigalian Chipola Formation, Alum Bluff, Florida (Gardner 1947; Akers 1972; Schmidt 1986; Vokes 1989).

*Order Notaspidea
Taxon: *Umbraculum sylvaerupis* (Harris) and *Eosinica elevata* (Aldrich).
Age and environment: Early Ypresian, middle shelf.
Unit: Bashi Marl Member, Hatchetigbee Formation, Woods Bluff, Alabama.
References: Palmer and Brann (1966), LaMoreaux and Toulmin (1959), Mancini (1984).
Notes: The umbraculid record in the Danian Brightseat Formation of Maryland (*Umbraculum*? sp. indet. of Govoni 1983) needs confirmation.

Class Bivalvia
Order Unionoida: Range extended to Devonian; see Gray (1988).

Arthropoda

Class Malacostraca
Order Amphipoda: Oldest records of this group, which has 'one of the most disappointing fossils records among any of the

Eumalacostraca' (Schram 1986, p. 175), are in terrestrial deposits of the Baltic Amber (Just 1974; Larsson 1978; Schram 1986).

Bryozoa

Class Gymnolaemata

Order Cheilostomata

Taxon: *Pyriporopsis portlandensis* Pohowsky.

Age and environment: Late Tithonian–Early Berriasian, nearshore and inner shelf.

Unit: Basal Shell Bed, Cockly Bed, Wockley Micritic Member, and Portland Roach Member of the Portland Stone, southern England, and Scallop Member of the Purbeck Limestone Formation, Dorset, England.

References: Townson (1975); Taylor (1981); Ensom (1985).

Echinodermata

Class Crinoidea

Order Isocrinida

Taxon: '*Isocrinus*' *smithi* Clark.

Age and environment: mid-Spathian, nearshore and inner shelf.

Unit: Thaynes Formation, many localities in Idaho, USA.

Taxon: '*Isocrinus* sp.' [= '*Isocrinus*' *smithi*, according to Klikushin 1987*b*].

Age and environment: mid-Spathian, nearshore and inner shelf.

Unit: Virgin Formation, many localities in Utah and Nevada, USA.

References: Kummel (1957); Carr and Paull (1983); Hagdorn (1986); Klikushin (1987*b*).

Order Millericrinida

Taxon: '*Millericrinus*' *silesiacus* (Beyrich).

Age and environment: Late Anisian, nearshore and inner shelf.

Unit: Upper part of Diplopora Dolomite, Piekary Slaskie, Upper Silesia, Poland.

References: Zawidzka (1975); Hagdorn (1985); Gluchowski and Boczarowski (1986); M.J. Simms (*in litt.*, 22 September 1988).

Notes: (1) Also recorded from similar environments in the Trochitendolomit and the Thörler Kalk of the eastern Alps, Austria, by Kristan-Tollmann and Tollmann (1967).

(2) *Dadocrinus*, included in by the *Treatise* in Millericrinida, appears to have closer affinities with the Encrinida (M.J. Simms, *in litt.* 1988).

Order Crytocrinida

Taxon: *Eudesicrinus* cf. *E. mayalis* of Jäger (1985).
Age and environment: Late Early Pliensbachian, middle shelf.
Unit: *davoei* Zone, Rottorf-am-Klei near Hannover, West Germany.
References: Hoffmann (1982); Jäger (1985).

Taxon: *Cotylederma lineati* Quenstedt.
Age and environment: Late Early Pliensbachien, middle shelf.
Unit: Upper Lias Gamma (probably *davoei* Zone), southern Germany.
References: Sieverts (1932); Brandt (1985); Geyer and Gwinner (1986).
Notes: (1) Oldest cyrtocrinid record in *The Fossil Record*, '*Cotylederma oppeli*' Torquem and Piette, does not belong in this order and is probably a millericrinid holdfast (M.J. Simms, *personal communication*, 8 August 1989; see also Sieverts 1932).
(2) Following Simms (personal communication, 1989), we place *Amaltheocrinus*, *Shroshaecrinus*, Cyclocrinidae, Hyocrinina in Millericrinida.

Order Bougueticrinida

Taxon: *Bourgueticrinus* sp.
Age and environment: Late Early Turonian, outer shelf.
Unit: unnamed pelagic chalk unit, Belbeck Valley, Crimea, USSR.
References: Klikushin (1982, 1985).
Notes: Early Cretaceous Bourgueticrinida cited by Klikushin (1975) are almost certainly the comatulid family Thollieri-crinidae (Klikushin 1987*a*).

*Order Comatulida

Taxon: *Paracomatula* n.sp.
Age and environment: Late Norian (or Rhaetian), inner/middle shelf.
Unit: Bouraké Formation, Ile Ducos, Baie de Saint Vincent, New Caledonia.
References: Campbell *et al.* (1985); Hagdorn and Campbell (in preparation; Hans Hagdorn *in litt.* 2 August 1989; Hamish J. Campbell *in litt.* 28 December 1989).
Notes: This species is considerably older than the more primitive *Eocomatula interbrachiatus* (Tate) (Lower Staithes Formation, early Pliensbachian [*davoei* Zone], Huntcliff, Yorkshire; see Simms 1988*a*, *b*). Also older than *Eocomatula* is an even more derived form, *Palaeocomaster styriacus* Kristan-Tollmann 1988,

from the mid-Hettangian Lower Allgäuschichten ('Fleckenmergel') near Lawinenstein, Austria. The marked lack of correspondence between stratigraphic occurrence and apparent phylogenetic position is another indication of the incomplete nature of the comatulid fossil record as presently known.

Order Uintacrinida—omitted as pelagic.

Order Roveacrinida—omitted as pelagic.

Class Echinoidea

Order Cidaroida — Now believed to be present in the late Palaeozoic (Kier 1984; Smith 1990; Smith and Hollingworth 1990).

*Order Echinothurioida

Taxon: *Echinothuria floris* Woodward.

Age and environment: *coranguinum* Zone, Early Santonian, outer shelf.

Unit: Broadstairs Member, Ramsgate Chalk Formation, Kent, England. Also probably in Santonian of Corneville (Eure), France (as the hoofed spines named *Phorosoma homoei* Lambert).

References: Smith and Wright (1989); Robinson (1986). Notes: The echinothuroids have a very poor fossil record with only three undoubted species, one Late Cretaceous and two Neogene (Smith and Wright 1989). Smith and Wright (1989) exclude '*Pelanodiadema*' *oolithicum* Hess from Echinothurioida on the basis of the ambulacral structure and lantern, which have features shared only with diadematoids. These authors say that this species is probably a *Pelanechinus*, and the type species of that genus, *P. corallinus*, is also not an echinothurioid.

Infraclass Acroechinoidea: The first occurrence of this monophyletic clade corresponds to the appearance of well-developed auricles, ambulacral processes that combine with the apophyses to provide a perignathic girdle that improves lantern muscle efficiency (Kier 1974; Smith 1984). This novelty marks the first appearance of the following paraphyletic order in traditional taxonomy.

Order Pedinoida

Taxon: *Hemipedina hudsoni* Kier.

Age and environment: Late Norian, inner shelf.

Unit: Sumra Formation of Hudson (1960; Hudson and Jefferies 1961) = lower part of middle member of Ghalialah Formation of Glennie *et al.* (1974), Elphinstone Group, Wadi Milaha, Musandam Peninsula, United Arab Emirates.

References: Hudson and Jefferies (1961); Kier (1977); Ricateau and Riché (1980); Searle *et al.* (1983).

Notes: The Late Carnian *Hemipedina? incipiens* Bather is better treated as indeterminate (Smith 1990).

*Order Diadematoida

Taxon: *Centrostephanus fragile* (Wiltshire).

Age and environment: *coranguinum* Zone, Basal Santonian, outer shelf.

Unit: Broadstairs Member, Ramsgate Chalk Formation, Thanet, England; also *quadrata* Zone of Salisbury, Wiltshire; and *coranguinum* Zone, Evreux, NW France (as spines named *Diadema ebroicensis* Caffin; Smith and Wright 1989).

Notes: As Smith and Wright (1989) point out, the fossil record of this order is extremely poor; the species listed above is the only one known from test and spine material that can definitely be attributed to the group. *Eodiadema* is a mixture of species of varying affinities, and *Pedinothuria cidaroides* Gregory is of uncertain affinities. The lone specimen of *Eosalenia miranda* Lambert is too poorly known for placement (Smith and Wright 1989).

*Order Micropygoida: only one fossil species is known that might belong in this order (Smith and Wright 1989).

Taxon: *Pedinothuria cidaroides* Gregory.

Age: Probably early Kimmeridgian, probably middle shelf.

Unit: Unknown, labelled as 'Weisser Jura, Germany'. Probably Weisser Jura Gamma, western Bavaria or Württemburg (Gregory 1897).

References: Gwinner (1976); Smith and Wright (1989).

Cohort Echinacea: This monophyletic clade is characterized by keeled teeth (among other features), which are mechanically stronger than those of more primitive euechinoids (Kier 1974, 1977; Smith 1981, 1984). This novelty first appears in the following paraphyletic order, using the traditional taxonomy.

Order Hemicidaroida

Taxon: *Pseudodiadema silbinense* Stefanini.

Age and environment: Rhaetian, inner shelf.

Unit: *Rhaetavicula contorta* beds, Monte Citona Formation, Selvena (Siena Province), Italy.

References: Passeri and Pialli (1973); Kier (1977); Ciarapica *et al.* (1982); Ciarapica (1985).

Order Salenioida

Taxon: *Acrosalenia lycetti* Wright.

Age and environment: *murchisonae* Zone, Middle Aalenian, nearshore.

Unit: Crickley Oncolite Member, and Cleeve Hill Oolite Member, Lower Inferior Oolite Formation, Crickley Hill and Cheltenham (Gloucestershire), England.

References: Richardson and Talbot Paris (1908); Mudge (1978).

Notes: This is the first species in which the diagnostic apical disc is preserved. Older possible representatives of this order include: *A. chartroni* Lambert from the Hettangian (or Sinemurian) near Saint-Cyr-en-Talmondais, France; *A. zararensis* Lambert from the Late Sinemurian of the Central High Atlas, Morocco; *A. cotteaui* Gauthier from the Late Pliensbachian of Bosc (Aveyron), France; *A. faurai* Lambert and *A. marratensis* Kier from the Toarcian of Spain and Saudi Arabia, respectively. All seem to be in onshore deposits.

Order Phymosomatoida

Taxon: *Jeannetia mortenseni* Mercier.

Age and environment: Early Hettangian, nearshore.

Unit: Calcaire de Valognes, Valognes (Manche), France.

References: Mégnien (1980); Smith (1990).

Notes: Smith (1984) includes the Arbacioida within this order.

Superorder Camarodonta: This monophyletic clade is characterized by further strengthening and bracing of the lantern, with expanded epiphyses sutured above the foramen magnum (Kier 1974; Smith 1981, 1984). This novelty marks the first appearance of the following paraphyletic order in traditional taxonomy.

Order Temnopleuroida

Taxon: *Hemidiadema neocomiense* (Cotteau).

Age and environment: Early Hauterivian, inner shelf.

Unit: Calcaire à Spatangues, Auxerre (Yonne), France.

References: C. Mégnien (1967); F. Mégnien (1980); Rat *et al.* (1987).

Taxon: *Glyptechinus montmollini* (Desor).

Age and environment: Early Hauterivian, middle shelf.

Unit: Marnes d'Hauterive, Villers-le-Lac (Doubs), France.

References: Guillaume (1966–67).

Notes: The Pliensbachian *Glyptodiadema cayluxense* (Cotteau), given in the *Treatise* as the oldest temnopleuroid, is in fact a member of the Triassic genus *Leurocidaris* Kier (A. B. Smith, personal communication, 25 August 1989).

Order Echinoida

Taxon: '*Psammechinus*' *biarritzensis* (Cotteau).

Age and environment: Late Lutetian, inner shelf?

Unit: Rochers de la Gourepe, Biarritz, France.

References: Castex and Lambert (1919); Alvinerie *et al.* (1964); Ducasse (1973); Pomerol (1980); Smith (1988).

Notes: *Echinometra thomsoni* Haime, from the Palaeogene of India, and thus potentially the oldest member of Order Echinoida, is in reality 'a slightly distorted phymosomatoid, possibly *Porosoma*', according to Smith (1988, p. 354).

Cohort Irregularia: This monophyletic clade (Smith 1981 and references therein) is characterized today by a large number of derived traits related to migration of the periproct outside of the apical system, related to an infaunal microphagous habit and unidirectional movement; these novelties accrued in stepwise fashion, and we treat them in turn. (See Kier 1974, pp. 50–1, on the advantages of unidirectional movement). The oldest representatives of this clade, recognized by the appearance of 'a very slightly elongated apical system' (Jesionek-Szymanska 1970), a flattened profile, and increased number and decreased size of tubercles and spines per plate, are placed by traditional taxonomy in the paraphyletic order Pygasteroida.

Order Pygasteroida

Taxon: '*Plesiechinus*' *hawkinsi* Jesionek-Szymanska.

Age and environment: Early Sinemurian, nearshore/inner shelf.

Unit: Sunrise Formation, New York Canyon, Gabbs Valley Range, Nevada.

References: Hallam (1965); Stanley (1971); Jesionek-Szymanska (1970); Imlay (1980); Rose (1982).

Taxon: *Atlasaster jeanneti* Lambert.

Age and environment: Early Sinemurian, middle shelf?

Unit: Unnamed, Oued Amdrhous, central High Atlas, Morocco.

References: Jenny (1988); Warme (1988).

Order Holectypoida

Taxon: *Holectypus hians* Lambert.

Age and environment: Late Early Pliensbachian (Late Carixian) (not Early Late Pliensbachian (Early Domerian) as originally described; see Du Dresnay 1966), inner shelf.

Unit: *margaritatus* Zone, Tounat-el-Mariz (= Tounate-el-Maariz), Middle Atlas, Morocco.

References: Du Dresnay (1966); Rose (1982); Rose and Olver (1985).

Superorder Microstomata: This monophyletic clade originated through loss of the lantern in adults (as indicated by the small peristome), development of weak oral phyllodes (an increased concentration of food-gathering tube feet around the mouth), and decreased size and denser packing of tubercles, particularly in the region of the mouth, all related to a shift to infaunal deposit-feeding. In traditional taxonomic terms, this represents the first occurrence of the order Cassiduloida. However, here we provide data on later species that comply to a more strictly cladistic definition of the order.

Taxon: *Eogalerophygus microstoma* (Lambert).
Age and environment: Middle Toarcian, nearshore/inner shelf.
Unit: Couches de Mibladène, near Dourirat on the east side of Djebel Ousilagh, Morocco.
References: Colo (1961–64); Jesionek-Szymanska (1978); Dagallier (1977); Benshili (1989).
Notes: '*Galeropygus*' *lacroixi* Lambert, from the Late Pliensbachian of Saint-Pons (Bouches du Rhone), France, may well be the oldest microstomate echinoid, but the specimen is crushed and lacks the apical disc. It occurs in sediments inferred to have been deposited in middle shelf environments (Arnaud and Monleau 1979).

Order Cassiduloida

Taxon: *Galeropygus agariciformis* Forbes, the first occurrence of strong phyllodes (A. B. Smith, personal communication, 8 August 1989).
Age and environment: Late Toarcian–Aalenian, inner and middle shelf.
Unit: Widespread in France, Germany, and Great Britian.
References: Mudge (1978); Smith (1984); Contini (1986).
Notes: Rose and Olver (1988) resurrect Order Galeropygoida Mintz (1968) for the Galeropygidae + Menopygidae. Pending further study, we prefer to start the Galeropygidae within the Cassiduloida, as suggested by Rose and Olver's (1988) cladogram (see also Smith 1981, 1984).

Order Pliolampoida: Included within Cassiduloida, following Smith (1984)

Order Oligopygoida

Taxon: Species of *Haimea* (and perhaps *Oligopygus*, though this needs confirmation), many of which are probably synonyms (Donovan 1987, 1988).

Age and environment: Early Lutetian, inner and middle shelf.

Unit: Font Hill Formation, Yellow Limestone Group, Jamaica; many units in western and central Cuba; unnamed Lower Lutetian unit, Senegal, West Africa.

References: Roman and Gorodiski (1959); Élouard (1962); Monciardini (1966); Kier (1967); Eva (1976); McFarlane (1977); Garcia *et al.* (1987); Robinson (1987); Donovan (1987, 1988, personal communication, 18 January 1990).

Notes: The Middle Eocene was a time of active tectonism in the Caribbean region, and deeper-water records may involve transported material.

Order Clypeasteroida

Taxon: *Togocyamus seefriedi* (Oppenheim), *T. alloiteaui* Roman and Gorodiski.

Age and environment: Thanetian, nearshore.

Unit: Ewekoro Formation and equivalents, Togo, Ghana, Gold Coast, Senegal, SW Nigeria, West Africa.

References: Roman and Gorodiski (1959); Chabaglian (1959); Adegoke (1977); du Chene *et al.* (1978); Kier (1982).

Series Atelostomata: This monophyletic clade represents the first occurrence of an elongate apical system—one response to the fabricational problem of posterior migration of the periproct in order to move the anus away from the feeding and respiratory areas (Kier 1974; Smith 1984). By the traditional taxonomy, this coincides with the first representative of the Order Disasteroida.

Order Disasteroida

Taxon: *Hyboclypus caudatus* Wright (= *H. harfordensis* Tablot Paris; A.B. Smith, personal communication, 31 August 1989).

Age and environment: Aalenian, inner shelf.

Units: Inferior Oolite and Harford Sands, Gloucestershire, England.

References: Talbot Paris (1908); Mudge (1978).

Notes: The Late Bajocian *Orbignyana ebrayi* Cotteau contains the first truly disjunct apical system, with migration of the periproct carrying genital 5 and the two posterior ocular plates backwards, and might be an alternative choice; it occurs in middle shelf environments near Le Guétin (Nievre), France (Mouterde 1953; Jesionek-Szymanska 1963; Mintz 1968; Mégnien 1980). *Cyclolampas killiani* Lambert, also Late Bajocian and perhaps one ammonite zone older than *O. ebrayi*, is considerably more derived than either of the preceding species (Clavel in Elmi 1984).

Order Holasteroida

Taxon: *Holaster cordatus* of Dubois (not of de Loriol).

Age and environment: *occitanica* Zone, Middle Berriasian - *pertransiens* Zone, Early Valanginian, inner shelf.

Unit: Pierre-Chatel Formation of Steinhauser and Lombard (1969) [= Calcaires de Thoiry of Mouty 1966] Fontanil (Isère), France.

References: Clavel *et al.* (1986); B. Clavel (*in litt.* 7 August 1987).

Notes: According to B. Clavel (*in litt.*, 7 August 1987), *Toxaster laffittei* Devriès is a synonym of *H. cordatus*. The former species, which is placed in *Eoholaster*, Holasteridae, by Solovjev (1971) (see Devriès 1973 for a different view), occurs in similar environments near Taghit Sidi Bel Kheir, northeast Algeria (Laffitte 1939; Devriès 1960).

Order Spatangoida

Taxon: *Toxaster africanus rochi* Lambert.

Age and environment: Berriasian, inner shelf.

Unit: Timsilline Formation, Cape Rhir, Morocco.

References: Petitot (1959); Wiedmann *et al.* (1982); Rey *et al.* (1987).

Taxon: *Toxaster granosus* (d'Orbigny) [may include *T. africanus*; B. Clavel, *in litt.*, 7 August 1987].

Age and environment: Middle Berriasian–Early Valanginian, inner and middle shelf.

Unit: (1) Calcaire Oolithique Inférieur of Rittener (1902) Sainte-Croix, Switzerland.

(2) Pierre-Chatel Formation of Steinhauser and Lombard (1969) [= Calcaires de Thoiry of Mouty 1966], Fontanil (Isère), France.

(3) Valtière Member, Fontanil Formation, Fontanil (Isère), France.

References: Morand (1913); Steinhauser and Charollais (1971); Arnaud *et al.* (1981); Clavel *et al.* (1986); B. Clavel (*in litt.* 7 August 1987).

Notes: The Late Jurassic *Allotoxaster tosaensis* (de Loriol) is based on a single indeterminate internal mould (A. B. Smith, *in litt.*, 10 June 1987).

Asteroida

*Order Trichasteropsida Blake (1987)

Taxon: *Trichasteropsis cilicia* (Quenstedt).
Age and environment: Late Anisian, inner and middle shelf.
Unit: Trochitenkalk, Gaismühle bei Craislheim, and Marbach, southern Germany; also recorded in Late Ladinian *nodulosus* Zone, Hauptterebratelbank, from several localities in northern and southern Germany.
References: Schöndorf (1910); Hagdorn (1985); Röhl (1986); Ockert (1988).

*Order Brisingida

Taxon: *Brisingella* sp.
Age and environment: Late Burdigalian, slope and deep basin.
Unit: uppermost Toyohama Formation, Morozaki Group, Chita Peninsula, Japan.
References: Shikama and Kase (1976); Hayashida (1986); Yamaoka (1987).
Notes: Early Oligocene material mentioned in Zullo *et al.* (1964) and in the *Treatise* is benthopectinid (Blake 1973).

*Order Forcipulatida

Taxon: *Germanasterias amplipapularia* Blake, *Hystrixasterias hettangiurnus* Blake.
Age and environment: Early Hettangian, inner shelf.
Unit: Gemünd echinoderm bed, Gemünd, south-west Germany.
References: Rosenkranz (1971); Seilacher *et al.* (1985); Blake (1990).

*Order Valvatida

Taxon: *Diclidaster gevreyi* de Loriol.
Age and environment: Late Hettangian, inner/middle shelf.
Unit: Couche granuleuse à *S. angulata*, Mercuer (Ardèche), France.
References: Elmi and Mouterde (1965); Mouterde (1984); Blake (1980).
Notes: May be represented in the Late Ladinian–Early Carnian Cassian Formation, northern Italy, by isolated ossicles described by Zardini (1973) (Gale 1987).

*Order Notomyoida

Taxon: *Plesioastropecten hallovensis* Blake.
Age and environment: Late Hettangian, middle shelf.

Unit: Psiloceratenschichten, Hallau (Kt. Schaffhausen), Switzerland.
References: Schlatter (1983); Blake (1984).

*Order Paxillosida

Taxon: *Tethyaster jurassicus* Blake.
Age and environment: Middle Bajocian, nearshore.
Unit: Judd Hollow Formation, Carmel Formation, St. George, Utah.

Taxon: *Advenaster inermis* Hess, *Pentasteria kelleri* Hess.
Age and environment: Late Bajocian, nearshore/inner shelf.
Unit: Upper Hauptrogenstein, Develier and Steinbruch, Switzerland.
References: Hess (1955, 1972, 1983); Imlay (1980); Blakey *et al.* (1983); Blake (1986).

*Order Velatida

Taxon: *Solaster murchisoni* (Williamson) and, probably, *Plumaster ophiuroides* Wright.
Age and environment: *davoei* Zone, Early Pliensbachian, nearshore.
Unit: Lower Staithes Formation, Robin Hood's Bay, Yorkshire, Great Britain.

Taxon: *Tropidaster pectinatus* Forbes.
Age and environment: Early Pliensbachian, middle shelf.
Unit: *davoei* Zone, Mickelton Tunnel, near Chipping Camden (Gloucestershire), England.
References: Wright (1863–80); J.F. Blake (1887); Hemingway (1974); Howard (1985).

*Order Spinulosida

Taxon: *Echinaster jacobseni* Rasmussen.
Age and environment: Late Thanetian, outer shelf.
Unit: Fur Formation, Silstrup Cliff near Tilsted, Fur, and Mors, northern Denmark.
References: Rasmussen (1972); Pedersen (1981); Pedersen and Surlyk (1983); Heilmann-Clausen *et al.* (1985).
Notes: *Henricia? venturana* Durham and Roberts, reported as a Late Cretaceous echinasterid, is a benthopectinid from rocks now considered Late Eocene in age (Blake 1981).

References

Adegoke, O. S. (1977). Stratigraphy and paleontology of the Ewekoro Formation (Paleocene) of southwestern Nigeria. *Bulletins of American Paleontology*, **71**, 5-379.

Aitken, W. G. (1961). Geology and palaeontology of the Jurassic and Cretaceous of southern Tanganyika. *Geological Survey of Tanganyika, Bulletin*, **31**, 1-144.

Akers, W. H. (1972). Planktonic Foraminifera and biostratigraphy of some Neogene formations, northern Florida and Atlantic Coastal Plain. *Tulane Studies in Geology and Paleontology*, **9**, 1-139.

Alvinerie, J., Mayeux, C. and Rechiniac, A. (1964). Esquisse sédimentologique de la coupe de Biarritz. *Mémoires du Bureau de Recherches Géologiques et Minières*, **28**, 407-24.

Arnaud, H., Gidon, M. and Thieuloy, J. -P. (1981). Les Calcaires du Fontanil des environs de Grenoble: Leur place dans la stratigraphie du Néocomien entre le Jura et le domaine vocontien. *Eclogae Geologicae Helvetiae*, **74**, 109-37.

Arnaud, M. and Monleau, C. (1979). Étude d'évolution d'une plateforme carbonatée: Exemple de la Provence au Jurassique (Hettangien-Oxfordien). *Travaux, Laboratoire de Géologie Historique et Paléontologie, Marseille*, no. 9 (two volumes).

Bate, R. H. (1975). Ostracods from Callovian to Tithonian sediments of Tanzania, East Africa. *Bulletin of the British Museum (Natural History), Geology*, **26**, 161-223.

Bayer, F. M. (1979). The correct name of the Helioporacean octocoral *Lithotelesto micropora* Bayer and Muzik. *Proceedings of the Biological Society of Washington*, **92**, 873-5.

Bayer, F. M. (1981). On some genera of stoloniferous octocorals (Coelenterata: Anthozoa), with descriptions of new taxa. *Proceedings of the Biological Society of Washington*, **94**, 878-901.

Beizel, A. L. (1977). *Khetella*, a new gastropod genus from the Upper Jurassic and Lower Cretaceous of northern Eurasia. *Paleontological Journal*, **11**, 136-7.

Beizel, A. L. (1983). *The Jurassic and early Cretaceous gastropods of the north of middle Siberia*. Nauka, Moscow. (In Russian).

Benshili, K. (1989). Lias-Dogger du Moyen-Atlas plissé (Maroc): Sédimentologie, biostratigraphie et évolution paléogéographique. *Documents des Laboratoires de Géologie, Université Claude-Bernard, Lyon*, **106**, 1-285.

Blake, D. B. (1973). Ossicle morphology of some Recent asteroids and description of some west American fossil asteroids. *University of California Publications in Geological Sciences*, **104**, 1-59.

Blake, D. B. (1980). On the affinities of three small sea star families. *Journal of Natural History*, **14**, 163-82.

Blake, D. B. (1981). The new Jurassic sea star *Eokinaster* and comments on life habits and the origin of the modern Asteroidea. *Journal of Paleontology*, **55**, 33-46.

Blake, D. B. (1984). The Benthopectinidae (Asteroidea: Echinodermata) of the Jurassic of Switzerland. *Eclogae Geologicae Helvetiae*, **77**, 631–47.

Blake, D. B. (1986). Some new post-Paleozoic sea stars (Asteroidea: Echinodermata) and comments on taxon endurance. *Journal of Paleontology*, **60**, 1103–19.

Blake, D. B. (1987). A classification and phylogeny of post-Palaeozoic sea stars (Asteroidea: Echinodermata). *Journal of Natural History*, **21**, 481–528.

Blake, D. B. (1989). Asteroidea: Functional morophology, classification and phylogeny. *Echinoderm Studies*, **3**, 179–223.

Blake, D. B. (1990). Hettangian Asteriidae (Echinodermata: Asteroidea) from southern Germany: Taxonomy, phylogeny and life habits. *Paläontologische Zeitschrift*, **64**, 103–23.

Blake, J. F. (1887). On a new specimen of *Solaster Murchisoni* from the Yorkshire Lias. *Geological Magazine*, New Series, Decade III, **4**, 529–31.

Blakey, R. C., Peterson, F., Caputo, M. V., Geesaman, R. C. and Voorhees, B. J. (1983). Paleogeography of middle Jurassic continental, shoreline, and shallow marine sedimentation, southern Utah. In *Mesozoic paleogeography of west-central United States* (ed. M. W. Reynolds and E. D. Dolly), pp. 77–100. Society of Economic Paleontologists and Mineralogists, Denver.

Borsuk, A. M. and Sholpo, V. N. (1983). Correlation of endogenous processes in the alpine cycle of the Caucasus. In *Profiles of orogenic belts* (ed. N. Rast and F. M. Delany), pp. 97–143. Washington, DC.

Brandt, K. (1985). Sea-level changes in the Upper Sinemurian and Pliensbachian of southern Germany. In *Sedimentary and evolutionary cycles* (ed. U. Bayer and A. Seilacher), pp. 113–26. Springer-Verlag, Berlin.

Cairns, S. D. (1983). A generic revision of the Stylasterina (Coelenterata: Hydrozoa). Part 1. Description of the genera. *Bulletin of Marine Science*, **33**, 427–508.

Campbell, H. J., Grant-Mackie, J. A., and Paris, J. -P. (1985). Geology of the Moindou-Téremba area, New Caledonia. Stratigraphy and structure of Téremba Group (Permian–Lower Triassic) and Baie de St. -Vincent Group (Upper Triassic–Lower Jurassic). *Géologie de la France*, **1985** (1), 19–36.

Carr, T. R. and Paull, R. K. (1983). Early Triassic stratigraphy and paleogeography of the Cordilleran miogeosyncline. In *Mesozoic paleogeography of west-central United States* (ed. M. W. Reynolds and E. D. Dolly), pp. 39–56. Denver.

Castex, L. and Lambert, J. (1919). Révision des échinides des falaises de Biarritz. *Actes de la Société Linnéenne de Bordeaux*, **71**, 117–200.

Chabaglian, J. (1959). Le Paléocène dans l'ouest du Sénégal. *Annales de la Faculté des Sciences, Université de Dakar*, **4**, 133–87.

Ciarapica, G. (1985). Il Trias dell'unita di Porto Venere e confronti con le coeve successioni Apuane e Toscane: Revisione degli 'Strati a *R. contorta*' auctt. dell'Appennino settentrionale. *Memorie della Societa Geologica Italiana*, **30**, 135–51.

Ciarapica, G., Cirilli, S. and Passeri, L. (1982). La Series Triassica del M. Cetona (Toscana Medirionale) e suo confronto con quella di La Spezia. *Memorie della Societa Geologica Italiana*, **24**, 155–67.

Clavel, B., Charollais, J., Busnardo, R. and Le Hegarat, G. (1986). Précisions stratigraphiques sur le Crétacé inférieur du Jura méridional. *Eclogae Geologicae Helvetiae*, **79**, 319–41.

Colo, G. (1961–64). Contribution à l'étude du Jurassique du Moyen Atlas Septentrionale. *Notes et Mémoires du Service Géologique, Maroc*, **139**, 5–226.

Contini, D. (1986). Notice explicative de la feuille Champlitte à 1/50 000. *Carte Géologique de la France a 1/50 000*, **XXXII-21**.

Cossmann, M. (1895–96). Contribution à la paléontologie française des terrains jurassiques. *Mémoires de la Société Géologique de France, Paléontologie*, **14**, 1–167.

Cox, L.R. (1965). Jurassic Bivalvia and Gastropoda from Tanganyika and Kenya. *Bulletin of the British Museum (Natural History), Geology, Supplement*, **1**, 213 pp.

Dagallier, G. (1977). Une série carbonatée littorale: Le Lias moyen (a Pb-Ba) de Mibladen (Maroc). *Sciences de la Terre*, **21**, 53–101.

Devriès, A. (1960). Contribution à l'étude de quelques groupes d'échinides fossiles d'Algérie. *Mémoires du Service de la Carte Géologique d'Algérie, Paléontologie*, **5**, 276 pp.

Devriès, A. (1973). Suite aux échindes d'Algérie. *Publications du Service de la Carte Géologique d'Algérie, Bulletin*, **44**, 45–100.

Donovan, S.K. (1987). A second B.W. Arnold collection of Jamaican fossil echinoids. *Journal of the Geological Society of Jamaica*, **24**, 36–43.

Donovan, S.K. (1988). A preliminary biostratigraphy of the Jamaican fossil Echinoidea. In *Echinoderm Biology: Proceedings of the 6th International Echinoderm Conference* (ed. R.D. Burke *et al.*), pp. 125–31. Balkema, Rotterdam.

Ducasse, O. (1973). Notes complémentaires à l'étude micropaléontologique (Ostracodes) des falaises de Biarritz. *Bulletin de l'Institut de Géologie du Bassin d'Aquitaine*, **14**, 89–102.

du Chene, R.E.J., Perch-Nielsen, K. and Petters, S.W. (1978). New stratigraphic data on the Paleogene Ewekoro and Akinbo Formations (southwestern Nigeria). *Archives des Sciences, Genève*, **31**, 203–12.

Du Dresnay, R. (1966). Observations stratigraphiques au sujet des gisements de Gastéropodes liasiques du Causse moyen-atlasique et du Haut Atlas oriental, étudies par R. Rourrouilh. *Notes et Mémoires du Service Géologique, Maroc*, **196**, 5–23.

Elmi, S. (coord.) (1984). Jurassique moyen: Dogger. *Mémoires du Bureau de Recherches Géologiques et Minières*, **125**, 177–221.

Elmi, S. and Mouterde, R. (1965). Le Lias inférieur et moyen entre Aubenas et Privas (Ardèche). *Travaux des Laboratoires de Géologie de la Faculté des Sciences de Lyon, Nouvelle Série*, **12**, 143–246.

Élouard, P. (1962). Étude géologique et hydrogéologique des formations sédimentaires du Guelba mauritanien et de la Vallée du Sénégal. *Mémoires du Bureau de Recherches Géologiques et Minières*, **7**, 274 pp.

Ensom, P.C. (1985). *Pyriporopsis portlandensis* Pohowsky 1973, a bryozoan from the Scallop Member, Purbeck Limestone Formation of Worbarrow Tout, Dorset. *Proceedings of the Dorset Natural History and Archaeological Society*, **106**, 167.

Eva, A.N. (1976). The palaeocology and sedimentology of Middle Eocene larger Foraminifera in Jamaica. *Maritime Sediments Special Publication*, **1**, 467–75.

Fan, J.S., Zhang, W., Ma, X., Zhang, Y.B. and Liu, H.B. (1982). The Upper Permian reefs in Lichuan district, west Hubei. *Scientia Geologica Sinica*, **1982** (3), 274–82. (In Chinese.)

Floris, S. (1979). Maastrichtian and Danian corals from Denmark. In *Cretaceous–Tertiary Boundary Events Symposium. I. The Maastrichtian and Danian of Denmark* (ed. T. Birkelund and R.G. Bromley), pp. 92–4. University of Copenhagen, Copenhagen.

Floris, S. (1980). The coral banks of the Danian of Denmark. *Acta Palaeontologica Polonica*, **25**, 531–40.

Galacz, A., Horvath, F. and Voros, A. (1985). Sedimentary and structural evolution of the Bakony Mountains (Transdanubian Central Range, Hungary): paleogeographic implications. *Acta Geologica Hungarica*, **28**, 85–100.

Gale, A.S. (1987). Phylogeny and classification of the Asteroidea (Echinodermata). *Zoological Journal of the Linnean Society*, **89**, 107–32.

Garcia, D., Gonzalez, R., Delgado, R., Brito, A. and Linares, E. (1987). Paleogeno de Cuba: Unidades en el mapa geologico (II Parte). *Serie Geologia—Centro de Investigaciones Geologica*, **1987** (3), 87–103.

Gardner, J. (1947). The molluscan fauna of the Alum Bluff Group of Florida. *United States Geological Survey Professional Paper*, **142**.

Geyer, O.F. and Gwinner, M.P. (1986). *Geologie von Baden-Württemberg (3rd edn).* E. Schweizerbart'sche Verlagsbuchhandlung, Stuttgart.

Glazunova, A.E. (1973). *Paleontological basis of the stratigraphical separation of the Cretaceous of the Volga region: Lower Cretaceous.* 'Nedra', Moscow. (In Russian.)

Glennie, K.W., Boeuf, M.G.A., Hughes Clarke, M.W., Moody-Stuart, M., Pilaar, W.F.H. and Reinhardt, B.M. (1974). Geology of the Oman Mountains. *Verhandelingen van het Koninklijk Nederlands geologisch mijnbouwkundig Genootschap*, **31**, 1–423.

Gluchowski, E. and Boczarowski, A. (1986). Crinoids from the Diplopora-Dolomite (Middle Muschelkalk) of Piekary Slaskie, Upper Silesia. *Bulletin of the Polish Academy of Sciences, Earth Sciences*, **34**, 189–96.

Govoni, D.L. (1983). Gastropod molluscs from the Brightseat Formation (Paleocene: Danian) of Maryland. Unpublished MS thesis, The George Washington University.

Gray, J. (1988). Evolution of the freshwater ecosystem: The fossil record. *Palaeogeography, Palaeoclimatology, Palaeoecology*, **62**, 1–214.

Gregory, J.W. (1897). On the affinities of the Echinothuridae; and on *Pedinothuria* and *Halikodiadema*, two new genera of Echinoidea. *Quarterly Journal of the Geological Society of London*, **58**, 112–22.

Guillaume, S. (1966–7). Le Crétacé du Jura français. *Bulletin du Bureau de Recherches Géologiques et Minières*, **1966** (1), 1–43; **1966** (2), 7–69; **1966** (3), 11–79; **1966** (6), 65–141; **1967** (1), 155–92.

Gwinner, M.P. (1976). Origin of the Upper Jurassic limestones of the Swabian

Alb (SW Germany). *Contributions to Sedimentology*, **5**, 75 pp.
Hagdorn, H. (1985). Immigration of crinoids into the German Muschelkalk Basin. In *Sedimentary and evolutionary cycles* (ed. U. Bayer and A. Seilacher), pp. 237–54. Springer-Verlag, Berlin.
Hagdorn, H. (1986). *Isocrinus? dubius* (Goldfuss, 1831) aus dem Unteren Muschelkalk (Trias, Anis). *Zeitschrift für Geologische Wissenschaften, Berlin*, **16**, 705–27.
Hallam, A. (1965). Observations on marine Lower Jurassic stratigraphy of North America, with special reference to the United States. *American Association of Petroleum Geologists Bulletin*, **49**, 1485–501.
Hanzlikova, E. and Roth, Z. (1963). Lithofacies, biofacies and sedimentary conditions in the Cretaceous beds of the flysch zone in the Czechoslovak Carpathians. *Geologicky Sbornik*, **14**, 83–102.
Hayashida, A. (1986). Timing of rotational motion of southwest Japan inferred from paleomagnetism of the Setouchi Miocene Series. *Journal of Geomagnetism and Geoelectricity*, **38**, 295–310.
Heilmann-Clausen, C., Nielsen, O.B. and Gersner, F. (1985). Lithostratigraphy and depositional environments in the Upper Paleocene and Eocene of Denmark. *Bulletin of the Geological Society of Denmark*, **33**, 287–323.
Hemingway, J. E. (1974). Jurassic. In *The geology and mineral resources of Yorkshire* (ed. D.H. Rayner and J.E. Hemingway), pp. 161–223. Yorkshire Geological Society, Leeds.
Hess, H. (1955). Die fossilen Astropectiniden (Asteroidea). *Schweizerische Paläontologische Abhandlungen*, **71**, 1–113.
Hess, H. (1972). Eine Echinodermen-fauna aus dem mittlern Dogger des Aargauer Juras. *Schweizerische Paläontologische Abhandlungen*, **92**, 1–87.
Hess, H. (1983). Ein neuer Seestern (*Testudinaster peregrinus* n.g.n.sp.) aus dem mittleren Dogger des Aargauer Juras. *Eclogae Geologicae Helvetiae*, **76**, 919–25.
Hoffmann, K. (1982). Die Stratigraphie, Paläogeographie und Ammonitenführung des Unter-Pliensbachium (Carixium, Lias gamma) in Nordwest-Deutschland. *Geologisches Jahrbuch*, **A55**, 3–439.
Howard, A.S. (1985). Lithostratigraphy of the Staithes Sandstone and Cleveland Ironstone Formations (lower Jurassic) of north-east Yorkshire. *Proceedings of the Yorkshire Geological Society*, **45**, 261–75.
Hudson, R.G.S. (1960). The Permian and Trias of the Oman Peninsula. *Geological Magazine*, **97**, 299–308.
Hudson, R.G.S. and Jeffries, R.P.S. (1961). Upper Triassic brachiopods and lamellibranchs from the Oman Peninsula, Arabia. *Palaeontology*, **4**, 1–41.
Imlay, R.W. (1980). Jurassic paleobiogeography of the conterminous United States and its continental setting. *United States Geological Survey Professional Paper*, **1062**.
Jacobshagen, V. (1965). Die Allgäu-Schichten (Jura-Fleckenmergel) zwischen Wettersteingebirge und Rhein. *Jahrbuch der Geologischen Bundesanstalt, Wien*, **108**, 1–114.

Jäger, M. (1985). Die Crinoiden aus dem Pliensbachium (mittlerer Lias) von Rottorf am Klei und Emplede (Süd-Niedersachsen). *Bericht der Naturhistorischen Gesellschaft zu Hannover*, **128**, 71–151.

Jenny, J. (1988). Carte géologique du Maroc au 1/000 000. Feuille Azilal (Haut Atlas Central), mémoire explicatif. *Notes et Mémoires du Service Géologique, Maroc*, **339** bis, 104 pp.

Jesionek-Szymanska, W. (1963). Échinides irréguliers du Dogger de Pologne. *Acta Palaeontologica Polonica*, **3**, 293–414.

Jesionek-Syzmanska, W. (1970). On a new pygasterid (Echinoidea) from the Jurassic (Middle Lias) of Nevada, U.S.A. *Acta Palaeontologica Polonica*, **15**, 411–23.

Jesionek-Syzmanska, W. (1978). On a new galeropygid genus (Echinoidea) from the Jurassic (Upper Lias) of Morocco. *Acta Palaeontologica Polonica*, **23**, 187–93.

Just, J. (1974). On *Palaeogammarus* Zaddach, 1864, with a description of a new species from western Baltic amber (Crustacea, Amphipoda, Crangonycidae). *Steenstrupia*, **3**, 93–9.

Kier, P.M. (1967). Revision of the oligopygoid echinoids. *Smithsonian Miscellaneous Collections*, **152**, (2), 147 pp.

Kier, P.M. (1974). Evolutionary trends and their functional significance in the post-Paleozoic echinoids. *Paleontological Society Memoir* **5**. (*Journal of Paleontology*, **48** (Supplement to No. 3), 95 pp.

Kier, P.M. (1977). Triassic echinoids. *Smithsonian Contributions to Paleobiology*, **30**.

Kier, P.M. (1982). Rapid evolution in echinoids. *Palaeontology*, **25**, 1–10.

Kier, P.M. (1984). Echinoids from the Triassic (St. Cassian) of Italy, their lantern supports, and a revised phylogeny of Triassic echinoids. *Smithsonian Contributions to Paleobiology*, **56**.

Klikushin, V.G. (1975). Mechanics of the column in the Bourgueticrinidae. *Paleontological Journal*, **9** (1), 121–4.

Klikushin, V.G. (1982). Cretaceous and Paleogene Bourgueticrinina (Echinodermata, Crinoidea) of the USSR. *Géobios*, **15**, 811–43.

Klikushin, V.G. (1985). Turonian, Coniacian and Santonian deposits of the Belbeck River, Crimea. *Byulleten Moskovskoe Obshchestvo Ispytatelei Prirody, Otdel Geologscheskii*, **60** (2), 69–82 [In Russian].

Klikushin, V.G. (1987*a*). Thiolliericrinid crinoids from the Lower Cretaceous of Crimea. *Géobios*, **20**, 625–65.

Klikushin, V.G. (1987*b*). Distribution of crinoid remains in the Triassic of the U.S.S.R. *Neues Jahrbuch für Geologie und Paläontologie, Abhandlungen*, **173**, 321–38.

Kollmann, H.A. (1982). Gastropoden-Faunen aus der höheren Unterkreide Nordwestdeutschlands. *Geologisches Jahrbuch*, **A65**, 517–51.

Kristan-Tollmann, E. (1988). *Palaeocomaster styriacus* n.sp., eine Schwebcrinoide (Comatulida) aus dem alpinen Lias des Salzkammergutes, Österreich. *Geologica et Palaeontologica*, **22**, 81–7.

Kristan-Tollmann, E. and A. Tollmann (1967). Crinoiden aus dem zentralalpinen Anis (Leithagebirge, Thörler Zug und Radstädter Tauern). *Wissenschaftliche Arbeiten aus dem Burgenland, Eisenstadt*, **36**, 1–55.

Ksiazkiewicz, M. (1975). Bathymetry of the Carpathian flysch basin. *Acta Geologica Polonica*, **25**, 309–27.
Kummel, B. (1957). Paleoecology of Lower Triassic formations of southeastern Idaho and adjacent areas. *Geological Society of America Memoir*, **67** (2), 437–68.
Kusmicheva, Y.I. (1975). Systematic composition and development of the family Helioporidae (Octocoralla). *Paleontological Journal*, **1975** (3), 278–86.
Laffitte, R. (1939). Étude géologique de l'Aurès. *Bulletin du service de la carte géologique de l'Algérie*, 2e série, **15**.
LaMoreaux, P.E. and Toulmin, L.D. (1959). Geology and ground-water resources of Wilcox County, Alabama. *Alabama Geological Survey County Report*, **4**.
Larsson, S.G. (1978). Baltic amber—a palaeobiological study. *Entomonograph*, **1**, 1–192.
Le Renard, J. (1983). Mise en évidence d'algueraies à *Caulerpa* par les Juliidae (Gastéropodes à 2 valves: Sacoglossa) dans l'Éocène du bassin de Paris. *Géobios*, **16**, 39–51.
Mahel', M. *et al*, (1968). *Regional geology of Czechoslovakia*. Part II. *The West Carpathians*. Stuttgart.
Malik, K. and Olszewska, B. (1984). Sedimentological and micropaleontological study of the Grodziszcze Beds at Zegocina (Flysch Carpathians). *Annales Societatis Geologorum Poloniae*, **54**, 293–334.
Mancini, E.A. (1984). Biostratigraphy of Paleocene strata in southwestern Alabama. *Micropaleontology*, **30**, 268–91.
Marcus, E. d. B. -R. (1972). On the Anaspidea (Gastropoda: Opisthobranchia) of the warm waters of the western Atlantic. *Bulletin of Marine Science*, **22**, 841–74.
McFarlane, N. (1977). Some Eocene and Oligocene faunas from central Jamaica. *Secondo Congreso Latinoamericano de Geologia, Memoria*, **3**, 1393–411.
Mégnien, C. (1967). Auxerre. *Carte Géologique de la France au 1/50 000*, **XXVI–20**.
Mégnien, F. (ed.) (1980). Synthèse géologique du Bassin de Paris. III. Lexique des noms de formation. *Mémoires du Bureau de Recherches Géologiques et Minières*, **103**.
Meyn, H. and Vespermann, J. (1987). Critical study of the type localities and type stages published by Koch 1851, Roemer 1841, 1842, and Reuss 1863 in the Lower Cretaceous of N.W. Germany. *Newsletters on Stratigraphy*, **18**, 7–19.
Mintz, L.W. (1968). Echinoids of the Mesozoic families Collyritidae d'Orbigny, 1853 and Disasteridae Gras, 1848. *Journal of Paleontology*, **42**, 1272–88.
Monciardini, C. (1966). La Sédimentation Éocène au Sénégal. *Mémoires du Bureau de Recherches Géologiques et Minières*, **43**.
Morand, M. (1913). Étude de la fauna des calcaires valanginiens du Fontanil (Isère). *Travaux du Laboratoire de Géologie, Faculté des Sciences, Université Grenoble*, **10**, 193–284.
Morycowa, E. (1988). Middle Triassic Scleractinia from the Cracow-Silesia region, Poland. *Acta Palaeontologica Polonica*, **33**, 91–121.

Morycowa, E. and Roniewicz, E. (1986). Class Anthozoa Ehrenberg, 1834. In *Geology of Poland.* Vol. III. *Atlas of guide and characteristic fossils.* Part 2a. *Mesozoic, Triassic* (ed. L. Malinowska), pp. 34–38. Wydawnictwa Geologiczne, Warsaw.

Mouterde, R. (1953). Études sur le Lias et le Bajocien des bordures nord et nord-est du Massif Central Francais. *Bulletin du Service de la Carte Géologique de la France*, **50** (236).

Mouterde, R. (coord.) (1984). Jurassique inférieur: Lias. *Mémoires du Bureau de Recherches Géologiques et Minières*, **125**, 119–76.

Mouty, M. (1966). Le Néocomien dans le Jura Méridional. Thèse 1369, Université Genève.

Mudge, D.C. (1978). Stratigraphy and sedimentation of the Lower Inferior Oolite of the Cotswolds. *Journal of the Geological Society of London*, **135**, 611–27.

Mutterlose, J. (1984). Die Unterkreide-Aufschlüsse (Valangin-Alb) im Raum Hannover–Braunschweig. *Mitteilungen aus dem Geologischen Institut der Technische Universität Hannover*, **24**.

Mutterlose, J. and Harding, I. (1987). Phytoplankton from the anoxic sediments of the Barremian (Lower Cretaceous) of north-west Germany. *Abhandlungen der Geologischen Bundesanstalt, Wien*, **39**, 177–215.

Nalivkin, D.V. (1973). *Geology of the U.S.S.R.* Oliver and Boyd, Edinburgh.

Ockert, W. (1988). Lithostratigraphie und Fossilführung des Trochitenkalks (Unterer Hauptmuschelkalk, mo1) im Raum Hohenlohe. In *Neue Forschungen zur Erdgeschichte von Crailsheim* (ed. H. Hagdorn), pp. 43–69. Gesellschaft für Naturkunde in Württemburg, Stuttgart.

Oravecz-Scheffer, A. (1987). Triassic foraminifers of the Transdanubian central range. *Geologica Hungarica, Series Palaeontologica*, **50**.

Palmer, K.V.W. and Brann, D.C. (1966). Catalogue of the Paleocene and Eocene Mollusca of the southern and eastern United States. *Bulletins of American Paleontology*, **48**.

Passeri, L. and Pialli, G. (1973). L'ambiente di sedimentazione dei Calcari a *R. contorta* dell' Umbria occidentale e del M. Cetona. *Geologica Romana*, **12**, 177–203.

Pedersen, G.K. (1981). Anoxic events during sedimentation of a Palaeogene diatomite in Denmark. *Sedimentology*, **28**, 487–504.

Pedersen, G.K. and Surlyk, F. (1983). The Fur Formation, a late Paleocene ash-bearing diatomite from northern Denmark. *Bulletin of the Geological Society of Denmark*, **32**, 43–65.

Petitot, M. -L. (1959). Contribution à l'étude des échinides fossiles du Maroc (Jurassique et Crétacé). *Notes et Mémoires du Service Géologique, Maroc*, **146**.

Pomerol, C. (1980). Biarritzien. *Mémoires du Bureau de Recherches Géologiques et Minières*, **109**, 224–6.

Rasmussen, H.W. (1972). Lower Tertiary Crinoidea, Asteroidea and Ophiuroidea from northern Europe and Greenland. *Biologiske Skrifter udgivet af Det Kongelige Danske Videnskabernes Selskab*, **19** (7), 1–83.

Rat, P., David, B., Magniez-Jannin, F. and Pernet, O. (1987). Le golfe du Crétacé inférieur sur le Sud-Est du Bassin parisien: Milieux (échinides, foraminifères) et évolution de la transgression. *Mémoires Géologiques de l'Université de Dijon*, **11**, 15–29.

Rey, J., Canérot, J., Peybernès, S., Taj-Eddine, K. and Thieuloy, (1988). Lithostratigraphy, biostratigraphy and sedimentary dynamics of the Lower Cretaceous deposits on the northern side of the Western High Atlas (Morocco). *Cretaceous Research*, **9**, 141–58.

Ricateau, R. and Riché, P. H. (1980). Geology of the Musandam Peninsula (Sultanate of Oman) and its surroundings. *Journal of Petroleum Geology*, **3**, 139–52.

Richardson, L. and Talbot Paris, E. (1908). On the stratigraphical and geographical distribution of the Inferior-Oolite echinoids of the west of England. *Proceedings of the Cotteswold Naturalists' Field Club*, **16**, 151–93.

Rigby, J.K., Racki, G. and Wrozlek, T. (1981). Occurrence of dyctyid hexactinellid sponges in the Upper Devonian of the Holy Cross Mts. *Acta Geologica Polonica*, **31**, 163–8.

Rittener, T. (1902). Étude géologique de la Cote-aux-Fées et des environs de Saints-Croix et Baulmes. *Matériaux pour la Carte Géologique suisse*, N.S., **13**, 1–116.

Robinson, E. (1987). Late Cretaceous and early Tertiary sedimentary rocks of the Central Inlier, Jamaica. *Journal of the Geological Society of Jamaica*, **24**, 49–67.

Robinson, R. N. (1986). Stratigraphy of the Chalk Group of the North Downs, southeast England. *Proceedings of the Geologists' Association*, **97**, 141–70.

Röhl, U. (1986). Feinstratigraphie und Mikrofazies des Oberen Muschelkalk im Hildesheimer Wald. *Neues Jahrbuch für Geologie und Paläontologie, Monatshefte*, **1986**, 489–511.

Roman, J. and Gorodiski, A. (1959). Échinides Eocènes du Sénégal. *Notes du Service de Géologie et de Prospection Minière, Dakar*, **3**, 1–91.

Rose, E. P. F. (1982). Holectypoid echinoids and their classification. In *Echinoderms: Proceedings of the 4th International Echinoderm Conference, Tampa Bay* (ed. J.M. Lawrence), pp. 145–52. Balkema, Rotterdam.

Rose, E. P. F. and Olver, J. B. S. (1985). Slow evolution in the Holectypidae, a family of primitive irregular echinoids. In *Echinodermata: Proceedings of the 4th International Echinoderm Conference, Galway* (ed. B.F. Keegan and B.D.S. O'Connor), pp. 81–9. Balkema, Rotterdam.

Rose, E.P.F. and Olver, J.B.S. (1988). Jurassic echinoids of the family Menopygidae: Implications for the evolutionary interpretation and classification of early Irregularia. In *Echinoderm Biology: Proceedings of the 6th International Echinoderm Conference* (ed. R.D. Burke *et al.*). pp. 149–58. Balkema, Rotterdam.

Rosenkrantz, D. (1971). Fossil-Lagerstätten Nr. 16: Zur Sedimentologie und Ökologie von Echinodermen-Lagerstätten. *Neues Jahrbuch für Geologie und Paläontologie, Abhandlungen*, **158**, 221–58.

Schlatter, R. (1983). Erstnachweis des tiefsten Hettangium Klettgau (Kanton Schaffhausen, Schweiz). *Mitteilungen, naturforschend Gesellschaft Schaffhausen*, **32**, 1–18.

Schmidt, W. (1986). Alum Bluff, Liberty County, Florida. In *Geological Society of America Centennial Field Guide, Southeastern Section* (ed. T.L. Neathery), pp. 355–7. Geological Society of America, Boulder, Colorado.

Schöndorf, F. (1910). Die Asteriden der deutschen Trias. *Niedersächsischer geologischer Verein, Hannover, Jahresbericht*, **3**, 90–116.

Schram, F. R. (1986). *Crustacea*. Oxford University Press, Oxford.

Searle, M. P., James, N. P., Calon, T. J. and Smewing, J. D. (1983). Sedimentological and structural evolution of the Arabian continental margin in the Musandam Mountains and Dibba zone, United Arab Emirates. *Geological Society of America Bulletin*, **94**, 1381–400.

Seilacher, A., Reif, W.-E. and Westphal, F. (1985). Sedimentological, ecological and temporal patterns of fossil Lagerstätten. *Philosophical Transactions of the Royal Society of London*, **B311**, 5–23.

Sepkoski, J. J., Jr. (1982). A compendium of fossil marine families. *Milwaukee Public Museum Contributions in Biology and Geology*, **51**.

Sheng Jin-zhiang, Rui Lin, and Chen Chu-zhen (1985). Permian–Triassic sedimentary facies and paleogeography of South China. In *The Tethys: her paleogeography and paleobiogeography from Paleozoic and Mesozoic* (ed. K. Nakazawa and J. M. Dickens), pp. 59–80. Tokai University Press, Tokyo.

Shikama, T. and Kase, T. (1976). Molluscan fauna of the Miocene Morozaki Group in the southern part of Chita Peninsula, Aichi Prefecture, Japan. *Science Reports, Yokohama National University*, Section 2, **23**, 1–25.

Sieverts, H. (1932). Kolonien von *Cotyloderma lineai* Quenstedt (Crinoidea) aus dem süddeutschen Lias. *Paläontologische Zeitschrift*, **14**, 96–107.

Simms, M. J. (1988*a*). The phylogeny of post-Palaeozoic crinoids. In *Echinoderm phylogeny and evolutionary biology* (ed. C. R. C. Paul and A. B. Smith), pp. 269–84. Oxford University Press, Oxford.

Simms, M. J. (1988*b*). An intact comatulid crinoid from the Toarcian of southern Germany. *Stuttgarter Beiträge zur Naturkunde*, **B140**.

Simms, M. J. (1990). Crinoids across the Triassic–Jurassic boundary. *Cahiers Scientifiques de l'Institut Catholique de Lyon*. (In press.)

Smith, A. B. (1981). Implications for lantern morphology for the phylogeny of post-Palaeozoic echinoids. *Palaeontology*, **24**, 779–801.

Smith, A. B. (1984). *Echinoid palaeobiology*. George Allen and Unwin, London.

Smith, A. B. (1988). Phylogenetic relationship, divergence times, and rates of molecular evolution for camarodont sea urchins. *Molecular Biology and Evolution*, **5**, 345–65.

Smith, A. B. (1990). Echinoid evolution from the Triassic to Lower Liassic. *Cahiers Scientifiques de l'Institut Catholique de Lyon*. (In press.)

Smith, A. B. and Hollingworth, N. T. J. (1990). Tooth structure and phylogeny of the Upper Permian echinoid *Miocidaris keyserlingi*. *Proceedings of the Yorkshire Geological Society*, **48**, 47–60.

Smith, A. B. and Wright, C. W. (1989). British Cretaceous echinoids. Part 2, Echinothuroida, Diadematoida and Stirodonta (1: Calycina). *Palaeontographical Society Monograph*, **143**, 101–98.

Solovjev, A. N. (1971). Late Jurassic and Early Cretaceous disasterid echinoids of the USSR. *Trudy Paleontologicheskogo Instituta, Akademiya Nauk SSSR*, **131**.

Solovjev, A. N. (1989). New echinoid genus *Eoholaster* (Spatangoida, Holasteridae) from the Berriasian. In *Fossil and Recent Echinoderm Researches*, pp. 148–55. Akademiya Nauk Estonskoi SSR, Tallinn.

Stanley, K. O. (1971). Tectonic and sedimentologic history of Lower Jurassic

Sunrise and Dunlap Formations, west-central Nevada. *American Association of Petroleum Geologists Bulletin*, **55**, 454–76.

Stanley, G. D., Jr (1988). The history of early Mesozoic reef communities: A three-step process. *Palaios*, **3**, 170–83.

Steinhauser, N. and Charollais, J. (1971). Observations nouvelles et réflexions sur la stratigraphie du 'Valanginien' de le région neuchateloise et ses rapports avec le Jura méridional. *Géobios*, **4**, 7–59.

Steinhauser, N. and Lombard, A. (1969). Définition de nouvelles unités litho-stratigraphiques dans le Crétacé inférieur du Jura méridional (France). *Compte Rendu des Séances de la Société de Physique et d'Histoire Naturelle, Genève*, **4**, 100–13.

Talbot Paris, E.T. (1908). Notes on some echinoids from the Lias of Worcestershire, Gloucestershire and Somerset. *Proceedings of the Cotteswold Naturalists' Field Club*, **16**, 143–50.

Taylor, J.D. and Morris, N.J. (1988). Relationships of neogastropods. *Malacological Review Supplement*, **4**, 167–79.

Taylor, P.D. (1981). Bryozoa from the Jurassic Portland Beds of England. *Palaeontology*, **24**, 863–75.

Thompson, T.E. (1976). *Biology of opisthobranch molluscs* Vol. I. Ray Society, London.

Thompson, T.E. and Seaward, D.R. (1989). Ecology and taxonomic status of the aplysiomorph *Akera bullata* in the British Isles. *Journal of Molluscan Studies*, **55**, 489–96.

Townson, W.G. (1975). Lithostratigraphy and deposition of the type Portlandian. *Journal of the Geological Society of London*, **131**, 619–38.

Vokes, H.E. (1989). An overview of the Chipola Formation, northwestern Florida. *Tulane Studies in Geology and Paleontology*, **22**, 13–24.

Warme, J.E. (1988). Jurassic carbonate facies of the Central and Eastern High Atlas rift, Morocco. In *The Atlas System of Morocco* (ed. V.H. Jacobshagen), pp. 169–199. Springer-Verlag, Berlin.

Wiedmann, J., Butt, A. and Einsele, G. (1982). Cretaceous stratigraphy, environments, and subsidence history at the Moroccan continental margin. In *Geology of the Northwest African continental margin* (ed. U. von Rad, K. Hinz, M. Sarnthen and E. Seibold), pp. 366–395. Berlin.

Wood, R. (1987). Biology and revised systematics of some Late Mesozoic stromatoporoids. *Special Papers in Palaeontology*, **37**.

Wright, T. (1863–80). British fossil Echinodermata of the Oolitic formations. Vol. 2. Asteroidea and Ophiuroidea. *Palaeontographical Society Monograph*.

Yamaoka, M. (1987). Fossil asteroids from the Miocene Morozaki Group, Aichi Prefecture, central Japan. *Kaseki No Tomo (Publication of the Tokai Fossil Society)*, **31**, 5–23.

Zardini, R. (1973). *Fossili di Cortina: Atlante degli echinodermi Cassiani Trias mediosuperiore della regione dolomitica attorno a Cortina d'Ampezzo*. Foto Ghedina, Cortina d'Ampezzo.

Zawidzka, K. (1975). Conodont stratigraphy and sedimentary environment of the Muschelkalk in Upper Silesia. *Acta Geologica Polonica*, **25**, 217–46.

Zullo, V., Kaar, R.F., Durham, J.W. and Allison, E.C. (1964). The echinoid genus *Salenia* in the eastern Pacific. *Palaeontology*, **7**, 331–49.

3. Major evolutionary radiations of the planktonic Foraminiferida

M.B. HART
Department of Geological Sciences, Polytechnic South West, Drake Circus, Plymouth, UK

Abstract

The planktonic Foraminiferida are passive floaters in the water column of the world's oceans and are capable only of small diurnal movements in response to feeding strategies. Appearing in the Triassic/Jurassic they evolved slowly through the Jurassic and Early Cretaceous, but became more diverse in the mid-Late Cretaceous. Following the K/T extinction event the group expanded rapidly, reaching an acme in the mid-Eocene. Following the Oligocene 'collapse' they have remained at or about the same level of diversity throughout the Neogene.

There are a limited number of strategies open to planktonic Foraminiferida in terms of test structure, morphology, and ornamentation. Studies of the test have shown the repeated appearance of certain characters, often many times over, in the post-Jurassic interval. Each evolutionary expansion appears to terminate at a distinct 'event', followed by re-colonization of the water column by surviving taxa. This pattern of development allows the planktonic Foraminiferida to be used as sensitive barometers of palaeoceanographic condition. There is, however, one major problem yet to be resolved. The evolution of the group determined from stratigraphic distribution, test morphology, and palaeoceanographic distribution appears to indicate that Jurassic, Cretaceous, and Neogene taxa follow the pattern observable in the modern ocean. In the Palaeogene, however, application of this same model results in a conflict with the published isotope data and this is still unresolved despite considerable effort.

Major Evolutionary Radiations (ed. P.D. Taylor and G.P. Larwood), Systematics Association Special Volume No. 42, pp. 59–72, Clarendon Press, Oxford, 1990.

Introduction

The planktonic Foraminiferida are extant, marine protozoans which originated from a benthic ancestor in the Triassic (Fuchs 1967, 1969; Oberhauser 1960; Kristan-Tollman 1964; Masters 1977; Hart 1980; Caron 1983*a, b*; Caron and Homewood 1983). Our knowledge of their biology is limited (for a major review see Bé 1977) and, while our understanding of their spatial distribution in the modern ocean is better, there are many areas where data are lacking. As a result, the geological interpretation of the group is severely hampered and much of the literature must be regarded as highly speculative.

Planktonic foraminifers are passive floaters capable only of small diurnal migrations (Rhumbler 1911; Bradshaw 1957; Parker 1960; Bé and Hamlin 1967; Tolderlund 1969) in the water column to facilitate feeding (on nannoplankton, small arthropods, etc.). Some species carry symbiotic zooxanthellae (Bé 1977, Pl. 1) and as such are restricted to the surface waters to a depth with appropriate light penetration. Other taxa have developed a life-style that is not dependent on such symbionts and are therefore free to exploit greater depths in the water column (down to 1000 m in the case of *Hastigerinopsis digitata* (Rhumbler)). During ontogeny the addition of calcareous test material causes a general sinking in the water column and, at the optimum depth for the individual taxon, they undergo gametogenesis. At this point the empty, adult, test is discarded (sinking to the sea floor) and the 'naked' juveniles tend to float upwards to begin their life cycle. The optimum depth of any taxon in the water column may be controlled by:

(1) the nature of the protoplasm (which may include the presence of vacuoles, fat bodies, etc.);

(2) the presence of long spines supporting the protoplasm as described in (1);

(3) the degree of calcification of the test and the presence of overgrowths during life (as distinct from those that may be produced immediately before, during, or after gametogesis);

(4) the presence of large apertures, accessory apertures, apertural modifications, enveloping overgrowths, etc;

(5) the limitations imposed by any symbiotic algae:

(6) the density, turbidity, and dynamics of the water column;

(7) the temperature stratification of the water column and the position of the thermocline;

(8) the presence, and intensity, of any oxygen minimum zone.

In the modern ocean (Bé 1977 and references therein) nine latitudinal zones can be recognized from plankton tow data, but only five latitudinally distributed zones can be identified from ocean floor sediments collected from above the carbonate lysocline. This is the direct result of mixing of the empty tests as they settle to the sea floor or become transported laterally by oceanic currents. For each of the latitudinal zones there is also a stratification within the water column (Bé 1977, table 4). The modern oceanic data implies three depth zones, as follows:

(1) *shallow water* — species living predominantly in the upper 50 m of the water column;

(2) *intermediate water* — species living in the upper 100 m of the water column but with adults predominantly between 50 and 100 m;

(3) *deep water* — species living in the upper few hundred metres of the water column and whose adult stages occur predominantly below 100 m

As indicated by Hart (1980), modern taxa identified within these zones often have characteristic morphological features, many of which can also be identified in fossil taxa. Because of all the factors outlined above, and the extremely simple genetic structure of foraminiferids, there are a limited number of things a planktonic foraminifer can do to its hard part morphology. The result of this is extensive homeomorphy, with many taxa (e.g. *Globorotalia* Cushman and *Rotalipora* Brotzen) having very few significant morphological differences. This feature of the planktonic Foraminiferida has been highlighted by many authors over a period of some 30 years.

Oceanic conditions

The planktonic Foraminiferida can be regarded as barometers of oceanic condition, passively responding to major changes in global circulation and climate. Some of the controls on the fauna can be summarized as follows:

(1) sea-level change: inundation of the shelves, increased erosion and changes in nutrient supply;

(2) climatic equitability: slow sluggish circulation;

(3) climatic extremes: active circulation, major oceanic turnovers;

(4) internal stratification: the result of vertical differences in temperature, salinity, and dissolved oxygen.

As always in palaeontological reconstructions, it is nearly impossible to identify the real cause of any faunal change in the geological record. Many of the changes are interconnected, thereby concealing the fundamental control. In the case of one parameter, temperature, isotope studies (Douglas and Savin 1978; Boersma and Shackleton 1981; Corfield 1987) have been used to indicate differences between surface waters and bottom sediments. The application of this data will be considered in a later section (p. 68).

Opportunities

The individual taxon, floating in the water column, can only migrate (or evolve) in three directions: horizontally within the same depth zone; downward into the zone below; and upward into the zone above. Studies to date appear to indicate that the latter option is at best uncommon or may not even be possible; subsequent evolutionary developments arise from taxa living in the shallow-water depth zone. Hart and Bailey (1979) highlighted the postulated downward migration of the *Praeglobotruncana* lineage, *Rotalipora* lineage, *P. helvetica* lineage, and the lineage to *Marginotruncana*. This was extended by Hart and Ball (1986) to embrace most Late Cretaceous taxa. The same conclusions were also arrived at by Caron (1983*a, b*) and have become generally accepted by most Cretaceous micropalaeontologists (e.g. Leckie 1989). In the Cainozoic Stainforth *et al.* (1975) highlighted similar patterns of evolution, especially those described by Jenkins (1971) and Toumarkine and Bolli (1970).

It seems logical that, if a taxon modifies its morphology by the addition of carbonate in the form of test thickening, ornamentation, keels, or other structures, this could cause a descent into the depth zone below. This process can be seen in studies of ontogeny where, for example, it is possible to see juvenile 'globigerine' chambers in the early whorl(s) which become more ornamented and keeled in the later whorls: a good example of this is *Contusotruncana fornicata* (Plummer). Banner and Desai (1988) have described a similar process operating in the Early Cretaceous with the development of muricocarinae and other forms of keel. The same downward migration with ontogeny has been recorded in the *Globigerinoides sacculifer* (Brady) to *Sphaeroidinella dehiscens* (Parker and Jones) transition by Bé (1965) and Bé and Hemleben (1970). It is also quite striking how, for example, the plano-convex morphotype appears several times in the Cretaceous to Recent interval. Hart and Ball (1986) attempted to show how such morphotypes (*Praeglobotruncana helvetica* (Bolli), *Dicarinella concavata* (Brotzen), *Globotruncanita calcarata* (Cushman), and *Gansserina gansseri* (Bolli)) follow each other in a form of evolutionary relay. In each of these quoted examples the ancestral taxa appear to occupy a shallower depth zone (as indicated both by morphological analogy to the Recent fauna and palaeoceanographic distribution).

The Cretaceous 'shallow water' fauna would appear to have been dominated by the genus *Hedbergella* during the Albian–Cenomanian, followed by *Whiteinella* and *Archaeoglobigerina* higher in the succession. This process of 'faunal replacement' was apparently quite slow, with no rapid evolutionary changes. This is in marked contrast to the 'intermediate water' and 'deeper water' faunas which change more frequently. Evolution within the 'shallow water' depth zone may therefore be the biological response to environmental pressures, slow climatic change, opening of new oceanic basins, opening of new oceanic connections, etc., and this only gives a rather subdued rate of speciation. Evolutionary changes involving migration into other water depth zones would be a response to stratification in the water column and this could produce more rapid changes affecting more taxa. It is postulated that intervals of increased speciation would probably be the result of such 'recolonization' events.

Patterns of evolution

Data are presented in Fig. 3.1 for the Albian to Recent interval. The eustatic curve and time-scale are based on the data of Haq *et al.* (1987). The number of species on the graph is the subjective view of the author using data from Blow (1969, 1979), Stainforth *et al.* (1975), Robaszynski and Caron (1979), Saito *et al.* (1980), Banner (1982), Kennett and Srinivasan (1983), Robaszynski *et al.* (1984), Caron (1985), Tourmarkine and Luterbacher (1985), Bolli and Saunders (1985), Jenkins (1985), Leckie (1989), Hart *et al.* (1989) and Premoli-Silva and Boersma (1988). The graph is not just a summation of the total number of published taxa but has involved an appraisal of all those included in the data base. As such it is a personal, but hopefully consistent, view.

The data in Fig. 3.1 show peaks in diversity in the late Cenomanian, late Turonian, mid-Maastrichtian, late Palaeocene, mid-Eocene, mid-Miocene, and early Pliocene. It is noticeable how much higher are the diversity levels in the Eocene. In contrast, the Cretaceous and Neogene have much lower diversity levels and remarkably similar patterns of change.

From this diversity data and the appearance of new taxa, it has been possible to calculate the rate of speciation. The overall Albian to Recent rate is over four species per Ma. Both the Cretaceous average and the Neogene average are less than this figure, while the Palaeogene average

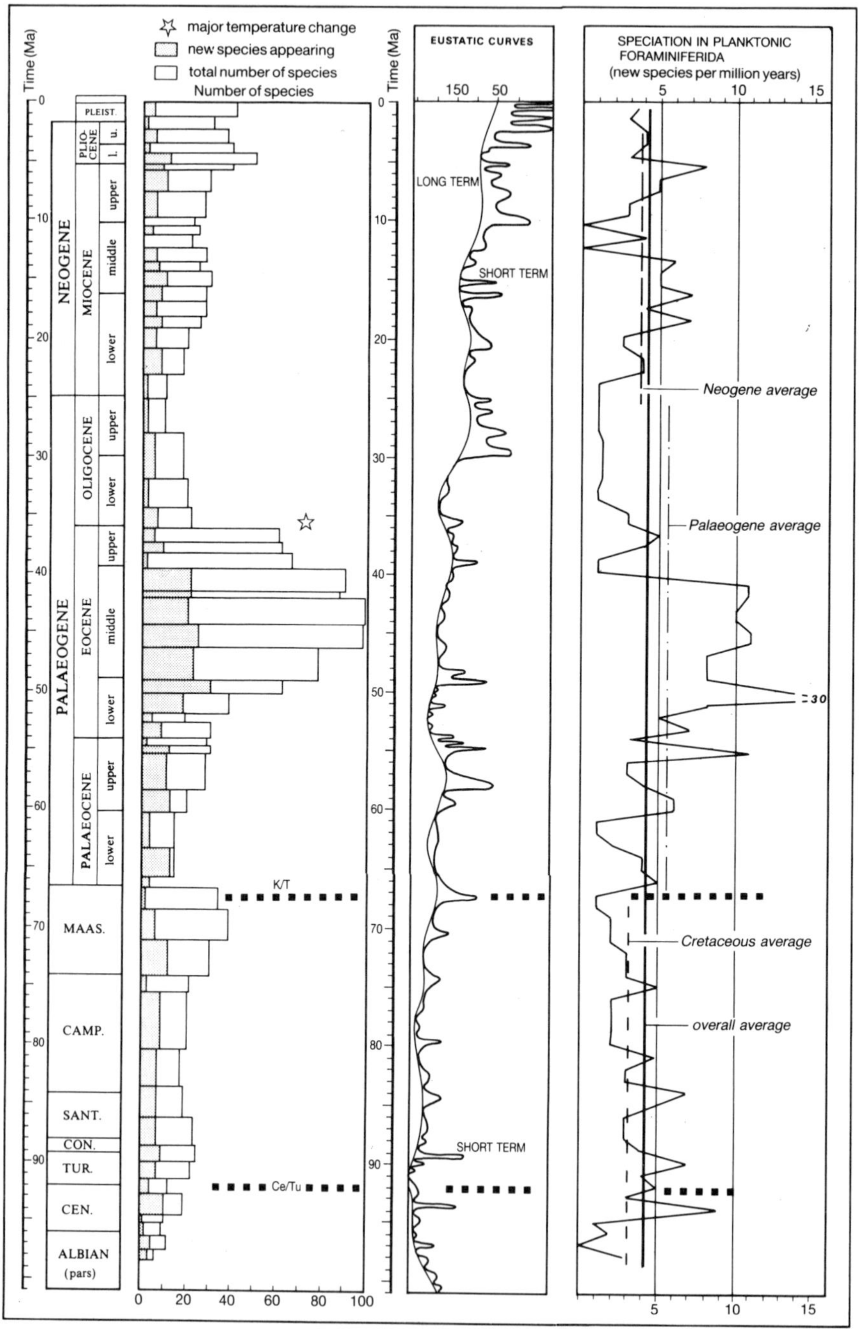
major temperature change
new species appearing
total number of species
Number of species
Time (Ma)
PLEIST.
PLIO-CENE
u.
l.
NEOGENE
MIOCENE
upper
middle
lower
PALAEOGENE
OLIGOCENE
EOCENE
PALAEOCENE
MAAS.
CAMP.
SANT.
CON.
TUR.
CEN.
ALBIAN (pars)
K/T
Ce/Tu
EUSTATIC CURVES
LONG TERM
SHORT TERM
SPECIATION IN PLANKTONIC FORAMINIFERIDA
(new species per million years)
Neogene average
Palaeogene average
Cretaceous average
overall average
30

is over five species per Ma. Indeed, if the Eocene were treated separately, the figure would be much higher, as the Palaeocene and Oligocene are both below average for the Palaeogene. The Oligocene, in particular, has a very low rate of speciation (about one species per Ma). While this graph is particularly interesting, an alternative view (Fig. 3.2) can be obtained by plotting the percentage of new species in any fauna (i.e. per capita rate of speciation). This immediately shows that the mid-Eocene diversity 'bulge' is in fact only producing a relatively small percentage of new taxa, while the very low diversity, early Palaeocene fauna is producing a very high percentage of new taxa. The average value is 29.9 per cent. This graph does show maxima in the late Albian, mid-late Cenomanian, early Palaeocene, late Palaeocene, early Eocene, early Oligocene, early Miocene, mid-Miocene, and late Miocene. All these 'events' are major colonization episodes, especially that following the K/T boundary; it is noticeable that this early Palaeocene peak hardly shows on the diversity plot. The early Paleocene and late Paleocene events have been documented in detail by Corfield (1987).

The major adaptive radiations therefore are:

(1) late Albian—appearance of keeled (?deeper water) taxa such as *Rotalipora, Planomalina, Praeglobotruncana*;

(2) late Cenomanian—further species of *Rotalipora* and *Praeglobotruncana*, and the first appearance of *Dicarinella*;

(3) early Palaeocene—appearance of the new Cainozoic fauna, including *Morozovella* and *Acarinina*;

(4) late Palaeocene/early Eocene—further species of the majority of extant lineages;

(5) early Oligocene—revival of the fauna after the end Eocene events;

(6) early Miocene—appearance of Neogene lineages after the Oligocene 'trough': especially species of *Globoquadrina* and *Globigerinoides*;

(7) mid-Miocene—arrival of many species of *Globorotalia*;

(8) late Miocene—renewed evolution within several lineages of *Globorotalia*.

Fig. 3.1. The Albian–Recent record of planktonic Foraminiferida, showing from left to right: number of species and new appearances; eustatic curve of Haq *et al.* (1987); and rate of speciation in terms of new species per Ma.

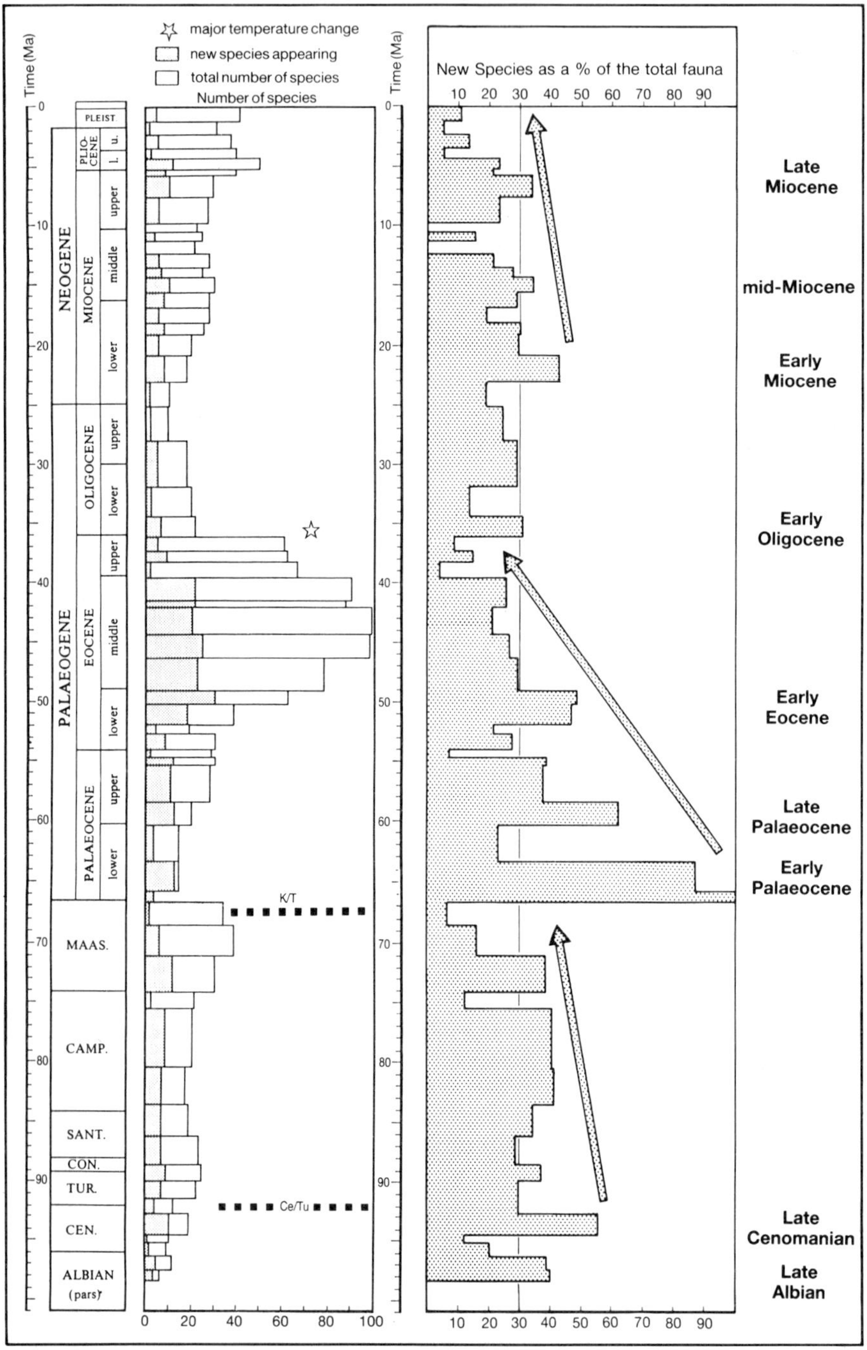
major temperature change
new species appearing
total number of species
Number of species
Time (Ma)
PLEIST.
PLIO-CENE
NEOGENE
MIOCENE
OLIGOCENE
PALAEOGENE
EOCENE
PALAEOCENE
upper
middle
lower
MAAS.
CAMP.
SANT.
CON.
TUR.
CEN.
ALBIAN (pars)
K/T
Ce/Tu
New Species as a % of the total fauna
Late Miocene
mid-Miocene
Early Miocene
Early Oligocene
Early Eocene
Late Palaeocene
Early Palaeocene
Late Cenomanian
Late Albian

Again, the Cretaceous and Neogene values of per capita speciation rate are nearly the same, although the Cretaceous is more uniform and the Neogene more varied. The oceanic stratification and highly developed O_2 minimum zone in the Cretaceous are clearly important influences (Jarvis *et al.* 1988), with depressed values being recorded in the early Turonian, Santonian, and late Campanian. The late Maastrichtian progressive decline, noted in many invertebrate groups (e.g. ammonites, Hancock 1969), is also detected. The Oligocene to Holocene interval is picked out by quite marked fluctuations and this may be attributable to the onset, and advance, of glaciation (both Antarctic and Arctic). There is a distinct drop in the percentage of new species recorded in the mid-late Miocene (10–12 Ma) with a similar 'event' being seen in the late Eocene (36–39 Ma). These two, added to the events at the K/T boundary (66 Ma) and the Cenomanian/Turonian boundary (92 Ma), provide a periodicity of 27, 28, and 26 Ma, respectively. This is extremely close to the predictions of Raup and Sepkoski (1984) and Sepkoski and Raup (1986) in their '26 million year' periodic extinction model. Figure 3.2 is, however, not measuring rates of extinction but levels of reduced speciation. Between these quite distinctive horizons there are other, less dramatic, levels of reduced speciation (Early Cenomanian, Late Campanian, latest Early Palaeocene, the Palaeocene/Eocene boundary, mid-Eocene, Early Oligocene, earliest Miocene, latest Early Miocene, Pliocene). The arrows on Fig. 3.2 indicate vague trends that might be significant. Again, the Cretaceous and the Neogene are similar. Following the initial exploitation of the water column by the single and double keeled taxa in the Cenomanian, rates of speciation decline gradually or remain nearly constant as fauna slowly replaces fauna (Hart and Ball 1986). In the Palaeogene the initial exploitation of the water column is followed by a progressive decline in the percentage of new species and, at the time of maximum diversity, the production of new taxa is running at a figure that is *below average*. This was a totally unexpected result in the compilation of this data. The 'stepwise' extinctions of Hut *et al.* (1987) and the near coincident temperature change recorded by Miller *et al.* (1987) are probably accentuated by this non-appearance of new taxa.

Fig. 3.2. The Albian–Recent record of planktonic Foraminiferida showing (on the right) a graph of the number of new species as a percentage of the total fauna in each zone. The average value of 29.9 per cent is indicated.

Current controversies

As indicated above, one current area of investigation involves the use of oxygen isotopes as temperature indicators (see, for example, Douglas and Savin 1978; Boersma and Shackleton 1981; Corfield 1987; Miller *et*

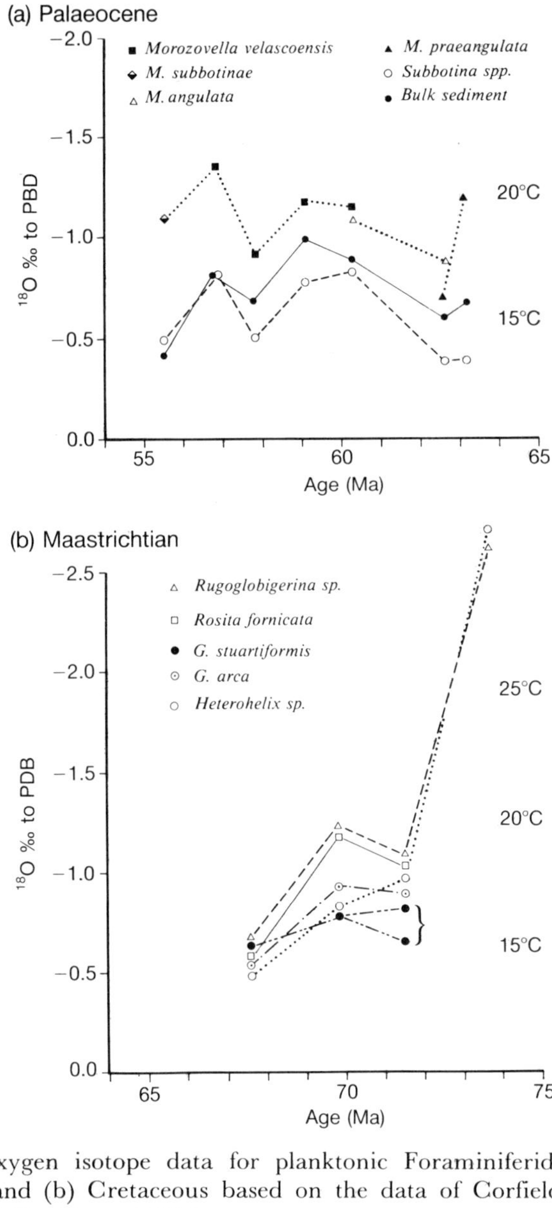

Fig. 3.3. Oxygen isotope data for planktonic Foraminiferida in the (a) Palaeocene and (b) Cretaceous based on the data of Corfield (1987) and Boersma and Shackleton (1981).

al. 1987). Several of the above authors have used their data (from Cretaceous and Palaeocene sediments) to determine the depth stratification of the planktonic Foraminiferida. In the case of the Cretaceous the *limited* data show a very reduced temperature range (Fig. 3.3(b)) but with the heavier, more calcified, keeled planktonic taxa inhabiting lower temperature environments than the non-keeled taxa. In the Palaeocene, however, Corfield's data (Fig. 3.3(a)) would indicate that the keeled *Morozovella* should have lived higher in the water column than the non-keeled *Subbotina*. This would reverse all the known evolutionary trends and run contrary to much of the perceived wisdom on morphology. It would also appear to be in conflict with the modern distribution pattern. What is clear at present is that in the Early Palaeogene the number of new species as a proportion of the whole fauna indicates a recolonization of the available water column following the K/T extinction event. What is also known is that it is the deeper-water Cretaceous taxa which succumb to the late Maastrichtian extinction, and the 'hedbergellid', surface-dwelling forms that survive to repopulate the Tertiary. That this form should suddenly, with little morphological change, migrate into deeper water leaving the surface waters to be repopulated by morozovellids seems unlikely and a resolution of the isotope/morphology dilemma must be sought. The *direction* of such repopulation events does not, however, affect the data shown in Figs 3.1 and 3.2.

Summary

A consideration of the mode of life of the planktonic Foraminiferida and an understanding of their distribution through time can be used successfully to investigate the 'health' of the world ocean. The main evolutionary radiations and major levels of decline have been delineated, as has the conflict in data (and interpretation) that confounds our use of the water depth model of Hart (1980) and Caron (1983 *a, b*) for the whole of the Triassic to Recent interval. Periods of reduced speciation appear to conform to the 26 Ma extinction periodicity model of Raup and Sepkoski (1984).

Acknowledgements

The author wishes to acknowledge the financial asistance of Polytechnic South West and the NAB/PCFC Research Initiative. Discussions with many colleagues have proved most fruitful; special thanks are extended to Drs S. M. Stanley and D. Jablonski for their interesting comments on the original lecture presentation. Mr John Abraham provided the cartographic expertise that produced the figures.

References

Banner, F. T. (1982). A classification and introduction to the Globigerinacea. In *Aspects of micropalaeontology* (ed. F.T. Banner and A.R. Lord), pp. 142–239. Allen and Unwin, London.

Banner, F.T. and Desai, D. (1988). A review and revision of the Jurassic–Early Cretaceous Globigerinina, with especial reference to the Aptian assemblages of Speeton (North Yorkshire, England). *Journal of Micropalaeontology*, **7**, 143–85.

Bé, A.W.H.(1965). The influence of depth on shell growth in *Globigerinoides sacculifer* (Brady). *Micropaleontology*, **11**, 81–97.

Bé, A.W.H. (1977). An ecological, zoogeographic and taxonomic review of Recent planktonic Foraminifera. In *Oceanic micropalaeontology* (ed. A.T.S. Ramsay), pp. 1–100. Academic Press, London.

Bé, A.W.H. and Hamlin, W.H. (1967). Ecology of Recent planktonic foraminifera: Part 3. Distribution in the North Atlantic during the Summer of 1962. *Micropaleontology*, **13**, 87–106.

Bé, A.W.H. and Hemleben C. (1970). Calcification in a living planktonic Foraminifera, *Globigerinoides sacculifer* (Brady). *Neues Jahrbuch für Geologie und Paläontologie, Abhandlungen*, **134**, 221–34.

Blow, W. (1969). Late Middle Eocene to Recent planktonic foraminiferal biostratigraphy. In *Proceedings of the 1st International Conference on Planktonic Microfossils, Geneva 1967*, Vol. 1 (ed. P. Bronnimann and H.H. Renz), pp. 199–422. Brill, Leiden.

Blow, W. (1979). *The Cainozoic Globigerinida*. Brill, Leiden.

Boersma, A. and Shackleton, N.J. (1981). Oxygen and carbon isotope variations and planktonic foraminifera depth habitats, Late Cretaceous to Paleocene, Central Pacific, Deep Sea Drilling Project Sites 463 and 465. In *Initial Reports of the Deep Sea Drilling Project, Leg 62*, pp. 513–26. US Government Printing Office, Washington, DC.

Bolli, H.M. and Saunders, J.B. (1985). Oligocene to Holocene low latitude planktic foraminfera. In *Plankton stratigraphy* (ed. H.M. Bolli, J.B. Saunders, and K. Perch-Nielsen), pp. 155–262. Cambridge University Press, Cambridge.

Bradshaw, J.S. (1957). Laboratory studies on the rate of growth of the foraminifera, *Streblus beccarii* (Linne) var. *tepida* (Cushman). *Journal of Paleontology*, **31**, 1138–47.

Caron, M. (1983*a*). La speciation chez les Foraminifères planctiques: une response adaptee aux constraintes de l'environment. *Zitteliana*, **10**, 671–6.

Caron, M. (1983*b*). Taxonomie et phylogenie de la famile des Globotruncanidae. *Zitteliana*, **10**, 677–81.

Caron, M. (1985). Cretaceous planktic foraminifera. In *Plankton stratigraphy* (ed. H.M. Bolli, J.B. Saunders, and K. Perch-Nielsen), pp. 17–86. Cambridge University Press, Cambridge.

Caron, M. and Homewood, P. (1983). Evolution of early planktic foraminifers. *Marine Micropaleontology*, **7**, 453–62.

Corfield, R.M. (1987). Patterns of evolution in Palaeocene and Eocene planktonic Foraminifera. In *Micropalaeontology of carbonate environments* (ed. M. B. Hart), pp. 93–109. Ellis Horwood, Chichester.

Douglas, R. G. and Savin, S. M. (1978). Oxygen isotopic evidence for the depth stratification of Tertiary and Cretaceous planktic Foraminifera. *Marine Micropaleontology*, **3**, 175–96.

Fuchs, W. (1967). Über Ursprung und Phylogenie der Trias 'Globigerinen' und die Bedeutung dieses Formenkreises für das echte Plankton. *Verhandlungen der Geologischen Bundesanstalt, Wien*, **1967**, 135–76.

Fuchs, W. (1969). Zur Kenntis des Schalenbaues der zu den Trias-''Globigerinen'' zählenden Foraminiferengattung *Praegublzinella. Verhandlungen der Geologischen Bundesanstalt,* Wien, **2**, 158–62.

Hancock, J. M. (1969). Some Cretaceous–Tertiary marine faunal changes. In *The fossil record* (ed. W. B. Harland *et al.*), pp. 91–104. Geological Society, London.

Haq, B. U., Hardenbol, J., and Vail, P. R. (1987). Chronology of fluctuating sea levels since the Triassic. *Science*, **235**, 1156–67.

Hart, M. B. (1980). A water depth model for the evolution of the planktonic Foraminiferida. *Nature*, **286**, 252–4.

Hart, M. B. and Bailey, H. W. (1979). The distribution of planktonic Foraminiferida in the mid-Cretaceous of N. W. Europe. *Aspekte der Kreide Europas, IUGS Ser. A*, **6**, 527–42.

Hart, M. B. and Ball, K. C. (1986). Late Cretaceous anoxic events, sea level changes and the evolution of the planktonic Foraminifera. In *North Atlantic palaeoceanography* (ed. C. P. Summerhayes and N. J. Shackleton), pp. 67–78. Geological Society, London.

Hart, M. B., Bailey, H. W., Crittenden, S., Fletcher, B. N., Price, R. J. and Swiecicki, A. (1989). Cretaceous. In *Stratigraphical atlas of fossil Foraminifera* (ed. D. G. Jenkins and J. W. Murray), pp. 273–371. Ellis Horwood, Chichester.

Hut, P., Alvarez, W., Elder, W. P., Hansen, T., Kauffman, E. G. Keller, G., Shoemaker, E. M. and Weissman, P. R. (1987). Comet showers as a cause of mass extinctions. *Nature*, **329**, 118–26.

Jarvis, I., Carson, G. A., Cooper, M. K. E., Hart, M. B., Horne, D., Leary, P. N., Tocher, B. A. and Rosenfeld, A. (1988). Microfossil assemblages and the Cenomanian–Turonian (late Cretaceous) oceanic anoxic event. *Cretaceous Research*, **9**, 3–103.

Jenkins, D. G. (1971). New Zealand Cenozoic planktonic Foraminifera. *New Zealand Geological Survey, Palaeontological Bulletin*, **42**, 1–278.

Jenkins, D. G. (1985). Southern mid-latitude Paleocene to Holocene planktic foraminifera. In *Plankton stratigraphy* (ed. H. M. Bolli, J. B. Saunders and K. Perch-Nielsen), pp. 263–282. Cambridge University Press, Cambridge.

Kennett, J. P. and Srinivasan, M. S. (1983). *Neogene planktonic Foraminifera.* Hutchinson Ross, Stroudsburg.

Kristan-Tollmann, E. (1964). Die Foraminiferen aus den rhätischen Zlambachimergelen der Fischerwiese bei Ausse im Salzkammergut. *Jahrbuch der Geologischen Bundesanstalt, Wien*, **10**, 1–189.

Leckie, R. M. (1989). A palaeoceanographic model for the early evolutionary history of planktonic Foraminifera. *Palaeogeography, Palaeoclimatology, Palaeoecology*, **73**, 107–38.

Masters, B. A. (1977). Mesozoic planktonic Foraminifera; a world wide review and analysis. In *Oceanic micropalaeontology* (ed. A. T. S. Ramsay), pp. 301–731, Academic Press, London.

Miller, K. G., Fairbanks, R. G. and Mountain, G. S. (1987). Tertiary oxygen isotope synthesis, sea-level history and continental margin erosion. *Paleoceanography*, **2**, 1–19.

Oberhauser, V. R. (1960). Foraminiferen und Mikrofossilien 'incertae sedis' der Ladinischen und Karnischen Stufe der Trias aus den Ostalpen und aus Persien. *Jahrbuch Geologischen Bundesanstalt, Wien*, **5**, 5–46.

Parker, F. L. (1960). Living planktonic Foraminifera from the equatorial and southeast Pacific. *Tohoku University, Science Reports, Ser. 2, (Geology) Special Volume*, **4**, 71–82.

Premoli-Silva, I. and Boersma, A. (1988). Atlantic Eocene planktonic foraminiferal historical biogeography and palaeohydrographic indices. *Palaeogeography, Palaeoclimatology, Palaeoecology*, **67**, 315–56.

Raup, D. M. and Sepkoski, J. J. (1984). Periodicity of extinctions in the geologic past. *Proceedings of the National Academy of Science*, **81**, 801–5.

Rhumbler, L. (1911). Die Foraminiferen (Thalamophoren) der Plankton Expedition; Teil 1. Die allgemeinen Organisationsverhaltnisse der Foraminiferen. *Plankton-Expedition, Humbolt-Stiftung, Ergebnisse*, **3**, (Lc), 1–331.

Robaszynski, F. and Caron, M. (1979). Atlas of Mid-Cretaceous Planktonic Foraminifera (Boreal Sea and Tethys). *Cahiers de Micropaléontologie*, Vols 1 and 2. Editions CNRS, Paris [Published in French and English.]

Robaszynski, F., Caron, M., Gonzalez Donoso, J. M. and Wonders, A. A. H. (1984). Atlas of Late Cretaceous Globotruncanids. *Revue de Micropaléontologie*, **26**, 145–345.

Saito, T., Hillman, N. S. and Janal, M. J. (1980). *Catalogue of planktonic Foraminifera*, 6 volumes. American Museum of Natural History, New York.

Sepkoski, J. J. and Raup, D. M. (1986). Was there a 26 Myr periodicity of extinctions? *Nature*, **321**, 533.

Stainforth, R. M., Lamb, J. L., Luterbacher, H., Beard, J. H., and Jeffords, R. M. (1975). *Cenozoic planktonic foraminiferal zonation and characteristics of index forms*. University of Kansas Paleontological Institute, Article 62. Lawrence, Kansas.

Tolderlund, D. S. (1969). Seasonal distribution patterns of planktonic Foraminifera at five ocean stations in the Western North Atlantic. Unpublished PhD Thesis, Columbia University, New York.

Toumarkine, M. and Bolli, H. M. (1970). Evolution de *Globorotalia cerroazulensis* (Cole) dans l'Eocene moyen et superieur de Possagno (Italie). *Revue de Micropaléontologie*, **13**, 131–45.

Toumarkine, M. and Luterbacher, H. (1985). Paleocene and Eocene planktic foraminifera. In *Plankton stratigraphy* (ed. H. M. Bolli, J. B. Saunders, and K. Perch-Nielsen), pp. 87–154. Cambridge University Press, Cambridge.

4. Early radiation of the Mollusca

N.J. MORRIS
Department of Palaeontology, Natural History Museum, London, UK

Abstract

During the last three decades more and more Cambrian Mollusca have been discovered, particularly among the small shelly faunas of the Lower and Middle Cambrian. Many of these have led to considerable discussion about their classification and the course of molluscan phylogeny. These early molluscs show an important radiation at the beginning of the Phanerozoic which led to the diversification into the recognized classes living today. This radiation seems to have been dependent on the origin of a dorsal exoskeleton and its subsequent calcification, and possibly its interference with the ancestral pattern of simple metamery. The apparent synchronous development of a skeleton in several independent phyla suggests there may be a link with changing composition of the atmosphere and sea.

Introduction

The development of a calcareous exoskeleton in the Mollusca took place at the beginning of the Phanerozoic, more or less at the same time as the development of a skeleton in a number of other phyla. These include brachiopods, arthropods, and a number of phyla of uncertain affinity.

In the view taken here, there are two monophyletic clades of molluscs: the Conchifera with their characteristic shell as a synapomorphy, and the chitons with a shell consisting of eight plates in the adult form. The two clades are here considered to be sister taxa, related by the presence of a radula. Rich earliest Cambrian faunas containing both chitons and conchiferans are now known from several parts of the world but are best preserved in South and Central China (a large number of references are

Major Evolutionary Radiations (ed. P.D. Taylor and G.P. Larwood), Systematics Association Special Volume No. 42, pp. 73–90, Clarendon Press, Oxford, 1990.

listed in Liu (1987), Yu (1987), and Qian and Bengtson (1989)), and in Northern Siberia (see Missarzhevsky 1982). The mollusc shells are thought to have originally consisted of aragonite with an organic matrix but have usually been replaced by a phosphatic mineral and are commonly preserved as small phosphatic shells or internal moulds. They usually occur in successions of calcarenites which suggest that they lived in an equable, shallow-water, marine environment.

The molluscan exoskeleton reflects only a modest amount of the animal's overall anatomy, so interpretation of the morphology of these early animals is both contentious and difficult. In this chapter I have tried to summarize the events of the earliest Phanerozoic evolution of the Mollusca which may be seen as an adaptive radiation following the acquisition of a dorsal exoskeleton and its subsequent calcification. The problems of relating earliest shells within a satisfactory phylogenetic classification of the Mollusca are discussed. However, the controversies that exist concerning the assignment of low-level taxa within higher taxa among Early Cambrian shells mean that any quantification of the radiation would be virtually meaningless at this stage.

Molluscan phylogeny

In spite of a number of differing views in the past, many aspects of the likely ancestor of the shelled Mollusca are now quite clear. The primitive form of limited metamerism was recognized by Wenz (1941) from the evidence of fossil taxa. It was supported by the discovery of living *Neopilina* and its subsequent description by Lemche and Wingstrand (1959). Although some reservations were initially expressed about the primitive nature of a monoplacophoran metameres, their plesiomorphy for the shelled Mollusca has been accepted by Salvini-Plawen and others in a series of critical papers (Pojeta and Runnegar 1976; Morris 1979; Salvini-Plawen 1981). Whether or not this apparently simple metamerism, as well as the limited development of coelomic cavities, is homologous with the segmentation in annelids remains an open question. In recent years a further and more comprehensive study of *Neopilina* and *Vema* by Wingstrand (1985) has shown many more details of the morphology of the Monoplacophora and the characters they share with the chitons.

Shared features of the shell-bearing Mollusca

Monoplacophora and Polyplacophora share an elongate body with a flattish, creeping foot and a calcified dorsal shell consisting of eight elements in mature chitons and just a single shell in the Neopilinidae.

The body is divided into eight metameres with similar muscle ennervation and blood supply. There is considerable similarity in the head and its radular feeding apparatus, the gonads, the nephridia, and the heart with its pericardium, and the arrangement of these organs within the simple metameric pattern. The gills are metamerically arranged in the Neopilinidae, but this feature is apparently lost in the chitons where a series of similar gills are concentrated in the pallial groove towards the posterior, possibly related to their proximity to the heart. During the ontogeny of the chitons, six trunk metameres apparently appear simultaneously while the head is at first only furnished with the second of its two head plates. This is matched by the ontogeny of the solenogaster *Nematomenia* where no anterior head plate is known and additional lines of scales appear on the flanks (Pruvot 1890). No details of the ontogeny on the Monoplacophora are known.

Many of these shared characters are also clearly shared by the late Lower Palaeozoic representatives of the monoplacophoran family Tryblidiidae. Along with the Neopilinidae this family has five trunk metameres in contrast to the six of chitons.

The eightfold metamerism as recognized by muscle scars or the ordering of organ systems is clearly reduced or obscured in all higher Mollusca. Even in the Tryblidiidae, amalgamation of the three head metameres can be seen to have taken place as far as the musculature is concerned. In the Lower Palaeozoic family Cyrtolitidae there is an obvious reduction to four pairs of body/foot muscles, and putative gill muscles are associated with only the two posterior pedal muscle scars. There may be further reduction in the Bellerophontacea. It is also apparent that a reduction took place to a single pair of retractor scars in the gastropods, and either one or two pairs in the cephalopods.

In the bivalves there seems to have been an entirely different pathway to the reduction of metamery. Pedal/body scars representing the trunk metameres are clearly preserved in a number of Ordovician bivalves and the subsequent evolutionary history of this musculature seems to be a matter of further specialization, sometimes atrophy. With the locomotory system relying on adduction in addition to the retractile foot, most features of the original metamery are lost in modern bivalves. Pojeta and Runnegar (1976) have given an account of the musculature of the Rostroconchia. In this case some multiple-paired retractor muscles were retained but they are reduced to one pair in scaphopods.

It is evident that a considerable amount of this reduction in metamery had taken place before the end of the earliest Cambrian. Although the muscle scars of the early Cambrian Meishucunian Mollusca are rather rarely and poorly preserved and their numbers very difficult to count, they do give a meagre but important amount of information. Whereas at least six pairs of scars are present in *Yangtzemerisma* (Yu 1987, plates 19, 20, and 22) and *Eoconus* (Yu 1987, plate 27, figure 2), in the tall planospirally coiled *Yangtzeconus* a single pair occurs high on the flank. *Eoconus*, like the similar Tremadocian genera *Kirengella* and *Romaniella* (Doguzhaeva 1972), could be interpreted as having an essentially eight-metamere morphology with coalescence of the head metameres. Although Yu (1987) has argued otherwise, the Merismoconchidae could be interpreted as early laterally compressed forms, like the ribeirioids, possibly with a full complement of muscles and metameres with an incipient pegma. *Yangtzeconus* clearly has reduced musculature and presumably reduced metamery.

A theory of molluscan metameric reduction

During the later ontogeny of trilobites and polychaetes, new trunk segments are added in front of the telson or its equivalent one by one. This might lead one to deduce that this is the primitive way to produce increasing numbers of segments. However, it is now apparent that in *Drosophila* and possibly the centipedes (Minelli and Borteletto 1988) increase in segment numbers is produced by binary division of pre-existing 'segments', and may be carried out by simple commands from control genes. Although it is necessary to make a special argument for the number of metameres that might be involved in a 'head', it would seem that the eightfold metamery and its reduction to sometimes four and sometimes two or one metameres that is evident in the Mollusca better fits the model of metamery in *Drosophila* and centipedes than the model of trilobites and polychaetes. The reduction deduced for molluscan evolution could have happened by successively earlier development of the shell which would 'freeze' the number of metameres because they are largely expressed, particularly in the Conchifera, by the number of pedal muscles. This theory avoids the necessity to experiment with successively fewer segments and could explain the virtually synchronous appearance of molluscs with several or very few metameres in the earliest Cambrian.

Chitons bound by the polyplacophoran shell

Before the recent discovery of chitons in the Meishucunian Stage they were not known prior to the Upper Cambrian. It was considered that chitons maintained a very low diversity until the end of the Mesozoic when a rapid diversification began, leading to the present-day total of about a thousand species. However, the Meishucunian contains a rich fauna of chitons, equal to any locality today.

Although the Meishucunian fauna includes a number of obvious chitons, the identification of other genera as polyplacophorans is less clear. Two of the more convincing are spiny genera with several conjoined plates. One series of 'genera' including *Canopoconus* consists of isolated cap-shaped plates with no sign of articulating surfaces. They first occur in the lowest fossiliferous horizon of the Meishucunian and are therefore the earliest known molluscs. Some of the specimens, kindly shown to me by Professor Yu Wen, have an outer layer represented by a phosphatic infilling of the aesthete canals. These are in the form of an outer zone, normal to the shell surface, and an internal zone, running parallel to the shell surface. The inner shell layer has no preserved structure. If *Canopoconus* is correctly identified as a chiton then it is an early primitive form in which the shell plates are simple, open conical structures that did not overlap.

The incorporation of the aesthetes into the outer shell layer of chitons led Haas (1981) to the conclusion that the chiton shell was not the exact homologue of the conchiferan shell and therefore that the Polyplacophora were not ancestral to the Conchifera. This is contrary to the view of Blumrich (1891) and Runnegar *et al.* (1979). The inner shell of the chitons, a crossed-lamellar layer, is unlike crossed-lamellar structure in the Conchifera. There is also little similarity of structure in the calcareous part of the outer layer (the tegmentum in the chitons) between the two groups.

Although Haas has described the presence of myostracum in polyplacophorans, in specimens of *Chiton* and *Stenochiton* that I have examined, the musculature seems to be attached to an inner part of the 'crossed lamellar' layer. There is no indication of the prismatic aragonite typical of conchiferan myostracum. The view that the calcification of a dorsal organic integument or shell took place separately in the Polyplacophora and the Conchifera is strongly supported here.

None of the early Cambrian chitons are preserved with a full complement of plates. However, following Rolfe's (1981) discovery that a small anterior eighth plate is present in the Upper Ordovician genus *Septemchiton*, there is no reason to suppose that any of them had other than eight plates (except during their larval development). In living polyplacophorans the incipient shells of six trunk metameres appear simultaneously with a single head plate; a second anterior head plate appears later.

Salvini-Plawen (1981) and Haas (1981) have argued that the classes of non-shelled chitons that used to be known as the Aplacophora, the Solenogasters and the Caudofoveata, are more primitive than the Polyplacophora and Conchifera. A scheme summarizing their classification is given by Salvini-Plawen (1981). Whereas this interpretation of relationships is possibly correct, it relies on a particular polarity of characters. A stumbling block that surely needs reinvestigation is provided by Pruvot's (1890) study of the development of the solenogaster genus *Nematomenia*. After 7 days in *Nematomenia* the test cells were shed, revealing flanks covered with numerous orbicular, imbricating scales illustrated as occurring in three rows. The dorsum bore a line of seven calcareous plates, each slightly overlapping the next posterior. Each plate was formed of simply juxtaposed calcareous spicules. There were no spicules present on the ventral surface. If Pruvot correctly interpreted what he saw, then it seems that the solenogasters go through an early chiton shell stage in their ontogeny, rendering much more likely the possibility of Pelseneer's (1906) hypothesis that the Aplacophora, or at least the Solenogasters, are derived from a polyplacophoran, shell-bearing ancestor.

If, on the other hand, Salvini-Plawen proves to have been right, the early Cambrian diversification of Mollusca that we see in the rocks is only part of the story. There is the possibility of a host of unpreserved Solenogasters and Caudofoviata contemporaneous with the small shelly fossils.

It seems to me that the ontogeny of *Nematomenia* is likely to be recapitulating its phylogeny and that the solenogasters are a later development from an eight-shelled ancestor. Put another way, the larval stage with seven dorsal plates is a synapomorphy of the Solenogasters plus the Polyplacophora, and together they are a sister-group of the Conchifera.

Range of form among the earliest Cambrian Conchifera

Many of the Early Cambrian Conchifera had shells with a simple form of logarithmic spiral growth that match the computer-generated shapes of Raup (1962). Even at the outset, however, they were constructed by regular incremental additions at the shell aperture usually not coinciding with the position of the theoretical generating curve, which lies in a plane that includes the coiling axis. The majority of the taxa do indeed have near planar apertures. Notable exceptions were the early representatives of the rostroconchs and possible bivalves where the shell became laterally compressed and the aperture curved when viewed from the side.

The planar apertures of the majority were rounded or ovoid and sometimes modified to accommodate earlier whorls of the shell spire. It seems that, in the case of shells that deviated from planospirality, the aperture was slightly angled to the plane normal to the coiling axis. Some genera, such as *Archaeospira*, include species with a generally planar aperture but with a distinctly arched notch on the lip closest to the coiling axis.

When the shells deviate from planospirality, they may have either right-or left-handed coiling when the apex is held uppermost. There seems to be a preponderance of left-handed spires although some right-handed forms clearly occur. In the Meishucunian some of these helically coiled species are well enough preserved and occur in sufficient numbers to allow reasonable certainty that each species coils in only one direction. As an example of this, the left-handed species *Archaeospira ornata* Yu, as interpreted by Qian and Bengtson (1989, p. 112), is represented by approximately 130 specimens. This species may be recognized by an arched lip close to the coiling axis; no such specimens have a right-handed shell. A second species, represented by only a very small number of specimens, has a much more planar aperture. This species, based on left-handed shells found in Hubei, is referred to as *Archaeospira* cf. *ornata* by Qian and Bengtson and is apparently conspecific with specimens referred to as *Bemella simplex* by Yu. These co-occur with right-handed shells described as *Hubeispira* by Yu (1987, plate 55, figs 1–3, plate 56, figs 5–8). These are either actual mirror images or, in the example on plate 56 of Yu, a mirror image with a slightly modified aperture. It seems that this early species could be either right-or left-handed although more evidence, in the form of additional specimens, is necessary before we can be certain. The importance of this possibility in understanding the evolution of gastropods will be considered further (p. 85).

Raup's (1962) computer-generated shell shapes broadly cover the whole field available to and occupied by univalved Conchifera and single valves of many bivalves. When the Lower Cambrian univalves with a near planar aperture are plotted on Raup's diagram it becomes immediately clear that only a small proportion of the field is exploited at this early time. In addition to the limited forms adopted by the generating curve, translation rate is limited to less than 1.0. Expansion rate may be as much as 10^5 in open, limpet-like forms but may also be very small. Dislocation of the generating curve from the axis of coiling is not great, apparently limited to 0.2.

The size of 'skeletonized' taxa at the beginning of the Cambrian has been a subject of some interest. Among the Mollusca, the earliest univalves all seem to be less than 3.5 mm in maximum shell dimension, and the chitons under 15 mm in total length. By the end of the early Cambrian, helcionellids are known that approach 10 mm in length. Early bivalves were particularly small, the mid-Lower Cambrian *Pojetaia* limited to 0.2 mm length, only the size of fairly early post-larval stages of post-Cambrian taxa. Runnegar and Bentley (1983) have emphasized small size to infer that bivalves of Lower Cambrian age must have relied on filter feeding; present-day juvenile bivalves of this size are filter feeders before the development of deposit feeding largely utilized by the adults. Whereas this seems a plausible argument in the context of the Bivalvia, contemporary representatives of other molluscan classes were equally small but did not lose their radula and are therefore unlikely to have been filter feeders as adults.

Although the small size of the Early Cambrian shelly fossils may reflect an overall small size for metazoans at this time, it is by no means certain that this is true. Certain Ediacaran fossils that appear to be segmented worms or arthropods are of considerable size and the Upper Sinian body fossils from Jiangsu and Anhui Provinces of China (Wang 1982) are all larger than the Meishucunian Mollusca. Although it is possible that all larger Metazoa suffered extinction before the beginning of the Cambrian, an alternative view would be that at the inception of calcareous skeletons only small individuals were possible.

The size of these small shelly fossils has given rise to the view that they may all have lived within the meiofauna. Some of the more tube-like forms do indeed have a modern analogue in the meiofaunal prosobranch gastropod genus *Caecum*. It seems highly unlikely, however, that the long-spined chitons were meiofaunal. Not only may their spines be interpreted as a defence against epifaunal predators, but they would also drastically impede progress within the sediment. An alternative explanation would be that ancestors of these small shelly fossils were meiofaunal. However, the flat molluscan foot that one would judge to be a primitive character for the Conchifera and Polyplacophora seems to be an adaptation for movement upon rather than within a substrate. This intriguing problem of size probably awaits a better explanation.

Although no measurements of shell thickness of these earliest molluscs, or assessment of their shell strength have been made, it is clear that many of them were relatively robust. They may well have been able to cope with minor amounts of surface sediment movement. On the other hand, it is very unlikely that similar organisms with purely organic and apparently very fragile conchiolin shells could have survived in the same environment. This would be particularly important when there was any sediment movement or if they were to any extent infaunal. It therefore seems likely that their ancestors, lacking a calcareous shell, migrated into the arenaceous or calcareous environment of the Lower Cambrian.

Infaunal life without adduction: the scaphopods and rostroconchs, half-way to the bivalves

The scaphopods share a number of characters with rostroconchs: (1) a bilaterally symmetrical shell with an elongate posterior aperture to the mantle cavity (in the scaphopods this is utilized by both inhalent and exhalent water currents); (2) a univalved larva and structural continuity of the shell across the dorsum; (3) a prominent anterior gape that is shared by the foot and feeding apparatus in the living scaphopods; (4) lack of a functional hinge like that of the bivalves and no adductor muscles.

Two main groups of rostroconchs are recognized—the Conocardioida and the Ribeirioida. They are distinguished by the nature of the shell; conocardioids usually have a shell that is ribbed at least within some of its layers, and the shell is bivalve-like with a ligament-like dorsal line developed to accommodate shell growth. In profile ribeiriod shells commonly resemble nuculoid bivalves in shape. Pojeta and Runnegar (1976) suggested the origin of scaphopods from the ribeirioid *Pinnocaris*. Although this is plausible, the particular ancestor they envisage has an apex near the anterior, and antero-posterior extension developed by hypertrophied growth in a posterior direction relative to the more primitive shape of *Ribeiria*. Whereas the larval shell of the Dentaliidae may have its apex to the anterior, the subsequent antero-posterior extension which follows ventral fusion of the shell material, takes place towards the anterior. This is partly a consequence of structural morphology—it is one way of producing a logarithmically curved conical tube. However, if it has developed from *Pinnocaris* other intermediate stages are necessary. It seems equally possible that one of the conocardioids may turn out to be the closest relative of the scaphopods. Later species of *Conocardium* certainly become very elongate and some lose their second, postero-ventral pallial opening. However, these later forms post-date the first scaphopods considerably and must be considered to have developed by convergence.

Many of the rostroconchs have a shell shape which parallels that of a number of different bivalves. This feature, together with additional musculature and inner shell impressions on some conocardioids (e.g. *C. longipennis*), suggests that they had paired ctenidia, perhaps quite similar to primitive bivalve ctenidia. If the relationship between the rostroconchs and scaphopods is correct, then the scaphopods lost the ctenidia early in their history. Their shell shape reflects the shape of the mantle cavity and the gently curved conical form leaves no room for gills.

The flat-fronted foot of some scaphopods, the food-collecting captaculae, the radula, and the salivary glands are all apparently primitive characters and it seems very likely that many of the rostroconchs also had these features. The rounded form of the anterior gape of many of the later conocardioids suggest that the morphology of the foot may have been rather similar. There are no known Cambrian scaphopods and few rostroconchs. The early form redescribed by Pojeta and Runnegar (1976), *Watsonella* (as *Heraultipegma*), does not differ a great deal in external form from the Tommotian genus *Anabarella* (see Missarzhevsky 1969, plate 4, figs 4–6), although in this compressed genus the apex is very close to the anterior margin.

Watsonella is also somewhat similar to the Meishucunian genera *Yangtzemerisma* (Yu 1987, plate 20, fig. 3) and *Minymerisma* (Yu 1987, plate 21), both belonging to Yu's family Merismoconchidae. These have a transverse shell bar on the dorsal inner surface a little behind the apex. This has been interpreted by Yu as a segmental boundary, possibly reflecting an evolutionary stage of shell-form between the Polyplacophora and the Conchifera. I would think it more likely that this structure is an incipent shell buttress or pegma such as occurs in the Ribeiriidae. In *Ribeiria apusoides* muscle scars are preserved on the anterior surface of this structure. The similarity of position and shape of these muscle scars to the transverse scar on metamere A of *Pilina* and *Tryblidium* suggests to me that these structures may be homologues. I prefer this interpretation to Yu's and I consider it unlikely that the conchiferan shell has evolved directly from a polyplacophoran shell. It seems more likely to me that the Merismoconchidae share this plesiomorphic character with the Ribeiriidae and that calcification took place separately in the two major groups.

Burrowing into the substrate: the origin of the bivalves

The bivalve shell is characterized by left and right calcareous valves which, at least in the early stages of growth, are equal in size and shape. They are rounded, veneriform, and joined at the dorsum by an elastic ligament. At early growth stages there is a sub-equal pair of adductor muscles that serve to keep the valves closed and form an effective counter to the ligament which opens the valves. Nearly all living bivalves have a single pair of ctenidia which are used to extract oxygen from the water and are commonly used to filter microplankton for food. A number of carnivorous forms without gills are an exception. There is no radular apparatus in the foregut and, by comparison with all other molluscan classes, this is considered to have been lost during an early stage of bivalve evolution.

The apparently primitive nuculoid bivalves feed by the intake of organically rich, fine sediment, using ciliated grooves and their palps. These palps are provided with proboscides which trail outside the shell within the substrate. Filter-feeding bivalves use the palps to take and to sort food particles captured by the ctenidia to the mouth. Some more advanced taxa have further used the ctenidial currents to become secondary deposit feeders, often using their siphons like a vacuum cleaner. Many of the septibranchs have lost their gills and have developed a muscular septum within the mantle cavity for much more efficient inflow of water. Some use this rapid water inflow to prey on small animals.

Crown group bivalves are first recognized in the very earliest Ordovician (Tremadoc), and a reasonable variety are known by Lower Arenig times. Some of these early taxa have multiple-paired pedal musculature, directly comparable to the pattern seen in chitons and monoplacophorans (Morris 1979).

The Cambrian record of bivalves is sporadic and preservation is often inadequate. Exceptions include the Lower and Middle Cambrian genera *Fordilla* and *Pojetaia* which have been described in considerable detail by Pojeta (1975) and Runnegar and Bentley (1983). Both *Fordilla* and *Pojetaia* have the veneriform shell and apparent ligament groove typical of modern bivalves. They also have a series of very well-impressed muscle scars. These scars are approximately in the position of the adductors and pallial line of living taxa but are really rather dissimilar. Indeed, in *Fordilla*, Pojeta and Runnegar (1976, p. 26, fig. 12) offer four different interpretations based on these scars, none of which is intermediate between the musculature of primitive bivalves and that of typical ribeirioids, their presumed sister-group. In mid-Ordovician nuculoids, the pallial line is usually invisible, whereas in both *Pojetaia* and *Fordilla* a series of well-impressed scars occur (see particularly Runnegar and Bentley 1983, fig. 8D). Since no other known Lower Cambrian Mollusca have such an obvious pallial line, there would seem to be a progressive evolution of this character, followed by a reversal. However, this pattern may relate to size as Runnegar and Bentley have pointed out. As can be seen in their fig. 8, both *Pojetaia* and *Fordilla* are approximately the same size as the very earliest growth stages known of the Lower Llanvirn nuculoid genus *Tironucula* (Morris and Fortey 1976, plate 1, figs 2–4). Although the nature of the pallial attachment is unknown in these early stages of early nuculoids, the hinge teeth and other aspects of shell morphology are virtually identical in equal-sized specimens of *Pojetaia* and *Tironucula*. One can hypothesize that the evolution of additional hinge teeth is related to size increase and also to physical protection of the hinge. The changes in the ontogeny of *Tironucula* recapitulate their evolution. An alternative hypothesis, which includes the interpretation of the muscle scars along the position of the pallial line as multiple paired pedal scars, is much less satisfactory, largely because it leaves such a small mantle cavity with virtially no room for any type of gill. Runnegar and Bentley rightly recognize the morphological similarity between *Pojetaia* and *Tironucula* and very reasonably suggest a close relationship. If they are also correct about the relationship between *Fordilla* and the Modiolopsidae, then increase in the number of hinge teeth did not occur in the Modiolopsidae.

Unlike rostroconchs, *Pojetaia* and *Fordilla* have anterior and posterior adductor scars. The dimyarian or heteromyarian condition is therefore a synapomorphy of the extant bivalves and these two Cambrian genera. However, Runnegar and Bentley (1983, pp. 83–5) suggest that this may have been preceded by the monomyarian condition with only a posterior adductor muscle.

The very earliest mollusc described as a bivalve, *Yangtzedonta* from the Meishucunian (Yu 1987), has a straight hinge with a groove where the ligament may have occurred. However, instead of having a veneriform shell typical of *Fordilla* and primitive for extant bivalves, the umbo is well to the anterior. *Yangtzedonta* more closely resembles members of the Pteriomorpha and does not have the shell shape thought of as being primitive. Unfortunately, in common with many Cambrian taxa, we have no muscle scars preserved and only a single valve is known. If *Yangtzedonta* is indeed correctly identified as a bivalve, it would seem to have a more primitive monomyarian condition; there is apparently no room for an anterior adductor. In this case the monomyarian condition seen in many Pteriomorpha is a later character reversal. However, *Yangtzedonda* does not seem to have the right shape to be in direct line of descent from a ribeirioid-like ancestor.

A change in orientation: the origin of the gastropods

All living gastropods go through an ontogenetic process known as torsion, where the lower body is twisted through 180° with respect to the shell and mantle cavity and the posterior part of the body. This is undertaken in the veliger larva stage, or its equivalent. It has been a commonly held view (Pelseneer 1906; Salvini-Plawen 1981) that torsion is something that happens during larval development and therefore evolved within the larval stage. Of course, it would have had major consequences for the adult stage and, when torsion first developed in the Cambrian, a larval shell had apparently not evolved.

However, it seems more likely that, at its inception, torsion was of adaptive significance to the adult shell and may well have first appeared at or following metamorphosis. Its initial development in an ontogenetic sense may have been moved into the larval stage when the gastropods first developed a veliger larva. Both processes, the development of a shell and torsion, would have been adopted through adultation of the larva, allowing a longer larval life with consequent wider geographical dispersal.

Lower Cambrian small shelly assemblages include a number of planospirally coiled shells that are often interpreted as exogastric monoplacophorans. In the Meishucunian, these are all coiled through less than 360°. Taxa which have turned through 360° or more have all deviated slightly from the planispiral form. These may have the apex deviating to the right (e.g. *Archaeospira* and *Yangtzespira*) or the left (e.g. *Cambroconus*) when viewed from above. *Hubeispira* more or less maintains its planospiral form but has altered the shape of the aperture by slight hypertrophy of its left anterior area and slight atrophy of the anterior right, thus accommodating the shell spire.

Several recent publications have sought to interpret the soft part anatomy of these early shells and hence their relationships and the early history of the gastropods and Monoplacophora. Only a limited number of possibilities exist. Shell coiling may be either left- or right-handed, the gastropod may or may not have undergone torsion, and the animal may have undergone a full 180° of torsion or less. The animal may also have turned clockwise or anticlockwise in relation to the shell and may have reduced its organ system on the pre-torsional left or right side. Finally the shells may be either endogastric or exogastric. The first of these possibilities is absolutely factual; many of the early Cambrian symmetrical shells are sinistral while others are dextral. The rest of the possibilities are a matter of deductive interpretation or pure speculation.

Yu (1987) has interpreted this group of early Cambrian Meishucunian shells as archaeogastropods. He considers them to be endogastric prosobranchs which by implication have undergone torsion. In the Archaeogastropoda the initial shell spire is turned to the pretorsional left (Fretter and Graham 1962, figs 225–9), i.e. they are right-handed spiral shells from the outset (this orientation is not so clear, however, in the Patellacea). Whether or not they are torted, neither *Archaeospira* nor *Yangtzespirira* resemble modern archaeogastropods in this respect. In fact, I see no reason to conclude that these are the shells of torted animals. The alternative interpretation—that the animal has only deviated by a small number of degrees from its shell—seems much more reasonable. *Cambroconus* is the only early Cambrian genus that resembles the pretortional veliger shell of an archaeogastropod.

The question remains whether several Cambrian shells with left-handed helicospirals belonged to a separate clade from modern gastropods. Dzik (1982) has pointed to the entirely left-handed nature of *Mimospira* and its relatives and has concluded that it is a descendant of Cambrian sinistral shells independently torted. Runnegar (1983) sees the majority of these shells as ultradextral rather than sinistral. He apparently bases this on a presumed relationship between the Cambrian left-handed Onychocheilidae and *Maclurites*, an Ordovician left-handed shell which is interpreted as ultradextral because of the curvature of its calcified operculum. He points to *Yuwenia* as an intermediate between *Maclurites* and the early Cambrian left-handed shells. I am inclined to agree with this interpretation but I am dissatisfied with his explanation of the Macluritidae which I also feel may belong to this left-handed group.

Among more recent molluscs, where the evidence for orientation is less ambiguous, sinistral shells are mirror images of dextral shells. The normally dextral whelk genus, *Neptunea*, evolved a left-handed species, *N. contraria*, which is common in the Pleistocene. This shell is a mirror image of right-handed species and both the larval and adult shell are left-handed. In the hyperstrophic architectonicacean *Heliacus*, the larval shell is left-handed but the animal inside is right-handed. Following metamorphosis, the shell reverses the direction of translation and normally becomes right-handed. Occasionally at metamorphosis this change of coiling direction does not occur; when such individuals survive into adulthood a very strange-shaped shell is produced ('*Torinia*') which is absolutely different from a mirror image. The one fossil group which has '*Torinia*'-like appearance is the archaeogastropod family Cirridae. I would suggest here that this family commonly has ultradextral shells. The flaw in Runnegar's (1981) proposal that early true gastropods are ultradextral lies in the problem that, when reversed images are made of species belonging to the left-handed family Onychochilidae, they seem to resemble right-handed taxa, i.e. unlike the Cirridae and '*Torinia*' they appear to be normal gastropods. This suggests that they may be real mirror image forms and truly sinistral.

If the majority of Yu's (1987) assignments to the gastropods rather than the Monoplacophora are correct, then there almost certainly must have been a planispiral, torted ancestor. This would apparently be in keeping with the views of Yochelson (e.g. 1978), who has supported the view of Knight (1940) that many Palaeozoic planispiral snails were indeed torted.

In the scenario offered here, all the Lower Cambrian genera I have discussed are exogastric, untorted snails normally referred to the paraphyletic group Monoplacophora. Many of these are planispiral, but some turn the apex to the right or the left in order to accommodate the earlier shell spire. For some reason, not necessarily related to the present argument, there was a tendency for these deviated shells to become slightly more high-spired. The body may then have twisted in relation to the shell in order to give a more satisfactory streamlining effect, of adaptive value for forward locomotion.

If the asymmetry of the gastropods is developed as a response to changes in orientation of the larva or adult, the die is cast at the time of early cleavage of the egg. This system and its control were obviously of a very complicated nature. In order to produce a true left-handed snail today the mirror image cleavage must take place at this early stage. If independent left- and right-handed snails developed early in the Cambrian we have no reason to suppose that the genetic controlling mechanism and the ontogeny it produced were in any way the same. As the supposedly left-handed snails are now extinct, we may never know how they developed.

The type of species that might have belonged to the ancestral forms deviating to both left and right is referred to the genus *Latouchella* by Yu (1987). Such a genus could have given rise to the cyrtolitids, sinuitids, and bellerophontids. These families, which I take to be untorted, all have more than one pair of symmetrically placed muscle scars holding the foot and body to the shell, but all have fewer pairs of scars than represented in the eight-metamere arrangement of the chitons, *Pilina* or *Neopilina*. Further reduction in the number of metameres is necessary to produce modern gastropods which usually retain only the pretorsional left, columellar muscle, the right being usually lost during ontogeny. According to Yu's interpretation only *Yangtzeconus* has just one pair of laterally placed muscle scars. The muscle attachment scars are, unfortunately, unknown in the majority of other comparable genera. Presumably this reduction of the muscle pairs in early gastropod evolution represents the atrophy of seven pairs and the retention of one of the more anterior sets because, during the torsion of *Haliotis* for example (Fretter and Graham 1962, p. 436), the retractor muscles lie directly over the head and run down and then backwards into the foot. These are clearly not the homologues of the posterior muscles which in the primitive monoplacophorans and polyplacophorans are associated with metameres incorporating the heart and nephridia etc. which are less torted.

Discussion

In the late Proterozoic, Upper Sinian (= Vendian), there is now good evidence of a major extinction of larger plankton which may correspond to a widespread and apparently equatorial glaciation (J.W. Schopf, personal communication). Preliminary assessment of the changing diversity of the Metazoa over this same interval suggests that there may also have been a decline in diversity, with a low at the beginning of the Cambrian (set arbitarily at 550 Ma). This diversity decline may be an artifact resulting from the lumping of all Ediacaran taxa at one time point and the lower diversity of the first Cambrian horizons bearing small shelly fossils. However, there are a number of biasing factors that should be considered. The non-survival of the Ediacaran taxa into the Cambrian may relate to their preservational potential. In the earliest Lower Cambrian sections of Yunnan Province, the low diversity of the first fossiliferous horizon may be affected by a detail of local environment rather than being a universal phenomenon. The preservation of soft-body fossils is now also known from higher in the Meishucunian in addition to three horizons of the Upper Sinian, and it seems that no account of this type of preservation is taken for the diversity at the earliest Cambrian low. It is to be hoped that a better picture of changing metazoan diversity will emerge as more information becomes available for this interesting and critical period.

Whether or not an extinction event affecting the Metazoa at the end of the Proterozoic was a reality, what is much more certain is that the Mollusca underwent a radiation in the first part of the Cambrian, which has continued to the present day. Development of a shell or dorsal integument in a member of the Bilateralia with simple metamerism, a flattish creeping foot, a limited amount of coelom formation, a simple head, and a complete gut together led to the development of a mantle cavity, some reduction of metamery, and development of completely new body organization. Some of these new body organizations evolved fairly quickly, within the first stage of the Cambrian, whereas the development of the full complement of characters defining some of the classes, including complete torsion in gastropods and float chambers in cephalopods (see Engeser, Chapter 6, this volume) apparently did not take place until the Late Cambrian. The major diversification occurred within the Conchifera rather than the Polyplacophora, and not until the latest Cambrian or earliest Ordovician.

The development of the shell, or at least its organic framework, may have been dependent on the changing marine environment, perhaps itself related to the amount of oxygen and carbon dioxide in the atmosphere. The shell does, however, seem to be essentially protective in that it shields the organism from its physical environment. Furthermore, the occurrence of spines in apparent polyplacophorans suggests that predation pressure may have been a significant selective force. The development of new shell forms by coiling and the production of compressed shapes are plausibly explained as adaptive to new life habits.

References

Blumrich, J. (1891). Das Integument der Chitonen. *Zeitschrift der Wissenschaftliche Zoologie*, **52**, 404–76.

Doguzhaeva, L. A. (1972). New Tremadocian monoplacophorans of the South Urals. *Paleontologicheskij Zhurnal*, **1972**, 24–8. (In Russian.)

Dzik, J. (1982). Larval development and relationships of *Mimospira*—a presumably hypostrophic Ordovician gastropod. *Geologiska Föreningens i Stockholm Förhandlingar*, **104**, 231–9.

Fretter, V. and Graham, A. (1962). *British prosobranch mollusc their functional anatomy and ecology*. Ray Society, London.

Haas, W. (1981). Evolution of calcareous hardparts in primitive Mollusca. *Malacologia*, **21**, 403–18.

Knight, J. B. (1940). Paleozoic gastropod genotypes. *Special Papers, Geological Society of America*, **32**, 1–510.

Lemche, H. and Wingstrand, K. G. (1959). The anatomy of *Neopilina galatheae* Lemche, 1957. *Galathea Reports*, **3**, 9–71.

Liu Diyong (1987). Brachiopods and tommotiids near Precambrian-Cambrian boundary in SW China. In *Stratigraphy and palaeontology of systemic boundaries in China. Precambrian-Cambrian boundary (1)*, pp. 345–413. Nanjing University Publishing House, Nanjing.

Minelli, A. and Borteletto, S. (1988). Myriapod metamerism and arthropod segmentation. *Biological Journal of the Linnean Society*, **33**, 323–43.

Missarzhevsky, V. V. (1982). Subdivisions and correlation of the Precambrian-Cambrian boundary beds based on some ancient groups of skeletal organisms. *Byulleten' Moskovskogo Obshchestva Ispytatelej Prirody, Otdel Geologicheskii*, **57**, 52–67. (In Russian.)

Morris, N. J. (1979). On the origin of the bivalves. In *The origin of major invertebrate groups*, Systematics Association Special Volume, no. 12, (ed. M. R. House), pp. 381–413. Academic Press, London.

Morris, N. J. and Fortey, R. A. (1976). The significance of *Tironucula* gen. nov. to the study of bivalve evolution. *Journal of Paleontology*, **30**, 701–9.

Pelseneer, P. (1906). Part 5, Mollusca In *A treatise on zoology* (ed. E. R. Lankester). A. and C. Black, London.

Pojeta, J. Jr. (1975). *Fordilla troyensis* Barrande and early pelecypod phylogeny. *Bulletin of American Paleontology*, **67**, 363–87.

Pojeta, J. Jr. and Runnegar, B. (1976). The paleontology of the rostroconch molluscs and the early history of the Phylum Mollusca. *Professional Papers of the United States Geological Survey*, **963**, 1–78.

Pruvot, G. (1890). Sur le developpement d'un Solenogastre. *Comptes Rendues de l'Academie des Sciences*, **111**, 689–92.

Qian Yi and Bengtson, S. (1989). Palaeontology and biostratigraphy of the Early Cambrian Meishucunian Stage in Yunnan Province, South China. *Fossils and Strata*, **24**, 1–156.

Raup, D. M. (1962). Computer as aid in describing form in gastropod shells. *Science*, **138,** 150–2.

Rolfe, W. D. I. (1981). *Septemchiton*—a misnomer. *Journal of Paleontology*, **55,** 675–8.

Runnegar, B., (1981). Muscle scars, shell form and torsion in Cambrian and Ordovician univalved molluscs. *Lethaia*, **14,** 311–22.

Runnegar, B. (1983). Molluscan phylogeny revisited. *Memoirs of the Association of Australasian Palaeontologists*, **1,** 121–44.

Runnegar, B. and Bentley, C. (1983). Anatomy, ecology and affinities of the Australian Early Cambrian bivalve *Pojetaia runnegari* Jell. *Journal of Paleontology*, **57,** 73–92.

Runnegar, B., Pojeta, J. Jr., Taylor, M. E., and Collins, D. (1979). New species of Cambrian and Ordovician chitons *Mathevia* and *Chelodes* from Wisconsin and Queensland. Evidence for the early history of polyplacophoran molluscs. *Journal of Paleontology*, **53,** 1374–94.

Salvini-Plawen, L. von (1981). A reconsideration of systematics in the Mollusca (phylogeny and higher classification). *Malacologia*, **19,** 249–78.

Wang Guixiang (ed.) 1982. *Research on the Upper Precambrian from the northern part of Jiangsu and Anhui Provinces*. Anhui Province Science and Technology Publishing House, Hefei.

Wenz, W. (1941). Gastropoda. In *Handbuch der Paleozoologie*, vol. 6(1) (ed. O. H. Schindewolf). Gebrüders Bornträger, Berlin.

Wingstrand, K. G. (1985). On the anatomy and relationships of Recent Monoplacophora. *Galathea Report*, **16,** 7–95.

Yochelson, E. L. (1978). An alternative approach to the interpretation of the phylogeny of ancient molluscs. *Malacologia*, **17**, 165–91.

Yu Wen (1987). Yangtze micromolluscan fauna in Yangtze Region of China with notes on Precambrian-Cambrian boundary. In *Stratigraphy and palaeontology of systemic boundaries in China. Precambrian–Cambrian Boundary* (*1*), pp. 19–344. Nanjing University Publishing House, Nanjing.

5. Adaptive divergence and taxonomic radiation in post-Palaeozoic bivalves

P.W. SKELTON
Department of Earth Sciences, The Open University, Milton Keynes, UK
J.A. CRAME
British Antarctic Survey, High Cross, Madingley Road, Cambridge, UK
N.J. MORRIS
Department of Palaeontology, Natural History Museum, London, UK
E.M. HARPER
Department of Earth Sciences, The Open University, Milton Keynes, UK

Abstract

Post-Palaeozoic marine bivalve families diversified with significant persistence. None of seven distinct clades dominated the expansion; most were involved. Only the pteriomorphs showed a decline, connected with the prolonged demise of exposed byssate taxa. Nine other autecological categories recognized by us recruited families more or less proportionately, although siphonate burrowers proliferated disproportionately at the expense of non-siphonate burrowers. Increasing predation pressure on bivalves from early in the Mesozoic, associated with the 'Mesozoic marine revolution', is identified as the main cause of these changes. The various adaptive pathways followed by different clades were highly conditioned by the constructional attributes of their body plans, a major role being played by the form of the ligament. Siphons evolved in several burrowing clades as a secondary adaptation.

While adaptive divergence can readily be demonstrated at lower taxonomic levels, the reasons for the widely differing extents of diversification even in closely related clades are often unclear. Taxonomic diversification is evidently 'driven' by many other factors besides the adaptive divergence of populations. Environmental context seems to be particularly important. Although lower taxa tend to proliferate in

Major Evolutionary Radiations (ed. P.D. Taylor and G.P. Larwood) Systematics Association Special Volume No. 42, pp. 91–117, Clarendon Press, Oxford, 1990.

relatively unstressed environments, there is a tendency for higher taxa to originate in stressed regimes, such as inshore and high-latitude environments. Inoceramids provide an example of a clade which probably originated in high latitudes but subsequently diversified in lower latitudes.

Introduction

Bivalves are widely considered to provide a good illustration of the concept of 'adaptive radiation'. Their living species show a broad repertoire of modifications to the simple body plan, which are clearly suited to different life habits (Stanley 1970; Yonge and Thompson 1976). The class also has a well recorded history of almost relentless increase in taxonomic diversity (Pease 1985; Fig. 5.1). Classically, the adaptive divergence is seen as essentially 'driving' the proliferation of taxa, although the influence of other, extrinsic factors, such as provinciality and global environmental change, is also acknowledged.

Recently, however, there has been much discussion of other circumstantial influences on diversity. In particular, it has been argued that certain emergent species-level properties (such as geographical and environmental range, and population size and structure), with some degree of 'heritability', might significantly influence rates of speciation and extinction, and hence patterns of diversification (cf. Vrba and Gould 1986; Levinton 1988). The implication of these arguments is that adaptive and taxonomic diversification might not, after all, be simply equated. We therefore distinguish between these two aspects in our analysis of bivalve taxa.

Taxonomic diversity

1. *Assembly of the data base*

The immense amount of data necessary and the biases of the fossil record prevent us from discovering how total bivalve species diversity has changed. Thus, higher taxa must serve as 'convenient proxies for species' (Sepkoski 1987). We have opted to analyse families, because of their greater manageability, in terms of systematic and stratigraphical checking and revision. We note that the post-Palaeozoic sequence of rises and falls in bivalve generic diversity shown by Miller and Sepkoski (1988, fig. 1A) is closely similar to that for families (Fig. 5.1), although the overall proportionate increase in the former is even greater.

For our analyses, we have revised and updated the listing of families, with stratigraphical ranges, published by Sepkoski (1982), and we have supplemented this with palaeoecological data. Our data base is deposited in the Palaeontology Library of The Natural History Museum, London.

We assume that the post-Palaeozoic marine fossil record of bivalves is reasonably comprehensive at the family level (see Pease 1985); only a handful of living families have not been documented as fossils. Regrettably, published records of their stratigraphical ranges are commonly not resolved below the stage (or epoch) level. This is the level of resolution adopted by Sepkoski (1982), and in only a few cases have we been able to improve on this, although we revise several of his stratigraphical ranges, based on a number of taxonomic revisions.

From these data it is a simple matter to estimate the number of families coexisting in each stage (the 'standing diversity') by summing the recruitment of new families with the survivors from the previous stage. This method overestimates the actual number of coexisting families in some stages since a few families which did not overlap in time are nevertheless summed. However, family durations tend to extend over many stages, and the gains and losses in each stage tend to be a small proportion of the total diversity. The amount of error involved is therefore small and probably outweighed by underestimates of the family ranges due to the incompleteness of the fossil record.

We have then normalized the data to equal time intervals of 5 Ma (Fig. 5.1), for statistical analysis, using the time-scale of Harland *et al.* (1982). In fact this involved little change since most of the stratigraphical divisions used by us (stages, epochs, or parts thereof) were of similar duration. Where at least 4 Ma of an interval occupied the bulk of a stage, the standing diversity of the interval was taken as that of the stage itself. Greater amounts of mismatch between stages and intervals were resolved conservatively, with artificial extension of the upper limits of family ranges into succeeding stages and of their lower limits into preceding stages being avoided. Where more precise range data were available the values for individual 5-Ma intervals were adjusted accordingly. These normalization procedures, together with uncertainties in the absolute time-scale, will inevitably have introduced further small errors. Obviously, any patterns of change within a time-scale of less than 5 Ma will have been smoothed over. Nevertheless, the longer-term patterns of change, which are the subject of this chapter, will not have been seriously distorted.

Perhaps the biggest source of problems for our analysis is the taxonomic status of the families. Bivalves usually have a rather limited range of taxonomically useful characters; evolutionary systematics suggests that convergence in these is relatively common. The monophyletic or paraphyletic status of established families is therefore frequently uncertain, although inferences may be drawn from stratigraphical considerations. Some authors have argued that paraphyletic taxa are artificial, and should be excluded from analyses of taxonomic diversity (e.g. Patterson and Smith 1987). It seems evident to us, however, that a paraphyletic taxon may represent a clear ecological grouping of related species, which is naturally distinct from that constituted by any monophyletic offshoot occupying a new adaptive zone. The survival or otherwise of such a paraphyletic taxon therefore registers valuable information about the fortunes of a certain adaptive type *vis-à-vis* its modified descendant stocks and so is not necessarily a mere taxonomic artefact. Skelton (1985), for example, recognized the (paraphyletic) grouping of all rudists with an external ligament as a naturally distinct assemblage of species. These were constrained to grow in a spirogyrate fashion because of this ligament, and so were restricted to spiral overgrowth of the substratum. In contrast, the derived (monophyletic) stock of rudists in which the ligament was invaginated were capable of 'uncoiled' shell growth, and thereby penetrated new adaptive zones.

We have attempted to eliminate polyphyletic taxa in our revision from our data base, as these are certainly not natural evolutionary entities.

The families were allocated among seven higher taxa so that the internal fabric of the overall radiation could be investigated. Each of these we judge to be a monophyletic clade, in terms of the post-Palaeozoic families allocated to it. The names applied to these clades, taken from conventional bivalve taxonomy, are only used as informal labels. In may thus be seen if the overall radiation was largely due to expansion of one or more particular clade(s) or was more generalized. The former pattern would imply the overriding importance of clade-specific features, such as 'key adaptations', while the latter would indicate some more general cause(s) affecting the entire class (at least).

2. *Analysis of the taxonomic data*

The long-term pattern of increase in taxonomic diversity of bivalves in post-Palaeozoic times is immediately apparent from Fig. 5.1, and has often been a source of comment (see references in Miller and Sepkoski 1988). The only checks to the increase were largely associated with episodes otherwise noted as times of mass extinction. Thus, Fig. 5.1 shows marked falls in diversity associated with the extinction episodes at the ends of the Triassic, Jurassic, and Cretaceous Periods, as well as lesser drops at some other times (Hallam and Miller 1988).

The overall trend for increase might seem obvious enough to warrant a search for general causal influences. However, stochastic explanations of clade growth (Raup *et al.* 1973) might postulate, in contrast, that the cumulative increase could simply have been randomly produced as in a 'lucky' random walk, with the originations and extinctions, due to myriads of indeterminate causes, equally probable in any given time interval. Such a proposal is readily tested by taking the random walk as a null hypothesis (H_0), against the hypothesis (H_1) that the pattern of increase is non-random.

In H_0 the probabilities of family origination and extinction in any given time interval t are equal. This being the case, the probability p of a net increase in family diversity from one time interval to the next is equal to that of a decrease. If q is the probability of diversity *either* decreasing *or* staying the same, then $p + q = 1$, and so $p < 0.5$ (since there is always a probability of diversity remaining the same). For the purposes of our test, we may nevertheless use $p = 0.5$ since this maximizes the scope for positive random walks and therefore biases the test against H_1. The standard deviation of the binomial distribution of random walks may be estimated as $\sqrt{(Npq)}$ for N steps (Bailey 1959). If we reject H_0 when the probability of the observed sequence of diversity changes being a random walk is less than 0.05, the critical region for identifying a significantly positive trend will be > 1.645 standard deviations above the level of equal numbers of increases and decreases (plus no changes). In this instance the test is one-tailed, since H_1 only concerns a trend for increasing diversity.

Figure 5.1 shows 50 time intervals of $t = 5$ Ma; thus the pattern of diversity change comprises 49 steps. The critical region may then be defined as that where the total number of steps of increase in diversity is greater than $0.5N + 1.645\sqrt{(Npq)} = 24.5 + 1.645\ \sqrt{(49 \times 0.5 \times 0.5)}\ =$ 30.26. As can readily be seen from Fig. 5.1, 31 positive steps are recorded. Thus H_0 may be rejected, and the acceptance of H_1 entitles us to seek some general cause or causes for the persistence with which originations tended to exceed extinctions through time.

3. *Interpretation of the results*

Turning now to how the radiation has been shaped by its seven constituent clades, two main points emerge from Fig. 5.1. It would not be fair to say that the radiation was largely fuelled by one or even a couple of the clades: most have shown proportionate increases with respect to their initial diversities. On the other hand, this effect is certainly uneven, and the pteriomorphs, in particular, have undergone a net decline in diversity. The heteroconchs, in contrast, show the most marked increase, especially with the Cretaceous 'bulge' generated by the rudists, sharply terminated at the close of the period, and with the proliferation of active burrowing families, especially in the Cainozoic.

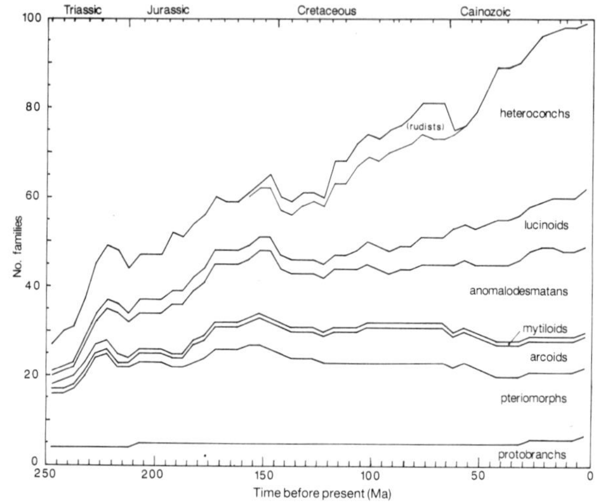

Fig. 5.1. Standing diversity of post-Palaeozoic marine bivalve families, in 5-Ma intervals. Stratigraphical divisions used in the data base are indicated along the top. Families are allocated among seven clades, following Morris (1978: heteroconchs and anomalodesmatans, including Myoida), Pojeta (1978: lucinoids, including leptonaceans and cyamiaceans; and protobranchs), and Waller (1978: mytiloids, arcoids and pteriomorphs).

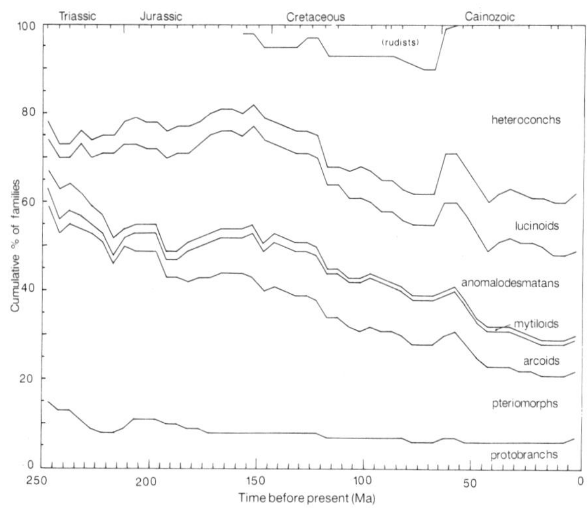

Fig. 5.2. Relative frequencies of families in the seven clades shown in Fig. 5.1. The data are as in Fig. 5.1, with diversities in each 5-Ma interval normalized to 100 per cent.

The relative contribution to the radiation of the seven clades can more clearly be seen in Fig. 5.2, where the diversity data for each interval have been normalized to 100 per cent. The pteriomorphs now stand out as the only clade which has shown a marked trend of decrease in relative frequency of families, though patterns of change at lower levels of taxonomic diversity within this clade show much variation, as will be discussed later (p. 99). Of the other clades, the protobranchs and mytiloids underwent slight declines (reflecting their numerical stability in Fig. 5.1), while the others all contributed to the compensatory 'wedge' of increasing relative frequency, though at rather different times. The most striking phase of disproportionate increase among the latter was that of the heteroconchs in the early Tertiary. Yet a one-tailed 'runs up and down' test on the sequence of changes in relative frequency of heteroconch families, designed to detect non-random clumping of steps of gain or loss (see Davis 1986), failed to reject the null hypothesis that these were randomly distributed, at the 0.05 level of significance. So even the heteroconchs diversified sporadically, relative to the others, rather than in any sustained episode(s).

Adaptive diversity

In order to study changes in the adaptive repertoire of post-Palaeozoic marine bivalves, we classified the families into 10 autecological categories, based on standard analogical and homological criteria (Fig. 5.3). Nearly all families wholly conformed to a given category. A few had mixed representation, and, where appropriate, these were double-scored in the categories concerned. A small number of aberrant families had to be arbitrarily placed in the nearest appropriate category (e.g. the Clavagellidae in the siphonate burrowing suspension-feeders). Data on numbers of families in each category were then normalized to 100 per cent for successive 5-Ma intervals, as before, and the results plotted in Fig. 5.3.

A pattern of changing frequencies somewhat reminiscent of Fig. 5.2 emerges. Here, it is the exposed byssate suspension-feeders which are shown to have undergone a marked decline in frequency, while all other categories more or less held their own or showed some expansion during at least a part of their history. The similarities with Fig. 5.2 are by no means coincidental; a large degree of overlap has always existed between pteriomorphs and the byssate category. Nevertheless, despite some recruitment of families from other clades, the relative frequency of byssate families declined even more markedly than did that of pteriomorph families. This can be explained by a cumulative 'escape' of several pteriomorph families into other autecological categories. In

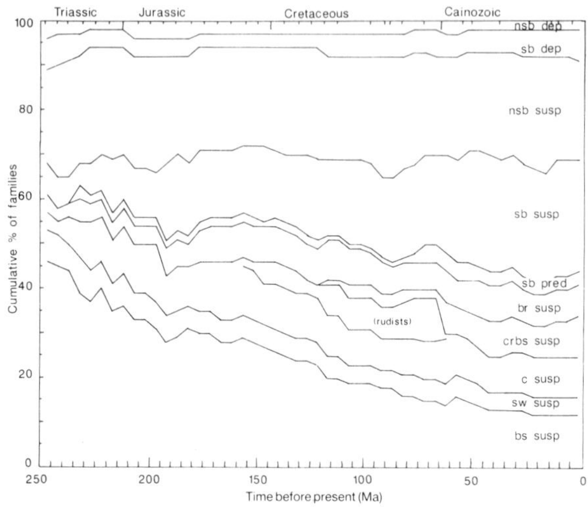

Fig. 5.3. Relative frequencies of post-Palaeozoic marine bivalve families in 10 autecological categories, through successive 5-Ma intervals (based on data of Fig. 5.1). The autecological categories are: (1) nsb dep, non-siphonate, and (2) sb dep, siphonate burrowing deposit-feeders; (3) nsb susp, non-siphonate, and (4) sb susp, siphonate burrowing suspension-feeders; (5) sb pred, siphonate burrowing (septibranch) predators to detritivores; (6) br susp, boring suspension-feeders; (7) crbs susp, cryptobyssate (cavity-dwelling to commensal) suspension-feeders; (8) c susp, cemented suspension-feeders; (9) sw susp, facultatively swimming (ex-byssate) suspension-feeders; and (10) bs susp, exposed byssate (endo-, to epibyssate and nestling) suspension-feeders.

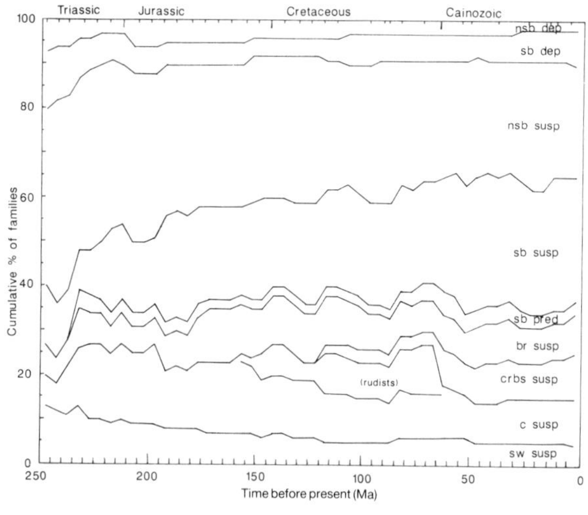

Fig. 5.4. Data of Fig. 5.3, with the 'bs susp' category (10) removed, and frequencies of the remaining categories renormalized to 100 per cent.

particular, facultative swimming was adopted by many scallops (Propeamussiidae, Pectinidae, and Syncyclonemidae), as well as some Limidae, and cementation evolved in several families, including the oysters, Plicatulidae and Spondylidae (see Waller 1978). Thus, out of 12 pteriomorph families registered from the early Triassic, 10 (83 per cent) were comprised entirely of exposed byssate suspension feeders. From among 15 Recent pteriomorph families, in contrast, only six (40 per cent) belong wholly to this category. Of these latter, moreover, some are greatly modified so as virtually to mimic cementation, with close, substrate-hugging growth (e.g. Anomiidae). Throughout the Mesozoic and Cainozoic, the exposed byssate life habit, which had been so common among Palaeozoic bivalves, substantially waned in prominence (Stanley 1977). They likely reasons for this are discussed later (p. 101).

Changes in the relative frequencies of the other autecological categories are more easily assessed by removing the distorting influence of the decline in exposed byssate suspension-feeders. In Fig. 5.4, the latter category has therefore been eliminated from the data set, and the remaining categories renormalized to 100 per cent. Apart from some erratic fluctuations in the first few intervals of the Triassic, when total numbers of families were low, the remaining categories can now be seen to have suffered remarkably little overall change in relative frequency with respect to each other: most appear on Fig. 5.4 as horizontal bands showing some fluctuations in width, with few consistent trends. A slight sustained increase in importance of siphonate burrowing suspension-feeders was closely matched by a decline in that of non-siphonate burrowing suspension feeders. Summed together, these two categories hovered closely around a mean relative frequency of 54.0 per cent, within a range of 40 per cent to 59 per cent, and with no consistent trend of change in joint frequency whatsoever. So, at the family level, burrowing suspension feeders (both siphonate and non-siphonate) did no better than the other categories shown on Fig. 5.4.

This is a rather surprising result, since much stress has been placed in the literature on the massive post-Palaeozoic radiation of siphonate burrowers (e.g. Stanley 1977). We do not doubt that the latter assertion is quite correct at lower taxonomic levels. Evidently, some of the families of siphonate burrowing suspension feeders have become very large. In view of the phenetic criteria employed by most classical taxonomists, this in turn implies only rather modest ecological differences between their many species. Returning to the family level, however, the adaptive zone of siphonate burrowing suspension-feeding was evidently just one of several, which were increasingly exploited by newly arising bivalve subclades as the exposed byssate taxa declined.

The broad enrichment of adaptive diversity in post-Palaeozoic bivalves stands in marked contrast to the more limited adaptive scope of most Palaeozoic bivalves. The great majority of the latter were either non-siphonate burrowing, or exposed byssate suspension feeders, or deposit feeders (see Stanley 1977; Pojeta 1978; Morris 1978). Nevertheless, most of the other life habits were also exploited in the Palaeozoic by just a few genera. Siphonate suspension-feeding taxa included the Permian Scaphinellidae (Trigoniacea) and, from the early Carboniferous onwards, various pholadomyoid anomalodesmatids, such as *Wilkingia* (Morris 1978). Boring has been documented from as early as the late Ordovician (Pojeta and Palmer 1976). Cementation is seen in the Carboniferous *Pachypteria* (Newell and Boyd 1970). Scallops with a morphology characteristic of swimmers, such as *Pernopecten* (Waller 1978), range from the early Carboniferous. Only the predatory septibranchs and cryptobyssate forms appear to be post-Palaeozoic innovations.

The many post-Palaeozoic families which exploited these other adaptive zones were not generally derived from the few Palaeozoic 'pioneers' cited above. Rather, most of them had diverse and independent origins. The proliferation of taxa in the expanding autecological categories of the post-Palaeozoic cannot therefore simply be ascribed to radiations sparked by the appearance of key adaptive innovations ('adaptive breakthroughs') in the Palaeozoic. It would seem that the balance of selection changed from early in the Mesozoic, channelling bivalve clades into several previously thinly tenanted adaptive zones. The adaptive responses of the different clades, however, were markedly different. This implies that preadaptive states were nevertheless of great importance in fashioning the eventual adaptive responses of the clades.

Apart from the inferred changes in CO_2 dissolved in sea water and its possible effect on calcite deposition (Sandberg 1983), we are unaware of any major permanent global change in physical circumstances, initiated at the close of the Palaeozoic, which might explain the changes in adaptive repertoire discussed here. A biotic explanation seems more likely. We believe that an outstanding candidate for such an explanation is the well-documented rise of many new predator groups, the 'Mesozoic marine revolution' (MMR) (Vermeij 1977). This involved a substantial diversification of durivores, and hence stepping-up of predation pressure upon shelly benthos, including the bivalves (Vermeij 1983). The adaptive responses of the major prey groups varied. Vermeij argued that bivalves, as a consequence of their generally poor capacity for rapid repair and survivorship of shell damage, relied chiefly on escape or avoidance. We see this as an explanation for most of the changes in the adaptive repertoire that we have discussed.

Our interpretation is certainly not new; the MMR has been independently implicated in several of these changes. The rise of siphonate burrowing, as a defensive adaptation against predators, has been argued notably by Stanley (1977, 1986), who has also attributed both the decline in exposed byssate forms and the adoption of swimming habits by some of these to predation pressure. Boring has also been interpreted as achieving some refuge against predator attacks (Carter 1978). The defensive value of cementation has been experimentally confirmed (Harper, in press). Various molluscivores (starfish and crabs) were offered a choice of byssally attached and cemented bivalve prey. Although it was possible for the predators to eat the latter, there was a marked, statistically significant preference for the byssate prey. Cementation appears to inhibit manipulation by the predator, and hence decreases the net energy gain. This places the cemented bivalves at a selective advantage over the byssate prey in terms of optimal foraging by the predators. This experimental evidence implies that the independent emergence of many new cemented taxa in early Mesozoic times may be causally linked with the inception of the MMR (Fig. 5.5). There is no evidence as yet, however, that cementation, by itself, affords any protection against drilling gastropods. Possibly, elaborations of shell structure (e.g. cellular and chalky fabrics, and conchiolin layers in oysters) and ornament (e.g. frills and spines in chamids, spondylids, etc.) help to inhibit gastropod attacks (cf. Vermeij 1983). Drilling gastropods first began to make a major impact from mid-Cretaceous times (Sohl 1969; Taylor *et al.* 1980; Taylor 1981). It is therefore interesting to note that the cemented category of Fig. 5.4 shows a slightly lowered relative frequency from this time, if the rudists are excluded. It is tempting also to speculate that cryptobyssate habits (e.g. in the leptonaceans) may also have evolved as an adaptation for predator avoidance.

We contend, then, that much of the adaptive divergence that has characterized post-Palaeozoic bivalves was instituted by the rise in predation pressure associated with the MMR. Only certain, rather bizzare innovations, such as uncoiled valve growth in rudists (Skelton 1985) and the adoption of predatory feeding habits in septibranchs, invite other explanations.

Even the apparent exceptions to this assertion—the remaining, sometimes even thriving taxa of exposed byssate forms—have alternative 'antipredator' adaptations. Though not particularly speciose, mussels are notably abundant the world over. They are relatively thin-shelled epibyssate forms, of a sort one might expect to fall easy prey to molluscivores. Yet their autecology also shows defensive adaptations. In addition to their successful exploitation of the intertidal refuge from at

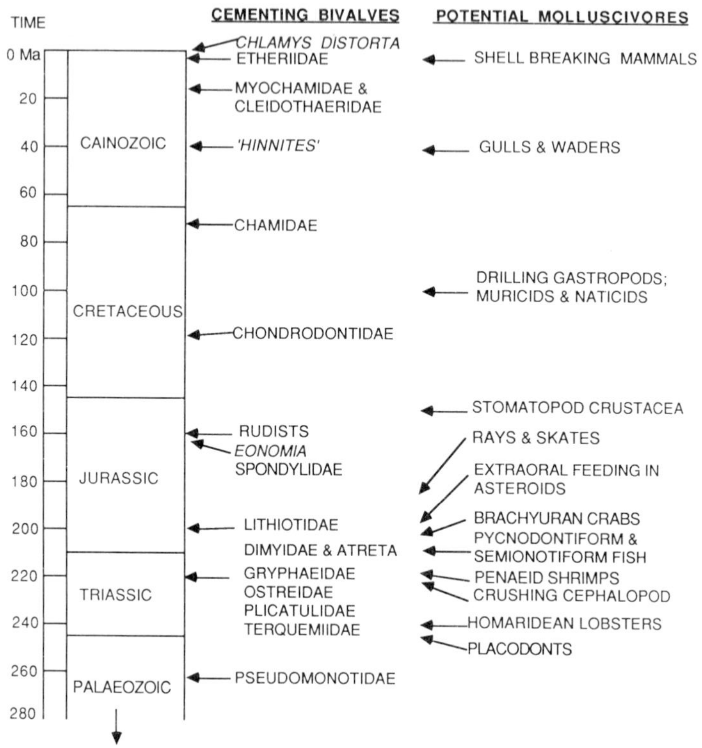

Fig. 5.5. Temporal distribution of the first appearances of cementing bivalves, and of molluscivory in various predator groups (modified from E.M. Harper, in press; key reference source, Vermeij 1987).

least some predators, it is now known that clusters of *Mytilus edulis* actively 'assault' their would-be raiders, drilling gastropods, by collectively trapping them with byssal threads and so exposing them to often fatal desiccation (Petraitis 1987). Vermeij (1983) noted that exceptions to his generalization about poor capability for shell repair are provided by the endobyssate *Pinna* and the free-swinging byssate pteriids, which have flexible valve margins that ensure hermetic sealing while repair is underway. Entrenched byssate nestling, sometimes leading to active endolithic habits (e.g. *Lithophaga* and the arcid *Litharca*), or as more passively acquired amidst growing skeleton-secreting hosts (e.g. *Tridacna* and the pectinacean *Pedum*), represents yet another defensive device.

If, as we argue, the MMR thus played the dominant role in the re-shaping of the adaptive repertoire of bivalves, how was the revolution in life habits distributed among clades, in terms of morphogenetic constraints (Seilacher 1985)?

Constructional controls on diversification

1. *Ligaments*

The exploitation of the autecological categories by the different bivalve clades was very uneven. One important control seems to have been ligamentary construction. The parivincular ligament system (Fig. 5.6), for example, was strongly associated with the burrowing clades. It comprises a posteriorly growing, arched hood behind the umbones. This permits the continuous generation of an external, lamellar (extensional) layer, and an inner lining of fibrous (compressional) ligament, by the contained mantle isthmus (Ansell 1961; Trueman 1969). The insertion surface in each valve naturally follows the spiral growth pattern of the valve, and so progressively twists somewhat with growth as it extends backwards (Fig. 5.6). The risk of consequent ventralward flexing (and rupture) of the older, anterior fibrous ligament is avoided by the co-adaptive growth of nymphs. These elongate ridges reflex the fibrous ligament insertion surfaces dorsally, so counteracting their progressive ventralward twisting towards the anterior (Fig. 5.6). The resulting elongate C-spring form of the ligament, with its biomechanically advantageous arrangement of the lamellar and fibrous layers, and only negligible rupturing of old ligament, is extremely efficient, as well as being compact. It is thus well predisposed to force the

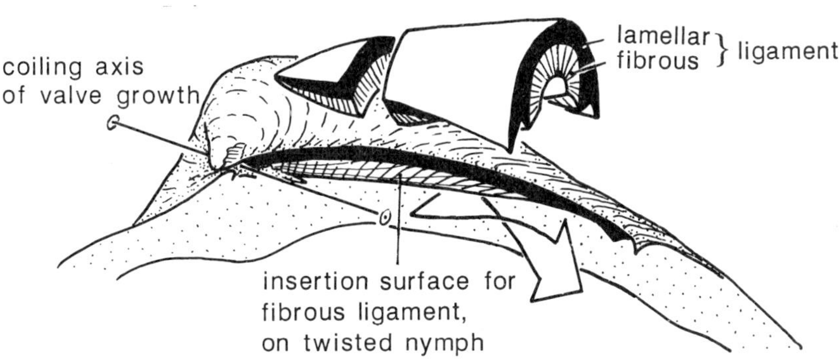

Fig. 5.6. Exploded diagram of a parivincular ligament and its insertion surface along the hinge of the right valve. The twisting during growth of the nymph is indicated by the broad arrow. The posterior part of the ligament has been cut away to show the arched lamellar and fibrous layers.

valves open against the surrounding sediment, while the foot probes during the burrowing sequence (Trueman 1966) without itself being an obstruction. In addition, the posteriorly tangential growth of the ligament has the constructional effect of making the umbones more or less prosogyral (Savazzi 1987, figs 2D and E), generating a broad flat surface in front of the umbones, the lunule. Stanley (1975) experimentally demonstrated the assistance this gives to the burrowing sequence. It is perhaps no surprise then that the heteroconch, lucinoid, and anomalodesmatan clades, which had acquired parivincular ligaments well before the Mesozoic, came to dominate as burrowers.

The arcoids, in contrast, despite relinquishing their primitive byssate habits for a burrowing existence in several post-Palaeozoic families, only made modest inroads in that realm, almost certainly because of their relatively weak and inefficient duplivincular ligament systems (Thomas 1978*a*,*b*.)

Curiously, the mytiloids also evolved parivincular-type ligaments with nymph-like supports in the Palaeozoic (Pojeta 1971). But their near total exclusion from active burrowing was presumably imposed by the pre-emptive constraints of the reduced foot and the anisomyarian adductors associated with their primitively byssate condition. A similar argument may be applied to the pteriomorphs.

2. *Siphons and mantle fusion*

The independent, multiple evolution of siphons in the heteroconchs, the lucinoids, and the anomalodesmatans can be interpreted as but a secondary adaptation, contingent upon their established status as efficient burrowers. Mantle cross-fusion did not represent such a fundamental threshold as is sometimes supposed, being only part of a series of changes. Certain of the siphonate clades did, however, become highly speciose.

Among the heteroconchs, for example, the evolution of siphons in the veneroids seems to have been polyphyletic, and the ensuing patterns of diversification were highly variable. The most parsimonious interpretation is the origin of three major clades of the veneroids in the Triassic: (1) the Cardiacea and their relatives; (2) the Arcticacea and their relatives; and (3) the Tellinacea and Solenacea group. These may have arisen from a late Palaeozoic common ancestor resembling *Astarte*. The first representatives of the three clades occur between the Carnian and Hettangian Stages. At this early time it is clear from the shell shapes and form of the pallial line that they have each developed a greater degree of mantle fusion than their common ancestor.

The three clades are separable in general shell shape and pedal muscle attachment patterns, which reflect their broadly differing burrowing. The Cardiacea have tumid and rounded shells with a truncated posterior margin that parallels the sediment surface. They are active burrowers with a powerful 'L'-shaped foot similar to the non-fused Trigoniidae. The early forms of the second clade are at present all included in the Arcticacea and are ovoid or rounded and also rather tumid. These, by comparison with living analogues, had a rather narrow blade-like foot but were apparently at first all rather sluggish burrowers. Even the earlier members of the third clade had narrow blade-like shells, with the umbones well back from the anterior margins, and seem well fitted for fast penetration of unconsolidated, sand-grade sediments. Two early families are recognized—the Tancrediidae, with normal radiating heterodont teeth, and the 'Quenstedtiidae'. Siphons are recognized by a deep pallial sinus in the 'Quenstedtiidae', and by characteristic gapes and later by a short sinus in the Tancrediidae. Shell shape and hinge teeth are characters which lead us to believe that both the living Tellinacea and Solenacea have evolved from members of the paraphyletic Tancrediidae.

The pattern of diversification of the Solenacea illustrates the highly uneven nature of speciation within different families. There are but two common genera of the Solenacea in the Cretaceous, *Senis* of the Pharidae and *Leptosolen* belonging to the paraphyletic Pharellidae. *Senis* was restricted to inshore silty facies and is of relatively regular occurrence between the Barremian and the Cenomanian, whereas *Leptosolen* occurred in more clayey offshore sediments, starting in the Albian. By late Campanian times the latter had grown larger and occurs in more sandy sediments. The living solenaceans share the advanced characters of an elongate shell and type 'B' mantle fusion along the posterior half of the venter, exposing the middle mantle fold with its papillate surface at the siphonal margins. An analysis of the remaining characters suggests that the only living descendant of *Senis*, *Pharus legumen*, has only altered the form of the anterior adductor, anterior part of the pallial line, and

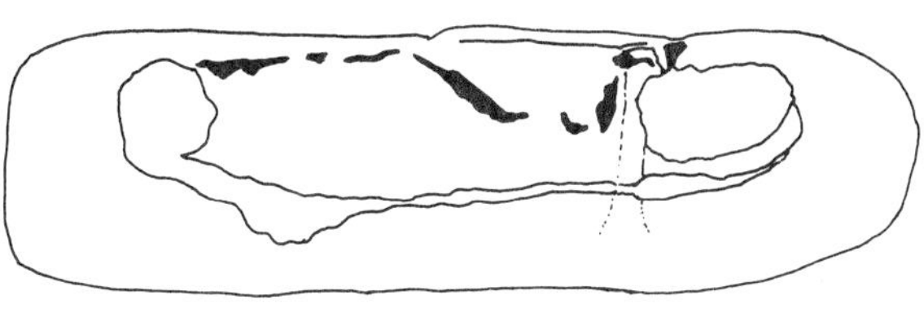

Fig. 5.7. Internal shell features of *Leptosolen biplicatens* (Conrad), Upper Campanian, USA, showing unmodified synapomorphic pedal muscle scar pattern (in black) of all Solenacea except in Pharidae.

shell anterior by a relative increase in shell length of nearly 30 per cent. *Leptosolen*, in contrast, developed a very particular pattern of pedal muscle attachment (Fig. 5.7), and a more anterior position of the hinge teeth and ligament. It separately underwent an increase in length and has given rise to all other living Solenacea. Why *Senis* and then *Pharus* remained at such remarkably low diversity, whereas *Leptosolen* underwent a considerable radiation that began very early in the Tertiary, is unclear. The Solenidae have been able to spread their habitat width from the high intertidal and estuarine, to relatively deep water and, concomitantly with this habitat expansion, exhibit a wide variation in siphon length and type, a wide variation of the amount and type of ventral mantle fusion, and a considerable diversity of type of pedal muscle attachment.

One part of the family Solenidae furnishes us with a clear-cut example of this increase in morphological diversity as well as of genera and species related to the invasion of new ecospace. The genus *Pharella* lives in inshore muddy sands in areas of fairly normal salinities. It has short siphons of type 'B' and a short pallial sinus, almost exactly as in *Leptosolen*. Some time during the Tertiary it invaded the thixotropic muds that occur in front of mangrove swamps and in estuaries from the Bay of Bengal to the coast of China. Most species have retained their razor shell shape. The species in the Ganges Estuary, *Novaculina*, and those of China, *Sinonovacula*, have developed a deep pallial sinus. The deep pallial sinus corresponds to very long siphons which enable it to slip deep into the thixotropic mud, perhaps driven by predation pressure. These siphons are not type 'B', but formed only of the inner mantle fold within the initially exposed middle fold, and are almost exactly convergent with the siphons of Tellinacea (type 'A'). However, they retain the pedal muscle scar attachment pattern of *Leptosolen*. A further significant change has taken place in the Malaysian area, where there has been a secondary reversal of shell shape, first to an oval form, then to the round shell of *Orbicularia*. This whole group would seem to indicate adaptation for living and moving in thixotropic mud, an environment that may not have existed before the Tertiary arrival of modern mangal community tree families.

Examples such as this illustrate how established species may diverge adaptively, according to differing environmental conditions. However, there remains the problem of how the species originate in the first place, and why some clades proliferate while others expand little, as with the descendants of *Senis* and *Leptosolen*. Factors other than adaptation must be considered.

Controls on differential rates of speciation and extinction

1. *Onshore–offshore patterns*

One of the most exciting developments in evolutionary studies in recent years has been the analysis of trends and rates within an ecological context: can we pinpoint certain types of environment that may have served consistently as the source of evolutionary novelties? This is particularly pertinent in the marine realm, where palaeoecological studies have suggested for some time that innovations may have occurred along environmental gradients (e.g. Berry 1974; Crimes 1974; Sepkoski and Sheehan 1983).

Such studies were formalized in a seminal paper by Jablonski *et al.* (1983), who showed that a succession of Cambro-Ordovician benthic communities first appeared in a nearshore setting and then expanded offshore. Each of Sepkoski's (1981) three great evolutionary faunas had obviously developed in this way. Certain 'adaptive types', as well as communities, were progressively displaced across the shelf. Perhaps the most important of these categories is that of the 'immobile epifaunal suspension feeders'. Once dominating Palaeozoic shelf seas, it became progressively isolated to the outer shelf and slope through the Mesozoic and survives today only in certain relict-type habitats.

There seems in fact to have been a persistent outward expansion of evolutionary innovations and new community types from a nearshore crucible into more offshore habitats (Jablonski *et al.* 1983; see summary in Conway Morris 1989). This is surprising for it has generally been accepted that there are somewhat higher rates of species-level evolution in offshore habitats (e.g. Jablonski 1980); if anything, a trend in precisely the reverse direction would be expected.

Subsequent research has revealed that, whilst some taxa have expanded offshore (i.e. have retained onshore representatives), others have retreated (i.e. have lost onshore representatives) (see Jablonski and Bottjer, in press). Examples of the former from within the Bivalvia would include the limopsids, and examples of the latter would include the parallelodontids (Bottjer and Jablonski 1988). The origin of lower taxa may in some way be diversity-dependent (Jablonski and Bottjer, Chapter 2, this volume). The more evolutionary experiments there are per unit time, the more novelties that are produced.

If these observations are correct, and are subsequently proved to be widespread within other marines clades, then there are some immediate macroevolutionary consequences. It may be that the origin of higher taxa is effectively decoupled from that of lower groups for, if the greater spatial and temporal heterogeneity of onshore environments was in some way promoting greater evolutionary divergence, then we would expect to find lower taxa (families and genera) consistently concentrated there. However, the fact that this is obviously not so for some groups means that we cannot account for the origin of higher taxa (e.g. orders) by a simple extrapolation of microevolutionary processes.

Some aspect of the onshore environment must be promoting the production and survivorship of morphological features that are sufficiently distinct to produce a taxon of higher rank: can we deduce what this may be? Abiotic factors seem unlikely (Jablonski and Bottjer, in press), and we are forced to review a variety of intrinsic biological mechanisms. Of these, there appear to be two distinct sets: one centred around the concept of differential origination and the other around that of differential extinction.

Nearshore environments are generally regarded as 'disturbed' or 'stressed', due to substantial fluctuations in parameters such as temperature, wave-energy, turbidity, and salinity; organisms that live in them are typically physiologically tolerant, have motile larval stages, and comprise widely dispersed populations. In direct contrast to this, stabler offshore environments are frequently characterized by much more specialized organisms with narrower geographical ranges (Jablonski and Bottjer 1983).

The same traits of physiological tolerance and broad distribution seem to confer on nearshore taxa somewhat lower speciation and extinction rates (Jackson 1974; Jablonski 1980). This means that, although it may take longer for an adaptive breakthrough to occur in such organisms, it is more likely to persist through time. It may well be the nearshore environments are not so much the sites of macromutations, but more the sorts of places where evolutionary innovations are prefentially maintained.

Once established, the subsequent spread of a clade across the shelf may be nothing more than a simple diffusion process (Gould 1988). It is logical to assume that most specialists are derived from generalists (rather than other specialists) and we would expect, *a priori*, an outward migration of taxa across the shelf. A logical corollary of this process is that speciation into stabler offshore environments will be more likely than speciation back into less stable environments.

2. *Polar–equatorial patterns*

As soon as the theory of onshore–offshore gradients of evolutionary origination became established, it was pertinent to ask whether there might not be equivalent polar–equatorial gradients. Polar regions (in both the marine and terrestrial realms) are typically high-stress ones and the tropics are frequently portrayed as environmentally stable (Lewin 1983; Sheehan 1986).

Such a polar–equatorial gradient runs contrary to the concept of the tropics as a major evolutionary centre (e.g. Darlington 1959; Fischer 1960; Stehli *et al.* 1969), but it is interesting to note that it is largely in agreement with Valentine's 'species diversity pump' model (1967, 1968, 1984). He suggested that stenothermal taxa contracted equatorward during cooling periods, vacating habitat space that could be occupied by forms with different temperature adaptations but otherwise similar niches. These, in turn, could subsequently be displaced equatorward, and, providing at least some of them adapted to warmer conditions, would lead to the species-enrichment of the lower latitudes. The model serves as a forceful reminder that at least some components of the equatorial high-diversity centres may have an extra-tropical origin.

Even though the polar palaeontological record is very incomplete, a group with considerable potential for testing this biogeographical hypothesis is that of the 'immobile epifaunal suspension feeders'; if component taxa within this category (bivalves, brachiopods, bryozoans) have been displaced progressively across the continental shelf, then is there any evidence of equivalent movements from high to low latitudes? Within Cretaceous shelf faunas certain pteriomorph bivalve groups may be particularly amenable to analysis, and in this study we have selected one very prominent group, the inoceramids. Members of the family Inoceramidae were prominent in mid-to outer-shelf assemblages throughout much of the Cretaceous, and ranged from low to very high latitudes (e.g. Jablonski and Bottjer 1983). Our analysis is a preliminary one of the evolutionary development of early Cretaceous inoceramids with particular reference to their palaeolatitudinal distribution (Figs 5.8 and 5.9).

After removal of some apparently unrelated convergent taxa, the rise of the inoceramids through the Early Cretaceous shows a steady pattern, punctuated by a marked decline in the Barremian–Aptian (Fig. 5.8). Reasons for the latter are unclear, but might involve any of sea-level fall, climatic change, an extinction event, or merely collection failure (unlikely in the Northern Hemisphere).

Although the total number of species involved is small, Berriasian inoceramids show a distinct preference for high-latitude occurrences in the Northern Hemisphere (i.e. 60°–90°N; Fig. 5.9A) (the mid-latitudes are here defined as 30°–60°N and the low latitudes as 0°–30°N). The high-latitude occurrences are due to the presence of three species in northern Siberia and four in the Far East of the USSR. The latter region is particularly important as it forms the core to the Northern Pacific Province, which was to dominate inoceramid development during the first three stages of the early Cretaceous (Pochialaynen

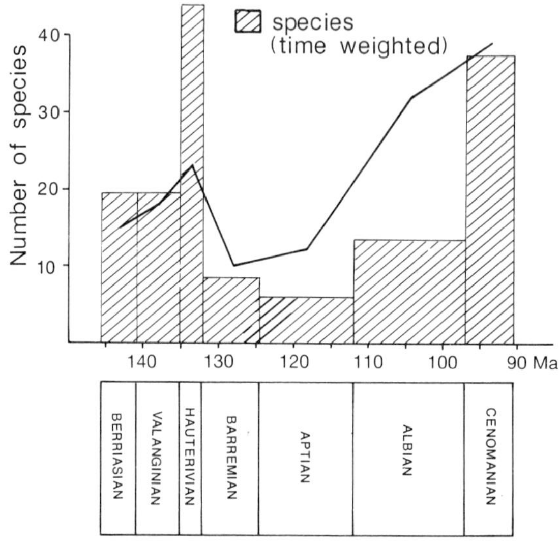

Fig. 5.8. Radiation of inoceramid bivalves through the early Cretaceous and into the Cenomanian. The solid line depicts global species diversity by stage. The histogram gives a time-weighted depiction of the species diversities of the stages. Data from Crame (1985), together with a card index file held by J. A. C. *Retroceramus* and *Anopaea* species are excluded.

1974). At times this province extended southwards to Sikhote Alin and Japan in one direction, and through Alaska to the Pacific coast of North America in the other.

The vast majority of Valanginian species occur within the North Pacific Province; however, records from southern Sikhote Alin and North America fall within the mid-latitude category (Fig. 5.9A). The same province reaches it acme in the Hauterivian, with the most diverse fauna being recorded from north-west Kamchatka (Pochialaynen 1969). This appears to have been a time of accelerated inoceramid evolution, with the appearance of a number of distinct taxonomic groups (subgenera?) centred on species such as *I. heteropterus*, *I. colonicus*, and *I. aucella* (Crame 1985). Some of these groups became widespread in the Northern Hemisphere.

From some 18 species in the Hauterivian of the Northern Hemisphere, there was an abrupt decline to just four in the Barremian (Fig. 5.9A). These were all forms that had arisen in the Hauterivian and it is apparent that there was virtually no evolutionary innovation in the family at this time. The crisis appears to have continued into the Aptian, with the only significant fauna being that which developed on the Russian Platform (Glazunova 1973).

The second main phase of inoceramid development began in the Albian. This coincided with the rise of a number of other benthic molluscan groups and can almost certainly be linked to a global increase in the area of shallow shelf seas (e.g. Sohl 1987). Inoceramids appeared at many new localities in Europe and North America, and it is apparent that the mid-latitudes were now their preferred site of development (Fig. 5.9A). In the mid-and late-Albian the first cosmopolitan inoceramids appeared (the *I. anglicus* and *Birostrina concentrica* groups).

This marked shift in the latitudinal development of Cretaceous inoceramids was further accentuated in the Cenomanian (Fig. 5.9A). Again, inoceramids are still present in significant numbers in the high latitudes but these forms are now overshadowed by those present in the 30°–50°N belt. Localities towards the southern end of the Western Interior Seaway (North America) and in central and southern Europe appear to have been the loci of inoceramid development during the initial stage of the late Cretaceous. The shift from high-to mid-latitude centres of development during the Berriasian–Cenomanian interval is statistically significant.

Remarkably few inoceramids are known from the Berriasian–Aptian of the Southern Hemisphere (Fig. 5.9B). This is almost certainly due in part to collection failure, but must also reflect the lack of shallow epicontinental seaways both around and across Gondwana. The small but persistent peak of taxa at 50°–60°S represents a fauna developed in the marginal basins of the Patagonia–Scotia arc region (Crame 1985).

The first extensive shallow-marine deposits in the Southern Hemisphere are Albian in age. They are frequently rich in inoceramids and largely, but not exclusively, endemic faunas are known from localities such as south-eastern Australia, New Zealand, and western Antarctica. These faunas can be taken to suggest that, although somewhat later in time, the initial inoceramid radiations in the Southern Hemisphere were again concentrated in the high latitudes. Cenomanian inoceramids are reasonably well known in the Southern Hemisphere, but show no preferential grouping in any palaeolatitudinal belt (Fig. 5.9B).

Although the data from the Southern Hemisphere are less rigorous, they may perhaps be taken to reinforce the high-to mid-latitude shifts established in the Northern Hemisphere. Many groups of late Cretaceous inoceramids seem to have reached their maximum development in the mid-latitudes (e.g. Dhondt 1983) and some even penetrated Tethyan carbonate environments (e.g. Kauffman 1968). Preliminary analyses suggest that other Mesozoic 'immobile epifaunal suspension

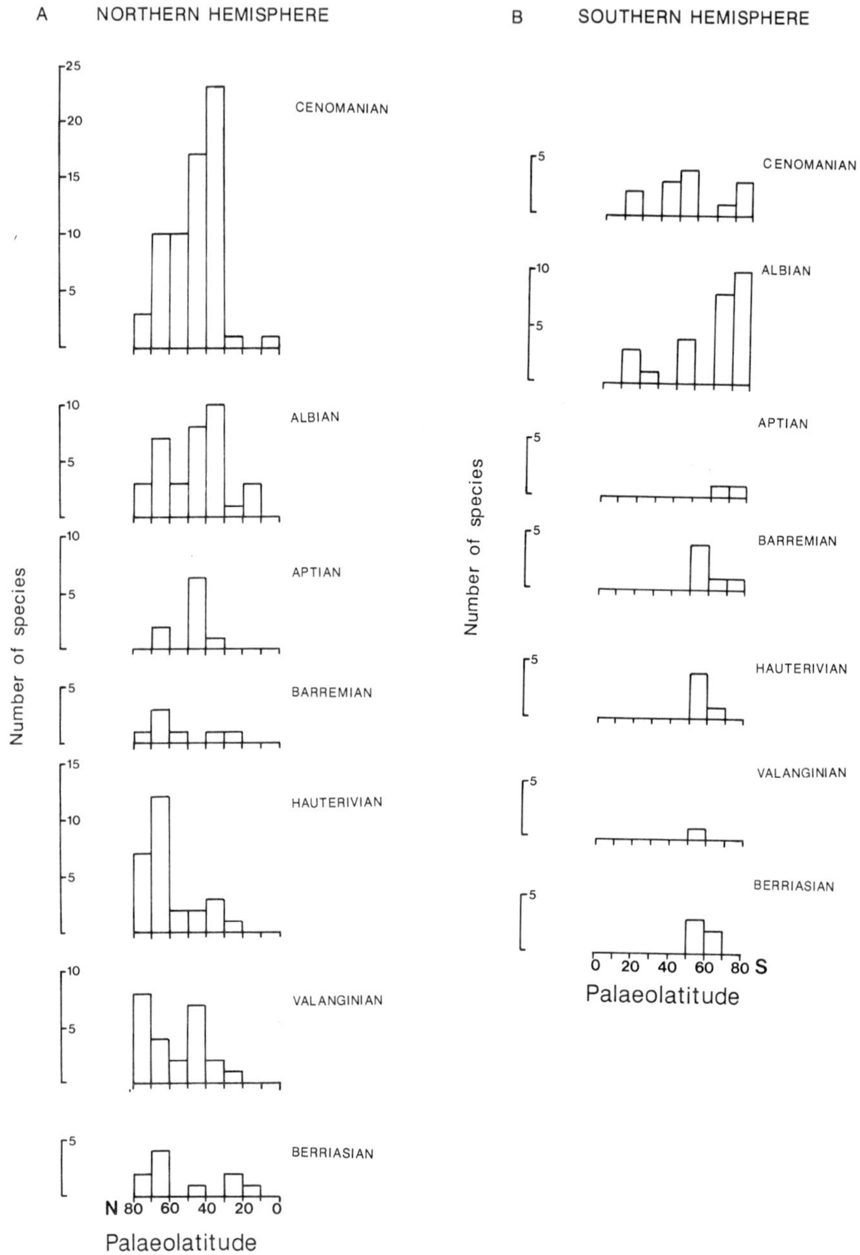

Fig. 5.9. Radiation of Cretaceous inoceramid bivalves in the (a) Northern and (b) Southern hemispheres. Histogram bars show numbers of species in 10° palaeolatitude belts (based on Smith and Briden 1977). These were divided into

feeder' bivalve groups showing similar latitudinal development trends include: retroceramids, buchiids, certain oxytomids (e.g. *Aucellina*), and some oyster clades.

Conclusions

We contend that the major post-Palaeozoic adaptations of bivalves were largely shaped by the rise in predation pressure associated with the Mesozoic marine revolution. Different clades showed strong tendencies to exploit different life habits as a response, according to the constraints and potentials of their constructional characters. Although such a picture emerges fairly clearly from the pattern of changes in the relative frequencies of families, it does not explain the highly uneven nature of lower-level taxonomic diversification within families.

Individual speciation events usually contain an element of selection pressure on individuals, leading to adaptive divergence, but, in addition, we look to macroevolutionary mechanisms such as increased high taxonomic innovation, related to species persistence, in stressed inshore and high-latitude environments, typified by our example of Cretaceous inoceramids.

Acknowledgements

E.M. Harper is in receipt of a NERC Studentship, and the manuscript was typed by Janet Dryden.

References

Ansell, A.D. (1961). The functional morphology of the British species of Veneracea (Eulamellibranchia). *Journal of the Marine Biological Association, UK*, **41,** 489–515.

Bailey, N.T.J. (1959). *Statistical methods in biology*. English University Press, London.

Berry, W.B.N. (1974). Types of Early Paleozoic faunal replacements in North

---

three categories for statistical testing: high (90°–60°), mid (60°–30°), and low (30°–0°) latitudes. Significant distributions (χ^2 test, $p < 0.05$) were found for the following stages in the Northern Hemisphere: Valanginian, Hauterivian, Albian, and Cenomanian. Significant differences exist there between the Hauterivian distribution and those obtained for both the Albian and the Cenomanian stages (Kolmogonov–Smirnov two-sample test, $p < 0.05$). The only significant distribution detected in the Southern Hemisphere was that for the Albian (χ^2 test, $p < 0.05$).

America: their relationship to environmental change. *Journal of Geology*, **82**, 371–82.

Bottjer, D.J. and Jablonski, D. (1988). Palaeoenvironmental patterns in the evolution of post-Paleozoic benthic marine invertebrates. *Palaios*, **3**, 540–60.

Carter, J.G. (1978). Ecology and evolution of the Gastrochaenacea (Mollusca, Bivalvia) with notes on the evolution of the endolithic habit. *Peabody Museum of Natural History, Yale University, Bulletin*, **41**, 1–92.

Conway Morris, S. (1989). The persistence of Burgess Shale-type faunas: implications for the evolution of deep water faunas. *Transactions of the Royal Society of Edinburgh (Earth Sciences)*, **80**, 271–83.

Crame, J.A. (1985). Lower Cretaceous inoceramid bivalves from the Antarctic Peninsula region. *Palaeontology*, **28**, 475–525.

Crimes, T.P. (1974). Colonisation of the early ocean floor. *Nature*, **248**, 328–30.

Darlington, P.J. (1959). Area, climate and evolution. *Evolution*, **13**, 488–510.

Davis, J.C. (1986). *Statistics and data analysis in geology* (2nd edn). Wiley, New York.

Dhondt, A.V. (1983). Campanian and Maastrichtian inoceramids: a review. *Zitteliana*, **10**, 689–701.

Fischer, A.G. (1960). Latitudinal variations in organic diversity. *Evolution*, **14**, 64–81.

Glazunova, A.E. (1973). *Paleontological evidence of the stratigraphic separation of Cretaceous deposits in the Volga region. Lower Cretaceous*. 'Nedra', Moscow. (In Russian.)

Gould, S.J. (1988). Trends as changes in variance: a new slant on progress and directionality in evolution. *Journal of Paleontology*, **62**, 319–29.

Hallam, A. and Miller, A.I. (1988). Extinction and survival in the Bivalvia. In *Extinction and survival in the fossil record*, Systematics Association. Special Volume, no.34 (ed. G.P. Larwood), pp. 121–138. Clarendon Press, Oxford.

Harland, W.B., Cox, A.V., Llewellyn, P.G., Pickton, C.A.G., Smith, A.G. and Walters, R. (1982). *A geologic time scale*. Cambridge University Press, Cambridge.

Harper, E.M. (in press). The role of predation in the evolution of cementation in bivalves. *Palaeontology*, **34**.

Jablonski, D. (1980). Apparent versus real biotic effects of transgressions and regressions. *Paleobiology*, **6**, 397–407.

Jablonski, D. and Bottjer, D.J. (1983). Soft-bottom epifaunal suspension-feeding assemblages in the late Cretaceous. In *Biotic interactions in recent and fossil benthic communities* (ed. M.J.S. Tevesz and P.L. McCall), pp. 747–812. Plenum Press, New York.

Jablonski, D. and Bottjer, D.J. (in press). Onshore–offshore trends in marine invertebrate evolution. In *Causes of evolution: a paleontologic perspective* (ed. R.M Ross and W.D. Allmon), 21–75. University of Chicago Press, Chicago.

Jablonski, D., Sepkoski, J.J., Bottjer, D.J. and Sheehan, P.M. (1983).

Onshore-offshore patterns in the evolution of Phanerozoic shelf communities. *Science*, **222**, 1123–5.

Jackson, J.B.C. (1974). Biogeographic consequences of eurytopy and stenotopy among marine bivalves and their evolutionary significance. *American naturalist*, **108**, 541–60.

Kauffman, E.G. (1968). The Upper Cretaceous *Inoceramus* of Puerto Rico. *Proceedings of the 4th Caribbean Geological Conference, Trinidad, 1965*, pp. 203–18. Caribbean Printers, Aruma, Trinidad and Tobago.

Levinton, J. (1988). *Genetics, paleontology and macroevolution*. Cambridge University Press, Cambridge.

Lewin, R. (1983). Origin of species in stressed environments. *Science*, **222**, 1112.

Miller, A.I. and Sepkoski, J.J. (1988). Modeling bivalve diversification: the effect of interaction on a macroevolutionary system. *Paleobiology*, **14**, 364–9.

Morris, N.J. (1978). The infaunal descendants of the Cycloconclidae: an outline of the evolutionary history and taxonomy of Heteroconchia, superfamilies Cycloconchacea to Chamacea. *Philosophical Transactions of the Royal Society of London*, **B284**, 259–75.

Newell, N.D. and Boyd, D.W. (1970). Oyster like Permian Bivalvia. *Bulletin of the American Museum of Natural History*, **143**, 221–81.

Patterson, C. and Smith, A.B. (1987). Is the periodicity of extinctions a taxonomic artefact? *Nature*, **330**, 248–51.

Pease, C.M. (1985). Biases in the durations and diversities of fossil taxa. *Paleobiology*, **11**, 272–92.

Petraitis, P.S. (1987). Immobilization of the predatory gastropod, *Nucella lapillus*, by its prey, *Mytilus edulis*. *Biological Bulletin*, **172**, 307–14.

Pochialaynen, V.P. (1969). Neocomian inocerams of the Anadyrskig-Koryakskiy folded region. *Trudy Severo-Vostochnogo Kompleksnogo Nauchno-Issledovatelskogo Instituta*, **32**, 124–62. (In Russian.)

Pochialaynen, V.P. (1974). Spreading of the Neocomian Pacific inoceramids. *Trudy Instituta Geolgii i Geodiziki Sibirskoe Otdelinie*, **80**, 174–87. (In Russian.)

Pojeta J. (1971). Review of Ordovician pelecypods. *US Geological Survey Professional Paper*, **695**, 1–46.

Pojeta, J. (1978). The origin and early taxonomic diversification of pelecypods. *Philosophical Transactions of the Royal Society of London*, **B284**, 225–43.

Pojeta, J. and Palmer, T.J. (1976). The origin of rock boring in mytilacean pelecypods. *Alcheringa*, **1**, 167–79.

Raup, D.M., Gould, S.J., Schopf, T.J.M. and Simberloff, D.S. (1973). Stochastic models of phylogeny and the evolution of diversity. *Journal of Geology*, **81**, 525–42.

Sandberg, P.A. (1983). An oscillating trend in Phanerozoic non skeletal carbonate mineralogy. *Nature*, **305**, 19–21.

Savazzi, E. (1987). Geometric and functional constraints on bivalve shell morphology. *Lethaia*, **20**, 293–306.

Seilacher, A. (1985). Bivalve morphology and function . In *Molluscs. Notes for a short course.* University of Tennessee, Department of Geological Sciences

Studies in Geology, no. 13 (ed. D.J. Bottjer, C.S. Hickman, and P.D. Ward), pp. 88–101.

Sepkoski, J.J. (1981). A factor analytic description of the Phanerozoic marine fossil record. *Paleobiology*, **7**, 36–53.

Sepkoski, J.J. (1982). A compendium of fossil marine families. *Milwaukee Public Museum, Contributions in Biology and Geology*, **51**, 1–125.

Sepkoski, J.J. (1987). Reply to Patterson and Smith, 1987. *Nature*, **330**, 252.

Sepkoski, J.J. and Sheehan, P.M. (1983). Diversification, faunal change, and community replacement during the Ordovician radiations. In *Biotic interactions in recent and fossil benthic communities* (ed. M.J.S. Tevesz and P.L. McCall), pp. 673–717. Plenum Press, New York.

Sheehan, P.M. (1986). Macroevolution and low diversity faunas. *Geological Society of America Abstracts with Programs*, **18**, 324.

Skelton, P.W. (1985). Preadaptation and evolutionary innovation in rudist bivalves. *Special Papers in Palaeontology*, **33**, 159–73.

Smith, A.G. and Briden, J.C. (1977). *Mesozoic and Cenozoic palaeocontinental maps*. Cambridge University Press, Cambridge.

Sohl, N.F. (1969). The fossil record of shell boring by snails. *American Zoologist*, **9**, 725–34.

Sohl, N.F. (1987). Cretaceous gastropods: contrasts between Tethys and the temperate provinces. *Journal of Paleontology*, **61**, 1085–111.

Stanley, S.M. (1970). Relation of shell form to life habits in the Bivalvia (Mollusca). *Geological Society of America, Memoir*, **125**, 1–296.

Stanley, S.M. (1975). Why clams have the shape they have: an experimental analysis of burrowing. *Paleobiology*, **1**, 48–58.

Stanley, S.M. (1977). Trends, rates and patterns of evolution in Bivalvia. In *Patterns of evolution* (ed. A. Hallam), pp. 209–50. Elsevier, Amsterdam.

Stanley, S.M. (1986). Population size, extinction and speciation: the fission effect in Neogene Bivalvia. *Paleobiology*, **12**, 89–110.

Stehli, F.G., Douglas, R.G. and Newell, N.D. (1969). Generation and maintenance of gradients in taxonomic diversity. *Science*, **164**, 947–9.

Taylor, J.D. (1981). The evolution of predators in the Late Cretaceous and their ecological significance. In *The evolving biosphere* (ed. P.L. Forey), pp. 229–40. British Museum (Natural History), London and Cambridge University Press, Cambridge.

Taylor, J.D., Morris, N.J. and Taylor, C.N. (1980). Food specialization and the evolution of predatory prosobranch gastropods. *Palaeontology*, **23**, 375–409.

Thomas, R.D.K. (1978*a*). Limits to opportunism in the evolution of the Arcoida (Bivalvia). *Philosophical Transactions of the Royal Society of London*, **B284**, 335–44.

Thomas, R.D.K. (1978*b*). Shell form and the ecological range of living and extinct Arcoida. *Paleobiology*, **4**, 181–94.

Trueman, E.R. (1966). Bivalve mollusks: fluid dynamics of burrowing. *Science*, **152**, 523–5.

Trueman, E.R. (1969). Ligament. In *Treatise on invertebrate paleontology. Part N. Mollusca 6. Bivalvia* (ed. R.C. Moore), Vol. 1, pp. N58–N64. Geological

Society of America and University of Kansas Press, Boulder, Colorado and Lawrence, Kansas.

Valentine, J.W. (1967). The influence of climatic fluctuations on species diversity within the Tethyan provincial system. In *Aspects of Tethyan biogeography* (ed. C.G. Adams and D.V. Ager), pp. 153–66. Systematics Association, London.

Valentine, J.W. (1968). Climatic regulation of species diversification and extinction. *Bulletin of the Geological Society of America*, **79**, 273–6.

Valentine, J.W. (1984). Neogene marine climate trends: implications for biogeography and evolution of the shallow-sea biota. *Geology*, **12**, 647–50.

Vermeij, G.J. (1977). The Mesozoic marine revolution: evidence from snails, predators and grazers. *Paleobiology*, **3**, 245–58.

Vermeij, G.J. (1983). Traces and trends of predation, with special reference to bivalved animals. *Palaeontology*, **26**, 455–65.

Vermeij, G.J. (1987). *Evolution and escalation, an ecological history of life*. Princeton University Press, Princeton, New Jersey.

Vrba, E.S. and Gould, S.J. (1986). The hierarchical expansion of sorting and selection: sorting and selection cannot be equated. *Paleobiology*, **12**, 217–28.

Waller, T.R. (1978). Morphology, morphoclines and a new classification of the Pteriomorphia (Mollusca: Bivalvia). *Philosophical Transactions of the Royal Society of London*, **B284**, 345–65.

Yonge, C.M. and Thompson, T.E. (1976). *Living marine molluscs*. Collins, Glasgow.

6. Major events in cephalopod evolution

T.S. ENGESER
Geologisch-Paläontologisches Institut der Universität Hamburg, Germany

Abstract

Previous classifications of the major groups of cephalopods are based mainly on features of the shell and the siphuncle and are phenotypic. In a new approach, the classification is reinvestigated by means of cladistic analysis. The Cephalopoda are a monophyletic group of molluscs. Despite obvious differences between, for example, *Nautilus* and the Coleoidea, there are some unique synapomorphies that are exclusively shared by these two distinct evolutionary lines. The latest common ancestor of the cephalopods will be considered. In contrast to previous studies which have considered only the fossil record for a suitable candidate, the hypothetical ancestral cephalopod (HAC) is here reconstructed using characters of Recent taxa. A plausible scenario for the evolutionary steps leading to HAC will be presented. This includes a discussion of the position of the Cephalopoda within the Mollusca and their probable sister-group. Previous workers used only characters of the shell (shell morphology, intrasiphuncular and intracameral deposits) for the classification of the Cephalopoda. Not used previously are different reproductive strategies and different developmental strategies of the buoyancy system. A new classification results from this approach.

Introduction

In today's seas, the coleoid or endococleate cephalopods are the most advanced molluscs and the only invertebrate group that can successfully compete with fishes. They have very effective locomotory organs (fins and funnel), manipulatory organs to catch and handle prey (arms), a powerful gnawing and cutting organ (beak), advanced camouflage organs (ink sac and chromatophores), an efficient vision organ

Major Evolutionary Radiations (ed. P.D. Taylor and G.P. Larwood), Systematics Association Special Volume No. 42, pp. 119–138, Clarendon Press, Oxford, 1990.

(adaptable lens eyes), and their highly developed brain correlates with a fairly high level of intelligence. The only grave disadvantage of the cephalopods is the possession of the blood pigment haemocyanin, only 20 per cent as effective as haemoglobin in binding free oxygen. This makes the coleoid cephalopods less perserving than vertebrates. About 700 extant species have been described and they are a very important link in the food chains of modern seas. Some scientists estimate that the biomass of coleoid cephalopods is as high as the biomass of fishes. They inhabit almost all parts of the sea from shallow water to the deep sea. The ectocochleate cephalopods are represented by only five or six species of the genus *Nautilus* which is restricted to the tropical Indo-Pacific. The *Nautilus* shell can sustain a pressure which corresponds to an approximately 700 m depth of water.

In the past, cephalopods were even more successful and had a much greater diversity. Until the end of the Mesozoic, the ectocochleate cephalopods seem to have been the more successful group in terms of species numbers and probably also in absolute numbers. It may be that the fossil record distorts this picture if the low fossilization potential of many coleoid cephalopod groups is taken into account. The aim of this chapter is to elucidate this evolutionary story of enormous success.

General considerations. The view of cephalopod evolution expressed in this chapter is based on the method known as 'phylogenetic systematics' or cladistics. The systematization of a major group normally starts with the extant groups because they show many more anatomical characters, allowing phylogenetic relationships to be more firmly established. Characters must be polarized into plesiomorphic ('primitive') and apomorphic ('derived'). In the case of the Cephalopoda the framework provided by the Recent groups is comparatively easy to establish. However, insertion of the many fossil groups is a much more difficult task.

The cladistic method also makes it possible to reconstruct the stem species ('latest common ancestor') of a major group. This reconstructed 'latest common ancestor' often has more credibility than a fossil ascibed the same status because of its age. The claim of many palaeontologists that phylogeny can only be reconstructed from the fossil record is here rejected because the fossil record is too incomplete. Even cephalopod groups that have a good fossilization potential often disappear from the fossil record for long periods of time, e.g. *Nautilus* from the Pliocene to the Holocene. However, only palaeontology can provide minimum ages and the sequence in which the characters evolved. This different view of the status of palaeontology for the reconstruction of phylogeny, and the consistent use of the cladistic method, have produced a different picture of cephalopod phylogeny to that in previous articles. These are the first preliminary results.

Definition of the Cephalopoda

At the time level of the present day there is no difficulty in defining the taxon Cephalopoda. However, to delimit the taxon in time is much more difficult. Usually the taxon Cephalopoda is defined by the first appearance of a phragmocone. This delimitation is, however, arbitrary. In the cladistic view the taxon Cephalopoda can be subdivided into the 'crown-group' and the 'stem-lineage' (Fig. 6.1; Ax 1984). The crown-group representatives possess the full set of cephalopodean apomorphies (see p. 124). The stem-lineage representatives must show at least one autapomorphy of the taxon Cephalopoda. This procedure may enlarge the concept of a taxon beyond the point intended by the original author. It may therefore be useful to restrict an existing taxon to the crown-group. Here the taxon Cephalopoda is expressly defined as the total group (i.e. crown-group + stem-lineage). The first recognizable cephalopodean apomorphy is not the phragmocone, as is the common view, but a more or less regular septation of a high conical shell. This regular septation stems from a tight attachment of the body mass to the shell (see

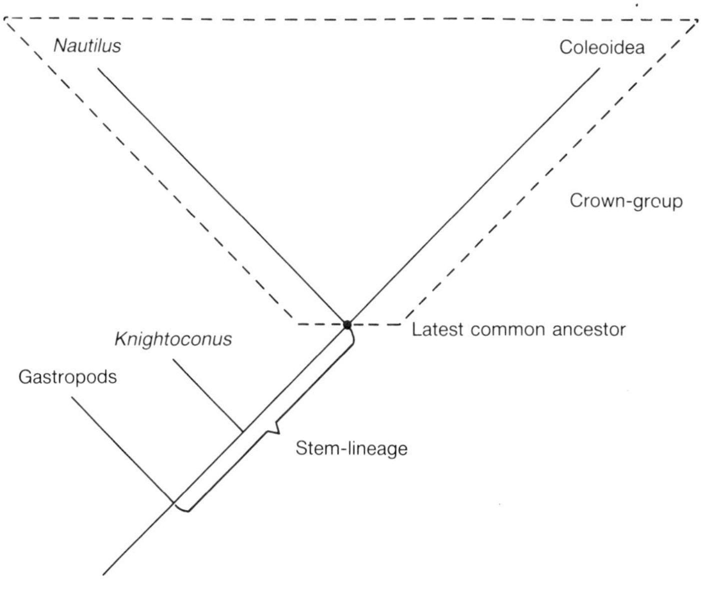

Fig. 6.1. Schematic phylogenetic tree of the Cephalopoda, including an explanation of the cladistic terms used in the text.

also p. 123). Therefore, *Knightoconus*, a Cambrian 'septate monoplacophoran', should be included in the taxon Cephalopoda as a stem-lineage representative (Fig. 6.1).

The ancestry of cephalopods

Most palaeontologists have assumed that cephalopods derived from monoplacophorans. There are two reasons for refuting this hypothesis. The first reason is a formal one. A group of extant organisms like the monoplacophorans cannot be the ancestor of another group of extant organisms like the cephalopods. They may share the same ancestor (however, not in this case; see below) and the common ancestor may have looked more like a representative of the more 'primitive' group, but an evolution from one extant group into another extant group is impossible. The second reason is that gastropods and cephalopods share some characters which can be interpreted as apomorphies, such as:

1. Reorientation of the body from a longitudinal axis into a dorsoventral direction. This reorientation is combined with a high conical shell and probably with a change in protective strategy as well. Instead of the plesiomorphic strategy of pulling the shell on to the substrate to protect the soft parts, the body mass was probably retracted into the shell. The shell opening could then be closed by an operculum. It has to be postulated that this operculum was lost in the Cephalopoda.
2. Epipodial protrusions. These belong to the ground plan of the Gastropoda and might be homologous with the arms of the Cephalopoda. If so they would be a synapomorphy of Gastropoda + Cephalopoda.
3. Head and body clearly separated and connected through a neck. The plesiomorphic character state is a body with no clearly separated head.

The Gastropoda and Cephalopoda are probably sister-groups. They are united in a supraordinal taxon, Rhacopoda Hennig, 1980.

The ground plan of the Cephalopoda

The Cephalopoda show a mosaic of plesiomorphic and apomorphic molluscan characters. They have, however, a number of unique autapomorphies which demonstrate their monophyly.

1. The development of a chambered, gas-filled shell (phragmocone) with a marginal siphuncular strand. The primitive cephalopod shell probably was slight endogastrically coiled (cyrtoconic). The phragmocone with its siphuncular strand allows buoyancy by gas. After the completion of a new chamber, the tissue of the siphuncular strand begins removing salt ions from the chamber liquid by active pumping. Chamber liquid is secreted by the mantle and its salinity is close to that of seawater. In *Nautilus* the salt-removing process comes to an end when the chamber liquid has reached 7–8 per cent of the salinity of sea water. This produces an osmotic pressure between chamber liquid and body liquid. Chamber liquid is now removed from the chamber and replaced by gas. When the chamber liquid falls below a level at which it is in direct contact with the siphuncular strand, the remaining liquid is removed by decoupled pumping, i.e. the chamber liquid is transported to the siphuncular strand through a hydrophilic membrane that covers the interior of the chamber. In *Nautilus* decoupled pumping is also used to adjust minor weight changes after the phragmocone is completed. The phragmocone is reduced or strongly modified in the majority of Recent cephalopods.

2. Transformation of the epipodial protrusions into 'many' (probably around 20 to 40) arms around the oral opening. The arms are arranged in two circles, and each arm has a adhesive surface. The number of arms in the hypothetical ancestral cephalopod is uncertain. Since *Nautilus* shows some signs of peratomorphosis (i.e. hypertrophy of some organs, e.g. doubling of the gill number), the arm number of 92 to 96 in *Nautilus* is probably greater than the primitive condition. In coleoid cephalopods there are indications that the arm number might have been reduced; a second, much smaller arm circle with 6 to 10 armlets is present in some coleoid groups. Moreover, some groups reduce their arm number during ontogenesis. In Recent coleoid cephalopods the arms are equipped with suckers, whereas in *Nautilus* each arm has an adhesive surface with transverse ridges. The character state of the arms of *Nautilus* is thought to be the 'primitive' condition.

3. Transformation of the foot into the funnel apparatus. All Recent Cephalopoda possess a funnel apparatus which makes possible locomotion by jet propulsion. Two character states of the funnel occur in Recent cephalopods: a funnel consisting of two folds with a valve (*Nautilus*); and a closed funnel pipe with or without a valve. The former state is considered as plesiomorphic within the Cephalopoda.

4. Transformation of simple chitinous jaws into beak-like jaws. All Recent cephalopods are carnivorous, and are variably active predators. Most of them have powerful beaks inside the buccal mass. The prey is caught by the arms, brought close to the buccal mass and reduced to small pieces by the sharp cutting edges of the beak before swallowing. The beak-like jaws can be evaluated as apomorphous.

5. Direct development of a meroblastic germ without true larval stages, discoidal cleavage of the yolk-rich eggs, and, therefore, a comparably large embryonic shell. The yolk content of cephalopod eggs is high in comparison with other molluscan groups. Cleavage of the eggs is incomplete. All cephalopods have a direct development and no larvae (see also discussion p. 128). These features are partly documented in the protoconch and are considered apomorphous (see also Berthold and Engeser 1987; Arnold *et al.* 1988).

In addition to these principal autapomorphies there are some further autapomorphies of the soft parts such as an increase in the size of the brain, an almost closed blood vessel system, hypotrophy of the coelomic cavity, and the development of pericardial and nidamental glands (Berthold and Engeser 1987).

In order to round off the ground plan of the taxon Cephalopoda, the most important plesiomorphies must also be specified.

1. *Marine habitat.* No extant cephalopod penetrates fresh water and the fossil record provides no evidence that any of the numerous fossil groups lived in fresh water.

2. *Possession of a radula.* The radula, the rasping tongue, is a typical molluscan autapomorphy (lost in Bivalvia). *Nautilus* with its 13 elements per row probably represents the plesiomorphic character state within the Cephalopoda.

3. *Pin-hole eye.* A simple pin-hole eye belongs to the ground plan of the Rhacopoda, i.e. Gastropoda + Cephalopoda. It is replaced by a lens eye in coleoid cephalopods.

4. *One pair of retractor muscles. Nautilus* still preserves the plesiomorphic rhacopodean condition of one pair of retractor muscles, whereas the Coleoidea and most Ammonoidea show a single retractor muscle.

5. *The typical conchiferan shell consisting of aragonite.* All ectocochleate and many endocochleate cephalopods have an aragonitic conchiferan shell with organic compounds, which is secreted by the shell gland. Only the belemnoids used calcite to form a secondary rostrum, and

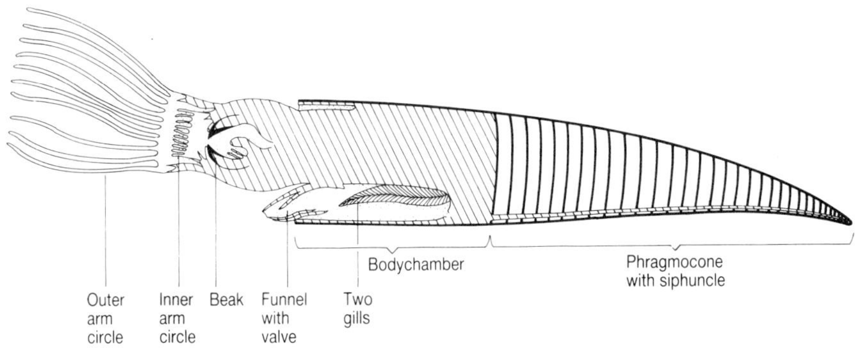

Fig. 6.2. A reconstruction of the hypothetical ancestral cephalopod (latest common ancestor of *Nautilus* and the Dibranchiata).

the female argonautids secrete a secondary calcitic brood shell using glands of the two dorsal arms. Mineralization is lost entirely in many decabrachian groups, and the shell is reduced remarkably in many octopod groups.

Evaluation of gill number, which is two in the Coleoidea but four in *Nautilus*, is controversial. If we assume a metameric molluscan ancestor, two gills would be an apomorphy of the Coleoidea. Conversely, if we assume a monomeric molluscan ancestor, the two-gilled condition would be a molluscan symplesiomorphy, and the four gills of *Nautilus* an apomorphy. Personally, I think that the plesiomorphic character state is two gills because no molluscan group shows any signs of true metamerism.

This mosaic of plesiomorphic and apomorphic characters is presented in a reconstruction of the latest common ancestor of the genus *Nautilus* and the Coleoidea (Fig. 6.2).

A model for the evolution of the Cephalopoda

As already indicated above, the Cephalopoda and the Gastropoda can be interpreted as adelphotaxa. The stem-species of the two groups, however, probably looked more like a gastropod. It had a high conical and slightly endogastric shell, two gills, two pin-hole eyes, a creeping foot provided with epipodial prostrusions, and a body mass which could be retracted into the shell by using two retractor muscles. The shell opening could be closed by an operculum secreted by a gland on the foot. The embryogenesis was probably yolk-poor and included a larval stage (Veliger larvae). The embryonic shell was small (about 0.1 to 0.2 mm).

In the first step towards the cephalopodean bauplan, the apical part of the shell was sealed off by septa. Septal formation in itself is not unique in the Mollusca; it is also known from Palaeozoic and Recent gastropods. In the case of the protocephalopod, however, the septal spacing was very regular. To achieve such a regular septation, the aptical part of the body mass must have been connected to the shell (as in Recent *Nautilus* and *Spirula*). In the course of the formation of a new septum, the visceral mass migrated into the exact position of the new septum and then became attached to the wall. The new septum reflects the form of the posterior body mass.

The most important step towards the latest common ancestor of *Nautilus* and the Coleoidea was the development of a strand of mantle tissue that passed through all of the septa and into the first chamber. This strand of mantle tissue, the siphuncular strand, was situated on the coiled side of the conch. Such an animal might have been capable of pumping some liquid out of the chambers. All molluscs are able not only to secrete but also to resorb calcium carbonate. During times when the animal needed calcium carbonate, the siphuncular strand absorbed ions from the chamber liquid. This produced an osmotic pressure between chamber liquid and body liquid, and gas entered into the chambers. Osmotic pressure was insufficient to remove all liquid from the chambers, and this mechanism worked only in very shallow water because water pressure is antagonistic to the osmotic pressure. The resulting slight buoyancy effect triggered an evolutionary pressure towards a more efficient control of the buoyancy apparatus.

After the protocephalopod had achieved a much higher mobility through a controlled buoyancy apparatus, the development of a more efficient locomotion organ was 'necessary'. The foot was transformed into the funnel apparatus, and the operculum reduced. Both characters were probably developed almost simultaneously. The reconstruction of *Plectronoceras* with a creeping foot and a phragmocone in Bandel (1982) and Teichert (1988) is unlikely and therefore rejected. Respiration was no longer achieved by ciliatory action, as in the molluscan ancestor, but through the mantle pump. Water was 'sucked' into the mantle cavity and expelled during withdrawal of the soft body by muscular action. The planktonic larval phase which serves as a vehicle of dispersal was suppressed since the adults were more mobile. The suppression of the larval stage was achieved by adding much more yolk to the egg. Moreover, the protocephalopods changed their dietary habit from omnivory to carnivory. They developed a powerful beak; to catch and hold prey effectively, the epipodial protrusions of the rhacopodean ancestor were transformed into grasping organs. Each arm had an adhesive surface. The scenario outlined here probably happened during the early to mid-Cambrian. The first cephalopods with the full set of cephalopodan autamorphies appeared by the late mid-Cambrian (Chen and Teichert 1983).

Buoyancy

The latest common ancestor of *Nautilus* and the coleoid cephalopods had already achieved neutral buoyancy. The floating position of the very early cephalopods was probably head-down, and the tip of the shell pointed obliquely upward, a typical position of hatchlings of Recent coleoids. This position is not very effective for horizontal locomotion; a horizontal position of the shell is much better. In order to accomplish this horizontal attitude, weight in the form of various kinds of calcareous deposits (intrasiphuncular and intracameral deposits) was added to certain parts of the shell without, however, losing the almost neutral buoyancy. The horizontal position of the shell could also be achieved by removal of the siphuncular strand from the early chambers with subsequent flooding of these chambers, or by coiling the shell.

Shell forms. The Cephalopoda developed a wide variety of shell forms, from straight elongated (longicone), short and thick (brevicone), more or less bent (cyrtocone), to loosely or tightly coiled. Shells may also be coiled in a three-dimensional spiral (torticone) or may develop irregular forms. They may be ornamented by striae, ribs, knots, and spines. The shell cross-section can vary from almost circular to broadly expanded laterally. The opening of the body chamber can be contracted in many ways, and a few groups have parts of the shell truncated. The adult shell, which ranges from less than 1 cm up to more than 10 m, can differ considerably from the juvenile shell.

Septal structures. The septa are generally concave forward. The distance between successive septa varies wildly. They may be orientated obliquely within the shell, or restricted to particular parts of the shell. The contact between septum and shell is called the suture, and may be straight, sinuous, or highly crenulated (e.g. in Mesozoic ammonoids). The inner sides of the chambers are covered by a mucous layer in which various kinds of intracameral deposits can be secreted. Forms with highly crenulated sutures may exhibit organic sheets which subdivide the chamber space. Where the siphuncular strand penetrates the septa various forms of backward or forward directed septal necks can be developed. They may be short (achoanitic) or extend backward through more than one chamber (macrochoanitic). The siphuncular strand is covered in many cephalopod groups by so-called connecting rings which consist of bundles of acicular aragonite crystals. These also show many variations, ranging from simple and thin to multilayered and thick. They may show longitudinal lamellae or they may be greatly thickened. The connecting rings may be greatly inflated so that they take up as much as one-half of the diameter of the camera. In many fossil cephalopods there are also various deposits on the inside of the septal necks (annuli, parietal deposits, etc.).

Siphuncular structures. The siphuncle of many groups contains organically deposited aragonitic structures, e.g. central rods, conical aragonitic deposits (Endocerida), or a complex tube system (Actinocerida). In some Cambrian (and a few Ordovician) groups, the organic siphuncular strand leaves the attachment site in the first chamber. The posterior siphuncular tube is then sealed off by transverse diaphragms.

Cameral deposits. The chambers of an adult *Nautilus* are almost empty. However, a certain amount of chamber liquid is stored in the hydrophilic membrane that covers the interior of the chambers. Small droplets of liquid may also be held in place between the connecting rings and the septal faces due to surface tension forces (Ward 1987). The hydrophilic membrane may contain clusters of organically deposited aragonite. This demonstrates that *Nautilus* is able to deposit calcium carbonate actively into the chambers. In many fossil cephalopods the camerae contain layers of aragonite and organic substances, probably the remains of the hydrophilic membranes. They were deposited on the anterior or posterior sides of the septa, and on the inside of the shell wall. Cameral deposits of the Devonian Lamellorthoceratidae are developed as radial lamellae. The cameral deposits usually increase in mass towards the apical end of the shell.

Reproductive strategies. As already stated, the eggs of cephalopods are generally large and yolk-rich compared with other groups of molluscs. There are no larval stages — the young hatch as miniature adults. Within modern cephalopods there are two character states: the moderate yolk content found in the eggs of most Recent cephalopods represents the plesiomorphic character state, whereas the extremely yolk-rich eggs of *Nautilus* can clearly be evaluated as apomorphic. The moderate yolk content of the eggs of most cephalopods is usually combined with high egg number, whereas the yolk-rich eggs of *Nautilus* are few in number. The hatchlings of the first group are small (about 1–2 mm), whereas those of the second group are greater than 3 cm in diameter. The development time of an egg in the first group may be less than 20 days under favourable temperatures (e.g. *Loligo*, see Hewitt 1988), but development may take one year in *Nautilus* (Arnold 1988). Males and females of the first group usually mature within 1 or 2 years and die after mating and spawning, whereas *Nautilus* may live for 20 years or more and survive more than 10 or even 15 mating and spawning cycles. The reproductive strategy of most coleoids may be characterized as *r*-selected. That of *Nautilus* is *K*-selected (Landman 1988). These completely different reproductive strategies are also reflected in the size of the embryonic shell (i.e. the shell formed inside the egg capsule). They can normally be determined in well-preserved fossil material.

Classification of the Cephalopoda

Fossil cephalopods are almost exclusively known from their hard parts, whereas most coleoids have only unmineralized hard parts (organic gladii) or no hard parts at all (e.g. *Octopus*). It is therefore most important to make full use of the information provided by the shells of fossil cephalopods. This must be interpreted in a functional context.

Major groups of fossil cephalopods were formerly recognized by the form of their shells (e.g. Orthocerida), siphuncle position and structure (e.g. Endoceratoidea and Actinoceratoidea), and truncation of the shell (Ascocerida), etc. Dzik (1984) grouped all non-bactritoid, non-ammonoid, and non-coleoid cephalopods in a supraordinal taxon Nautiloidea. Teichert (1988) proposed an equivalent systematic position for the Endoceratoidea, Actinoceratoidea, and Nautiloidea. A phylogenetic viewpoint dictates that the 'nautiloid' groups should have at least one autapomorphy present in the Recent genus *Nautilus*, or an autapomorphy of the Coleoidea, if they are more closely related to this latter group. If they have apomorphies of neither group they are stem-lineage representatives of the taxon Cephalopoda or paraphyletic groupings. Within the Recent Cephalopoda we can recognize two sister-groups, the Coleoidea and the genus *Nautilus*. Both are characterized in what follows. The different fossil groups are discussed briefly, and their position in a phylogenetic scheme for the Cephalopoda is suggested (Fig. 6.3).

The ground plan of Nautilus. The only extant representatives of the Nautilidia are species of the genus *Nautilus*. The monophyletic status of *Nautilus* can be demonstrated by the following autapomorphies.

1. A calcified beak. The central parts of the beak of *Nautilus* are heavily mineralized by high-Mg calcitic and aragonitic layers (Lowenstam

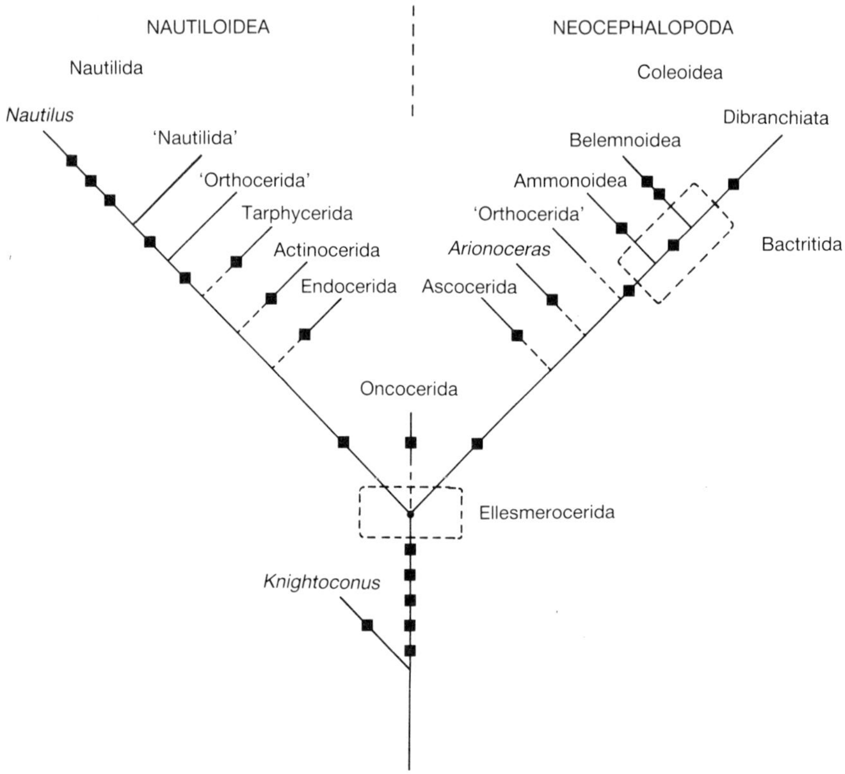

Fig. 6.3. Phylogenetic tree of the Cephalopoda. The black squares indicate apomorphies (see text for explanation). The positions of groups with hatched lines are uncertain.

et al. 1984). In other molluscan groups the beaks, if present, are entirely horny. The calcification of lower beaks of the Ammonoidea (aptychi) has to be regarded as an independent development (Lehmann 1985). Calcified nautiloid beaks appeared for the first time in the early Triassic (Dzik 1984).

2. A coiled phragmocone. Within the taxon Cephalopoda a slightly cyrtoconic phragmocone is plesiomorphic (see p. 123). A coiled phragmocone may have developed several times within the Cephalopoda.

3. Development of extremely yolk-rich and large eggs (up to 4 cm in length), and large embryonic shells (up to 31 mm) (Arnold 1988). All Cephalopoda have yolk-rich eggs. Within the taxon

Cephalopoda, the species of *Nautilus* show particularly yolk-rich eggs. These extremely yolk-rich eggs are regarded as apomorphic.

Further autapomorphies of the soft parts of *Nautilus*, which have no fossilization potential, are listed in Berthold and Engeser (1987).

Ellesmerocerida. The Ellesmerocerida generally have variably curved cyrtoconic shells with the siphuncle positioned marginally. They show no synapomorphies with *Nautilus* or the Coleoidea. Neither do they have an autapmorphy that would demonstrate their monophyly. They probably represent a grade group from which both major lineages arose. The Ellesmerocerida are placed as Cephalopoda *incertae sedis*.

Actinocerida. The Actinocerida comprise cephalopods with large and mostly straight shells. The most conspicious autapomorphies of the taxon are the central or subcentral siphuncle with strongly inflated connecting rings, and the intrasiphuncular deposits with a complicated tube system. The first chamber seems to be very large, implying a *K*-selected reproductive strategy. They are therefore placed as stem-lineage representatives of the genus *Nautilus*. Stratigraphical range: Middle Ordovician–Upper Carboniferous.

Endocerida. Most Endocerida have large and straight shells. The siphuncle is always strictly marginal and comparatively thick. The apical parts are filled with conical layers of intrasiphuncular deposits which can be interpreted as an autapomorphy of the taxon. The first chamber is very large, and it can be assummed that the Endocerida had yolk-rich development and large hatchlings. They are therefore grouped in the stem-lineage of the genus *Nautilus*. Stratigraphic range: Lower Ordovician–Middle Silurian.

Tarphycerida. Representatives of the Tarphycerida have loosely coiled and planispiral shells. The body chamber diverged from the shell when the animal was adult and the siphuncle was situated centrally. Intracameral deposits may occur. The inital chamber is cup-shaped and the siphuncle is attached to the shell in the form of a shoe. This curious mode of siphuncle attachment might be evaluated as an apomorphy. Development was clearly yolk-rich and the order is therefore placed in the stem-lineage of *Nautilus*. Stratigraphic range: Lower Ordovician–Middle Devonian.

Discosorida. The Discosorida include cephalopods with straight or cyrtoconic shells. The siphuncle is positioned centrally or marginally, the connecting rings are strongly inflated, and the siphuncle contains layered intrasiphuncular deposits in its apical part. This mode of forming intrasiphuncular deposits can be evaluated as apomorphic. However, the general mode of forming siphonal deposits is similar to the Actinocerida and can therefore be evaluated as synapomorphic. Ontogenesis is poorly known but seems to have been yolk-rich with a large and shallow initial chamber. Discosorids are placed in the Actinocerida. Stratigraphic range: Middle Ordovician–Upper Devonian.

Oncocerida. The Oncocerida comprise cephalopods with cyrtoconic and breviconic shells. Specialized forms may even be straight or torticonic. The siphuncle is positioned on the ventral side. A clear autapomorphy of the oncocerids has yet to be demonstrated. Most oncocerids contract the opening of the body chamber. In some groups the shell opening is almost completely closed apart from a few slits. Oncocerids in this stage of development could not have fed and it seems possible that these shells were of mature females that kept their eggs in the mantle cavity until hatching of the young before dying. It may also indicate a *K*-selected reproductive strategy. A similar brooding habit is known from Recent *Octopus*. After spawning the animals cease feeding, guard the egg clutches and die after the young are hatched. Oncocerid ontogenesis is unknown. They are placed as Cephalopoda *incertae sedis*. Stratigraphic range: Middle Ordovician–Lower Devonian.

Orthocerida. The Orthocerida contain all ectocochleate cephalopods with an almost straight and a centrally positioned siphuncle with simple septal necks and connecting rings, but typically with various kinds of intracameral deposit. The central siphuncle may have developed more than once based on the distribution of other characters. At least three different embryonic shell types are recognizable. The initial part of the shell and the mode of attachment of the siphuncle in *Pseudorthoceras* are very similar to the early shell of *Nautilus*. They are therefore interpreted as stem-lineage representatives of the 'order' Nautilida. Two other types of embryonic shells are discussed under the Neocephalopoda (p. 133). Because of the polyphyletic status of the Orthocerida, the taxon has to be rejected and the various 'orthocerid' groups must be distributed among other cephalopod groups.

Nautilida. The Nautilida have curved-to-coiled shells with a more or less central siphuncle which can be interpreted as an apomorphic character. Some advanced forms have shifted the siphuncle towards the margin. The nautilid ground plan includes a horn-shaped and large (about 15–20 mm) embryonic shell which may be even larger in derived forms. The Nautilida include stem-lineage representatives of the genus *Nautilus*. Stratigraphic range: Lower Devonian-Recent.

The Neocephalopoda. The taxon Coleoidea would be enlarged beyond the original intention if the stem-lineage representatives were to be included. The total group is here called Neocephalopoda. There are no good autapomorphies for this group, except the reduction of the arm number down to 10 and the radula elements down to 9. These characters are known only from a few species and specimens. Plesiomorphic characters are *r*-selected reproductive strategies, small embryonic shells, and marginal siphuncles.

Orthocerida. As already stated above (p. 132), the Orthocerida are clearly polyphyletic; the character 'almost straight shell with a central siphuncle' evolved more than once. Orthocerid genera with a spherical first chamber and a small embryonic shell (e.g. *Michelinoceras*, *Sphaerorthoceras*, *Parasphaerorthoceras*, and *Sikhanoceras*) certainly belong to the Neocephalopoda. *Michelinoceras* has been shown to have 9 radula elements (Mehl 1984). Stürmer's (1985) small coleoid from the early Devonian is probably a spherorthoceratid (Mehl, personal communcation, 1988). It has 9 to 10 arms.

Genera like *Arionoceras*, *Kaliceras*, *Vericeras*, and *Psilorthoceras* have a comparatively large and slighty trigonal first chamber (Ristedt 1968). Individuals in these groups hatched, however, with only one chamber (Hewitt 1988). Their peculiar embryogenesis can be regarded as an apomorphy of this group. They can only be placed in the stem-lineage of the Neocephalopoda. They are known from the Devonian.

Ascocerida. Members of the 'order' Ascocerida underwent a shell truncation during ontogenesis; the adult shell is very different from the juvenile shell. The truncated shell form of the adult can be inferred as an apomorphy. The size of the embryonic shell is approximately 5 mm. The Ascocerida might be a specialized off-shoot of the Neocephalopoda. Stratigraphical range: Middle Ordovician–Upper Silurian.

Bactritida. Shells of the Bactritida show a camerate phragmocone of a low angle and a comparatively large body chamber. The first chamber is globular and clearly demarcated. The siphuncle is situated marginally. The size of the embryonic shell was small, implying an *r*-selected reproductive strategy. 'Bactritids' with distinct lobes are excluded from the group as here understood. The Bactritida represent a typical grade group with no autapomorphy. They probably include the ancestors of both the Ammonoidea and the Coleoidea. Stratigraphical range: Ordovician–Permian.

Ammonoidea. The Ammonoidea probably arose from the Bactritida. Their main autapomorphy is the development of a lobe line, a feature which must be viewed in connection with its function. The interior walls of the chambers of *Nautilus* are covered by a hydrophilic mucous layer which serves as a storage organ for cameral liquid that can be removed for buoyancy adjustment. The folding of the septal attachment sites allows more storage of fluid for this purpose. Moreover, ammonoids have on average two to three times the number of chambers as *Nautilus*. In Triassic ammonoids Weitschat and Bandel (in press) have observed organic sheets subdividing the chambers. Many of these sheets are connected to the siphuncle. The compartments formed by the sheets allow the storage of large amounts of cameral liquid that is available for rapid removal. Whereas the pumping rate of *Nautilus* is too low for density-powered vertical movement, this process seems possible for the ammonoids. The reproductive strategy of the ammonoids was *r*-selected. The ammonitella (i.e. the embryonic shell of the ammonoids) was only 0.5–1.6 mm in size (Landman 1988). It had, however, one to seven functional chambers. Bactritids which possess distinct lobes (e.g. *Lobobactrites* and *Cyrtobactrites*) are included in the taxon Ammonoidea because they possess the apomorphy of the Ammonoidea.

Coleoidea. The Coleoidea (syn. Endocochleata) comprise cephalopods whose shells (or shell rudiments) are covered by the mantle. The taxon Coleoidea comprises the extinct Belemnoidea and the extant Dibranchiata. Some papers (e.g. Jeletzky 1966) have recommended abandonment of the term Dibranchiata for reasons such as 'the character two gills is plesiomorphic' or 'the character two gills cannot be proven in fossils' (cf. Reitner and Mehl 1989). However, the taxon must fulfil only one requirement—it must be monophyletic—even if monophyly is demonstrated by a character other than that which gave the taxon its name. The ground plan of the Coleoidea is characterized by the following autapomorphies.

1. The shell is covered by a mantle fold. This allows attachment of the fins to the outside of the shell and therefore a much higher manoeuverability. The formation of a counterbalance (rostrum) on the outside of the shell can now be controlled directly by the mantle. In the event of injury the phragmocone can be repaired to a certain degree. The tissue on the outside of the shell can also bear chromatophores which have enabled coleoids to evolve a

changeable, multicoloured skin camouflage and signalling abilities (House 1988). This was certainly a major advance compared to the unchangeable camouflage pattern on the shells of ectocochleate cephalopods.

2. Development of an ink sac. A character certainly closely-related with the increase of manoeuverability is the ink sac. The ink can be expelled in the event of an attack and works as a phantom picture or shape replacement of the cephalopod body. Meanwhile, the cephalopod escapes by using its jet propulsion system. Moreover, the ink of some coleoids (e.g. *Octopus*) paralyses, or at least irritates, the olfactory nerves of some predators of coleoids. Belemnoids apparently possessed an ink sac, whereas ectocochleate cephalopods were too slow to benefit from such a character and had no ink sac (Lehmann 1985).
3. Development of adaptable lens eyes. Most coleoids (except for a few specialized forms) have adaptable lens eyes which are as highly developed as vertebrate eyes. Distinct circular impressions on the lateral parts of the head region of two belemnoid species, *Belemnoteuthis antiqua* from the Callovian of England, and *Chondroteuthis wunnenbergi* from the Toarcian of southern Germany (personal observations), imply that the extinct belemnoid cephalopods already possessed eyes with lenses. It is less likely that these distinct impressions were left by simple pin-hole eyes. The lens eye produced much sharper images and was certainly closely associated with the development of the optic lobe of the brain in order to process the much more numerous data coming from the retina of the eye. It is unknown whether the ammonoids possessed lens eyes.

To complete the picture of the ancestral coleoid cephalopod, the following plesiomorphies should be added:

1. Ten almost equal arms, each equipped with two rows of cirri. This character is most likely not an apomorphy of the Coleoidea but was inherited from much earlier (ectocochleate) ancestors. Stürmer's (1985) so-called 'loliginid' from the early Devonian Hunsrück Shale (West Germany) is an ectocochleate cephalopod (*Sphaerorthoceras* sp.) (personal observation). The specimen, nevertheless, displays clearly nine arms, and may have had 10 arms. Neither suckers nor hooks are visible. The arms of the ancestral coleoid were probably equipped with two rows of cirri.
2. A ventrally closed body chamber. All Recent coleoid cephalopods use the mantle pump technique for respiration. This is achieved by

two antagonistic muscle layers in the mantle muscle. However, the ancestral coleoid cephalopod must still have had the plesiomorphic chamber pump technique, i.e. a rhythmical contraction of the two retractor muscles which resulted in a withdrawal of the body mass and thus an expulsion of water from the mantle cavity. Within the Coleoidea we recognize two sister-groups: the extinct Belemnoidea and the Dibranchiata. The first appearance of the Coleoidea was in the early Devonian (Bandel *et al.* 1984).

Belemnoidea. The Belemnoidea are characterized by at least two autapomorphies.

1. The closing membrane (a first organic septum).
2. The five conothecal layers (nacreous septum, inner prismatic layer, middle organic layer, outer prismatic layer, and an outer organic layer) (for data see Bandel *et al.* 1984). The 10 to 40 pairs of belemnoid hooks on all 10 arms mentioned by Berthold and Engeser (1987) as an autapomorphy of the Belemnoidea may not have been present in the most primitive groups of the Belemnoidea, the Aulacocerida, and this character is therefore considered an autapomorphy of the higher belemnoids. The Belemnoidea have a stratigraphic range from the early Devonian to the end of the Cretaceous. The so-called 'Eocene belemnoid' *Bayanoteuthis* is a representative of the Decabrachia and therefore excluded from the Belemnoidea.

Dibranchiata. This name is applied to the coleoid cephalopods minus the Belemnoidea. An autapomorphy of the Dibranchiata is the possession of suckers on the arms. All Recent coleoid cephalopods have suckers on their arms. Two sucker types occur within the Dibranchiata: the sessile suckers of the Octopoda, and the pedunculate suckers with horny sucker rings of the Decabrachia.

Further autapomorphies based on the soft parts are listed in Berthold and Engeser (1987). The Dibranchiata split into two lines. The Octopodiformes are characterized by differentiation (or loss) of the second arm pair, the sessile sucker type, and reduction of the phragmocone; and the Decabrachia by transformation of their fourth arm pair into tentacles. Within the Decabrachia the phragmocone was lost in the Oegopsida as well as in some groups of the Myospida. The Coleoidea which reduced the phragmocone developed three methods to achieve neutral buoyancy: ammonium ion replacement of denser metallic ions in the soft tissue; low-density oil storage in the liver; and dense sulphate ion replacement by chlorine ions (Clarke 1988). The so-called fossil 'teuthids' (late Triassic to late Cretaceous) are now transferred to the Vampyromorpha because they possessed only eight arms (Bandel and Leich 1986; Engeser 1988*b*). The oldest true octopod is *Palaeoctopus newboldi* (Woodward) from the late Cretaceous of the Lebanon (Engeser 1988*a*).

Acknowledgements

I wish to thank Professor Dr K. Bandel and Dr W. Weitschat (both Universität Hamburg) for extensive discussion of various aspects of this chapter. Dr P. D. Taylor kindly corrected the English.

References

Arnold, J. M. (1988). Reproduction and embryology of *Nautilus*. In *Nautilus, the biology and paleobiology of a living fossil* (ed. B. C. Saunders and N. H. Landman), pp. 353–72. Plenum Press, New York.

Arnold, J. M., Landman, N. H., and Mutvei H. (1988). Development of the embryonic shell of *Nautilus*. In *Nautilus, the biology and paleobiology of a living fossil* (ed. B. C. Saunders and N. H. Landman), pp. 373–400. Plenum Press, New York.

Ax, P. (1984). *Das Phylogenetische System*. Fischer Verlag, Stuttgart.

Bandel, K. (1982). Morphologie und Bildung der frühontogenetischen Gehäuse bei conchiferen Mollusken. *Facies*, **7**, 1–198.

Bandel, K., and Leich, H. (1986). Jurassic Vampyromorpha. *Neues Jahrbuch für Geologie und Paläontologie Monatshefte*, **1986**, 129–48.

Bandel, K., Engeser, T., and Reitner J. (1984). Die Embryonalentwicklung von *Hibolithes* (Belemnitida, Cephalopoda). *Neues Jahrbuch für Geologie und Paläontologie Abhandlungen*, **167**, 275–303.

Berthold, T. and Engeser, T. (1987). Phylogenetic analysis and systematization of the Cephalopoda (Mollusca). *Verhandlungen des Naturwissenschaftlichen Verein Hamburg, Neue Folge*, **29**, 187–220.

Chen, J.-Y. and Teichert, C. (1983). Cambrian cephalopods from China. *Palaeontographica*, **A181**, 1–102.

Clarke, M. R. (1988). Evolution of buoyancy and locomotion in Recent cephalopods. In *Paleontology and neontology of cephalopods* (ed. M. R. Clarke and E. R. Trueman), pp. 203–13. Academic Press, San Diego, California.

Dzik, J. (1984). Phylogeny of the Nautiloidea. *Palaeontologica Polonica*, **45**, 3–203.

Engeser, T. S. (1988*a*). Fossil 'octopods'—a critical review. In *Paleontology and neontology of cephalopods* (ed. M. R. Clarke and E. R. Trueman), pp. 81–7. Academic Press, San Diego, California.

Engeser, T. S. (1988*b*). Vampyromorpha ('Fossile Teuthiden'). *Fossilium Catalogus Animalia*, **130**, 1–165.

Hewitt, R. A (1988). Significance of early septal ontogeny in ammonoids and other ectocochliates. In *Cephalopods—present and past* (ed. J. Wiedmann and J. Kullmann), pp. 207–14. Schweizerbart, Stuttgart.

House, M. (1988). Major features of cephalopod evolution. In *Cephalopods—present and past* (ed. J. Wiedmann and J. Kullmann), pp. 1–16. Schweizerbart, Stuttgart.

Jeletzky, J. A. (1966). Comparative morphology, phylogeny, and classification of fossil Coleoidea. *University of Kansas Paleontological Contributions, Mollusca Article*, **7**, 1–162.

Landman, N. H. (1988). Early ontogeny of Mesozoic ammonites and nautilids. In *Cephalopods—present and past* (ed. J. Wiedmann and J. Kullmann), pp. 215–28. Schweizerbart, Stuttgart.

Lehmann, U. (1985). Zur Anatomie der Ammoniten; Tintenbeutel, Kiemen, Augen. *Paläontologische Zeitschrift*, **59**, 99–108.

Lowenstam, H. A., Traub, W., and Weiner, S. (1984). *Nautilus* hard parts: a study of the mineral and organic constituents. *Paleobiology*, **10**, 268–79.

Mehl, J. (1984). Radula und Fangarme bei *Michelinoceras* sp. aus dem Silur von Bolivien. *Paläontologische Zeitschrift*, **58**, 211–29.

Reitner, J. and Mehl. J. (1989). Ein besonderes Fossil. *Paläontologische Zeitschrift*, **63**, 3–4.

Ristedt, H. (1968). Zur Revision der Orthoceratidae. *Akademie der Wissenschaften und der Literatur Mainz Abhandlungen der Mathematisch-Naturwissenschaftlichen Klasse*, **1968**, 211–87.

Stürmer, W. (1985). A small coleoid with soft parts from the Lower Devonian discovered by using radiography. *Nature*, **318**, 53–5.

Teichert, C. (1988). Main features of cephalopod evolution. In *Paleontology and neontology of cephalopods* (ed. M. R. Clarke and E. R. Trueman), pp. 11–79. Academic Press, San Diego, California.

Ward, P. (1987). *The natural history of Nautilus*. Allen and Unwin, Boston.

Weitschat, W. and Bandel, K. (1990). Organic components in phragmocones of Boreal Triassic ammonoids: their implications for ammonoid paleobiology. *Paläontologische Zeitschrift*, (In press.)

7. Evolutionary radiations in the Trilobita

R.A. FORTEY
Department of Palaeontology, Natural History Museum, London, UK
R.M. OWENS
Department of Geology, National Museum of Wales, Cardiff, UK

Abstract

Radiations have been understood in several ways; the trilobites have a good fossil record which can be used to examine these different usages. Supposed radiations in the sense of accelerated cladogenesis (especially involving higher taxa) are discussed at two important times: the basal Cambrian radiation; and the basal Ordovician radiation at the inception of the 'Palaeozoic evolutionary fauna'. At the Precambrian–Cambrian boundary a cladistic analysis of trilobite relationships shows that the fossil record inadequately represents the early history of the Trilobita, both with regard to the relationships of the group to other arthropods, and with regard to early vicariant or migration events resulting in faunal provinces. If the analysis of relationships is correct there is no record of crucial evolutionary events in the Tommotian or earlier, which may call into question the correctness of regarding the Precambrian–Cambrian stratigraphic record as a historical narrative. The early Ordovician marked the appearance of the major clades that dominate subsequent trilobite history, but our understanding is constrained by phylogenetic problems. For example, the Cambrian sister groups of characteristic Ordovician (and later) families are in dispute, and the earliest representatives of some groups already appear to be differentiated biogeographically and morphologically. Quantification of 'adaptive radiation' in trilobites is hampered by our lack of knowledge of trilobite life habits. One approach to the problem is to recognize distinctive morphotypes (repeated evolutionary trends), which are likely responses to similar life strategies, the appearance of which *can* be plotted against time. For a set

Major Evolutionary Radiations (ed. P.D. Taylor and G.P. Larwood), Systematics Association Special Volume No. 42, pp. 139–164, Clarendon Press, Oxford, 1990.

of eight such morphotypes the Ordovician appears to have been the acme of trilobite radiation, because it was the time when the maximum number of families contributed the maximum number of morphotypes. Subsequent extinction of taxa (especially at the end of the Ordovician) meant that the morphotypes, when they reappeared, were recruited from different phylogenetic sources. Minor radiations took place in the Devonian, Carboniferous, and Permian. In general, in the post-Ordovician, a declining range of morphotypes was derived from progressively fewer families.

'Radiation' has also been applied to describe the evolution of clades within defined palaeogeographic 'provinces'. The examples of Cambrian 'biomeres' and Malvinokaffric phacopides are cited. Finally, there are ecological aspects of radiation which can be investigated from the stratigraphical record. There is evidence that the earliest members of eight out of 10 major trilobite clades appeared in inshore sites, and subsequently spread into a wider range of shelf sites.

Introduction

Radiations in the fossil record have been construed in several ways. This chapter briefly considers how some of these concepts can be examined using trilobites as evidence. The fossil record of the group is good, and trilobites span some of the important thresholds in the history of life. Hence, they are a useful group for attempting to describe historical events from the stratigraphic sequence. There is no occasion here to discuss different possible definitions of 'radiations' in detail. But it is necessary to state the different concepts which can be addressed by trilobite evidence. These can be summarized in three categories, which will make up the major divisions of this chapter.

1. Periods of accelerated cladogenesis. Cladistic analysis of the relevant taxa is important in this case.
2. Adaptive radiation or deployment. This is evolutionary radiation in the sense of Simpson (1944) or Stanley (1979).
3. History of particular clades in terms of ecology, or defined geographic pattern, directly testable from stratigraphic succession.

The classical use of the concept of radiation is probably that of Simpson (1944), especially with regard to the early Tertiary history of the mammals. This involves both the rapid appearance of higher taxa (accelerated cladogenesis) *and* the appearance of a range of morphologies available to fill vacant niches. The notion of adaptive radiation is linked to the taxonomic history. The explanation of the radiation is predicated by the notion that 'Nature abhors a vacuum', although this is expressed in ways which attempt to avoid teleology. The recent use of cladistic approaches to character distribution may change this. Cladistic classification is based upon synapomorphies (derived characters) alone—without regard to adaptive theories or statigraphic order. The refined methods of cladistic analysis may identify a primitive sister taxon of a large clade which does not differ greatly in gross aspect from the sister taxon of another major clade. Their gross similarity may be because the bulk of their characters are symplesiomorphic and, in traditional taxonomy, they would have been accommodated together within a paraphyletic 'grade group'. Both may be primitive organisms, early in the history of their respective groups, and as yet lacking the whole suite of the derived characters of their clades, which are the prime expression of the 'adaptive' part of radiation. Cladistic analysis seeks to dismantle paraphyletic groups, consigning primitive and often stratigraphically early members to their correct clades. Cladistic analysis may thus decouple the 'events' of cladogenesis from those of adaptation, as expressed in the taxonomy.

Another major aspect of radiation is the investigation of the ecological or other pattern shown by an episode of diversification of a clade. In this category would belong, for example, Jablonski and Bottjer's (1990; Chapter 2, this volume) contention that major marine clades generally originated in inshore biofacies and subsequently tracked out across the shelf and into deeper biofacies. In the same category also we examine diversification within a defined palaeogeographic area or province. We consider as examples the radiations claimed to occur at the base of each Cambrian 'biomere' (Stitt 1975), and the diversification of Phacopina within the Devonian Malvinokaffric Province (Eldredge and Branisa (1980).

Accelerated cladogenesis

1. *The reality of the early Cambrian 'explosive' radiation*

The basal Cambrian is usually considered to coincide with the most rapid radiation in the fossil record, with the appearance of many different kinds of skeletonized animals within a short time interval. Trilobites were not the earliest animals with hard parts, which were the small shelly fossils of the oldest Cambrian, the Tommotian, but they appeared very early in most sections spanning the Cambrian-Precambrian boundary, in the succeeding Atdabanian Stage. There are problems in the correlation of the earliest Cambrian rocks (see reviews in Cowie and Brasier 1989), and the basal Cambrian boundary has yet to be formally defined. Correlation problems are of particular importance with regard to certain sections in China. In most Cambrian–Precambrian sections a single genus appears first, and is rapidly joined by several more up-section (Brasier in Cowie and Brasier 1989). The question is: does this rapid appearance of Trilobita—and their apparently rapid diversification—reflect radiation, or is some taphonomic factor as important?

In sections in North America and Scandinavia the earliest trilobites encountered are olenellids. These have been excluded from the Trilobita (Lauterbach 1980), but more characters support the concept of the Trilobita as a natural group including olenellids, which are the most primitive trilobites with a calcareous exoskeleton (Fortey and Whittington 1989). One cannot treat the appearance of the calcified Trilobita in isolation from that of the many other uncalcified arthropods, a variety of which are known from the Burgess Shale in the Middle Cambrian, and in the Lower Cambrian of south-west China (Chiungchussu Formation, Hou 1987) and Greenland. The relationships of the Trilobita to these other arthropods has been assessed cladistically by Briggs and Fortey (1989). This analysis reveals that the trilobites are a highly derived clade relative to almost all the 'Cambrian type' arthropods (Fig. 7.1). The best candidate for the sister taxon of calcified Trilobita is *Naraoia*, which is also known early in the Cambrian in China. If the characters used by Briggs and Fortey are correct, the inference is that before the appearance of the first olenellid there was a considerable history of cladogenesis *prior to the initiation of the 'hard part' fossil record of the Arthropoda*, as first exemplified by the trilobites. This involves at least 12 major evolutionary steps, according to Briggs and Fortey (1989). Even given the 'burst' model of cladogenesis which has been proposed for this time interval (Conway Morris 1985), it may not be correct to assume that all this happened within the short timespan of the Tommotian (estimated at about 10 Ma). It does seem possible to us that the appearance of hard parts—undoubtedly important though this was—was but one feature in the differentiation of the arthropods, and one which coincided, as with so many groups, with the basal Cambrian. But many uncalcified sister taxa of the trilobites existed before the

Fig. 7.1. A, Cladogram illustrating inferred relationships of the Trilobita and major evolutionary steps in the earliest history of the Arthropoda (after Briggs and Fortey 1989). If this is correct, a complex Tommotian or earlier history is assured. B, Summary of relationships of trilobite clades present in the early Cambrian (after Fortey 1990), which approximates to the stratigraphic order of their appearance.

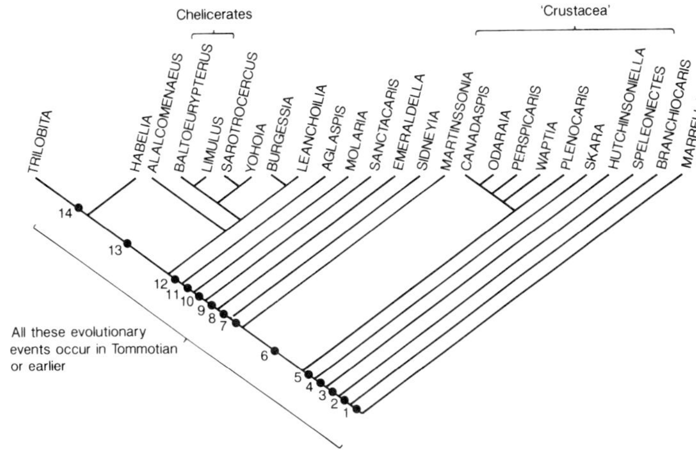

Chelicerates
'Crustacea'
TRILOBITA
HABELIA
ALALCOMENAEUS
BALTOEURYPTERUS
LIMULUS
SAROTROCERCUS
YOHOIA
BURGESSIA
LEANCHOILIA
AGLASPIS
MOLARIA
SANCTACARIS
EMERALDELLA
SIDNEYIA
MARTINSSONIA
CANADASPIS
ODARAIA
PERSPICARIS
WAPTIA
PLENOCARIS
SKARA
HUTCHINSONIELLA
SPELEONECTES
BRANCHIOCARIS
MARRELLA
14
13
12
11
10
9
8
7
6
5
4
3
2
1
All these evolutionary events occur in Tommotian or earlier

A

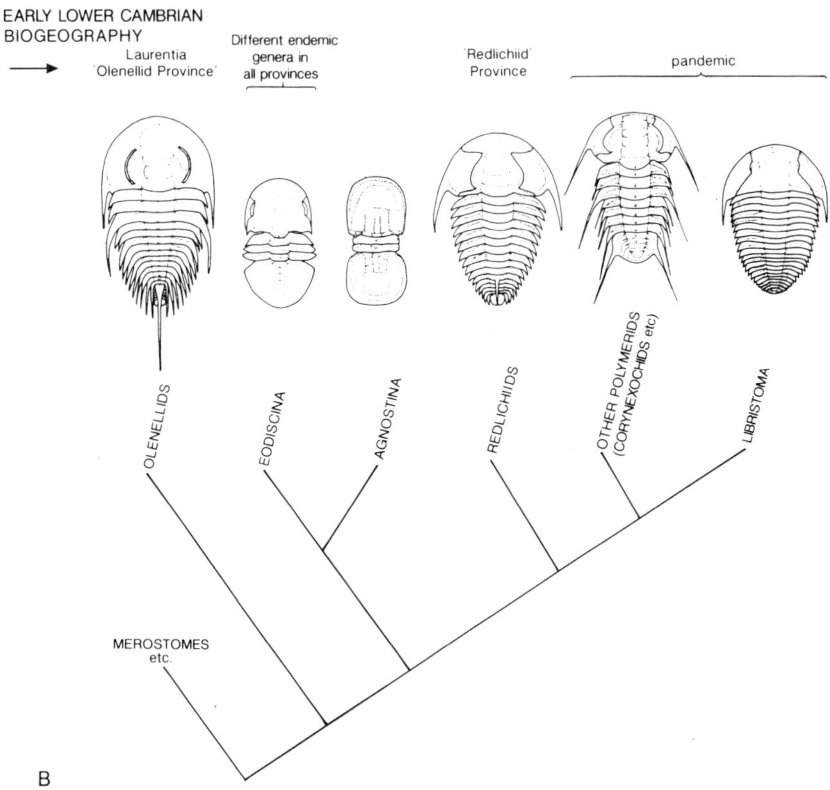

EARLY LOWER CAMBRIAN BIOGEOGRAPHY
Laurentia 'Olenellid Province'
Different endemic genera in all provinces
'Redlichiid' Province
pandemic
OLENELLIDS
EODISCINA
AGNOSTINA
REDLICHIIDS
OTHER POLYMERIDS (CORYNEXOCHIDS etc)
LIBRISTOMA
MEROSTOMES etc.

B

Cambrian (as defined on the appearance of trilobites) and survived to coexist successfully with them. This could be regarded as in support of the notion of a metazoan origin going back to 1000 Ma, as suggested by some molecular evidence (e.g. Runnegar 1982), controversial though this is, rather than the shorter time-scale 'explosive' alternatives.

The history of the Trilobita after their first appearance can be studied in numerous sections. Although there is evidence of increase in variety of taxa in each important section, these are not the same taxa in each succession, and neither is there a consistent order of appearance from one section to another (Brasier in Cowie and Brasier 1989). The relationships of the major groups in the trilobites is not yet agreed—for example, many authors are unsure about where to place the Agnostida. Figure 7.1B is a summary of views based on the characters and cladograms in Fortey and Whittington (1989) and Fortey (1990). This shows the redlichiids as advanced compared with the olenellids. This is encouraging in that some authors believe that the first appearance of olenellids in North America pre-dated the appearance of redlichiids in China (though the contrary view has been advanced by Chinese workers). Eodiscoids appeared very early in several sequences, which is not inconsistent with their position accorded on the cladogram. Libristoma includes the ptychopariids of authors (see Fortey 1990). All the groups indicated on Fig. 7.1B appeared within the Lower Cambrian, and this accounts for the main cladogenesis event in the Trilobita. If the stratigraphic position and position on the cladogram of olenellids is correct, then this constitutes a taxonomic radiation in any view. Furthermore, *within* any one group in any one biogeographic region there is a proliferation of genera upwards through the Lower Cambrian (e.g. S. China; Zhang *et al.*. 1980), and this likely records cladogenesis in progress (e.g. by heterochronic change; McNamara 1978).

However, it has long been recognized that from the earliest Cambrian there is clear biogeographic differentiation (Richter and Richter 1941; Cowie 1971). The Olenellid Province or Realm included much of Europe, Siberia, and North Ameica; the Redlichiid Realm included eastern Asia, China, and Australia. In the latter olenellids are not known. Each region has endemic genera, belonging to all the major trilobite groups, representing local radiation in the epeiric seas. If it is correct to assert the monophyly of the trilobites, there has, therefore, to be a phase of vicariance and evolution (and, presumably, migration) to allow for this geographic differentiation (Fig. 7.2). Of this phase we know nothing—the trilobites appear as suddenly in the Chinese sections as they do elsewhere. Because the Chinese forms (eodiscoids and redlichiids) are more advanced than olenellids (if Fig. 7.1B is correct)

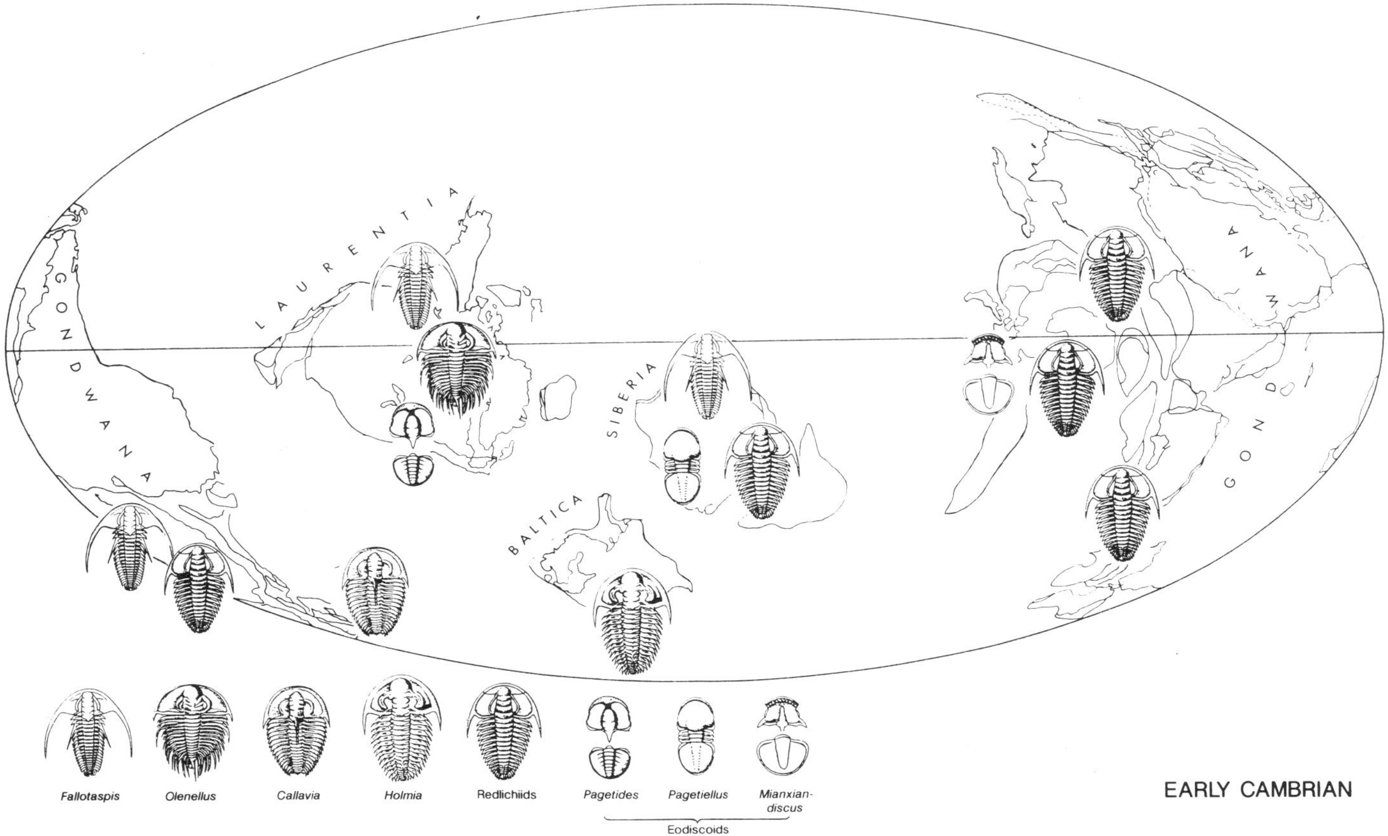

Fig. 7.2. World-wide distribution of some trilobite genera in the early Lower Cambrian, showing how provinciality was established from the beginnings of trilobite history, with implications for the completeness of their fossil record.

there is a crucial piece of evidence of their derivation which is lacking from the record. In summary, there was no doubt an evolutionary radiation in the early Cambrian history of the trilobites. However, if the cladistic analysis of relationships is correct, this may have been comparatively late in the cladogenesis of marine arthropods. Furthermore, some critical events in the geographic differentiation of Cambrian trilobite faunas are not preserved in the record. A simple reading of the Precambrian–Cambrian fossil record as a 'historical narrative' is clearly naïve.

2. *Early Ordovician radiation*

The top of the Cambrian has received a lot of attention as an extinction event (e.g. review in Westrop 1989), but less as a time of origination. It is, however, well-known that the Ordovician marked an increase in overall familial diversity (Sepkoski 1981). Many of the major marine clades (of cephalopods, articulate brachiopods, and graptolites, for example, as well as trilobites) have first records close to the Cambrian–Ordovician boundary. Fortey (1989) pointed out that increased cladogenesis near the origin of major groups elevated rates of taxonomic pseudoextinction by virtue of the preponderance of paraphyletic taxa, and that this may serve to elevate the apparent extinction at this horizon. Radiation and extinction rates may be correlated in this way.

The Tremadoc–Arenig interval marks the recorded origins of many of the higher taxa which dominate subsequent trilobite history (Fig. 7.3). Families that have their first records in this interval nearly all survived for at least 50 Ma (Ordovician) and many survived more than 100 Ma. Foote (1988) has proved that (using present taxonomy) the survivorship of groups in the Ordovician and younger is much longer than comparable taxonomic units in the Cambrian. Sloan (1990) recorded a steady climb in the standing diversity of genera on the Laurentian palaeocontinent throughout most of the Ordovician. He also calculated a regular increase in mean generic duration through the Ordovician (Fig. 7.4). Hence the Ordovician records rapid cladogenesis followed by a decline at the Ordovician–Silurian extinction event; in these respects the trilobite pattern is not unlike that of the Metazoa as a whole at this time, as described by Sepkoski (1981). This is a classical radiation pattern.

However, a critical point to be made about the Ordovician radiation is that the Ordovican clades at family level and above are both well-defined and long-lived, but their Cambrian sister taxa are in many cases not identified with certainty. Deeper-water taxa, such as Agnostida and Ceratopygidae, pass through the Cambrian-Ordovician boundary

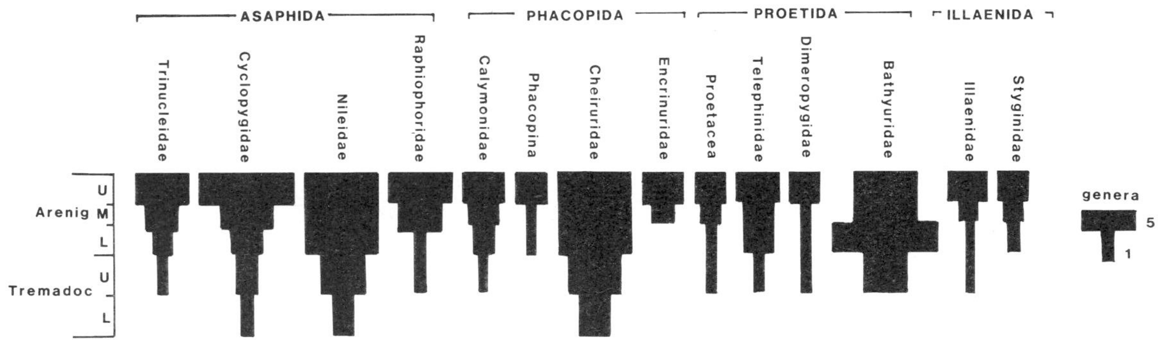

Fig. 7.3. Numbers of genera of major clades of post-Cambrian families through the first two series of the Ordovician (with approximately stage level subdivision) on a world-wide basis. This shows general increase in all stocks through this interval in a classical radiation pattern.

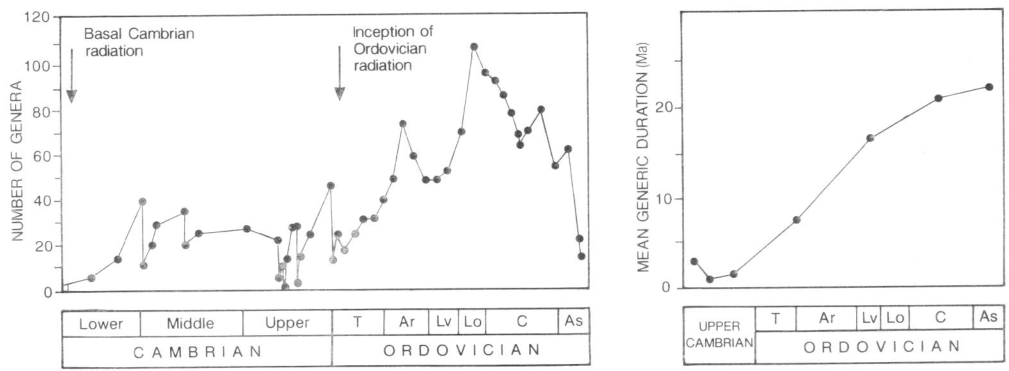

Fig. 7.4. Changing numbers of trilobite genera present in the Cambrian and Ordovician, with early Cambrian and early Ordovician radiations indicated (left). Increase in mean generic duration (right) in the Upper Cambrian and throughout the Ordovician (after Sloan 1990).

relatively unscathed (Fortey 1983; Westrop 1989), but these do not seem to be the close relatives of 'Ordovician' superfamilies and families, such as Cheiruracea, Calymenacea, Phacopacea, Cyclopygacea, Trinucleidae, Raphiophoridae, Illaenidae, and Styginidae. Whittington (1981, p. 594) described this Ordovician cladogenesis as '. . . a cryptogenetic renewal of trilobites, the steps in which are largely unknown,' and even considered the possibility that '. . . some of the new Ordovician groups may have arisen from ancestors that were unknown because they were not preserved, having had an uncalcified exoskeleton.' This is to invoke another basal-Cambrian type of radiation at this level and is subject to the same problems as described above. Nor is it likely. The Ordovician trilobite families listed above all share more characters with other trilobites (Fortey and Whittington 1989) than with other arthropods—even those like *Naraoia* without a clacified cuticle. Hence polyphyletic calcification as a solution to cryptogenesis in the Ordovician 'radiation' is scarcely parsimonious. Rather than a failure in finding plausible ancestors it is likely that there has been a failure in the phylogenetic analysis (Fortey and Chatterton 1988). The Cambrian sister taxa may be rather rare. For example, a true odontopleurid was described in the late Cambrian by Bruton (1983) (see Thomas and Holloway 1988); a possible late Lower Cambrian odontopleurid, *Eodontopleura*, was described by Zhang *et al.* (1980, p. 284), from a very few specimens, but there are enough to remove the Odontopleurida as an Ordovician innovation. It is not possible at the moment to predict the effects of greater phylogenetic resolution on the basal Ordovician radiation as perceived, except that if Cambrian sister taxa *are* found it may separate the time of cladogenesis (at family level) from that of the adaptive radiation represented by the increased array of morphological types in the Ordovician (see p. 151). However, the climb in numbers of genera *within* the monophyletic groups through the earlier part of the Ordovician happened in all groups (Fig. 7.3), and this may reflect the 'adaptive component' in the Ordovician radiation.

Another approach to the same problem is to look at the earliest representatives of known Ordovician monophyletic groups. If there are several different early forms this implies an unknown earlier history. We might take as an example the Phacopina (Fig. 7.5), which is a distinctive monophyletic higher taxon. Early representatives of this group are Arenig to early Llanvirn in age.

1. *Ormathops*. This genus occurs only in the high latitude, western part of the Gondwana palaeocontinent.

2. *Gyrometopus*. Confined to the Baltica palaeocontinent (Jaanusson

Fig. 7.5. World-wide distribution of the oldest known members of the Suborder Phacopina, in the Arenig and earliest Llanvirn series of the Ordovician. Different genera occur in different 'provinces' implying an earlier history of which there is no known record.

1975), this genus is dissimilar to *Ormathops*. A little known form from south-east Siberia, *Prodalmanitina*, somewhat resembles *Gyrometopus*.

3. *Prosopiscus*. This is a peculiar, blind trilobite confined to the tropical, eastern part of the Gondwana palaeocontinent. It was interpreted as a dalmanitacean by Fortey and Shergold (1984).

4. *Calyptaulax*. This genus probably does not appear until the earliest Llanvirn on the Laurentian palaeocontinent, within the tropical carbonate belt. It is again quite different from the three genera named above.

It is obvious that there has to have been a considerable history of the clade Phacopina before the first records of these genera. This is needed to account both for the vicariant and/or dispersal events producing distinctive taxa on climatically and geographically separate palaeocontinents; and for evolutionary events sufficient to account for descent of these highly different taxa from a common ancestor. It would be wrong to claim that the first appearance of these phacopines is representative of the early Ordovician 'radiation': clearly, if there was a radiation we have, as yet, no direct evidence for it. Similar criticisms can be applied to other Ordovician clades: for example, odontopleurids are already differentiated by the Arenig into two major morphological types (represented by *Selenopeltis* and *Ceratocephala*) on different palaeocontinents and palaeolatitudes. These must have had an earlier phylogenetic history in the later Cambrian of which virtually nothing is known. Future work should assess to how many of the Ordovician and later clades this criticism can be applied.

Despite the real problems of phylogenetic resolution, there *is* a change between Cambrian and post-Cambrian trilobite faunas, which is apparent in the gross composition of the faunas, with the increasing dominance of groups such as those in Fig. 7.3. This change coincides with radiations claimed in other invertebrate groups, and may be a phenomenon of co-evolution. There are many speculations available, most of them hard to test, and they cannot be discussed in detail here. These include: the appearance of new predators—notably nautiloids—serving to 'channel' morphology into a limited range of adaptive types; increases in biogeographic and climatic differentiation because of widely-dispersed palaeocontinents in the early Ordovician; that oceanic oxygenation had been generally low in the Cambrian, and changed in the Ordovician towards present levels; the appearance and evolution of groups such as bryozoans and corals which permitted new biological interactions, especially as bioherms become re-established; and increasing colonization of infaunal habitats.

Adaptive deployment

This aspect of radiation requires some method of assessing the adaptive range of the trilobites concerned. Because we know only the rudiments of a few trilobite life habits this cannot be approached directly. However we *do* know that certain distinctive morphological types recurred in the history of the group (as 'evolutionary trends'). It is reasonable to suppose that these morphotypes were related to specific life habit(s). Hence, a time of radiation could be defined as being when the maximum number of morphotypes can be recognized as having originated from the maximum number of phylogenetic sources. Note that one needs phylogenetics in this type of analysis to the extent that the homoplasous characters can be recognized and the taxon ascribed to its correct phylogenetic source (especially family); it is not necessary to have a completely resolved phylogeny. But it is necessary to be careful to select 'trends' which describe a similar adaptation. For example, one of the commonest 'trends' in trilobite evolution is effacement, in which the dorsal furrows are lost, so that eventually the trilobite becomes smoothed out and featureless (as smooth as an egg in the case of some agnostid cephala). This does not seem to be an equivalent trend in all cases where it occurs. It can happen with equal facility in inner-shelf trilobites (*Leiocoryphe*), deep-water benthics such as olenids and raphiophorids (*Leiobienvillia*, *Ellsaspis*), or pelagics (as claimed for Agnostida). It seems improbable that the reason for effacement is similar in all these disparate examples. Some eight distinctive morphotypes of trilobites which appeared iteratively in the fossil record were recognized by Fortey and Owens (1990), who give a more detailed discussion. Briefly, these are:

1. *Pelagic*. Trilobites having hypertrophied eyes with an exceptionally wide field of view, and often found in deep-water facies, or independent of benthic biofacies type.

2. *Phacomorph*. *Phacops*-like trilobites with tumid and forward-expanding glabellas, with modified glabellar furrows, and strongly furrowed pygidia, often found in epeiric limestones.

3. *Illaenimorph*. A distinctive type of 'smoothed out' trilobite, like *Illaenus*, often found in association with reefal habitats.

4. *Atheloptic*. 'Shrunken eyed' trilobites (blind or with the visual field much reduced) with close relatives having normal eyes, and associated with deeper-water sites. Eyes can be lost or reduced in several different ways.

5. *Trilobites with an anterior array of marginal cephalic spines.*

6. *Miniaturized trilobites.* Usually such forms arose progenetically; they are not 'stunted'. A miniature trilobite has a mature size of a few millimetres.

7. *Pitted fringe.* A structure like the trinucleid fringe is unique to the Trilobita; despite several suggestions its function is not known.

8. *Olenimorph.* Olenimorphs were adapted to coping with a dysaerobic environment. They characteristically have a thin cuticle (50 μm), numerous narrow thoracic segments, and are caecate.

Figure 7.6 illustrates the history of these morphotypes, calibrated at approximately series level. A thick black block indicates that a number of families are contributing examples of the morphotype during the time interval concerned. To the right we sum the number of families contributing the morphotypes, and also show the curve for the total number of families known through the same intervals, using the *Treatise* (Moore 1959), supplemented by subsequent major compilations (e.g. Lu *et al.* 1965) to derive the curve. On this definition of radiation there is no doubt that the Ordovician marks the acme of trilobite success. At this time a large number of trilobite families contributed a large number of individual genera to the morphotypes selected. If it is correct to assume that each morphotype is also an adaptive type, then the Ordovician was a time of adaptive radiation. Note also that the Silurian was a low, and the Devonian (and even the late Permian) marks a modest subsequent radiation. The effect of the extinction event at the end of the Ordovician was marked. No subsequent example of the pelagic morphotype is known to us, and it is presumed that this life habit was eliminated from the trilobite repertoire. More types again disappeared within the late Devonian faunal crisis.

This method of presentation shows that there is some decoupling of radiation in the adaptive sense from the taxonomic diversity as measured by number of families. Cambrian families were as numerous as Ordovician ones, but this is not reflected in morphotype recruitment, which takes time. This may also reflect the 'generalized' nature of Cambrian trilobites, if not the taxonomic practices of Cambrian specialists. Conversely, in the Devonian and later a range of morphotypes had to be recruited from fewer and fewer families, until ultimately only proetaceans survived. This accounts for the observation that late proetaceans 'mimic' genera of Ordovician to Devonian age.

It is possible to criticize this approach to adaptive radiation on several grounds. First, trilobites obviously had many more life habits than is

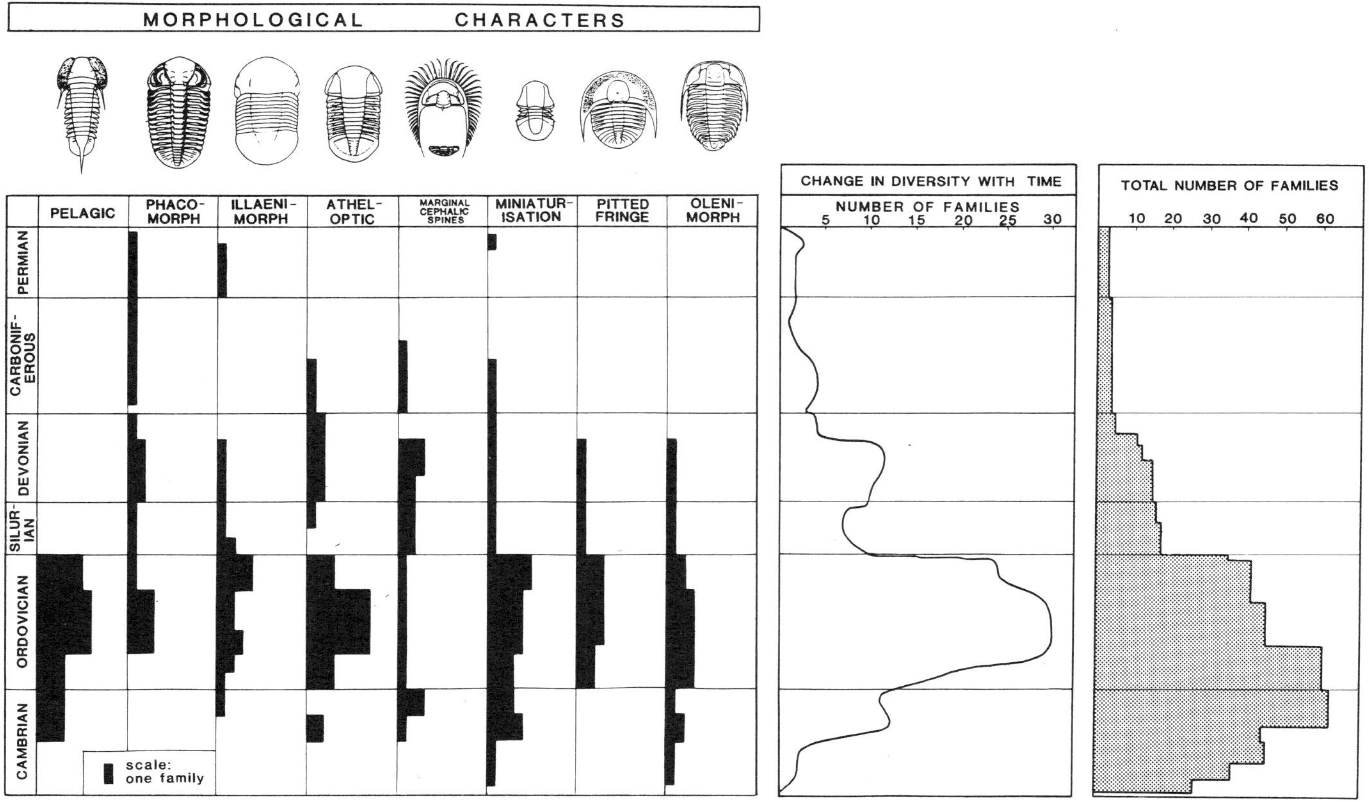

Fig. 7.6. Distribution of eight morphotypes throughout the history of the Trilobita, the change in diversity with time, and the number of families from which they were recruited. Note the basal Ordovician pseudoextinction (after Fortey and Owens 1990).

summarized by these morphotypes. This deficiency can only be addressed by more work on functional morphology and comparative anatomy, and it will be very interesting to find out whether the introduction of more examples changes, or reinforces the pattern. Several additional categories occurred to us as we prepared this chapter (e.g. 'fan-tailed' trilobites, or those that developed long, anterior 'noses') and were excluded because of doubts about whether they were truly comparable in all cases. Second, emphasizing repeated morphological trends does not allow for the appearance of unique morphologies which are some reflection of adaptive radiation also. To give one of many possible examples, the Ordovician trilobite *Bulbaspis* developed a bladder-like structure on the end of a median cephalic spine, which is known on no other trilobite. But for all these deficiencies this approach provides one way to look at adaptive deployment in a quantifiable way.

Radiation of monophyletic clades

1. *Radiation of clades within discrete geographical regions*

There are several examples which fall into this category, the problem being how to distinguish what might be thought of as a radiation from background evolution. Should one exclude, for example, the kind of evolutionary history portrayed for Trinucleidae in the Ordovician by Hughes *et al.* (1975, fig. 120), which undoubtedly shows an increase in generic number and vicariance through time (accompanied by morphological and presumably adaptive change)? So far as we know these authors nowhere described trinucleid evolution as a radiation. The kind of radiation that should be considered is, perhaps, based on the Hawaiian model, whereby *in situ* evolution in Insecta took place rapidly within a comparatively isolated geographical region. In the case of the trilobites this has to apply to areas larger than islands, for example on a palaeocontinental scale, but isolated climatically or by a palaeo-ocean from neighbouring areas. A requirement of the concept is that the trilobites concerned were those that inhabited cratonic areas, which tended towards the endemic, or were otherwise inhibited in dispersal. Obviously, free passage between separate geographical blocks would complicate the discrete vicariance required by the theory. Once more, a high standard of taxonomic resolution is crucial to analyse phylogenesis.

Cambrian 'biomere' pattern. Enough has been written about the biomeres to make repetition superfluous here (reviewed in Palmer 1984). The term radiation (or adaptive radiation) has been explicitly applied to biomeres (Stitt 1975). Such a radiation is supposed to have happened early in a given biomere following the extinction event at the end of the preceding biomere. The biomere was a North American (strictly, Laurentian) phenomenon, this palaeocontinent having had an isolated palaeoequatorial position in the Cambrian. The great majority of trilobites concerned were described as North American endemics at the generic or even familial level; most were also confined to a single biomere, which was thus a discrete evolutionary package. The radiation in each biomere is treated in the literature as if it happened *in situ* in Cambrian Laurentia rather than, for example, by invasion of taxa from elsewhere, and so it does conform to an 'island' model. As with Hawaiian Drosophilidae, a few pioneer species are supposed to have given rise to diverse taxa in each radiation. There is no doubt of the reality of the biomere as a stratigraphic pattern in the distribution of species, which is supported by superb fieldwork by Palmer and his colleagues. However, as a taxonomic radiation, the concept depends on the taxonomy, and the problem is that the comparative taxonomy on a world-wide basis lags behind that of North American palaeontologists. For example, if the taxa prove to have been more widespread, then the biogeographic discreteness breaks down, and with it the 'Hawaiian' style of radiation. Also it can be claimed that the definition of many Cambrian families is imperfect, being based on stratigraphy and with incomplete analysis of morphology, and a sceptic might wonder whether the biomere concept controls the taxonomy to the extent that it becomes self-fulfilling. If this were the case the biomere would become an indubitable stratigraphic phenomenon, but a disputable 'radiation' phenomenon. For example, in one account (Fortey and Chatterton 1988), the Pterocephaliidae were considered to be plesiomorphic Asaphina without good characters for familial recognition, whereas Palmer (1965) regarded the same family as well-defined within the limits of one biomere.

Malvinokaffric Devonian. This example is related to strong Devonian climatic zonation; cold-water palaeolatitudes had a range of endemic trilobites, which radiated within the Malvinokaffric region to produce a spectacular range of morphologies (Eldredge 1980; Eldredge and Branisa 1980), and a variety of endemic genera (Fig. 7.7) but from relatively few (presumably cold adapted) families. Systematic works on Devonian trilobites from South Africa and South America, especially Bolivia, confirm the geographic discreteness of the Malvinokaffric province. This is adaptive radiation by most definitions, even if we do not know the function of most of the morphotypes produced.

Such climatic zonation existed at other times in the history of trilobites. For example, in the early Ordovician southern Europe and North Africa lay at high palaeolatitudes, and there again one finds a

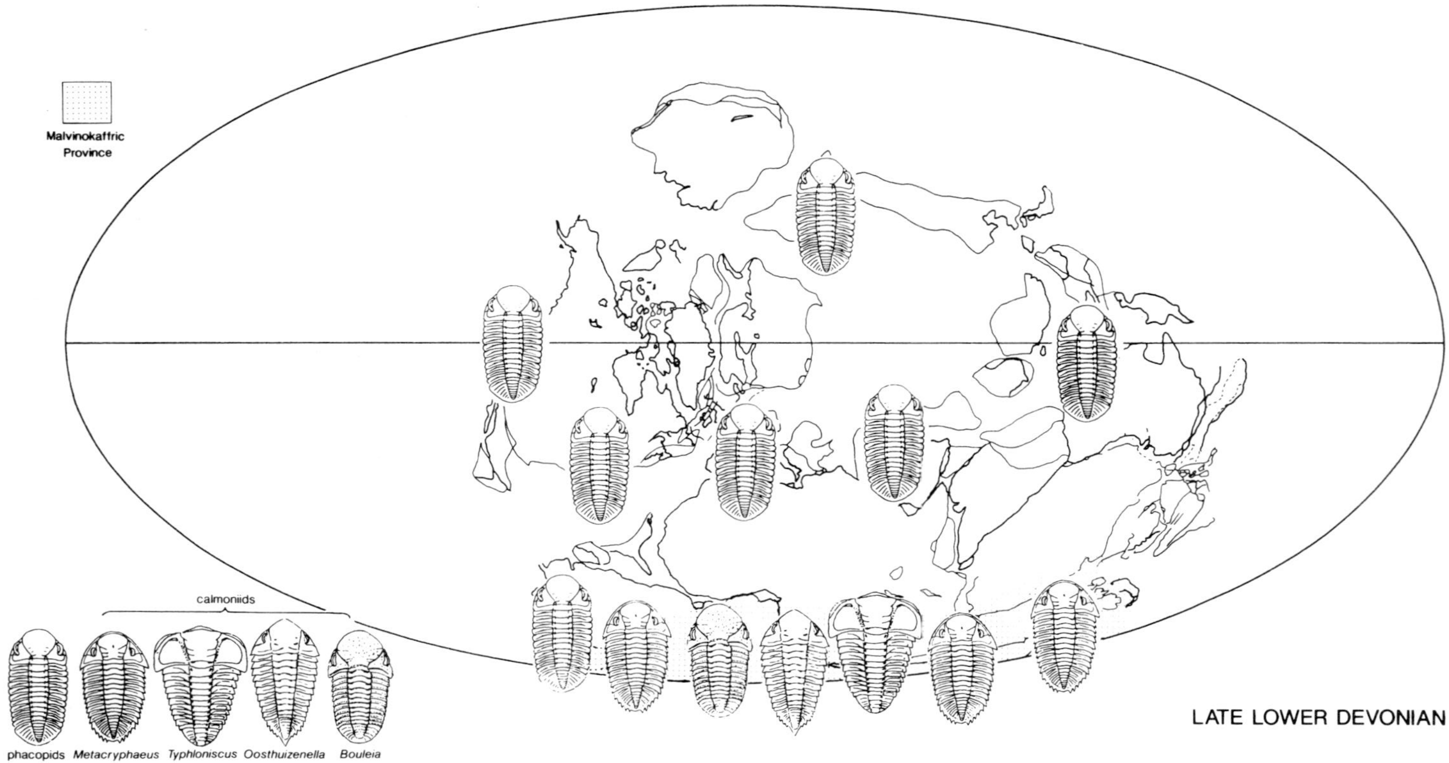

Fig. 7.7. World-wide distribution of selected phacopide trilobites in the mid-Devonian. Note the distinctive Malvinokaffric Province with its endemic ('radiated') calmoniids, lacking *Phacops* and its allies that are widely distributed elsewhere.

large suite of endemics, mostly referable to Calymenacea and Dalmanitacea, well described by Henry (1980). This is not generally described as a radiation, but it is difficult to see any difference in principle from the Malvinokaffric story, unless it be that the morphological distance covered is somehow deemed to be less, or that the rapidity of origination of endemic taxa is considered to be slower (for which the evidence is unclear). It may be simply that nobody has elected to describe the early Ordovician boreal endemics as the product of radiation.

2. *Patterns of environmental origination of major clades*

Jablonski and Bottjer (1990: Chapter 2, this volume) have described a pattern in the environmental range of the early history of major post-Palaeozoic clades (at order level), in which the first recorded members of the clades were often confined to inshore sites, with an expansion subsequently across-shelf, and, in some cases, eventual restriction to deep-water sites. This is a narrative for radiation, and one that should be testable for older taxa with a good fossil record, including trilobites. A good, phylogenetically based high-level classification is required, and it is doubtful whether we have that in the Trilobita. The biofacies is also not known in all cases desirable. But it is probably worthwhile ignoring these problems to look for indications that the pattern could apply. At the least this would require evidence of first appearance of the clade concerned in an inshore environment and, subsequently, in a wider range of habitats. Additionally, it might be shown that the last survivors of a clade tended to be in offshore habitats. Fortey (1980) has already shown that trilobite *generic* longevity was longest in more offshore environments.

Class Trilobita. Accepting the olenellids as the most primitive trilobites (above), the earliest occurrences of the group are in shallow-water deposits in cratonic successions in (*inter alia*) Siberia, Europe, western USA, and Morocco. The olenellids occur above Tommotian shelly faunas which are thought of as even shallower. These inshore facies and their possible (or likely) diachronism lie at the root of the problems with defining and correlating the Cambrian–Precambrian boundary (Conway Morris 1987). We can find no unequivocal record of trilobites in deep-water facies until close to the end of the Lower Cambrian. This makes it likely that the clade as a whole took several Ma to occupy the whole width of the shelf and upper slope. Deeper-water facies are commoner in the mid-Cambrian (Conway Morris and Rushton 1988). At the class level the theory of inshore to offshore migration is supported.

Trilobite orders. At this taxonomic level there are problems, because a phylogenetic classification of trilobites has yet to be fully worked out. *Treatise* (Moore 1959) orders are given, with modifications from Fortey and Owens (1975), Fortey and Chatterton (1988), and Fortey (1990), and are simply listed alphabetically in the preliminary and brief account which follows.

Order Agnostida. The Suborder Eodiscina is included in the Agnostida in our understanding of the Order Agnostida. Eodiscids antedate agnostids, and the earliest pagetiid eodiscoids retain many primitive features which link them with other trilobites. This has led some authors to doubt whether eodiscoids are truly Agnostina. The oldest and most primitive forms are found in China, and these eodiscids seem to have been inhabitants of shallow-shelf habitats (Fig. 7.8A; Zhang *et al.* 1980, text-fig. 2), but not the shallowest, shoreward biofacies, where redlichiids were typically present. Agnostid faunas representing the pelagic realm and occupying exterior sites in the onshore–offshore gradient were not established until near the end of the Lower Cambrian, at least in the North Atlantic region (Theokritoff 1979, p. 288). A rich diversity of mid to late Cambrian agnostids favoured outer-shelf habitats (Robison 1972) and were rare in the inner shelf (e.g. inner detrital belt biofacies). If agnostids were pelagic as Robison maintained this is scarcely surprising, but Jago (1973) claimed to have recognized benthic agnostid 'communities', and the occurrence of agnostids was also related to benthic biofacies patterns in the Ordovician (Fortey 1975). The history of Agnostida is generally compatible with the onshore–offshore pattern.

Order Asaphida. This major group was tentatively regarded as a suborder of Ptychopariida by Fortey and Chatterton (1988), but as an order by Fortey (1990). Primitive Asaphida were placed in the superfamily Anomocaracea. Because the recognition of the group depends on the presence of the ventral median suture—which has been described on very few Middle Cambrian forms—it is probably premature to identify the earliest Asaphida.

Order Corynexochida. Lower Cambrian corynexochids appear in shallow-shelf faunas, both in North America and Siberia. Middle Cambrian corynexochids are familiar as part of deeper-shelf faunas, as in the celebrated Burgess Shale. At least one genus appears to have become pelagic in mid-Cambrian times.

Order Lichida. Thomas and Holloway (1988) have reviewed this order. The oldest and most primitive lichids are placed in the family Lichakephalidae, which is first recorded from the Middle Cambrian. Lichakephalids are rare, and the early ones are rarer still, only recorded from the Soviet Union. We are uncertain of the detailed ecology of their

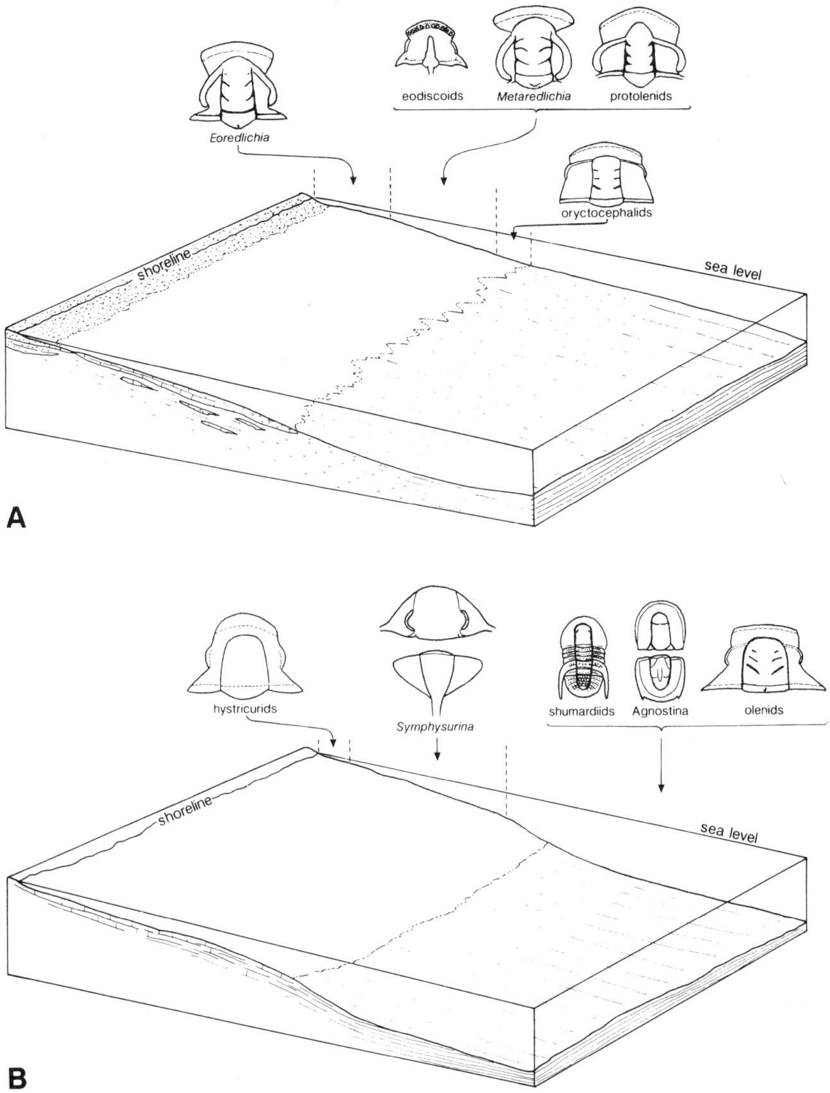

Fig. 7.8. Inshore origins of trilobite clades. A, Early Lower Cambrian of southern China with redlichiids in the shallowest water, joined by other polymerids and eodiscoids farther offshore. At this time trilobites had not yet occupied deeper-water habitats (based on Zhang *et al.* 1980). B, Early Ordovician in the tropical carbonate belt of Laurentia. New clades originated in shallow water on the open shelf; older ones, of Cambrian origin, had become adapted to more external sites.

occurrence there, but they do appear to be associated with platform limestones typical of Siberia at this time. Lichids are rare until about the mid-Ordovician, but from this time until the Devonian they occupied inner to outer shelf sites, although apparently uncommon in deeper-water biofacies (e.g. Whittard (1955–67), recorded a mere three lichid specimens among thousands of trilobites from his deeper-shelf Ordovician faunas of Shropshire).

Order Odontopleurida. The identification of *Acidaspides praecurrens* Lermontova as an odontopleurid (Bruton 1983; see also Thomas and Holloway 1988, p. 239) gives an Upper Cambrian record of the order. This appears to have been an inhabitant of inshore biofacies. Fortey (1990) suggested that Damesellacea might prove to be the sister-group of Odontopleurida; this group was typical of shallow shelf facies in China. By the Ordovician odontopleurids appear to have occupied a wide range of shelf habitats. Although Siluro-Devonian occurrences include some genera, such as *Miraspis*, which are found in comparatively deep-water shales, it does not seem to be the case that odontopleurids became progressively restricted to deeper-shelf sites; they were terminated by the same late Devonian extinction event that affected other shelf taxa.

Order Phacopida. The most primitive phacopides belong the Calymenacea. The earliest undoubted calymenid is *Neseuretus* (Ordovician, early Arenig), which was a genus characteristic of the most inshore trilobite biofacies in the early Ordovician. However, as was noted in discussing the Phacopina above, the early record of Phacopida is incomplete. Tremadocian Phacopida, such as *Pharostomina* and *Bavarilla*, do not fit obviously into a phacopide phylogeny, nor are they shallow-shelf inhabitants (both were probably mid- to outer-shelf forms). Hence Phacopida do not, at the present state of knowledge, conform to an onshelf–offshelf pattern.

Order Proetida. As conceived at present the earliest Proetida belong to the early Ordovician Subfamily Hystricurinae (Fortey and Owens 1975). Hystricurines were among the shallowest shelf trilobites inhabiting Ordovician Laurentia (Fig. 7.8B) and other sections of the former tropical carbonate belt. In the field they are often associated with interbedded stromatolites, rip-up conglomerates, and other indicators of extremely shallow deposition—the shallowest in which trilobites occurred at this time. More derived Proetida include: Bathyuridae, which occupied Ordovician shallow- to mid-shelf environments, but never reached deeper shelf habitats; and the proetid-aulacopleurid clade which during a long subsequent history occupied all habitats available to the Trilobita, as we noted above in discussing iterative trends, eventually as the last survivors of the group. The last of the trilobites were *not* deep-shelf inhabitants, however.

Order Ptychopariida. Limited in scope by Fortey (1990), the Ptychopariida includes the most primitive libristomate trilobites, but is a paraphyletic group. Although Ptychopariida certainly occur in the early Cambrian, there are problems in defining the earliest member of the group, and more work needs to be done on this before a potential pattern can be described. Nonetheless, it is interesting that Zhang *et al.* (1980, text-fig. 4) show what must be early Ptychopariida occupying the shallowest biofacies in south-west China after this site has been vacated by the Redlichiida. By later mid-Cambrian times they occupied the entire environmental gradient. On Fortey's (1990) understanding of Ptychopariina, olenids were the last surviving ptychopariines, and they inhabited deeper-water biofacies.

Order Redlichiida. As noted above, the earliest redlichiids are recorded in the shallowest biofacies in the earliest Cambrian of Southwest China by Zhang *et al.* (1980) (see Fig. 7.8). A redlichiid diversification into some 11 families must have occupied a variety of habitats but detailed appraisal of this remains to be done.

Order Illaenida. The status of this group is not yet decided, and here it is used to mean a clade comprising Illaenidae + Scutelluidae. The earliest known members of this group have not yet been described, but occur in shallow shelf limestones of latest Tremadoc or earliest Arenig age from western Newfoundland (Illaenidae) and Spitsbergen (Scutelluidae), collected by R.A.F. By the end of the Arenig some illaenids (*Ectillaenus*) were flourishing in outer-shelf environments in Wales and the Lake District, and scutelluids were common in similar environments in mariginal Laurentia.

Preliminary though this account is, eight out of the 10 major clades considered show some evidence of origination in inshore environments, and in several cases it is known that the clade started in the shallowest end of a described onshore-to-offshore environmental gradient. Clades which do not support the hypothesis (Asaphida, Phacopida) have problems with the recognition of their earliest members, and may yet prove to conform to the pattern. An expansion of the range across-shelf during 'radiation' is likely in all the cases considered, but not all the clades reached, or survived exclusively in the deeper-water environments available to trilobites. Detailed 'mapping' of the environmental history of these trilobite groups would be profitable to gain an understanding of the historical pattern of radiation.

References

Briggs, D.E.G. and Fortey, R.A. (1989). The early radiation and relationships of the major arthropod groups. *Science, New York*, **246**, 241–3.

Bruton, D.L. (1983). Cambrian origins of the odontopleurid trilobites. *Palaeontology*, **26,** 875–85.

Conway Morris, S. (1985). The Ediacaran biota and early metazoan evolution. *Geological Magazine*, **122,** 77–81.

Conway Morris, S. (1987). The search for the Precambrian–Cambrian boundary. *American Scientist*, **75,** 156–67.

Conway Morris, S. and Rushton, A.W.A. (1988). Precambrian to Tremadoc biotas in the Caledonides. In *The Caledonian–Appalachian orogen*, Geological Society Special Publication No. 38 (ed. A.L. Harris and D.J. Fettes), pp. 93–109. Blackwell, Oxford.

Cowie, J.W. (1971). Lower Cambrian faunal provinces. In *Faunal provinces in space and time* (ed. F.A. Middlemiss, P.F. Rawson and G. Newall), pp. 31–46. *Geological Journal*, Special Issue No. 4.

Cowie, J.W. and Brasier, M.D. (ed.) (1989). *The Precambrian–Cambrian boundary*. Clarendon Press, Oxford.

Eldredge, N. (1980). An extravagance of species. *Natural History*, **89 (7),** 45–51.

Eldredge, N. and Branisa, L. (1980). Calmoniid trilobites of the Lower Devonian *Scaphocoelia* Zone of Bolivia with remarks on related species. *Bulletin of the American Museum of Natural History*, **165,** 185–289.

Foote, M. (1988). Survivorship analysis of Cambrian and Ordovician trilobites. *Paleobiology*, **14**, 258–271.

Fortey, R.A. (1975). Early Ordovician trilobite communities. *Fossils and Strata*, **4,** 331–352.

Fortey, R.A. (1980). Generic longevity in Ordovician trilobites: relation to environment. *Paleobiology*, **6,** 24–31.

Fortey, R.A. (1983). Cambrian–Ordovician trilobites from the boundary beds in western Newfoundland and their phylogenetic significance. *Special Papers in Palaeontology*, **30**, 179–211.

Fortey, R.A. (1989). There are extinctions and extinctions: examples from the Lower Palaeozoic. *Philosophical Transactions of the Royal Society of London*, **B325,** 327–55.

Fortey, R.A. (1990). Ontogeny, hypostome attachment and trilobite classification. *Palaeontology*, **33**, 529–576.

Fortey, R.A. and Chatterton, B.D.E. (1988). Classification of the trilobite suborder Asaphina. *Palaeontology*, **31**, 165–222.

Fortey, R.A. and Owens, R.M. (1975). Proetida: a new order of trilobites. *Fossils and Strata*, **4,** 227–39.

Fortey, R.A. and Owens, R.M. (1990). Trilobita. In *Evolutionary trends* (ed. K.J. McNamara) pp. 121–142. Belhaven Press, London.

Fortey, R.A. and Shergold, J.H. (1984). Early Ordovician trilobites, Nora Formation, central Australia. *Palaeontology*, **27,** 315–66.

Fortey, R.A. and Whittington, H.B. (1989). The Trilobita as a natural group. *Historical Biology*, **2,** 125–38.

Henry, J.-L. (1980). Trilobites ordoviciens du Massif Armoricain. *Mémoires de la Société géologique et minéralogique de Bretagne*, **22,** 1–250.

Hou, X.-G. (1987). Three new large arthropods from Lower Cambrian, Chengjiang, eastern Yunnan. *Acta Paleontologica Sinica*, **26**, 286–98.

Hughes, C.P., Ingham, J.K., and Addison, R. (1975). The morphology, classification and evolution of the Trinucleidae (Trilobita). *Philosophical Transactions of the Royal Society of London*, **B272**, 537–607.

Jaanusson, V. (1975). Evolutionary processes leading to the trilobite Suborder Phacopina. *Fossils and Strata*, **4**, 209–18.

Jablonski, D. and Bottjer, D.J. (1990). The ecology of evolutionary innovation: the fossil record. In *Evolutionary Innovations* (ed. M.H. Nitecki), University at Chicago Press.

Jago, J.B. (1973). Cambrian agnostid communities in Tasmania. *Lethaia*, **6**, 405–21.

Lauterbach, K.-E. (1980). Schlüssereignisse in der Evolution des Grundplans der Arachnata (Arthropoda). *Abhandlungen des Naturwissenschaflichen Vereins in Hamburg*, **23**, 163–327.

Lu, Y.-H. *et al.* (1965). *Trilobites of China*. Academia Sinica, Beijing.

Moore, R.C. (ed.) (1959). *Treatise on invertebrate paleontology, Part O. Arthropoda 1*. Geological Society of America and University of Kansas Press, Boulder, Colorado, and Lawrence, Kansas.

McNamara, K.J. (1978). Paedomorphosis in Scottish olenellid trilobites (early Cambrian). *Palaeontology*, **21**, 53–92.

Palmer, A.R. (1965). Trilobites of the late Cambrian Pterocephaliid Biomere in the Great Basin, western United States. *Professional Papers, US Geological Survey*, **493**, 1–105.

Palmer, A.R. (1984). The biomere problems: evolution of an idea. *Journal of Paleontology*, **58**, 599–611.

Richter, R. and Richter, E. (1941). Das Kambrium am Toten Meer und die alteste Tethys. *Abhandlungen von der Senckenbergischen Naturforschenden Gesellschaft*, **460**, 1–50.

Robison, R.A. (1972). Mode of life of agnostid trilobites. *24th International Geological Congress*, session 7, pp. 33–40. Montreal.

Runnegar, B. (1982). A molecular-clock date for the origin of the animal phyla. *Lethaia*, **15**, 199–205.

Sepkoski, J.J. (1981). A factor analytic description of the fossil record. *Paleobiology*, **7**, 36–53.

Simpson, G.G (1944). *Tempo and mode in evolution*. Columbia University Press, New York.

Sloan, R.E. (1990). A chronology of North American Ordovician trilobite genera. *Bulletin of the Geological Survey of Canada*. In press.

Stanley, S.M. (1979). *Macroevolution: pattern and process*. Freeman and Co, San Francisco.

Stitt, J.B. (1975). Adaptive radiation, trilobite palaeoecology, and extinction, Ptychaspidid Biomere, late Cambrian of Oklahoma. *Fossils and Strata*, **4**, 381–90.

Theokritoff, G. (1979). Early Cambrian provincialism and biogeographic boundaries in the North Atlantic region. *Lethaia*, **12**, 281–96.

Thomas, A. T. and Holloway, D. J. (1988). Classification and phylogeny of the trilobite order Lichida. *Philosophical Transactions of the Royal Society of London*, **B321,** 179–262.

Westrop, S. R. (1989). Trilobite mass extinctions near the Cambrian-Ordovician boundary in North America. In *Mass extinctions* (ed. S.K. Donovan), pp. 89–103. Belhaven Press, London.

Whittington, H. B. (1981). Paedomorphosis and cryptogenesis in trilobites. *Geological Magazine*, **118,** 591–602.

Whittard, W. F. (1955–67). Ordovician trilobites of the Shelve Inlier, West Shropshire. *Palaeontographical Society [Monographs]*, 1–352.

Zhang, W.-T. *et al.* (1980). Cambrian trilobite faunas of southwestern China. *Palaeontologia Sinica*, n.s., **616,** 1–497.

8. The late Palaeozoic radiation of malacostracan crustaceans

D.E.G. BRIGGS
Department of Geology, University of Bristol, Bristol, UK
E.N.K. CLARKSON
Grant Institute of Geology, University of Edinburgh, Edinburg UK

Abstract

The record of Palaeozoic shrimps is strongly influenced by taphonomic factors. The vast majority of shrimps lacked a mineralized exoskeleton. They are underrepresented in fully marine environments. The conditions required for their preservation, however, are prevalent in environments transitional between marine and fresh-water which were widespread during the Carboniferous. The earliest shrimps are middle Devonian in age, but it is likely that their range is taphonomically curtailed and that they had undergone a significant radiation (rapid cladogenesis) by this time. The recorded diversity of shrimps increases markedly in the early Carboniferous, and remains in excess of 10 families through most of the Visean to Westphalian. It reaches a maximum in the Westphalian, when 15 genera are recorded and at least 16 families must have been present (three are without representatives), and declines in the Stephanian. All but four of the shrimp families recorded in the Palaeozoic disappear by the end of the era, only some families of isopods and decapods surviving into the Mesozoic. Other Recent orders, Stomatopoda, Syncarida, Lophogastrida, and Hemicaridea, originate in the Carboniferous but are unrepresented during part or all of the Permian, reappearing in the Mesozoic (Lazarus taxa). A shrimp carapace from the Visean of China is the only example known from outside the Laurussian Province. It suggests that a substantial diversity of shrimps remains to be discovered in the Angaran Province. In spite of taphonomic biases it is likely that the fossil record of shrimps reflects general trends in diversity. The absence of the families characteristic of

Major Evolutionary Radiations (ed. P. D. Taylor and G. P. Larwood), Systematics Association Special Volume No. 42, pp. 165–186, Clarendon Press, Oxford, 1990.

the Palaeozoic in similar environments in the Mesozoic confirms that their decline and disappearance represents a real extinction.

The majority of Palaeozoic shrimp assemblages are dominated by rapacious carnivores and scavenger/low-level carnivores. Other feeding strategies—filter and detritus-algal feeding—cannot be identified with confidence. Thus there is little evidence to sustain the concept of a persistent trophic structure in late Palaeozoic crustacean communities. There are only minor variations in the recorded diversity of shrimps in marine environments during the Palaeozoic. The Carboniferous radiation is represented mainly in environments transitional between marine and fresh-water, reflecting the widespread occurrence of broad coastal plains and low-lying swamp environments at this time.

Introduction

The term 'shrimp' is used here as a convenient way of referring to two groups of malacostracans—the subclasses Hoplocarida (the stomatopods and their extinct relatives) and the Eumalacostraca (including the syncarids, eucarids, and peracarids). The fossil record of Palaeozoic shrimps is now much more completely known than it was when Schram (1969*a*) first considered the stratigraphic distribution of the Eumalacostraca. Palaeozoic shrimps have received considerable attention recently and both their taxonomy (e.g. Schram 1986) and their taphonomy and environmental distribution (e.g. Briggs and Clarkson 1989) have been reassessed. The shrimps underwent a significant radiation in the late Palaeozoic, but the majority of these families became extinct before the end of the Permian. The late Permian record of those groups which survived to provide the basis for a more substantial diversification in the Mesozoic and Tertiary is very incomplete. In this chapter we document and interpret the late Palaeozoic radiation of the shrimps in the context of their distribution in different environments, and biases in their fossil record.

Origin of shrimps

The earliest known shrimps, *Eocaris oervigi* and *Devonocaris cuylerensis*, from the middle Devonian of West Germany and New York, respectively, are based on single poorly-preserved specimens (Brooks 1962*a*). Although they are clearly shrimps, their affinities cannot be determined. There is no fossil evidence that shrimps arose before the middle Devonian (Schram *et al.* 1978, p. 1385), although other crustaceans (e.g. phyllocarids, ostracodes) occur in the Cambrian. The stratigraphic range of shrimps, however, may be taphonomically curtailed; it is likely that earlier forms existed and await discovery where rare circumstances allowed their preservation.

The major morphological gap between the shrimps and other crustacean groups has left plenty of scope for speculation about their likely origins. Most authors have inclined to the view that the Phyllocarida, usually regarded as a subclass of the malacostracans, were a stem-group or primitive sister-group of the shrimps (Rolfe 1981). As a result of phylogenetic analysis, however, Schram (1986) grouped the phyllocarids with the cephalocarids and branchiopods, returning to an older concept of 'Phyllopoda', characterized by polyramous limbs with a leaf-like form. Furthermore, Schram (1986, p. 527) regarded his taxon Malacostraca (restricted to the Hoplocarida and Eumalacostraca) as 'the most primitive of the three "higher crustacean classes" [Malacostraca, Phyllopoda, Maxillopoda] in that they possess appendages on all segments of the body' (Schram 1986, p. 526, fig. 43-1).

Recent data on one of the most completely known fossil phyllocarids, *Nahecaris* from the exceptionally preserved pyritized fauna of the Devonian Hunsrück Slate of West Germany, support a return to their more traditional taxonomic placement within the Malacostraca (Bergström *et al.* 1987, 1989; Dahl 1987). *Nahecaris* does not share the synapomorphies which characterize the clade (combining the Phyllopoda and Maxillopoda) into which Schram (1986, fig. 43-1) placed the phyllocarids, i.e. a lack of most or all abdominal limbs, a uniramous first antenna, and loss of the mandibular palp. *Nahecaris* also lacks the polyramous phyllopodous limb which characterizes Schram's (1986) Phyllopoda; it has a biramous, stenopodous appendage like that in other Malacostraca. Further support for a malacostracan affinity of the phyllocarids comes from possible homologies between the carapace grooves in some Devonian phyllocarids and those in decapods (Rolfe 1981). Thus the affinities of some fossil phyllocarids may lie within the Malacostraca, as a primitive sister-group of the shrimps (hoplocarids and eumalacostracans). Other fossils assigned to the Phyllocarida, however, are poorly known, and further evidence may show that the affinities of these forms lie elsewhere (as demonstrated for the Concavicarida: Briggs and Rolfe 1983; Rolfe 1985). In addition, the relationship of the Recent Leptostraca to the fossil orders included in the Phyllocarida requires further consideration.

Late Palaeozoic radiation

1. *The nature of the data*

The decapods are the only shrimps which have a partially calcified exoskeleton, and this may have been absent in some of their early representatives. The exoskeleton of the majority of shrimps is organic

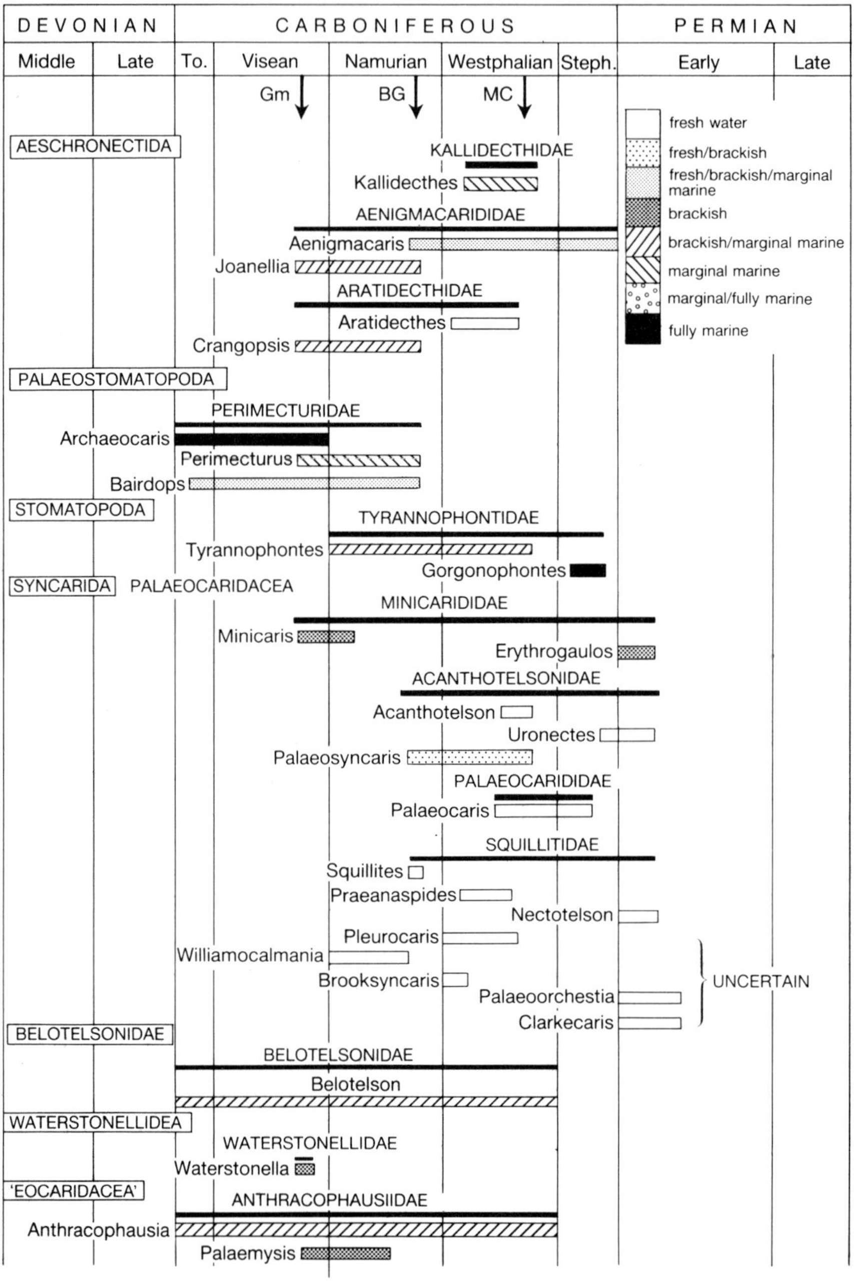

Fig. 8.1. Stratigraphical ranges (genera and families) and sedimentary environments (genera) of Upper Palaeozoic shrimps (Hoplocarida and Eumalacostraca). The arrows indicate the position of the three main

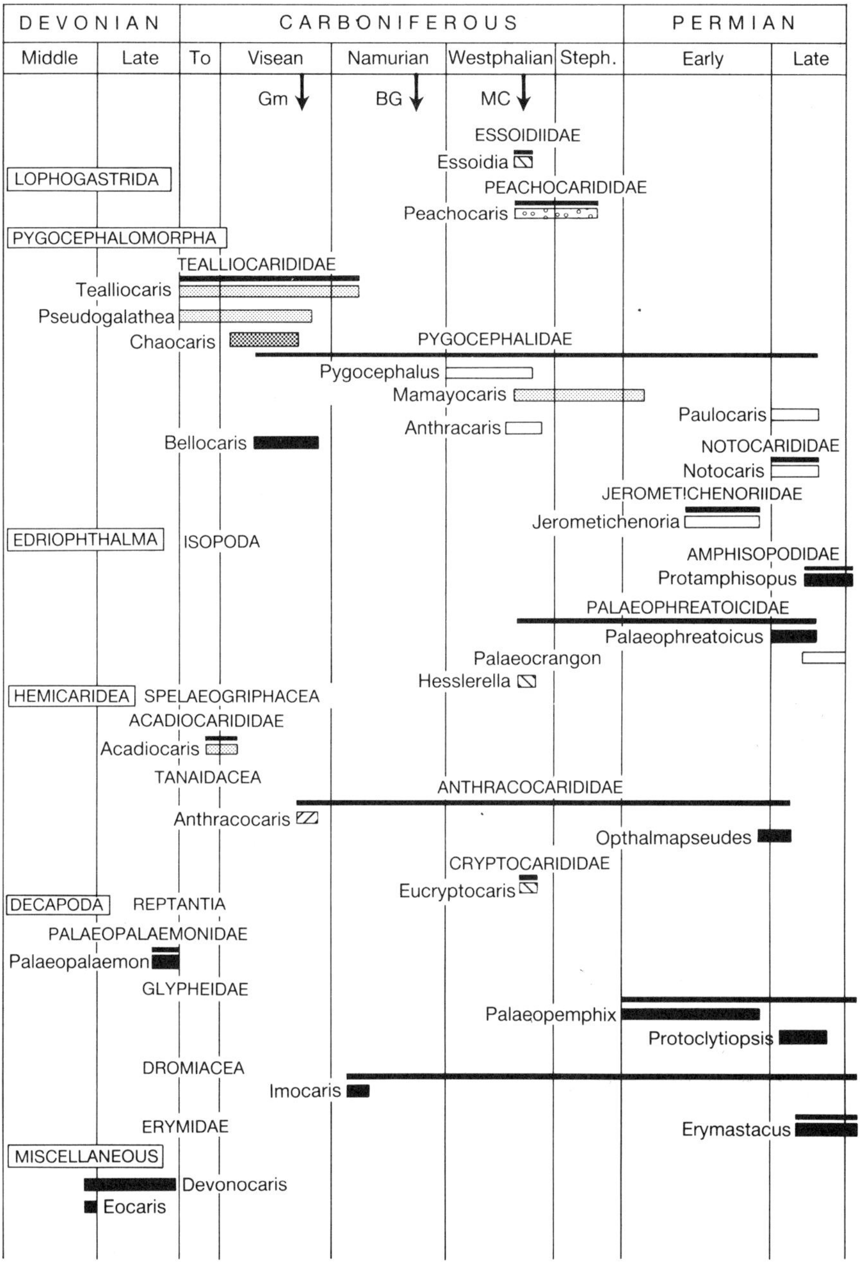

Konservat-Lagerstätten: Glencartholm (Gm), Bear Gulch (BG), and Mazon Creek (MC). Some of the ranges are exaggerated due to uncertainties about stratigraphic horizon.

(chitinous) and provides only limited resistance to decay. Carcasses disintegrate, even under quiet anaerobic conditions, within about 25 weeks (Allison 1988*a*). Thus shrimps are normally preserved only where the effects of decay are nullified by early diagenetic mineralization. Such exceptional preservation requires a precise combination of circumstances including, for example, quantity and timing of input of organic material, ion concentrations, grain size, and rate of sedimentation, salinity, and position of the oxic/anoxic boundary (Allison 1988*b*; Cater *et al.* 1989). Shrimps may be preserved in phosphate, as is commonly the case in the early Carboniferous shrimp beds of northern Britain (Cater *et al.* 1989), in carbonate—calcium carbonate in the Namurian Bear Gulch of Montana (Horner and Lund 1985), siderite in the Westphalian Mazon Creek of Illinois (Woodland and Stenstrom 1979)—or, very rarely, in pyrite (Schram 1978).

The analysis here is based on genera (Fig. 8.1). Species are too incompletely represented to provide a reliable indication of diversity patterns. Because unusual circumstances are required for their preservation, occurrences of shrimps are rare, but specimens usually retain relatively complete morphological data and are often found in large numbers. Over 80 per cent of the 55 genera include at least one species which is based on tens of specimens, and which preserves details of the limbs. Only 7 per cent of genera are known on the basis of the carapace alone. The associated biota is also usually preserved. There is likely to be a reasonably complete record of shrimp higher taxa and, while species and genera are doubtless underrepresented in the fossil record compared with shelly organisms, there is often a better basis for assessing their taxonomic affinity, mode of life, and ecology.

2. *Konservat-Lagerstätten*

The Carboniferous includes three major Konservat-Lagerstätten with significant shrimp faunas: Glencartholm in the Visean of southern Scotland (Schram 1983); Bear Gulch in the Namurian of Montana (Schram 1985); and Mazon Creek in the Westphalian (D) of Illinois (Schram 1979*b*). Other notable examples include Granton in the Visean of southern Scotland (Briggs and Clarkson 1983; Cater 1987; Briggs *et al.* 1990), and Blanzy/Montceau in the Stephanian of France (Rolfe *et al.* 1982). Any locality yielding shrimps can be regarded as a Konservat-Lagerstätte, as their preservation indicates a level of decay inhibition such that chitin survives. There is a gradation in examples which reach this taphonomic threshold, however, from occurrences of a few scrappy specimens to major Konservat-Lagerstätten.

The frequent occurrence of shrimps in the Carboniferous, in spite of the narrow range of circumstances necessary for their preservation, reflects the conditions which prevailed (Cater *et at.* 1989). The broad coastal delta plain provided a setting in which restricted water bodies, such as interdistributary bays and lagoons, could readily form. These were sites of low or episodic sedimentation rates, and variable salinity (due to mixing of normal marine and fresh-water). The low-latitude coastal plain setting allowed high rates of organic production and accumulation. Conditions in these transitional environments resulted in the preservation of a large number of occurrences of Carboniferous shrimps: in Britain alone there were at least 12 in the Dinantian (Briggs and Clarkson 1989; Cater *et al.* 1989), and over 40 in the Westphalian (Schram 1979*a*). This 'taphonomic window' suggests that at least the Carboniferous record should retain the major features of shrimp diversification. Indeed, the record of brackish-water shrimps may be relatively complete, although shrimps are doubtless underrepresented in fully marine environments.

It is widely acknowledged that Konservat-Lagerstätten introduce distortions into compilations of diversity based on the normal fossil

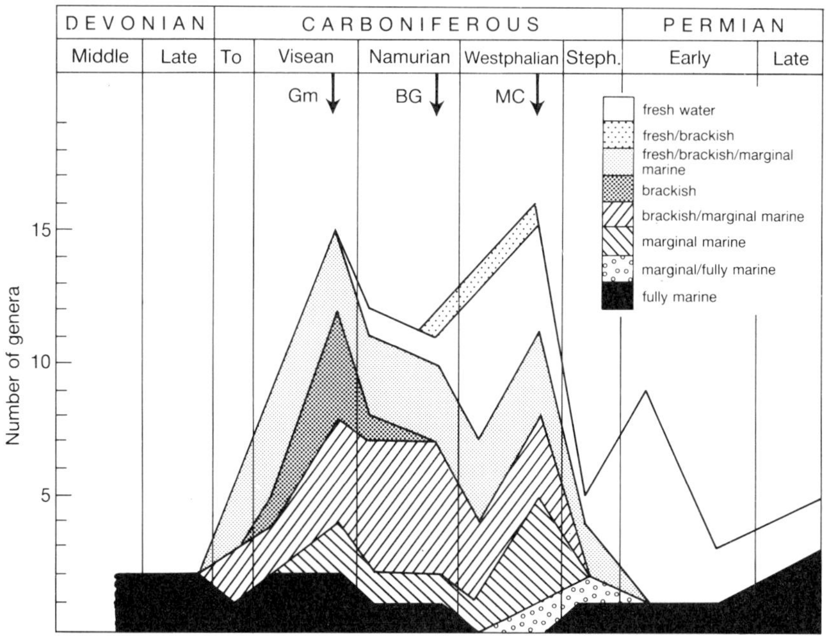

Fig. 8.2. Total recorded shrimp genera (Hoplocarida and Eumalacostraca) through the Upper Palaeozoic subdivided according to the sedimentary environments in which they occur. The arrows indicate the position of the three main Konservat-Lagerstätten, Glencartholm (Gm), Bear Gulch (BG), and Mazon Creek (MC). Data from Fig. 8.1.

record (Hoffman and Ghiold 1985; see also the spike that represents the Burgess Shale in Sepkoski's (1979, figs 4, 7) plot of Phanerozoic familial diversity). However, less than 20 per cent of the shrimp genera are known from only one locality, and less than 10 per cent are confined to one of the major Konservat-Lagerstätten: *Essoidia*, *Hesslerella*, and *Eucryptocaris* to Mazon Creek, and *Waterstonella* (Fig. 8.3) to Granton. This limits the distortion introduced by these major Konservat-Lagerstätten into the record of shrimp diversity. A similar conclusion was reached by Schram (1985, p. 470) using rarefaction techniques. The peaks in diversity (Fig. 8.2), however, do coincide with Glencartholm and Mazon Creek.

3. *Palaeozoogeography*

Before the discovery of a shrimp in the Carboniferous of China (Shen Yan-bin 1983), known Devonian and Carboniferous shrimps were confined to Laurussia, and essentially to tropical latitudes (Schram 1977). Schram (1977) argued that this is a real reflection of endemism. It should be noted, however, that broad coastal plains in equatorial latitudes are essentially confined to Laurussia in both Early and Late Carboniferous times (e.g Scotese *et al.* 1979). Thus the transitional environments in which shrimp communities are most likely to be preserved (Briggs and Clarkson 1989) may have been widely prevalent only in Laurussia. The occurrence of *Chaocaris*, represented by a single carapace in the Visean of Anhui Province of China, extends the range of the late Palaeozoic shrimps to the Angaran Province (Shen Yan-bin 1983). *Chaocaris* is associated with *Lingula* and bivalves, perhaps indicating conditions that were not fully marine, but its occurrence suggests that a substantial diversity of shrimps may remain to be discovered outside Laurussia.

4. *The pattern of diversification*

A radiation can be monitored on the basis of the rate at which higher taxa appear—i.e. the evolution of disparity. A meaningful calibration of the diversity of higher taxa through time relies on a rigorous taxonomic scheme. In many traditional taxonomies the older, more primitive members of a clade tended to be designated a separate taxon which became 'extinct' (a pseudoextinction) when it was replaced by a new taxon comprising the more advanced members (a pseudoorigination; see Briggs *et al.* 1988). The taxonomy of the shrimps has been refined recently, however, and the relationships of higher taxa investigated cladistically (e.g. Schram 1984*b*, 1986).

The earliest well-preserved shrimp, *Palaeopalaemon* (Fig. 8.3) from late Devonian localities mainly in the US midwest, is interpreted as a

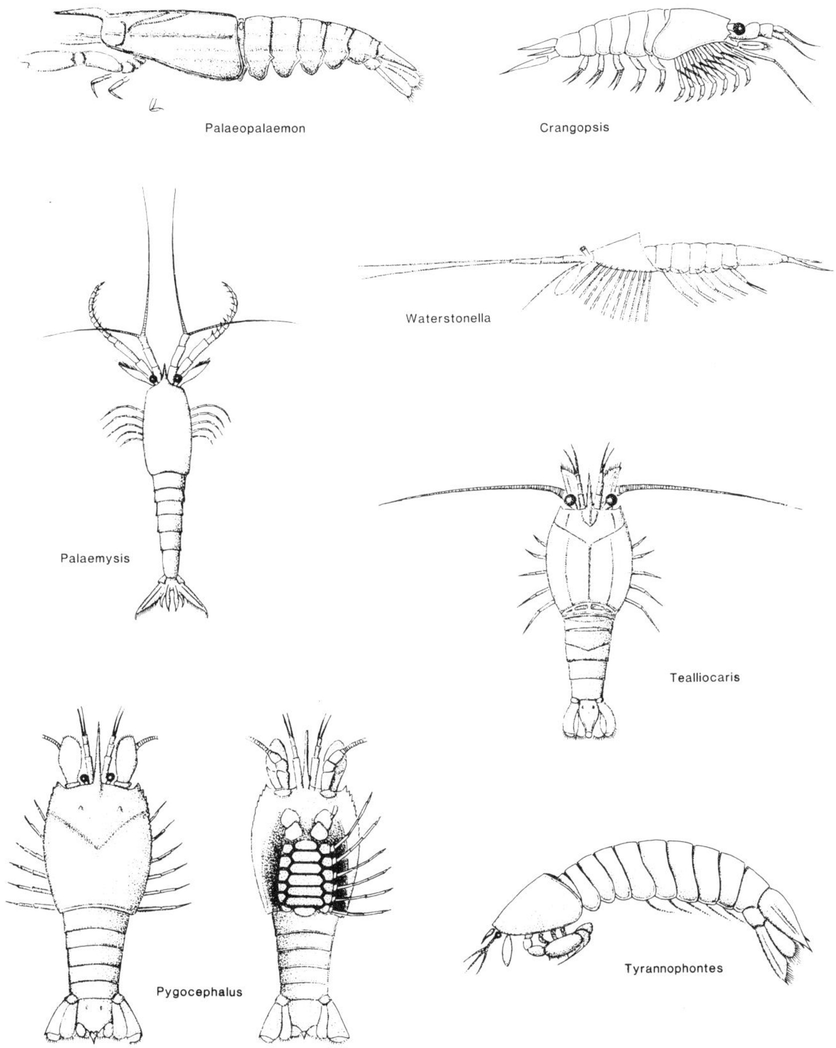

Fig. 8.3. Some representative Upper Palaeozoic shrimps: *Palaeopalaemon newberryi* Whitfield, 1880, Upper Devonian, USA (from Schram *et al.* 1978); *Palaemysis dunlopi* Peach, 1908, Visean-Namurian, Scotland (after Clarkson 1985); *Pygocephalus cooperi* (Huxley, 1857), Upper Carboniferous, England (based on a restoration by Jill Rankine); *Crangopsis socialis* (Salter, 1861), Visean, Scotland (after Peach 1908); *Waterstonella grantonensis* Schram, 1979, Visean, Scotland (after Briggs and Clarkson 1983); *Tealliocaris woodwardi* (Etheridge, 1877), Visean, Scotland (after Briggs and Clarkson 1985); *Tyrannophontes theridion* Schram, 1969, Namurian, USA (after Factor and Feldmann 1985).

decapod (Schram *et al.* 1978) and therefore belongs to one of the more derived taxa within the Malacostraca (Schram 1984*b*, 1986). If the results of cladistic analysis (Schram 1986, p. 530) reflect the timing of origins of eumalacostracan groups, other orders must also have appeared by the late Devonian even though fossil examples of this age have yet to be discovered. This conclusion is borne out by the occurrence of representatives of at least four additional orders in Tournaisian rocks (Fig. 8.1). Thus the fossil record of the shrimps and their relationships indicate a period of rapid cladogenesis in the Devonian.

The diversification of fossil groups is considered more usually in terms of genus and family records. The predicted pattern of an adaptive radiation (Paul 1989)—an exponential increase until equilibrium is reached, following which the diversity curve levels off into a plateau broken only by major extinctions—is rarely followed. This may be because the equilibrium model is inappropriate (see Benton, Chapter 18, this volume), or because the fossil record is too biased by taphonomic filters. In addition, it is unrealistic to consider the radiation of one clade in isolation; shrimps, for example, are just one element in communities which include a range of other organisms.

Only three genera of shrimps are represented in the late Devonian (Figs 8.1 and 8.2). Fourteen genera in 10 families are known from the late Visean. Although recorded generic diversity drops in the Namurian, 12 families were present (two, the Anthracocarididae and Pygocephalidae, occur in the Visean but have no representatives in the Namurian—they are Lazarus taxa, reappearing in the Westphalian; the Spelaeogriphacea, which first appear in the Tournaisian, must also have been present). The peak of diversity occurred in the late Westphalian when 15 genera in 13 families are known, and at least three additional families (Minicarididae, Anthracocarididae, and one of the Spelaeogriphacea) were present but lack representatives. Only six genera are known from the Stephanian, but at least 11 families must have been present. By the late Permian only eight genera in eight families are known, but Stomatopoda, Lophogastrida, and Spelaeogriphacea must also have been present.

Four Recent orders, Stomatopoda, Syncarida, Lophogastrida, and Hemicaridea, appear in the Carboniferous but have no representatives during part or all of the Permian (Fig. 8.1), reappearing in the Mesozoic. Of the 27 families (a conservative figure, omitting those taxa which are not assigned to a family) represented in the Palaeozoic, all but four became extinct before the end of the era. The Erymidae occur in the Triassic–Cretaceous. The Amphisopodidae and Glypheidae have Triassic (to Recent) representatives, although the assignment of the

Palaeozoic material to these families is somewhat uncertain, at least in the latter case (Schram and Mapes 1984). The dromiacean decapods, represented by *Imocaris* from the Namurian of Arkansas (Schram and Mapes 1984), also occur in the Mesozoic (Lower Jurassic) to Recent. Only one of the Palaeozoic genera, *Erymastacus*, is known to be represented in the Mesozoic, and the Upper Permian specimen (from the Zechstein of Germany) consists only of a chelipede (Bachmeyer and Malzahn 1983), the fragmentary nature of the material making it 'difficult to compare with other known Palaeozoic body fossils' (Schram and Mapes 1984, p. 167). Indeed, it is identified to the genus with a query. Thus the most important post-Palaeozoic groups, the decapods and edriophthalmids (isopods and amphipods), appeared in the late Palaeozoic but did not radiate substantially until the Mesozoic and, more significantly, the Tertiary. The low number of representatives of the decapods between their first appearance in the late Devonian and their major radiation in the Jurassic and later is an example of a 'macroevolutionary lag' (*sensu* Jablonski and Bottjer, Chapter 2, this volume). Their radiation may have awaited the opportunity afforded by the demise of the other late Palaeozoic shrimps.

Clearly the record of Palaeozoic shrimps is strongly influenced by taphonomic factors. However, it is likely that the pattern reflects general trends in diversity, particularly in transitional environments. Similar Mesozoic environments, such as the Triassic Grès à Voltzia (Briggs and Gall 1990) that yields only isopods and decapods among malacostracans, confirm that the disappearance of the families characteristic of the Palaeozoic represents a real extinction.

5. *Life strategies*

As a number of the late Palaeozoic shrimp taxa are remarkably well preserved, a consideration of their radiation can go beyond a simple calibration of diversity shifts through time, or the origination of major clades. The adaptive radiation should also be reflected in the appearance of a range of feeding strategies. Detailed interpretation of feeding in fossil crustaceans is difficult. Diagnostic evidence (of the proximal parts of appendages, for example) may not be preserved. Feeding may be unspecialized—'many crustaceans are deposit feeders, herbivores, omnivores, and scavengers' (McLaughlin 1982, p. 229).

The feeding type that can be inferred most reliably on the basis of appendage morphology is rapacious carnivory. The palaeostomatopods, *Perimecturus* and *Bairdops*, and the stomatopod, *Tyrannophontes* (Fig. 8.3), have raptorial subchelae on the thoracopods ('maxillipedes') like those in the living stomatopods. A raptorial appendage is also preserved in material of *Palaemysis* (Fig. 8.3) from Granton (Briggs and

Clarkson 1983) and Bearsden (Wood 1982, fig. 4). This appendage has been interpreted as a second antenna (Rolfe *in* Wood 1982; Briggs and Clarkson 1983) but evidence on specimens from Bearsden (Namurian of Scotland) suggests to N. D. L. Clark that it may have been a mandibular palp. Schram (1979*a*) synonymized *Palaemysis* with *Anthracophausia*, but new material (N. D. L. Clark, in preparation) indicates that they are distinct. Although definitive evidence has yet to be found, it is likely that *Anthracophausia* was also a predator.

While appendage morphology may provide evidence of a predatory mode of life, other modes of feeding are much more difficult to infer directly. With well preserved material, however, the likely mode of feeding usually can be constrained. *Tealliocaris woodwardi*, for example, has anteriorly converging thoracopods which do not meet along the midline, conical sternal processes, very fine setae on the thoracic exopods, and large powerful mandibles (Briggs and Clarkson 1985*b*). There is, therefore, clearly no food groove and the limb bases could not pass material anteriorly to the mouth. It seems likely that *Tealliocaris* (Fig. 8.3) was some kind of 'sweep feeder', using rhythmic grasping movements of the thoracopods to bring material directly to the region of the mouth, where it could be broken up by the mandibles. It is therefore best categorized as a scavenger/low-level carnivore or detritus feeder.

Schram (1981) identified communities of malacostracan crustaceans, based on consistent associations of taxa and feeding types, through the late Palaeozoic (see also Factor and Feldmann 1985). He identified four feeding types: (1) rapacious carnivores, which actively seek prey; (2) low-level carnivores or scavengers, which generally feed on carcasses; (3) filter-feeders, which generate currents to remove suspended food from the water; and (4) grazing detritus-algal feeders, which live on organic material in the sediment, or on algae. A further category, parasitism, was not noted in the fossil shrimps. (Feeding strategies in crustaceans can be categorized in a number of ways. Grahame (1983), for example, groups scavenging with deposit feeding, distinguishing the latter (detritus feeding in Schram's scheme) from grazing.) Schram's interpretations of feeding categories were based partly on the evidence of appendages, but also relied heavily on comparisons with living relatives.

Schram (1981) considered that the three major Carboniferous Konservat-Lagerstätten (Glencartholm, Bear Gulch, and the Essex fauna of Mazon Creek) represent the nearshore marine environment. Although they include the most diverse Carboniferous shrimp assemblages and come closest to representing normal marine conditions, all three show evidence of salinity fluctuations below normal marine (Briggs and Clarkson 1989, p. 298). Schram (1981, table 1) divided the shrimps into the four major feeding categories, thus demonstrating that these strategies persisted in related taxa through time. Evidence accumulated since the publication of Schram's (1981) seminal paper allows a number of the original interpretations of feeding type to be updated and suggests that others are more uncertain than previously thought (Table 8.1).

1. *Rapacious carnivores*. Schram (1981, table 1) assigned only palaeostomatopods and stomatopods (*Perimecturus*, *Bairdops*, and *Tyrannophontes*) to this category. New evidence suggests that anthracophausiids (see above), at least some phyllocarids, and tanaidaceans also belong here. The large anterior appendage on the phyllocarids *Sairocaris* and *Kellibrooksia* has been reinterpreted as a raptorial limb (Rolfe 1981, p. 20; Schram 1986, pp. 330–1), although it could have been fossorial (Schram, personal communication). A similar limb has been recorded in *Ceratiocaris* (Rolfe, personal communication). The feeding appendages in the tanaidaceans *Anthracocaris* and *Eocryptocaris* (= *Cryptocaris*, see Schram 1989) are poorly preserved in both taxa. Modern tanaids, however, are generally raptors (Schram 1986, p. 555), as is the Permian *Opthalmapseudes rhenanus* (Schram *et al.* 1986), and *Anthracocaris scotica* from Glencartholm preserves an enlarged anterior thoracopod which may be a chelipede (Schram 1979*a*, p. 103; Schram *et al.* 1986).

2. *Scavenger low-level carnivores*. Here Schram (1981) included *Belotelson*, and the pygocephalomorphs *Pseudogalathea*, *Pseudotealliocaris*, and *Mamayocaris*. The limb bases of *Belotelson* are poorly preserved (Schram 1974, 1979*a*) but the posterior cephalic appendages appear to have large plate-like endites, presumably gnathic in function (Schram 1986). There is no evidence of setose exites or other filtering apparatus. The arguments for including *Tealliocaris* in this category are given above, and *Pseudotealliocaris* is similar (synonymous in the view of N.D.L. Clark, personal communication). Like *Tealliocaris*, *Mamayocaris* (Schram 1974) has trunk limbs which do not interact in the midline, and robust mouth parts. *Pseudogalathea* is included, in the absence of evidence of the trunk limb morphology, on the grounds that a feeding strategy similar to closely related forms is likely. The phyllocarid *Dithyrocaris* probably also should be assigned to this category. The mandible is 'very large and well sclerotized' (Schram and Horner 1978, p. 395), and therefore more indicative of a carnivore or scavenger than a detritus-algal feeder.

3. *Filter feeders*. None of the shrimps interpreted by Schram (1981, table 1) as filter feeders can be assigned with confidence to this

Table 8.1. Shrimp assemblages from the three major Carboniferous Konservat-Lagerstätten divided into feeding categories (compare Schram 1981, table 1; Factor and Feldmann 1985, table 8, which differs from Schram in some of the taxonomic determinations of the Bear Gulch shrimps, and in the inadvertent omission of the labels for some of the feeding categories)

Trophic level	Dinantian Glencartholm	Mississippian Bear Gulch	Pennsylvanian Essex fauna
Rapacious carnivores	*Perimecturus parki*	*Perimecturus rapax*	
	Bairdops elegans		
		Tyrannophontes theridion	*Tyrannophontes theridion*
	?*Anthracophausia dunsiana*	?*Anthracophausia* sp.	?*Anthracophausia ingelsorum*
	Sairocaris elongata	*Sairocaris centurion*	*Kellibrooksia macrogaster*
			Ceratiocaris sp.
	Anthracocaris scotica		*Eucryptocaris hootchi*
			Palaeosyncaris micra
		Concavicaris sinuata	
Scavenger/ low-level carnivores	*Belotelson traquairi*	*Belotelson magister*	*Belotelson magister*
	?*Pseudogalathea macconochiei*		
	Tealliocaris etheridgei	*Tealliocaris* sp.	*Mamayocaris jaskoskii*
		Dithyrocaris rolfei	*Dithyrocaris* sp.
Feeding type uncertain	*Crangopsis eskdalensis*	*Crangopsis eskdalensis*	*Kallidecthes richardsoni*
			Peachocaris strongi
			Essoidia ipiceron
			Hesslerella shermani
	Joanellia elegans	*Aenigmacaris cornigerum*	

category. The filter feeders included aeschronectids and anthracophausiids. *Kallidecthes* is the most completely preserved of the aeschronectids; little detail of the appendages of *Crangopsis* (Fig. 8.3) is known. Schram (1969*b*) concluded that *Kallidecthes* was most likely nektonic, and that the setiferous thoracic limbs and possible food groove suggest filter feeding. Detritus feeding could not, however, be ruled out (Schram 1969*b*, p. 250). Schram (1986, p. 553) later introduced the possibility of a predatory habit, based on the suggestion (Kunze 1981, 1983) that all hoplocaridans were predatory. The anthracophausiids were apparently rapacious carnivores (see above). The appendages of the two other Mazon Creek taxa listed by Schram (1981, table 1) as filter feeders—*Peachocaris* and *Essoidia*—are poorly-preserved, and there is little on which to base an interpretation of feeding. They are therefore transferred, like *Kallidecthes*, to the category 'Feeding type uncertain' (Table 8.1).

4. *Detritus-algal feeders*. The phyllocarids and tanaidaceans included by Schram (1981, table 1) with the detritus-algal feeders have been reinterpreted as either rapacious or low-level carnivores (above). The well-preserved appendages of the only shrimp remaining in this category, the earliest isopod, *Hesslerella*, provide no clear evidence that it was a detritus-algal feeder. Later phreatoicids have a subchelate second thoracopod. It is therefore designated here 'Feeding type uncertain' (Table 8.1).

This review of feeding types shows that only two categories can now be recognized with confidence: rapacious carnivores, and scavenger low-level carnivores. Thus there is only limited evidence to sustain Schram's (1981) concept of a persistent community in near-normal marine conditions through the Carboniferous. The degree of taxonomic continuity recognized by Schram (1981) remains, but there is little evidence of trophic structure. The proportion of rapacious carnivores is higher than previously recognized, and no taxon can be unequivocally identified as a filterer or detritus-algal feeder (although some of those designated 'Feeding type uncertain' in Table 8.1 may represent these categories). Surprisingly, no fewer than 80 per cent of the shrimp species in Glencartholm and Bear Gulch, and 67 per cent in the Mazon Creek Essex fauna, fall into the category of rapacious carnivores or scavenger/low-level carnivores (and, in each case, the rapacious carnivores are dominant). Where data are available for numbers of specimens at Glencartholm (Schram 1983) and Bear Gulch (Schram 1985) this dominance of carnivores is almost as pronounced. At Glencartholm 67 per cent of individuals are carnivores (made up of 42 per cent rapacious carnivores, 25 per cent scavengers/low-level carnivores). These proportions may be relatively undistorted by exuviae; Schram (1983, p. 11) reported that 'the crustaceans . . . are usually virtually intact.' At Bear Gulch 67.3 per cent of individuals are carnivores (49.3 per cent rapacious carnivores, 18 per cent scavengers/low-level carnivores).

Rapacious carnivores make up a smaller percentage of other crustacean assemblages. *Bairdops* is only 6 per cent of individuals at Foulden (Briggs and Clarkson 1985*a*,*b*), but the remainder of the shrimp assemblage consists of *Belotelson* and very rare *Tealliocaris*, both scavengers/low-level carnivores. At Granton (Briggs and Clarkson 1983; Briggs *et al.* 1990) *Waterstonella* (Fig. 8.3), which is probably a filter or detritus feeder, is far more abundant than any other shrimp, but 63 per cent of the shrimp species present are carnivores (rapacious carnivores—*Anthracocaris*, *Bairdops*, and *Palaemysis*; scavengers/low-level carnivores—*Pseudogalathea* and *Tealliocaris*).

Schram (1981, table 2; see also Schram 1984*a*, p. 217, table 3) also identified recurring shrimp assemblages in brackish to freshwater environments. The assemblage consists of a pygocephalomorph (Fig. 8.3) and syncarids and is best represented in the Braidwood fauna at Mazon Creek, and in the British Coal Measures. The lack of a food groove or filtering setae, and the morphology of the robust mouthparts, support Schram's (1981, table 2) interpretation of the pygocephalomorphs as scavengers/low-level carnivores (e.g. *Tealliocaris*—see above, *Anthracaris*—Brooks 1962*b*), although it is possible that some were grazers (Schram 1986, p. 555).

Schram's (1981) interpretation of the syncarids as detritus-algal feeders relied mainly on a comparison with living representatives. In 1984, however, he grouped a number of palaeocaridacean syncarids, including *Acanthotelson stimpsoni* and *Palaeosyncaris micra*, a new species from Mazon Creek (Schram 1984*a*), in the Acanthotelsonidae, based on their possession of raptorial thoracopods. *Palaeocaris*, which occurs in the Braidwood fauna and in the Coal Measures, has a massive mandible which projects below the margin of the cephalic shield. It may have been a scavenger/low-level carnivore. The other syncarids known from these brackish to freshwater assemblages—*Praeanaspides*, *Pleurocaris, Minicaris*, and *Clarkecaris*—lack details of the appendages which might permit an interpretation of their feeding type. As Schram (1986, p. 553) pointed out, with the exception of the raptors, 'nothing is known with certainty about how these Late Palaeozoic syncarids may have fed.'

Hence, just as in the major Konservat-Lagerstätten which represent a near normal marine environment (Schram's (1981) nearshore marine assemblage), carnivores are a much more significant component of shrimp assemblages in brackish water (Table 8.2). In this environment

Table 8.2. Shrimp assemblages from Carboniferous brackish-water environments divided into feeding categories (compare Schram 1981, table 2; Schram 1984*a*, table3)

Trophic level	Coal Measures	Pennsylvanian Braidwood fauna
Rapacious carnivores		*Acanthotelson stimpsoni* *Palaeosyncaris micra*
Scavenger/ low-level carnivores	*Palaeocaris retracta* *Pygocephalus cooperi*	*Palaeocaris typus* *Anthracaris gracilis*
Feeding type uncertain	*Praeanaspides praecursor* *Pleurocaris annulatus*	

there is even less evidence to support the idea of a persistent community structure, with related taxa occupying parallel roles.

The majority of shrimp faunas are preserved, and presumably lived, on fine-grained organic-rich muddy soft-bottom substrates. This suggests that they should occur in communities dominated by deposit-feeding organisms. The conspicuous lack of deposit-feeding shrimps suggests either that there was a diversity of deposit-feeding infaunal groups (other than crustaceans) which are not preserved, or that the sediment was permanently anoxic.

The earliest well-preserved shrimps are apparently rapacious carnivores (late Devonian *Palaeopalaemon* (Fig. 8.3); Tournaisian *Archaeocaris*, *Bairdops*, and ?*Anthracophausia*), or scavengers/low-level carnivores (Tournaisian *Belotelson*, *Tealliocaris*, and *Pseudogalathea*). Due to the difficulty of identifying other feeding types, it is not clear when they first appear. *Waterstonella* (Fig. 8.3), which was probably a filterer or detritus/algal grazer, occurs in the late Visean.

6. *Environments*

In considering the distribution of shrimps in different environments through time we have plotted the stratigraphic ranges of all known genera, together with their habitat preferences (based on Briggs and Clarkson 1989 and data from the literature). It is evident that diversity has fluctuated markedly throughout the later Palaeozoic and that different environments have been invaded and occupied at different times. Four major environmental categories are recognized: freshwater, brackish water, marginal marine, and fully marine. A number of shrimps are euryhaline and occur in more than one of these categories. Although interpretations of environment are limited by uncertainties, when the data are plotted as a composite diversity curve a basic pattern emerges.

Fully marine forms are doubtless underrepresented due to taphonomic factors, but are present from the Middle Devonian onwards (Fig. 8.1). The earliest known eumalacostracans (*Devonocaris*, *Eocaris*) and hoplocaridans (*Archaeocaris*) are marine. This is consistent with physiological evidence which suggests a marine ancestry for these groups (see Briggs and Clarkson 1989).

From early Carboniferous times, shrimps invaded a range of marginal marine to brackish habitats (Figs 8.1 and 8.2). Those genera with the longest recorded stratigraphic ranges occurred in such environments. These predominantly euryhaline taxa were presumably less subject to directional selection and are characterized by lower rates of speciation and extinction than more narrowly adapted stenohaline taxa (see discussion in Briggs and Clarkson 1989). The major diversification in environments transitional between fully marine and freshwater occurs in the Carboniferous.

Although some genera occurring in the Tournaisian and Visean are found in environments which were apparently freshwater, no eumalacostracan genera confined to freshwater habitats are known before the Namurian. *Aratidecthes* is the only hoplocaridan confined to freshwater. Most of the freshwater taxa in the Westphalian and Permian are pygocephalomorphs and syncarids. *Erythrogaulos* (Schram 1984*a*) is one of the very few Permian shrimps from environments transitional between marine and freshwater (Figs 8.1 and 8.2). In the early Permian freshwater shrimps dominated but by the late Permian marine forms were in the ascendancy. The major radiation of the decapods in the Mesozoic, and particularly the Tertiary, was essentially in marine environments. The fossil record indicates that the origins of shrimps were marine and that they subsequently colonized environments of intermediate salinity before finally invading fresh water.

Schram (1981) argued that the 'near-shore marine environment' represented by Glencartholm, Bear Gulch, and the Essex fauna of Mazon Creek (now known to include conditions of reduced salinity), provided a reservoir from which shrimps radiated into other environments during the Carboniferous. Although some details of the model have been reinterpreted (above; Briggs and Clarkson 1989), the possibility remains that euryhaline taxa in these diverse assemblages did provide such a reservoir. Unfortunately, the detailed data on the relationships, environmental preferences, and stratigraphic occurrence of the shrimps, which would be required to test the hypothesis, are not available at present.

Environmental factors control not only the distribution of shrimps, but their preservation. Chitin has a relatively low preservation potential in marine environments; the recorded diversity of marine shrimps contributes a relatively insignificant background through the Palaeozoic, increasing in the Mesozoic with the appearance of calcified cuticles. Shrimps have a higher fossilization potential in transitional environments. The major radiation of shrimps in the Carboniferous reflects the widespread occurrence of broad coastal plains and low-lying swamp environments in equatorial latitudes at this time. The demise of Palaeozoic forms in the Permian is paralleled by a general increase in continental relief and much drier climatic conditions in tropical latitudes following the suturing of Pangaea (see Scotese *et al.* 1979).

Acknowledgements

We are grateful to N.D.L. Clark, R.A. Fortey, and F.R. Schram for discussion and comments on the text. The research was funded by Natural Environment Research Council Grant GR3/5610, awarded to the authors.

References

Allison, P.A. (1988*a*). The decay and mineralization of proteinaceous macrofossils. *Paleobiology*, **14**, 139–54.

Allison, P.A. (1988*b*). Konservat-Lagerstätten: cause and classification. *Paleobiology*, **14**, 331–44.

Bachmeyer, F. and Malzahn, E. (1983). Der erste Nachweis eines decapoden Krebses im miederrheinischen Kupferschiefer. *Annalen des Naturhistorischen Museums in Wien*, **85A**, 99-106.

Bergström, J., Briggs, D.E.G., Dahl, E., Rolfe, W.D.I. and Stürmer, W. (1987). *Nahecaris stuertzi*, a phyllocarid crustacean from the Lower Devonian Hunsrück Slate. *Paläontologisches Zeitschrift*, **61**, 273–98.

Bergström, J., Briggs, D.E.G., Dahl, E., Rolfe, W.D.I. and Stürmer, W. (1989). Rare phyllocarid crustaceans from the Devonian Hunsrück Slate. *Paläontologisches Zeitschrift*, **63**, 319–333.

Briggs, D.E.G. and Clarkson, E.N.K. (1983). The Lower Carboniferous Granton 'shrimp-bed', Edinburgh. *Special Papers in Palaeontology*, **30**, 161–78.

Briggs, D.E.G. and Clarkson, E.N.K. (1985*a*). Malacostracan Crustacea from the Dinantian of Foulden, Berwickshire, Scotland. *Transactions of the Royal Society of Edinburgh: Earth Sciences*, **76**, 35–40.

Briggs, D.E.G. and Clarkson, E.N.K. (1985*b*). The Lower Carboniferous shrimp *Tealliocaris* from Gullane, East Lothian, Scotland. *Transactions of the Royal Society of Edinburgh: Earth Sciences*, **76**, 173–201.

Briggs, D.E.G. and Clarkson, E.N.K. (1989). Environmental controls on the taphonomy and distribution of Carboniferous malacostracan crustaceans. *Transactions of the Royal Society of Edinburgh: Earth Sciences*, **80**, 293–301.

Briggs, D. E. G. and Gall, J.-C. (1990). The continuum in soft-bodied biotas from transitional environments: a quantitative comparison of Triassic and Carboniferous Konservat-Lagerstätten. *Paleobiology*. (In press.)

Briggs, D. E. G. and Rolfe, W. D. I. (1983). New Concavicarida (new order: ?Crustacea) from the Upper Devonian of Gogo, Western Australia, and the palaeoecology and affinities of the group. *Special Papers in Palaeontology*, **30**, 249–76.

Briggs, D. E. G., Clark, N. D. L. and Clarkson, E. N. K. (1990). The Granton 'shrimp-bed', Edinburgh—a Lower Carboniferous Konservat-Lagerstätte. *Transactions of the Royal Society of Edinburgh: Earth Sciences*. (In press.)

Briggs, D. E. G., Fortey, R. A. and Clarkson, E. N. K. (1988). Extinction and the fossil record of the arthropods. In *Extinction and survival in the fossil record*, Systematics Association Special Volume, no. 34 (ed G. P. Larwood), pp. 171–209. Clarendon Press, Oxford.

Brooks, H. K. (1962*a*). Devonian Eumalacostraca. *Arkiv for Zoologi*, **15**, 307–15.

Brook, H. K. (1962*b*). The Paleozoic Eumalacostraca of North America. *Bulletins of American Paleontology*, **44**, 163–335.

Cater, J. M. L. (1987). Sedimentology of part of the Lower Oil-Shale Group (Dinantian) sequence at Granton, including the Granton 'shrimp-bed'. *Transactions of the Royal Society of Edinburgh: Earth Sciences*, **78**, 29–40.

Cater, J. M. L., Briggs, D. E. G. and Clarkson, E. N. K. (1989). Shrimp-bearing sedimentary successions in the Lower Carboniferous (Dinantian) Cementstone and Oil Shale Groups of northern Britain. *Transactions of the Royal Society of Edinburgh: Earth Sciences*, **80**, 5–15.

Clarkson, E. N. K. (1985). Carboniferous crustaceans. *Geology Today*, (**1985**), 11–15.

Dahl, E. (1987). Malacostraca maltreated—the case of the Phyllocarida. *Journal of Crustacean Biology*, **7**, 721–6.

Factor, D. F. and Feldmann, R. M. (1985). Systematics and paleoecology of malacostracan arthropods in the Bear Gulch Limestone (Namurian) of Central Montana. *Annals of the Carnegie Museum*, **54**, 319–56.

Grahame, J. (1983). Adaptive aspects of feeding mechanisms. In *The biology of Crustacea*, vol. 8 (ed. D. E. Bliss), pp. 65–107. Academic Press, New York.

Hoffman, A. and Ghiold, J. (1985). Randomness in the pattern of 'mass extinctions' and 'waves of origination'. *Geological Magazine*, **122**, 1–4.

Horner, J. R. and Lund, R. (1985). Biotic distribution and diversity in the Bear Gulch Limestone of Central Montana. *Compte Rendu, Neuvième Congrès International Stratigraphie Géologie Carbonifère*, **5**, 437–42.

Kunze, J. C. (1981). The functional morphology of stomatopod Crustacea. *Philosophical Transactions of the Royal Society of London*, **B292**, 255–328.

Kunze, J. C. (1983). Stomatopoda and the evolution of the Hoplocarida. In *Crustacean phylogeny*, Crustacean Issues, no. 1 (ed. F. R. Schram), pp. 165–88. Balkema, Rotterdam.

McLaughlin, P. A. (1982). Comparative morphology of crustacean appendages. In *The biology of Crustacea*, vol. 2 (ed. D. E. Bliss), pp. 197–256. Academic Press, New York.

Paul, C. R. C. (1989). Patterns of evolution and extinction in invertebrates. In *Evolution and the fossil record* (ed. K.C. Allen and D.E.G. Briggs), pp. 99–121. Belhaven Press, London.

Peach, B. N. 1908. A monograph of the higher Crustacea of the Carboniferous rocks of Scotland. *Memoirs of the Geological Survey of Great Britain, Palaeontology*, 82 pp., 12 pls.

Rolfe, W.D.I. (1981). Phyllocarida and the origin of the Malacostraca. *Geobios*, **14**, 17–27.

Rolfe, W. D. I. (1985). Form and function in Thylacocephala, Conchyliocarida and Concavicarida (?Crustacea): a problem of interpretation. *Transactions of the Royal Society of Ediburgh: Earth Sciences*, **76**, 391–9.

Rolfe, W.D.I., Schram, F.R., Pacaud, G., Sotty, D., and Secretan, S. (1982). A remarkable Stephanian biota from Montceau-les-Mines, France. *Journal of Paleontology*, **56**, 426–8.

Schram, F.R. (1969*a*). The stratigraphic distribution of the Paleozoic Eumalacostraca. *Fieldiana Geology*, **12**, 213–34.

Schram, F. R. (1969*b*). Some Middle Pennsylvanian Hoplocarida (Crustacea) and their phylogenetic significance. *Fieldiana Geology*, **12**, 235–89.

Schram, F. R. (1974). The Mazon Creek caridoid Crustacea. *Fieldiana Geology*, **30**, 9–65.

Schram, F.R. (1977). Paleozoogeography of late Paleozoic and Triassic Malacostraca. *Systematic Zoology*, **26**, 367–79.

Schram, F.R. (1978). *Jerometichenoria grandis* n. gen., n. sp. (Crustacea: Mysidacea) from the Lower Permian of the Soviet Union. *Journal of Paleontology*, **52**, 605–7.

Schram, F. R. (1979*a*). British Carboniferous Malacostraca. *Fieldiana Geology*, **40**, 1–129.

Schram, F.R. (1979*b*). The Mazon Creek biotas in the context of a Carboniferous continuum. In *Mazon Creek fossils* (ed. M.H. Nitecki), pp. 159–90. Academic Press, New York.

Schram, F.R. (1981). Late Paleozoic crustacean communities. *Journal of Paleontology*, **55**, 126–37.

Schram, F. R. (1983). Lower Carboniferous biota of Glencartholm, Eskdale, Dumfriesshire. *Scottish Journal of Geology*, **19**, 1–15.

Schram, F. R. (1984*a*). Fossil Syncarida. *Transactions of the San Diego Society of Natural History*, **20**, 189–246.

Schram, F.R. (1984*b*). Relationships within the Eumalacostracan Crustacea. *Transactions of the San Diego Society of Natural History*, **20**, 301–12.

Schram, F. R. (1985). The Bear Gulch crustaceans and their bearing on Late Paleozoic diversity and Permo-Triassic evolution of Malacostraca. *Compte Rendu, Neuvième Congrès International Stratigraphie Géologie Carbonifère*, **5**, 468–72.

Schram, F. R. (1986). *Crustacea*. Oxford University Press, New York.

Schram, F.R. (1989). Designation of a new name and type for the Mazon Creek (Pennsylvanian, Francis Creek Shale) tanaidacean. *Journal of Paleontology*, **63**, 536.

Schram, F. R. and Horner, J. (1978). Crustacea of the Mississippian Bear Gulch Limestone of Central Montana. *Journal of Paleontology*, **52**, 394–406.
Schram, F. R. and Mapes, R. H. (1984). *Imocaris tuberculata*, n. gen., n. sp. (Crustacea: Decapoda) from the upper Mississippian Imo Formation, Arkansas. *Transactions of the San Diego Society of Natural History*, **20**, 165–8.
Schram, F. R., Feldman, R. M. and Copeland, M. J. (1978). The late Devonian Palaeopalaemonidae and the earliest decapod crustaceans. *Journal of Paleontology*, **52**, 1375–87.
Schram, F. R., Sieg, J. and Malzahn, E. (1986). Fossil Tanaidacea. *Transactions of the San Diego Society of Natural History*, **21**, 127–44.
Scotese, C. R., Bambach, R. K., Barton, C., Van der Voo, R. and Ziegler, A. (1979). Paleozoic base maps. *The Journal of Geology*, **87**, 217–77.
Sepkoski, J. J. Jr. (1979). A kinetic model of Phanerozoic taxonomic diversity II. Early Phanerozoic families and multiple equilibria. *Paleobiology*, **5**, 222–51.
Shen, Yan-bin (1983). A new pygocephalomorph genus (Eumalacostraca) of Lower Carboniferous from Anhui. *Acta Palaeontologica Sinica*, **22**, 663–8.
Wood, S. P. (1982). New basal Namurian (Upper Carboniferous) fishes and crustaceans found near Glasgow. *Nature*, **297**, 574–7.
Woodland, B. G. and Stenstrom, R. C. (1979). The occurrence and origin of siderite concretions in the Francis Creek Shale (Pennsylvanian) of Northeastern Illinois. In *Mazon Creek fossils* (ed. M. H. Nitecki), pp. 69–103. Academic Press, New York.

9. Major insect radiations

ROBIN J. WOOTTON
Department of Biological Sciences, University of Exeter, Exeter, UK

Abstract

The principal aspects and events in the evolution of insects, and the nature of their fossil record are reviewed. A Palaeozoic fauna of extinct orders and suborders, diverse in size, habitat, habit, and diet, was progressively replaced in the Permian by the stem-groups and early representatives of extant orders. The distribution in time of the Palaeozoic orders, and the first appearances of superfamilies of most of the remaining orders, are figured and discussed. There is no very strong evidence for explosive supraordinal radiations following significant evolutionary events, and the first appearances of important groups are not always immediately followed by major diversification. Within orders rapid expansions can be recognized following the presumed appearance of new types of larva, and particularly in association with the rise of the angiosperms, which provided the impetus for massive radiations at several trophic levels.

Introduction

Measured in terms of diversity, distribution, and sheer numbers, insects are an extraordinarily successful group. There are relatively few marine forms, most confined to the littoral zone; but insects are crucial components of almost every ecosystem on land and in fresh water—and they seem to have been so for a substantial proportion of Phanerozoic time.

With hindsight, it is possible to recognize several components of their ground plan which must particularly have contributed to their success. The *cuticle*, a spectacularly versatile composite material whose thickness, hardness, compliancy, and permeability are under fine local control,

Major Evolutionary Radiations (ed. P. D. Taylor and G. P. Larwood), Systematics Association Special Volume No. 42, pp. 187–208, Clarendon Press, Oxford, 1990.

provided support, enclosure, waterproofing, an efficient respiratory system, and a wide range of sensors and effectors (Neville 1978; Vincent 1980). Of the last, *jointed legs* allowed efficient, stable substrate locomotion, and their reduction to *three pairs*, which could be spread out so that their fields of movement did not overlap, gave manoeuvrability. *Three pairs of jaws* gave scope for great trophic diversity. Today insects variously use them to chew living and dead plants and animals, to seize and manipulate prey, to pierce and suck plant and animal fluids, to lap and sip plant exudates and other organic-rich fluids, to scrape encrusting algae, and to filter-feed; contrast the uniformly predatory spiders, with a single pair of jaws which can only pierce, inject, and tear.

Beyond the ground plan we can identify three enormously significant later developments. The *origin of wings* gave insects mobility in three dimensions and must greatly have helped their exploitation of tall plants, besides giving them new scope for dispersal and migration. If, as is disputed but probable, the first winged insects had outspread wings like modern dragonflies, their mobility on the ground must have been impaired; but this was restored by the development of *wing folding*, which also allowed the animals to penetrate vegetation and small spaces. The evolution of *internal wing development* (endopterygoty) combined with *holometabolous development via a pupa* brought new possibilities for physical and physiological divergence between juveniles and adults, for resource partitioning, and for precise 'keying-in' to seasonal changes, and to the life cycles of food organisms.

A further significant factor is the capacity of insects to evolve to be *small*. This brings with it the potential for occupying the wide range of microhabitats available in any ecosystem, with consequent niche diversification, and the opportunity—by no means always exploited —of *rapid reproduction*. The primitively wingless Collembola (springtails) include some of the most minute modern insects—as little as 0.2 mm in length. In winged insects, however, small size necessitates high wing-beat frequencies, which can only be achieved if the flight muscles are of the special *fibrillar type*, which is capable of contracting several times for each nerve impulse. Fibrillar muscle seems to have evolved independently many times within the Class (Pringle 1976).

The insect fossil record

Everything listed so far can be deduced from extant insects without reference to the fossil record, but only fossils can indicate when these landmarks were reached and show the patterns of diversification which accompanied these and other significant events. Since most palaeontologists have a limited knowledge of the insect record, it may be helpful to outline its characteristic features and limitations. The rest of this review should be read with these in mind.

The fossil record of insects is unusual. It is extensive, and fairly well-documented, but fossils make up less than 1 per cent of described species (Muller and Campbell 1954; F.M. Carpenter, personal communication)—a remarkably high proportion when one considers the enormous modern fauna and the scarcity of palaeoentomologists, but contrast over 40 per cent for the molluscs and chordates. Many species are known from single specimens. The record is probably therefore rather unrepresentative. Beds with abundant well-preserved insects are uncommon, and many are certainly overlooked. In consequence, a high proportion of the known species in any major geological division will have come from perhaps three or four really productive and well-studied localities, and often from a restricted part of the globe: we still know comparatively little about non-tropical insect faunas from the Palaeozoic, and of tropical faunas from the Mesozoic and Cainozoic. It follows that the discovery of a major new deposit, such as that from the early Cretaceous of Australia (Jell and Roberts 1986), can drastically transform our view of the abundance and distribution of insect groups in time.

The fossilization process is heavily selective. Insects are surprisingly resistant to decomposition (see Whalley 1985 for a useful discussion), but only preserve well in fine-grained sediments and in amber. Body and head characters are frequently hard to decipher, and taxonomy therefore relies substantially on wing characters; indeed isolated wings are often all that remain. Wings provide plenty of characters, but their use is full of pitfalls. Their variability itself varies greatly from group to group, and it is sometimes hard to tell whether differences between similar wings represent individual variation, or define species, genera, or higher taxa. Furthermore, the functional significance of most aspects of wing design has until recently been quite unknown, so that it has been very difficult to gauge the probability of characters arising in different groups by convergence, either singly or in correlated assemblages.

Not surprisingly, there are major differences of opinion in fossil insect systematics, and the position of many groups, especially in the Palaeozoic, is still best regarded as uncertain. Quantitative treatment of the record is therefore difficult. The analysis which follows draws extensively, and fairly uncritically, on Rohdendorf and Rasnitsyn (1980), the most recent major summarizing account of the fossil insects as a whole, and on later Soviet monographs where these supersede the earlier work. Useful summaries of the distribution of families in time are given in Ponomarenko (1988). The collected data in the imminent insect volumes, by F.M. Carpenter, of the *Treatise on invertebrate paleontology*, will allow far more detailed analyses than are now practicable.

With so many uncertainties it is not surprising that the impact of cladism has been rather slight, and paraphyletic taxa are widely used. This practice has been defended by Rasnitsyn (1988). I have followed traditional classification, but indicate paraphyletic taxa where appropriate.

Major radiations

Our knowledge of the relationships between the principal groups of insects is still very incomplete. Kristensen (1981) discusses the principal areas of uncertainty. Figure 9.1 indicates my own views and reservations on the evolution and interrelationships of the major subdivisions.

1. *The apterygote radiations*

We know very little of the initial radiations of the primitively wingless insects. The earliest known are the possibly Siegenian springtails (O. Collembola) (Whalley and Jarzembowski 1981) and the problematical

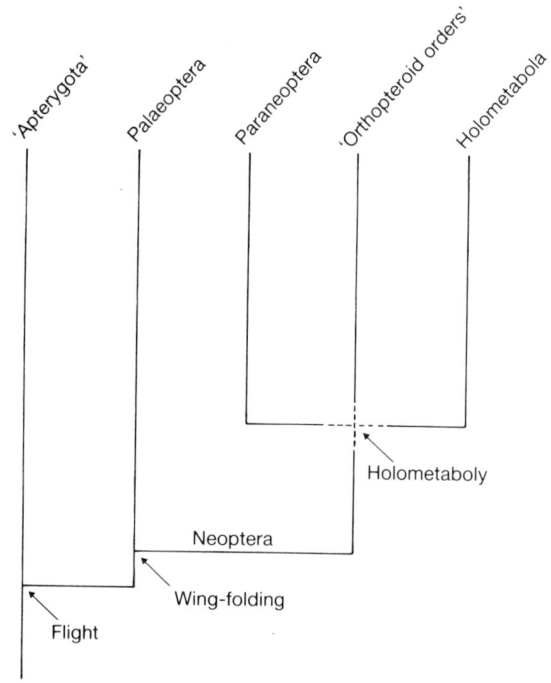

Fig. 9.1. The probable interrelationships of the major insect groups.

Rhyniognatha, an isolated pair of biting mandibles; all from the Rhynie Chert. Apparent bristletails (Archaeognatha), are known from the Emsian of Quebec and the Givetian of Gilboa (Labandeira *et al.* 1988). Kukalová-Peck (1987) has described a japygid dipluran and figured a thysanuran (O. Zygentoma), both from the Upper Carboniferous (Westphalian D) of Mazon Creek, Illinois, and Monura are known from several Upper Carboniferous and Lower Permian localities.

It is clear that considerable radiation had already taken place by the mid-Devonian; Kukalová-Peck (1987) reviews the interrelationships of these groups, and concludes more radically that the basic division into major ordinal lineages had already taken place by the early Silurian. There is good evidence for early trophic diversity. Most modern Collembola feed on a variety of small-scale plant debris, some on plant cell contents. Most Archaeognatha and Zygentoma are detritivores. Japygidae are predators.

2. *The initial pterygote, palaeopterous, and neopterous radiations and the Upper Carboniferous/Lower Permian fauna*

The circumstances of the origin of insect flight continue to be discussed (for recent reviews see Quartau 1986; Wootton 1986). Precisely when flight evolved is still unknown. Winged insects appear first in the Namurian; and our knowledge of these early forms has recently been significantly improved by the description of many new insects from the Namurian B of Hagen-Vorhalle, West Germany (Brauckmann 1988). It is clear that by this stage not only flight but also wing-folding had evolved, and that both the palaeopterous (non-folding) and neopterous (folding) lines had undergone significant radiations. Less clear is which if any of the known fossils belong to the stem-group of Pterygota, rather than to the monophyletic Palaeoptera or Neoptera; and how they held their wings at rest. The Carboniferous Paoliidae were believed by Sharov (1966) to be stem Pterygota, and to have had at least partly folding wings, and this is supported by Rasnitsyn (1976, 1981). I find the evidence for their stem status unconvincing, and at present prefer to treat Paoliidae as *incertae sedis*.

Figure 9.2 shows the number of families, in each of four subdivisions of the Upper Carboniferous and Permian, of the principal definable groupings which resulted from the early radiations of the Palaeoptera and Neoptera. The Upper Carboniferous is here divided at the top of Westphalian A—approximately the end of the Bashkirian.

By the Namurian B, Palaeoptera had already separated into at least two clades: Odonatoidea—the dragonfly group—and Palaeodictyopteroidea—an exclusively Palaeozoic group of orders characterized by piercing and sucking mouthparts. As the odonatoids appear to be the

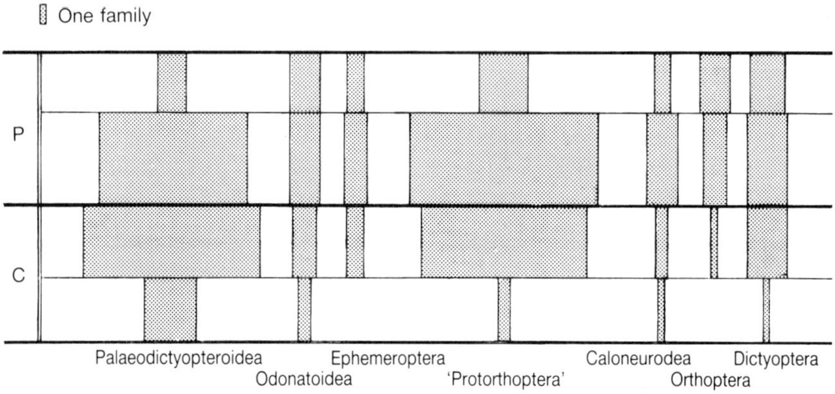

Fig. 9.2. The number of families, in each of four late Carboniferous and Permian subdivisions, of the principal groups of Palaeoptera and Neoptera which first appear in the Carboniferous.

sister-group of Ephemeroptera (mayflies) (Riek and Kukalová-Peck 1984) the latter may also have been present, but the earliest known mayfly fossils are Westphalian C (Kukalová-Peck 1985). All Palaeoptera, fossil and extant, belong to one of these three groups. They continued to diversify through the Upper Carboniferous and Lower Permian, but the palaeodictyopteroids were in decline by the mid-Permian, and vanish by the end of the period. Most Palaeozoic dragonflies and mayflies are quite distinct from Mesozoic and later forms, and have been placed in separate suborders. The most archaic odonatoid group—which may have extended into the Mesozoic—is often treated as an order, the Protodonata, separate from Odonata.

The Palaeodictyopteroidea appear to have radiated rapidly before and during the Upper Carboniferous and Lower Permian. They are usually divided into three, or sometimes four orders, peaking at different times: Palaeodictyoptera in the late Carboniferous and Megasecoptera, Diaphanopterodea, and Permothemistida in the Lower Permian. All were extinct by the end of the Permian. They form a distinct assemblage of fluid-feeding insects which strikingly parallel the neopterous Hemiptera in their piercing/sucking mouthparts. Their size range is remarkable: wing spans vary between 10 and *c* 560 mm. The Diaphanopterodea developed the ability to fold their wings, quite independently from the true Neoptera.

Carboniferous odonatoids are all Protodonata (= Meganisoptera), and are only moderately diverse. Brauckmann and Zessin (1989) recognize four families, of which two are probably paraphyletic. Three are already present in the Namurian. Two continue into the Permian (Carpenter 1960), where also occur smaller dragonflies, usually regarded as true Odonata. These are known only from wings, and their relationship to later Odonata is still controversial, but the wing shapes of some are strikingly similar to those of modern damselflies (Suborder Zygoptera) while others parallel those of small modern dragonflies of the Suborder Anisoptera. Together the Palaeozoic odonatoids show a huge size range with wing spans from 44 to 710 mm.

All modern odonatoids are predators, and the grasping forelegs of *Namurotypus sippeli* (Brauckmann and Zessin 1989) and of related Stephanian and Lower Permian forms show that they too had predatory adults. The nymphs of Palaeozoic odonatoids are still hardly known.

Ephemeroptera are very rare in Carboniferous rocks, with one certain family and two which may well be (Kukalová-Peck 1985; but see Carpenter 1987 and Kluge 1989). The few nymphs which are known appear to be aquatic, like all modern forms. Mayflies are fairly abundant in the Lower Permian, where both adults and aquatic nymphs are known (Carpenter 1979; Hubbard and Kukalová-Peck 1980).

The classification of Carboniferous and Permian Neoptera is still disputed. In the Carboniferous, four orders are clear-cut: Orthoptera, with one family; Dictyoptera, represented here by innumerable cockroaches in perhaps five families; and the extinct Caloneurodea and Miomoptera. Beyond these lie perhaps 30 families (Carpenter and Burnham 1985) of insects with chewing mouthparts, some expansion in the anal area of the hind wing, and abdominal cerci, representing a burst of radiation of rather generalized neopterous forms. Following Carpenter (1966 and elsewhere) and Burnham (1983), these forms are treated here as a single, unquestionably artificial Order Protorthoptera. The same five orders are found in the Permian, where cockroaches are far less common, 'Protorthoptera' far more so, with well over 30 families (Carpenter 1935, 1950, 1966, 1976). A new order, the Protelytroptera, is restricted to the Permian.

In this assemblage of extinct orders, and extinct suborders of extant orders, we can distinguish a clearly-marked Palaeozoic fauna, rich in species and diverse in size, form, and habit, which had almost entirely disappeared by the end of the Permian. Their ecology is still poorly-known, but a variety of adaptations were discussed by Carpenter (1971), and evidence of plant–arthropod interactions has been reviewed by Scott and Taylor (1983), Taylor and Scott (1983) and Scott *et al.* (1985). It seems certain that the palaeodictyopteroids were predominantly liquid-feeders, probably sucking sap and in some cases the liquid endosperm of seed-fern and gymnosperm megaspores (Sharov 1971; Taylor and Scott 1983) and perhaps feeding directly on microspores (Smart and Hughes 1972). Lycopod spores have been confirmed in gut contents of the Carboniferous protorthopteran *Eucaenis ovalis* (Taylor and Scott 1983) and there is good evidence, mainly from coprolites, that spore-feeding was widespread among Palaeozoic arthropods, many of which will have no doubt been Protorthoptera (Scott *et al.* 1985). Evidence for feeding on living plant leaves and stems is still scanty, but cockroaches and doubtless some Protorthoptera will have been scavengers and detritivores. Other Protorthoptera were certainly raptorial predators (Carpenter 1971). So were adult odonatoids, and the similarity of the wings of Permian forms to those of extant dragonflies indicates a high degree of flight adaptation, and hints that they might already have been capable of taking prey on the wing. Some adult Permian mayflies had substantial jaws, and may also have been predatory. The broad size-range of both herbivores and predators suggests a complex web of predator–prey interactions in a wide variety of terrestrial habitats.

The extent of Palaeozoic insect radiation in fresh water is less clear; see Wootton (1988) for a recent review. Several prominent workers (e.g. Štys and Soldán 1980; Kukalová-Peck 1983, 1987; Toms 1984) believe that the first insects were aquatic at all life-stages, and that the aquatic juveniles of many groups are therefore primitively so. If this is the case, the larvae of odonatoids and Ephemeroptera were aquatic from the first. More evidence is needed. Apparently aquatic mayfly-like nymphs are known from the Westphalian of Illinois and Czechoslovakia (Kukalová-Peck 1985), and the excellent mayfly nymphs from the Lower Permian (Kukalová 1968; Hubbard and Kukalová-Peck 1980) were certainly aquatic, as apparently were some Lower Permian Protorthoptera (Carpenter 1935; Kukalová-Peck 1978). Some Permian aquatic larvae were certainly predatory; but we know little else about the range of insect feeding habits in Palaeozoic freshwater systems.

3. *Radiation patterns of the extant orders*

From the Lower Permian onwards, initially alongside the Palaeozoic faunal components just described, the foundations of the modern fauna were being laid. Orthoptera, with five Permian families, were expanding; so were the stoneflies (Plecoptera), which appear here for the first time. Furthermore, by the Artinskian it is clear that three more major landmarks had been passed. The radiation of Paraneoptera—a monophyletic cluster of exopterygote orders of which the Hemiptera are by far the most important—was underway; holometabolous development had evolved; and stem Mecopteroidea, a significant monophyletic group of holometabolous orders, were present. Of Paraneoptera, the orders Psocoptera, Thysanoptera, and Hemiptera are all present in the

Artinskian. Of Holometabola, beside stem mecopteroids we find Neuroptera, Coleoptera, and a small extinct order, the Glosselytrodea, whose systematic position is unclear, and which is unknown after the Jurassic.

These insects, and those which follow, fall into two categories. We may first distinguish a group of orders and suborders which appear to have exploited a range of resources which were available from an early stage, and to have undergone relatively moderate radiations thereafter, achieving fair, rather than extraordinary diversity. Beyond these we can identify five immense orders, which together make up some 93 per cent of all modern described insect species—and about 70 per cent of all extant species of animals. These are the Hemiptera, which are exopterygote Paraneoptera, and the Lepidoptera, Diptera, Coleoptera, and Hymenoptera, all of which are endopterygote Holometabola, the first two being mecopteroids.

Figure 9.3 illustrates, for the majority of these orders, the numbers of superfamilies appearing *for the first time* within each of a series of subdivisions of the geological column. These subdivisions are unequal, but none is more than 41 Ma long. The figure hence shows something of the major periods of diversification *at superfamily level*. Two limitations should be borne in mind. First, the superfamily category is more commonly used in large orders than in small, but the differences between families in small orders may be as great as those between superfamilies in large orders. Second, the uneven information from different periods can lead to wrong conclusions. When a period with relatively few good localities, like the Triassic, precedes a more productive one, like the Jurassic, radiations which actually took place in the former may not appear until the latter.

Dealing first with the more moderately-sized groups: three exopterygote orders—Odonata, Ephemeroptera, and Plecoptera—typically have aquatic juveniles and terrestrial/aerial adults. In each group the juveniles are significantly different in appearance from the adults; Odonata in particular show that a considerable degree of metamorphosis is sometimes possible at a single moult without need for a pupal stage.

Odonata show moderate diversification throughout. They are predatory as both larvae and adults. Suitable prey will have been available from the first, and the evolution of the group seems to have consisted of adaptive refinement and modest adjustment to changing conditions and prey.

Ephemeroptera show a rather similar pattern, though the diagram shows breaks, which may be spurious, in the Triassic and Cretaceous. Alone among extant insects, mayflies have two winged stages, both

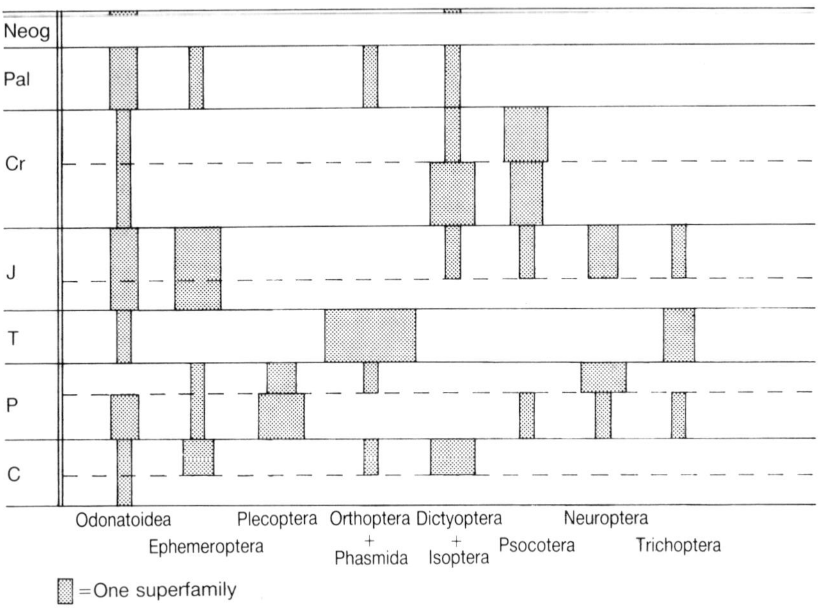

Fig. 9.3. First appearance through time of superfamilies of a selection of insect orders.

short-lived. Today only the larvae feed, and it is the larvae which show most evidence of spectacular adaptive radiation. They occur in most kinds of freshwater bodies, and in and on a variety of substrates; and their feeding habits range from filtering to active predation. I have discussed elsewhere (Wootton 1988) the time-ranges within which these habits probably arose.

Plecoptera, which are neopterous, make an interesting contrast. There are today perhaps half as many species of stoneflies as of mayflies. Their habitats are more restricted—most are confined to running water—and they are much less diverse in form. Their radiation at superfamily level appears restricted to the Permian, though there is evidence for a burst of new families in the Triassic and Jurassic (Sinichenkova 1987).

Orthoptera and Dictyoptera are rather generalized exopterygote, neopterous, terrestrial—and paraphyletic—orders, with their origins in the Carboniferous. The saltatory Orthoptera show a notable radiation of superfamilies in the Triassic, where also appears their offshoot, the Order Phasmida — stick- and leaf-insects. No new superfamilies appear after the Triassic, but later radiation at and below family level

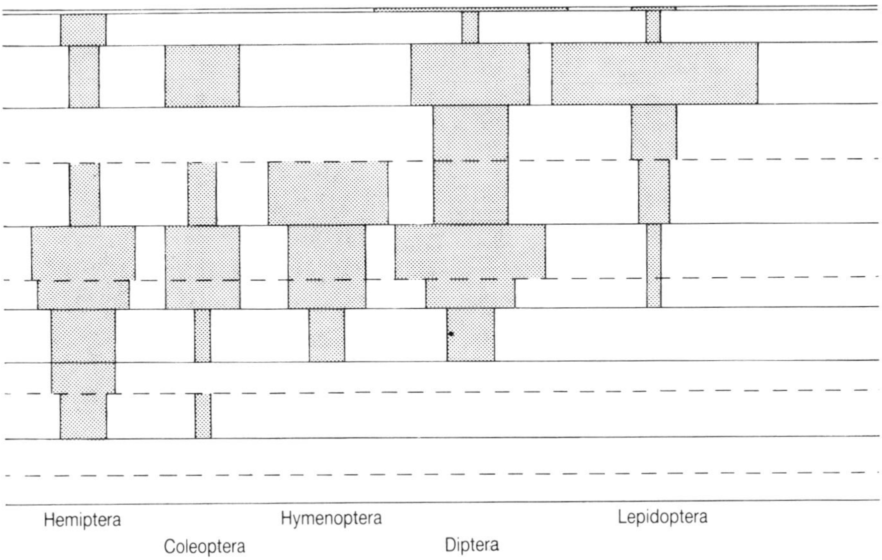

Fig. 9.3. *continued.*

is considerable. Today over half the species in this large order belong to a single family, the Acrididae, the whole of whose expansion may have taken place since the late Cretaceous, in association with angiosperm radiation, and particularly with the development of open grasslands. Tettigoniidae, which today provide a third of the total, seem also to have radiated within this time.

Dictyoptera show a very different pattern. Carboniferous cockroaches (Suborder Blattodea) are very numerous, very variable, and very hard to classify. They are placed in up to seven families, in a separate division from all modern forms, from which some at least differ in having a long ovipositor. All extant cockroaches have short ovipositors, and almost all lay their eggs in cuticular packages (oothecae). Cockroaches are scarcer in Permian and Triassic beds, and their principal superfamily radiations are Jurassic and Cretaceous. Oothecae are known from the Upper Cretaceous.

The remarkable eusocial, cellulose- and wood-feeding termites (O. Isoptera) first appear in the Cretaceous (Jarzembowski 1981) and are clearly derived from cockroaches. The predatory mantises (Dictyoptera, Suborder Mantodea) are unknown before the Tertiary.

The highly significant mecopteroid Holometabola comprise five orders. The Siphonaptera (fleas), are hardly known as fossils, but first appear in the Lower Cretaceous (Jell and Roberts 1986). Two orders, the Mecoptera and Trichoptera, have been of moderate importance: the latter far more so today. Two orders, the Lepidoptera and Diptera, are among the five largest orders, and are discussed below.

Mecopteroids appear first in the Artinskian, and undergo significant radiation in the Permian. Permian forms have conventionally been placed in Mecoptera and Trichoptera, both paraphyletically defined, the former order being ancestral to the other four. In cladistic terms, however, Mecoptera are first recognizable in the Upper Permian, alongside the stem of Lepidoptera + Trichoptera, and stem Diptera (Willmann 1989; Wootton and Ennos 1989), and are a relatively minor group, never sufficiently diverse to be divided into superfamilies.

The wings of primitive Trichoptera (caddises) are nearly identical to those of primitive Lepidoptera, and the fossils are almost impossible to tell apart. In the diagram Permian and Triassic 'Trichoptera' may in fact include members of both orders and their common stem, and their larval habitats are unknown. Modern Trichoptera have aquatic larvae. The diagram for this order is based on Sukacheva (1982), and shows modest radiation, with all extant superfamilies established by the beginning of the Cretaceous. Only then do the larval cases, which characterize most modern species, appear; and there is evidence for a substantial Cretaceous/Palaeogene radiation of families, particularly within the Superfamily Phryganeoidea: perhaps accompanying the adoption of detritus-shredding in slow-flowing and still water (Wootton 1988).

The radiation patterns of the five great orders show clear differences from one another. The only one with exopterygote development is the Order Hemiptera (bugs)—the dominant order of the Division Paraneoptera. Paraneoptera are first found in the Lower Permian, where Hemiptera occur beside the earliest members of the related but far smaller Orders Psocoptera and Thysanoptera.

Hemiptera all have piercing and sucking mouthparts. They radiated rapidly in the Permian, as the similarly-endowed palaeodictyopteroids declined and became extinct, an apparent example of ecological replacement. Radiation continued throughout the Triassic, where first appear members of a major clade, the Suborder Heteroptera, which contributes significantly to later radiations. Many of the early fossil Heteroptera are aquatic, with clear predatory adaptations (Popov 1971). Today the only predatory Hemiptera are Heteroptera. Whether predation was the primitive habit of Heteroptera is disputed (see Schaefer 1981 for a review) but there is no doubt that from the beginning it played an important role in their evolution.

The Jurassic saw extensive development of new superfamilies, most of them Heteroptera. Others continue to appear in the Cretaceous and

Tertiary; but the most significant post-Jurassic radiations were at the level of family or below. While members of several other orders feed on animal juices and plant exudates, Hemiptera and the closely related Thysanoptera remain the only groups significantly to exploit plant sap. Many of these are oligo- or monophagous, and the radiation of angiosperms was certainly accompanied by massive generic and specific diversification.

Coleoptera similarly first appear in the Lower Permian (Ponomarenko 1969), though many Permian fossils previously thought to be beetles now appear to belong to the extinct Protelytroptera (Kukalová 1969). They are abundant throughout the Mesozoic (Ponomarenko 1977, 1985) and Tertiary. Figure 9.3 suggests that the principal periods of radiation were the Jurassic and the Palaeogene—but this is certainly misleading. Fossil beetles frequently consist of single elytra, or of paired elytra with a pronotum, and are particularly hard to classify. The radiation diagram is in the early stages biased towards the few groups—like water-beetles, click-beetles, and weevils—which can confidently be recognized using these and a few other gross body characters. The apparent superfamily explosion in the Palaeogene is certainly largely due to the Baltic Amber, in which many beetles are preserved in fine, systematically useful detail. It seems probable that Coleoptera were actually radiating vigorously from the Triassic onwards, and that most or all modern superfamilies were established by the Palaeogene. However, since several of the largest families consist exclusively of phytophagous insects, many of which are host-specific, there can be no doubt that there was extensive late radiation below the superfamily level, associated with the development of the modern flora.

Coleoptera are an enormous group, which today form over one-third of all described animal species. They have chewing mouthparts, and their success reflects their capacity for exploiting an immense range of food sources, plant and animal, living and dead, in a great variety of terrestrial and aquatic habitats throughout the world.

The Hymenoptera present an unusually compact radiation pattern, which in part reflects their particularly intensive study over some 25 years by A.P. Rasnitsyn (Rasnitsyn 1988). Whereas Coleoptera and Diptera are both magnificent general-purpose orders, Hymenoptera are notable for doing a few things extremely well. Most of the Suborder Symphyta (sawflies) parallel the Lepidoptera in having phytophagous, caterpillar-like larvae, variously attacking leaves, fruits, and wood. The Suborder Apocrita include a great range of specialized predators, including parasitoids, on insects and other arthropods, most of the predation being carried out by specialized larvae hatching in or next to the prey with which their parent has supplied them. Other Apocrita are bees—solitary and social pollen and nectar-feeders; and ants— versatile foragers which combine a wide food range with highly complex eusocial behaviour. The fossil record indicates that these ways of life evolved sequentially. Symphyta appear in the Triassic, and their superfamilies diverged in the Jurassic. Parasitoid Apocrita appear in the Lower Jurassic and diversify through the Jurassic. The predatory wasps appear in the Jurassic and radiate extensively in the Cretaceous, where the first ants occur. The bees, which depend heavily on angiosperm flowers, are now represented in the late Cretaceous by a specimen belonging to a eusocial subfamily (Michener and Grimaldi 1988). Social organization seems to have been established in several distinct Cretaceous groups (Burnham 1978).

The two giant mecopteroid orders, Diptera and Lepidoptera, contrast sharply. Stem Diptera were probably present in the late Permian (Willmann 1989; Wootton and Ennos 1989). They are known from very few Triassic localities, but the initial major radiation of the Suborder Nematocera seems to have taken place in the Triassic and early Jurassic, that of the Brachycera in the mid-and late Jurassic, and of the Cyclorrhapha in the Cretaceous and Palaeogene (Kovalev 1987). The key to the success of the Diptera may well have been the remarkably versatile apodous larva (Keilin 1944), which has shown itself adaptable to life in a wide spectrum of aquatic, semiaquatic, and terrestrial habitats, including many highly anoxic situations where no other insects are found. In particular the Mesozoic expansion of nematoceran and brachyceran Diptera must have altered the structure and functioning of all freshwater ecosystems beyond recognition (Wootton 1988), and the Cretaceous radiation of Cyclorrhapha must have transformed decomposition processes.

The Lepidoptera were the last of the major orders to expand, and are perhaps the least versatile: with very few exceptions the larvae are phytophagous, feeding on leaves, stems, roots, or fruits, and the adults sip nectar and other organic-rich liquids. Whalley (1986) has reviewed the Mesozoic records, which are still scanty. The great majority of superfamilies and families first appear in the Palaeogene, and their spectacular radiation is evidently linked with the rise of the modern, angiosperm-dominated flora.

Discussion

1. *Supraordinal radiation patterns*

Smart (1963) and Smart and Hughes (1972) described insect evolution in terms of a series of six explosive radiations, each leading to a recognizable assemblage of orders. They illustrated each as a burst of lines radiating from a single point at a single moment in time, actually representing (Smart and Hughes 1972) a series of close-coupled dichotomies occurring within a relatively brief period. They are:

(1) a radiation of winged (pterygote) forms, which immediately followed the origin or flight;

(2) the radiation of Palaeoptera (winged forms, primitively incapable of folding the wings);

(3) the radiation of Neoptera, following the evolution of wing-folding;

(4) the radiation of Paraneoptera;

(5) the radiation of 'Oligoneoptera' (= Holometabola = 'Endopterygota'), following the origin of holometabolous development;

(6) the radiation of the mecopteroid orders.

Inasmuch as Pterygota, Palaeoptera, Neoptera, Paraneoptera, Holometabola, and Mecopteroidea are probably all monophyletic groups (Hennig 1981; Kristensen 1981; Carpenter and Burnham 1985), these six radiations were probably real events, but their dating and 'explosiveness'—in the sense of shortness of duration and compactness of branching pattern—need to be examined.

Radiations 1, 2, and 3, as well as the initial radiations of the apterygote groups, took place before the Namurian, when the insect fossil record begins in earnest. If, as seems most probable, flight evolved through parachuting and gliding from tall plants (Wootton and Ellington 1990), it is unlikely to have originated before the late Devonian when tree-like forms became widespread, but this still allows an ample 40 Ma for the varied and highly adapted Namurian fauna to have developed. We have no means of knowing whether the origin of flight, of wing-folding, and of piercing and sucking mouthparts were followed by rapid radiation, or gradual, progressive diversification.

We do not yet know precisely when the stems of Paraneoptera and Holometabola became distinct, but their radiations (points 4 and 5 above) have taken place well within the period from which we have fossil information. In each case, one group seems to have undergone fairly rapid expansion—in Paraneoptera the Hemiptera, in Holometabola the stem mecopteroids—while others (Psocoptera and Thysanoptera, Coleoptera and Neuroptera) have developed more slowly. Interestingly, in view of the present dominating position of Holometabola, the origins of internal wing development and the pupal stage (usually assumed to be closely linked) do not seem to have led to an immediate remarkable burst of evolutionary activity; the enormous potential of the holometabolous developmental pattern appears to have been realized progressively, over a long period of time. Even now the larvae and adults of many Coleoptera feed side by side in the same habitats, like Orthoptera and Hemiptera; and show far less resource partitioning between stages than do the hemimetabolous dragonflies, mayflies, stoneflies, and cicadas.

We must conclude that the fossil record does not particularly support Smart's model of explosive radiations following his six landmarks, though they cannot all be ruled out. For examples of rapid radiations one needs instead to look within individual orders.

2. *Intraordinal radiation patterns*

Attention has already been drawn to an array of orders—Odonata and Ephemeroptera, Plecoptera, Orthoptera and Dictyoptera, Psocoptera, Thysanoptera and Neuroptera—which became established in the Carboniferous or Lower Permian, exploiting broad resources which were already available and have continued to be so. These orders have undergone evolutionary change and expansion as the resources have changed and expanded, and in some cases—notably Ephemeroptera and Dictyoptera—have extended their habitat range and feeding capabilities.

Some of these generalizations also apply to two of the great orders, Hemiptera and Coleoptera. Both were present in the Permian, each possibly exploiting a single type of food: plant sap in the former, perhaps wood in the latter—though this last is far from certain. Both, however, underwent spectacular later radiations after breaking into new adaptive zones: in the Hemiptera suctorial predation on land, and in and on water; in Coleoptera terrestrial and aquatic predation, and a great range of variations on the themes of detritophagy and phytophagy—including, importantly, spore and pollen-feeding.

Diptera, Hymenoptera, and Lepidoptera were all later arrivals, and each broke new ground, making full use of the possibilities of holometaboly. The caterpillar-like larvae of Hymenoptera Symphyta (sawflies), and much later those of Lepidoptera, became the most important chewers of living plants, along with some Orthoptera and Coleoptera. Parasitoidy, in which a specialized apodous larva hatches and feeds in or on its prey, appears to have been a hymenopteran innovation (later adopted by several groups of Diptera) and was followed by a considerable radiation of families in the Jurassic (Rasnitsyn 1988), and of genera and species thereafter. A radiation of predatory wasps, rather similarly adapted, can be seen in the Cretaceous. The apodous larvae of Diptera allowed widespread colonization of aquatic and semi-aquatic habitats, including in due course decomposing materials, the interior of living animals and plants, and oxygen-poor substrates of many kinds, to which successive Mesozoic radiations of dipterous clades probably bear witness.

The adults of Diptera, Lepidoptera, and many Hymenoptera feed mainly on organic liquids: animal and plant exudates, decomposition products, nectar, and, in the case of Diptera, arthropod and vertebrate blood. Several groups in each order take spores and pollen; and it is this habit, which they share with many Coleoptera, which seems to have led from the Cretaceous onwards to the extraordinary co-evolution of insects with the flowering plants.

It is clear from palaeontology and from our knowledge of extant ecosystems that the importance to insect evolution of the rise of the angiosperms cannot be overemphasized. It led to the radiation of the bees (Hymenoptera, Apoidea), which depend entirely on angiosperms and are unknown before the late Cretaceous, provided a new—or at least greatly enlarged—energy-rich source of liquid food for other Hymenoptera, for Lepidoptera, and for many groups of Diptera and Coleoptera, and created a new, widely distributed, very diverse flora in an extensive range of new habitats, which phytophagous insects were able to exploit. The effects can be seen in many different groups, some of them already long-established. The late rise of Acrididae and Tettigoniidae, the two dominant modern families of Orthoptera, has already been mentioned. Weevils (Coleoptera, superfamily Curculionoidea) had been present since the Jurassic, but their enormous later radiation (they are now by far the largest superfamily of animals) must have been greatly stimulated by the angiosperm revolution. So must many families of Hemiptera and Hymenoptera-Symphyta.

Other groups, today heavily dependent on angiosperms, seem to have radiated extensively alongside them: leaf-beetles (Superfamily Chrysomeloidea), now second in number of species only to the weevils; cecidomyid, agromyzid, tephritid, psilid, and many other flies; gall-wasps (Hymenoptera, family Cynipidae). Most spectacularly Lepidoptera, which seem to have been inconspicously present since the early Jurassic, clearly underwent dramatic radiation in the late Cretaceous and Palaeogene—incidentally providing a notable example of a macroevolutionary lag (Jablonski and Bottjer, Chapter 2, this volume). Measured in described species, Lepidoptera are now the second largest order.

The enormous diversification of species reflects the fact that very many of these phytophagous insects are mono- or oligophagous, and they in turn led to a notable radiation in host-specific, or host-limited parasitoids, particularly, though not exclusively, in the hymenopteran superfamilies Ichneumonoidea, Evanioidea, Proctotrupoidea, and Chalcidoidea, all of which had first appeared in the late Jurassic or early Cretaceous, and the dipteran family Tachinidae, which is first certainly recorded in the Palaeogene.

Summary

In a taxon as large as the insects, it is not surprising that a wide spectrum of different radiation patterns can be recognized. Examples of rapid expansion into new adaptive zones are widespread. Many follow adaptations—apparent in the fossil record or deduced from other evidence—which are not obviously linked to external events: piercing and sucking mouthparts in palaeodictyopteroids and later in Hemiptera; parasitoidy in Hymenoptera; the apodous larva of Diptera; social organization in termites and several lines of Hymenoptera; asynchronous flight muscle, along with small size, many times in different groups; and perhaps the origin of flight. Other radiations follow the colonization of new habitats: fresh water, by Mesozoic Coleoptera, Heteroptera, and Diptera; leaf mesophyll by agromyzid Diptera and nepticulid moths. Each of these seems to have been limited more by chance or by evolutionary rates rather than by external opportunity. Others, however, clearly follow expansions in other groups: most notably bees and other nectar-feeding Hymenoptera, Lepidoptera, Diptera, and Coleoptera, following the angiosperm expansion; and the three linked tiers of major Cretaceous and Tertiary radiations—flowering plants, phytophagous insects, and their specialized predators and parasitoids.

The first appearances of significant insect groups are not always followed immediately by evidence for major radiations: Coleoptera and Lepidoptera in particular show pronounced macroevolutionary lags. Rather surprisingly, even holometabolous development does not seem to have led immediately to radiation on the scale which we might expect from the present dominance of Holometabola.

At out present level of palaeoecological knowledge it is difficult to recognize clear examples of ecological replacement of one group by another. The most convincing is that of palaeodictyopteroids by Hemiptera in the Permian. Whalley (1985) has drawn attention to the significant wave of extinctions at family level which seems to have taken place between the early Jurassic and the early Cretaceous. Many Cretaceous families—most of which survive, can perhaps be considered to have 'replaced' related Jurassic families.

In conclusion, it is essential again to stress the need for caution. The known fossil record is a tiny, and a profoundly and irregularly biased sample of insect biota through time. Tiny samples may, however, give valuable information: the presence of a single representative of an extant group at a particular horizon is unequivocal evidence that both the group and its sister-group were established by that time. Furthermore, although our knowledge of the nature of the bias is very incomplete, it can be significantly improved in any particular case by studying *all* the geological evidence, biogenic and non-biogenic, and can then indicate danger areas. For example, the angiosperm-linked insect radiations first become apparent in the late Cretaceous, predominantly in high-latitude localities. However Crane and Lidgard (Chapter 17, this volume) show that angiosperms first became important in low palaeolatitudes, whose insect faunas are hardly known. The Cretaceous insect radiations may well have taken place appreciably earlier than we realize.

References

Brauckmann, C. (1988). Hagen-Vorhalle, a new important Namurian insect-bearing locality (Upper Carboniferous; F.R. Germany). *Entomologia Generalis*, **14**, 73–9.

Brauckmann, C. and Zessin, W. (1989). Neue Meganeuridae aus dem Namurium von Hagen-Vorhalle (BRD) und die Phylogenie der Meganisoptera. *Deutsche entomologische Zeitschrift*, **36**, 177–215.

Burnham, L. (1978). Survey of social insects in the fossil record. *Psyche*, **85**, 85–133.

Burnham, L. (1983). Studies on Upper Carboniferous insects: 1. The Geraridae (Order Protorthoptera). *Psyche*, **90**, 1–57.

Carpenter, F. M. (1935). The Lower Permian insects of Kansas. Part 7. The Order Protoperlaria. *Proceeding of the American Academy of Arts and Sciences*, **70**, 101–46.

Carpenter, F. M. (1950). The Lower Permian insects of Kansas. Part 10. The Order Protorthoptera: the family Liomopteridae and its relatives. *Proceedings of the American Academy of Arts and Sciences*, **78**, 185–219.

Carpenter, F. M. (1960). Studies on North American Carboniferous insects. 1. The Protodonata. *Psyche*, **67**, 98–110.

Carpenter, F. M. (1966). The Lower Permian insects of Kansas. Part 11, the Orders Protorthoptera and Orthoptera. *Psyche*, **73**, 46–88.

Carpenter, F. M. (1971). Adaptations among Paleozoic insects. *Proceedings of the North American Paleontological Convention, Chicago, 1969*, **1**, 1236–51.

Carpenter, F. M. (1976). The Lower Permian insects of Kansas. Part 12. Protorthoptera (continued), Neuroptera, additional Palaeodictyoptera and families of uncertain position. *Psyche*, **83**, 336–78.

Carpenter, F. M. (1979). Lower Permian insects from Oklahoma. Part 2. Orders Ephemeroptera and Palaeodictyoptera. *Psyche*, **86**, 261–90.

Carpenter, F. M. (1987). Review of the extinct family Syntonopteridae (order uncertain). *Psyche*, **94**, 373–88.

Carpenter, F. M. and Burnham, L. (1985). The geological record of insects. *Annual Review of Earth and Planetary Science*, **13**, 297–314.

Hennig, W. (1981). *Insect phylogeny* (translated and edited by Adrian Pont, with revisionary notes by Dieter Schlee). Wiley, Chichester.

Hubbard, M. D. and Kukalová-Peck, J. (1980). Permian mayfly nymphs: new taxa and systematic characters. In *Advances in Ephemeroptera biology* (ed. J. F. Flannagan and K. E. Marshall), pp. 19–31. Plenum, New York.

Jarzembowski, E. A. (1981). An early Cretaceous termite from southern England (Isoptera: Hodotermitidae). *Systematic Entomology*, **6**, 91–6.

Jell, P. A. and Roberts, J. (ed.) (1986). *Plants and invertebrates from the Lower Cretaceous Koonwarra Fossil Bed, South Gippsland, Victoria.* Association of Australian Palaeontologists, Memoir 3. Sydney.

Keilin, D. (1944). Respiratory systems and respiratory adaptations in larvae and pupae of Diptera. *Parasitology*, **36**, 1–66.

Kluge, N. J. (1989). A question of the homology of the tracheal gills and paranotal processes of mayfly larvae and wings of insects, with reference to the taxonomy and phylogeny of the Order Ephemeroptera. In *Chteniya Pamyati Nikolaya Aleksandrovicha Kholodskogo, 1988* (ed. E. P. Narchuk), pp. 48–77. Nauka, Leningrad. (In Russian.)

Kovalev, V. G. (1987). Classification of the Diptera in the light of palaeontological data. In *Dipterous insects: systematics, morphology, ecology* (ed. E. P. Narchuk), pp. 40–8. Nauka, Leningrad. (In Russian.)

Kristensen, N. P. (1981). Phylogeny of insect orders. *Annual Review of Entomology*, **26**, 135–57.

Kukalová, J. (1968). Permian mayfly nymphs. *Psyche*, **75**, 310–27.

Kukalová, J. (1969). On the systematic position of supposed Permian beetles, Tshekardocoleidae, with a description of a new collection from Moravia. *Sborník Geologkých Věd, Paleontologie*, **11**, 139–62.

Kukalová-Peck, J. (1978). Origin and evolution of insect wings and their relation to metamorphosis, as documented by the fossil record. *Journal of Morphology*, **156**, 53–125.

Kukalová-Peck, J. (1983). Origin of the insect wing and wing articulation from the arthropod leg. *Canadian Journal of Zoology*, **61**, 1618–69.

Kukalová-Peck, J. (1985). Ephemeroid wing venation based upon new gigantic Carboniferous mayflies and basic morphology, phylogeny and metamorphosis of pterygote insects. *Canadian Journal of Zoology*, **63**, 933–55.

Kukalová-Peck, J. (1987). New Carboniferous Diplura, Monura and Thysanura, the hexapod ground plan, and the role of thoracic side-lobes in the origin of wings. *Canadian Journal of Zoology*, **65**, 2327–45.

Labandeira, C. C., Beall, B. S. and Hueber, F. M. (1988). Early insect diversification: evidence from a Lower Devonian bristletail from Quebec. *Science*, **242**, 913–16.

Michener, C. D. and Grimaldi, D. A. (1988). A *Trigona* from the late Cretaceous amber of New Jersey (Hymenoptera, Apidae, Meliponinae). *American Museum Novitates*, **2917**, 1–10.

Muller, S.W. and Campbell, A. (1954). The relative number of living and fossil species of animals. *Systematic Zoology*, **3**, 169–70.

Neville, A.C. (1978). *The biology of the arthropod cuticle.* Carolina Biology Readers, no. 103. Carolina Biological Supply Company, Burlington, North Carolina.

Ponomarenko, A.G. (1969). The historical development of archostematan beetles. *Trudy Paleontologicheskogo Instituta, Akademii Nauk USSR*, **125**, 1–240. (In Russian.)

Ponomarenko, A.G. (1977). Mesozoic beetles. *Trudy Paleontologicheskogo Instituta, Akademii Nauk USSR*, **161**, 1–204. (In Russian.)

Ponomarenko, A.G. (1985). Beetles from the Jurassic of Siberia and Southern Mongolia. In *Jurassic insects of Siberia and Mongolia* (ed. A.P. Rasnitsyn). *Trudy Paleontologicheskogo Instituta, Akademii Nauk USSR*, **211**, 47–91. (In Russian.)

Ponomarenko, A.G. (ed). (1988). *The Cretaceous biocenotic crisis and the evolution of insects*. Nauka, Moscow. (In Russian.)

Popov, Y.A. (1971). The historical development of bugs of the Infraorder Nepomorpha (Heteroptera). *Trudy Paleontologicheskogo Instituta, Akademii Nauk USSR*, **129**, 1–228. (In Russian.)

Pringle, J.W.S. (1976). The muscles and sense organs involved in insect flight. In *Insect flight*, Symposia of the Royal Entomological Society of London, no. 7, (ed. R.C. Rainey), pp. 3–15. Blackwell, Oxford.

Quartau, J.A. (1986). An overview of the paranotal theory on the origin of the insect wings. *Publicacoes do Instituto de Zoologia 'Dr Augusto Nobre', Faculdade de Ciencias do Porto*, **194**, 1–42.

Rasnitsyn, A.P. (1976). On the early evolution of insects and the origin of Pterygota. *Zhurnal Obshchei Biologii*, **37**, 543–55.

Rasnitsyn, A.P. (1981). A modified paranotal theory of insect wing origin. *Journal of Morphology*, **168**, 331–8.

Rasnitsyn, A.P. (1988). An outline of evolution of the hymenopterous insects. *Oriental insects*, **22**, 115–45.

Riek, E.F. and Kukalová-Peck, J. (1984). A new interpretation of dragonfly wing venation based on Early Upper Carboniferous fossils from Argentina (Insecta: Odonatoidea) and basic character-states in pterygote wings. *Canadian Journal of Zoology*, **62**, 1150–66.

Rohdendorf, B.B. and Rasnitsyn, A.P. (ed.) (1980). Historical development of the class of insects. *Trudy Paleontologicheskogo Instituta, Akademii Nauk USSR*, **175**, 1–269. (In Russian.)

Schaefer, C.W. (1981). The land bugs (Hemiptera: Heteroptera) and their adaptive zones. *Rostria*, **33** (Suppl.), 67–83.

Scott, A.C. and Taylor, T.N. (1983). Plant–animal interactions during the Upper Carboniferous. *Botanical Review*, **49**, 259–307.

Scott, A.C., Chaloner, W.G. and Paterson, S. (1985). Evidence of pteridophyte-arthropod interactions in the fossil record. *Proceedings of the Royal Society of Edinburgh*, **86B**, 133–40.

Sharov, A.G. (1966). *Basic arthropodan stock.* Pergamon, London.

Sharov, A.G. (1971). Morphological specializations and the mode of life of the palaeodictyopterans. In *Chteniya Pamyati Nikolaya Aleksandrovicha Kholodskogo,*

1971 (ed. E.P. Narchuk), pp. 49-63. Nauka, Leningrad. (In Russian.)
Sinichenkova, N.D., (1987). Historical development of stoneflies. *Trudy Paleontologicheskogo Instituta, Akademii Nauk USSR*, **221**, 1-143. (In Russian.)
Smart, J. (1963). Explosive evolution and the phylogeny of insects. *Proceedings of the Linnaean Society of London*, **172**, 125-6.
Smart, J. and Hughes, N.F. (1972). The insect and the plant: progressive palaeoecological integration. In *Insect/plant relationships*, Symposia of the Royal Entomological Society of London, no. 6 (ed. H.F. van Emden), pp. 143-55.
Štys, P. and Soldán, T. (1980). Retention of tracheal gills in adult Ephemeroptera and other insects. *Acta Universitatis Carolinea. Biologica*, **1978**, 409-35. Blackwell, Oxford.
Sukacheva, I.D. (1982). Historical development of the caddis order. *Trudy Paleontologicheskogo Instituta, Akademii Nauk USSR*, **197**, 1-112. (In Russian.)
Taylor, T.N. and Scott, A.C. (1983). Interactions of plants and animals during the Carboniferous. *Bioscience*, **33**, 488-93.
Toms, R.B. (1984). Were the first insects terrestrial or aquatic? *South African Journal of Science*, **80**, 319-23.
Vincent, J.F.V. (1980). Insect cuticle: a paradigm for natural composites. In *The mechanical structure of biological materials*, Society for Experimental Biology Symposium, no. 34 (ed. J.F.V. Vincent and J.D. Currey), pp. 183-210. Cambridge University Press, Cambridge.
Whalley, P.E.S. (1985). The systematics and palaeogeography of the Lower Jurassic insects of Dorset, England. *Bulletin of the British Museum (Natural History), Geology Series*, **39**, 107-189.
Whalley, P.E.S. (1986). A review of the current evidence of Lepidoptera in the Mesozoic. *Biological Journal of the Linnaean Society*, **28**, 253-71.
Whalley, P.E.S. and Jarzembowski, E.A. (1981). A new assessment of *Rhyniella*, the earliest known insect, from the Devonian of Rhynie, Scotland. *Nature, London*, **291**, 317.
Willmann, R. (1989). Evolution und phylogenetisches System der Mecoptera (Insecta: Holometabola). *Abhandlungen der Senckenbergischen Naturforschenden Gesellschaft*, **544**, 1-153.
Wootton, R.J. (1986). The origin of insect flight: where are we now? *Antenna*, **10**, 82-6.
Wootton, R.J. (1988). The historical ecology of fresh water insects: an overview. *Palaeogeography, Palaeoclimatogy, Palaeoecology*, **62**, 477-92.
Wootton, R.J. and Ellington, C.P. (1990). Biomechanics and the origin of insect flight. In *Biomechanics and evolution*, Society for Experimental Biology, Seminar Series (ed. J.M.V. Rayner). Cambridge University Press, Cambridge. (In press.)
Wootton, R.J. and Ennos, A.R. (1989). The implications of function on the origin and homologies of the dipterous wing. *Systematic Entomology*, **14**, 507-20.

10. Major evolutionary radiations in the Bryozoa

P.D. TAYLOR
Department of Palaeontology, Natural History Museum, London, UK
G.P. LARWOOD
Department of Geological Sciences, University of Durham, UK

Abstract

Three principal phases of major radiation can be distinguished during the Ordovician to Recent history of marine bryozoans; an Ordovician stenolaemate radiation; a mid-Mesozoic cyclostome radiation; and a late Mesozoic cheilostome radiation. Each radiation entailed numerical diversification and morphological divergence within apparently monophyletic clades. Primitive taxa at the base of the three radiating clades had runner-like colony-forms characterizing a fugitive ecological strategy. The spectrum of more advanced colony-forms which appeared during each radiation provide striking examples of homeomorphy through parallel evolution. Comparisons between species-rich and species-poor sister-groups permit the identification of evolutionary innovations which might have played a role in the three radiations. Structural innovations include the acquisition of a calcified skeleton during the Ordovician stenolaemate and late Mesozoic cheilostome radiations, and of free-walled skeletal organization during the Ordovician stenolaemate and mid-Mesozoic cyclostome radiations. These innovations released structural constraints on colony-form thereby enlarging available morphospace. However, significant radiation of cheilostomes appears to have been delayed until the evolution of non-planktotrophic larvae, a larval type which is likely to have reduced gene flow within and between populations, promoting accelerated rates of speciation. Similar changes in larval type may have occurred during the other two radiations, although difficulty in inferring larval type from skeletal morphology in

Major Evolutionary Radiations (ed. P.D. Taylor and G.P. Larwood), Systematics Association Special Volume No. 42, pp. 209–233, Clarendon Press, Oxford, 1990.

stenolaemates means that good supportive evidence for this hypothesis is currently lacking.

Introduction

The Bryozoa are a phylum of colonial metazoans with a rich fossil record extending back to the early Ordovician. Most modern bryozoans are marine and have calcified skeletons which provide the majority of morphological characters used in their systematics. Although research on both fossil and living bryozoans has been limited, an impression of the broad pattern of bryozoan evolution is beginning to emerge (e.g. Morozova and Viskova 1977; Schopf 1977; Taylor and Larwood 1988; McKinney and Jackson 1989).

In this chapter we use data on the changing diversity of bryozoans through the Phanerozoic to identify three phases of major evolutionary radiation in which relatively rapid cladogenesis was combined with morphological divergence. These major radiations occurred during the Ordovician, mid-Mesozoic, and late Mesozoic, and involved stenolaemates, cyclostomes, and cheilostomes, respectively. Primitive taxa at the base of each radiation had runner-like encrusting colonies. From these a wide range of more advanced taxa with diverse colony-forms evolved in parallel, providing striking examples of homeomorphy. Comparison between groups which radiated, and primitive sister-groups which did not, allow us to suggest evolutionary innovations which may have played a role in the radiations. In particular, it is hypothesized that a switch from planktotrophic to non-planktotrophic larvae facilitated or even triggered radiation by accelerating rates of speciation.

Diversity history

1. *General pattern*

Although the fossil history of bryozoans is still imperfectly known, available data on diversity changes through time reveals a pattern which we deem unlikely to be altered greatly by future discoveries or taxonomic revisions. A modified version (Taylor and Larwood 1988) of Sepkoski's (1982) family-level data base had been used to construct Fig. 10.1. This pattern of family diversity includes the following prominent features: a steep Ordovician rise in diversity following the initial appearance of fossil bryozoans; a Palaeozoic plateau of roughly 20–30 families; a sharp drop in diversity at the end of the Palaeozoic; a minor diversity peak in the Triassic; a mid-Jurassic

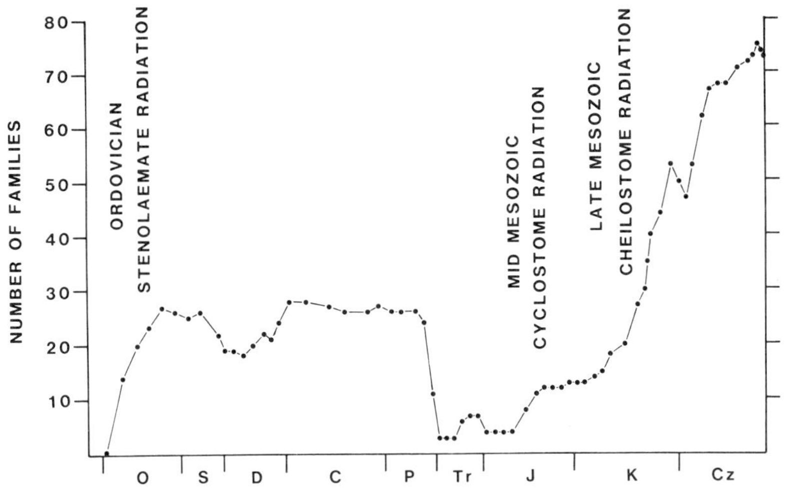

Fig. 10.1 Family diversity of fossil bryozoans (after Taylor and Larwood 1988).

increase in diversity; and a steep mid-Cretaceous to Recent climb in diversity to a present-day level of 70–80 fossilizable families.

Clearly, the details of this family-level pattern are highly dependent on taxonomic procedure. Methodological rigour has not been applied to the definition of bryozoan families and many currently used families are undoubtedly paraphyletic. As Smith and Patterson (1988) have emphasized, paraphyletic groupings are artificial constructs whose geological ranges are often truncated by pseudoextinctions rather than true extinctions. Nevertheless, Sepkoski (1989), using simulations, has shown that paraphyly does not necessarily invalidate the use of such loosely-defined families as a proxy for species diversity.

Evidence that the broad family-level changes in bryozoan diversity are not simply a taxonomic artefact comes from data on changes in species diversity in bryozoan assemblages through the Phanerozoic. In our experience these changes track one another fairly closely, i.e. times of high global family diversity coincide with times when typical diversities in species assemblages was high, and low global family diversity with low species assemblage diversities. Therefore, it is reasonable to equate rises in family diversity, especially pronounced rises, with times of species diversification when origination rates exceeded extinction rates.

Three significant diversifications are evident in the geological history of the Bryozoa. The first, during the Ordovician, entailed the radiation of several orders of stenolaemates following the first appearance of bryozoans in the fossil record. This will be referred to subsequently as the Ordovician stenolaemate radiation. A less dramatic radiation of cyclostomatous stenolaemates occurred from the late Triassic into the early Cretaceous, and will be referred to as the mid-Mesozoic cyclostome radiation. The final bryozoan group to radiate were cheilostomes which first appeared in the latest Jurassic but failed to diversify significantly until the mid-Cretaceous. Their radiation is here termed the late Mesozoic cheilostome radiaton. It should be noted that although cheilostome radiation continued into the Cainozoic (ascophoran grade cheilostomes underwent radiation mainly in the Palaeogene, see Cheetham 1971), only the Mesozoic portion of cheilostome radiation will be considered here.

2. *Ordovician stenolaemate radiation*

The early fossil history of the bryozoans is poorly known and has been confused by several fossils incorrectly determined as bryozoans, e.g. *Marcusodictyon* (see Taylor 1984), *Archaeotrypa* (see Kobluk 1984). Unequivocal bryozoans first appear in the Lower Ordovician. The earliest currently known are two genera of putative trepostomes from

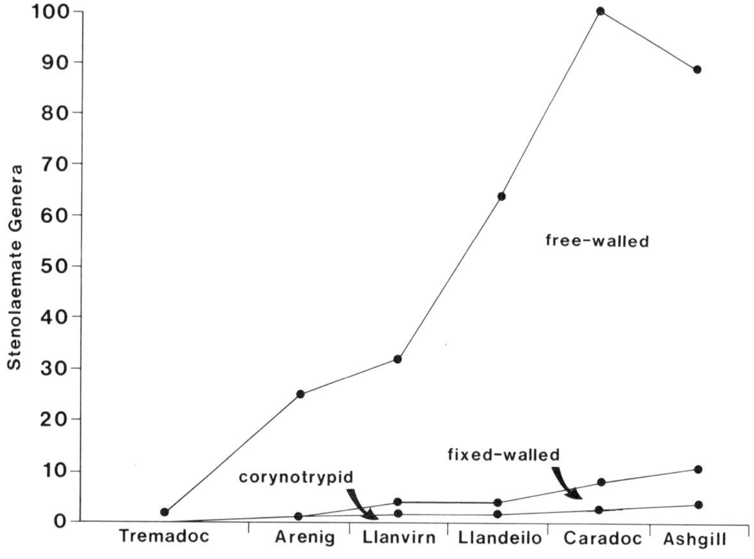

Fig. 10.2. Generic diversification of stenolaemate bryozoans during the Ordovician. Compiled using data mainly from Astrova (1978), Dzik (1981), Boardman *et al.* (1983), Lavrenteva (1985), Ross (1985), and Goryunova (1985).

the Fenshiang Fm. (mid–late Tremadoc) of Hubei Province, China (Hu and Spjeldnaes, in press). In the Arenig all five stenolaemate orders (Cyclostomata, Trepostomata, Cystoporata, Cryptostomata, and Fenestrata) are represented and 25 bryozoan genera have been recorded (Fig. 10.2). Arenig bryozoans have been reported from several geographical areas (see Taylor and Curry 1985; Taylor and Cope 1987), and moderately diverse assemblages are known in the Volkhov horizon of Esthonia which has yielded 15 genera (Goryunova 1986). Ordovician genus diversity appears to have peaked at about 100 genera in the Caradoc, before declining slightly into the Ashgill. However, high-diversity assemblages persist into the Ashgill, notably in the Cincinnati Province of North America (see Anstey 1986). It should be noted here that gymnolaemates make their first appearance as fossils in the late Ordovician where borings attributed to ctenostomes are found in calcareous substrata (Pohowsky 1978). Penetrant habit is regarded as a relatively specialized mode of life, pointing to a prior history of soft-bodied ctenostomes which has left no known fossil record.

3. *Mid-Mesozoic cyclostome radiation*

A major turnover in bryozoan faunas occurred after the end of the Palaeozoic in association with Permian and possibly Triassic extinction events. Most, if not all, of the dominant stenolaemate orders of the Palaeozoic failed to survive into the Jurassic (see Taylor and Larwood 1988). However, cyclostomes, a very minor order during the Palaeozoic, did survive and underwent a relatively modest radiation which began in the mid-Jurassic and lasted until at least the mid-Cretaceous (Fig. 10.3). The oldest cyclostomes recorded from the post-Palaeozoic come from the Upper Triassic Cassian Formation (Carnian) of the Italian Dolomites (Bizzarini and Braga 1981, 1985; note, however, that supposed cerioporine cyclostomes from this formation and elsewhere in the Triassic are calcified demosponges, see Engeser and Taylor 1989). Cyclostome diversity remains low until the Middle Jurassic where assemblages exceeding 10 species appear for the first time in the Aalenian of the Anglo-Paris Basin (Walter 1970). Overall genus diversity peaks at about 30 in the Bathonian. However, the apparent diversity decline in the late Jurassic (Fig. 10.3) is almost certainly an artefact caused by the incompleteness of the fossil record, as indicated by Upper Jurassic gaps in the ranges of many genera known from the Middle Jurassic and Lower Cretaceous. Diversity rises again in the Lower Cretaceous to attain level of 40 genera by the Albian. Assemblage diversities may reach 50 species (Pitt and Taylor 1990) in the Lower Cretaceous. It must be stressed that this pattern of radiation is based almost entirely on data from Europe; only a small number

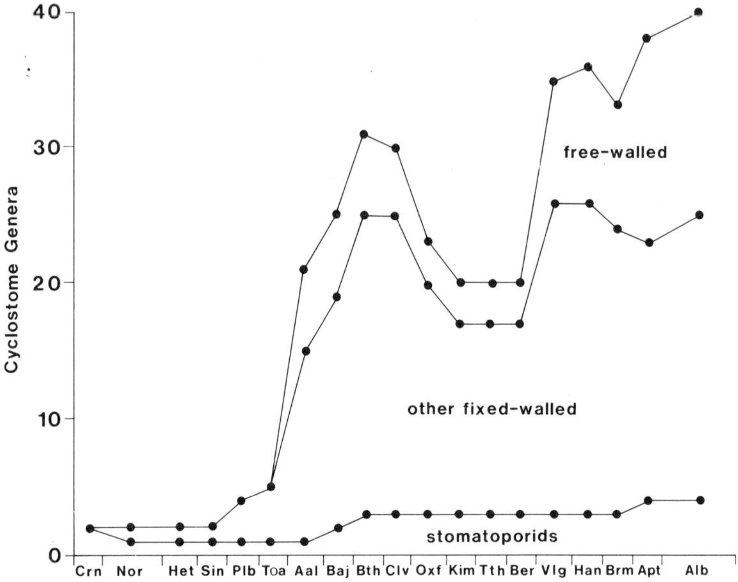

Fig. 10.3. Generic diversification of cyclostome bryozoans from late Triassic to mid-Cretaceous.

of mid-Mesozoic cyclostomes have been reported from elsewhere in the world.

4. *Late Mesozoic cheilostome radiation*

The majority of modern marine bryozoans belong to the Cheilostomata. The oldest known cheilostome, *Pyriporopsis portlandensis*, is from the late Jurassic Portland Beds (Tithonian) of southern England (Pohowsky 1973). Very few cheilostome genera have been recognized in deposits older than late Albian (Taylor 1986). Significant radiation of cheilostomes commenced in the late Albian or early Cenomanian (Fig. 10.4). Family-level global diversity rose dramatically from an Aptian value of one to a late Eocene value of about 45, after which there was an apparent levelling-off. This rise is matched by changes in species assemblage diversities: while no pre-late Albian assemblages contain more than two species of cheilostomes, assemblages of over 50 species are commonly found in the Maastrichtian chalks and associated deposits of north-western Europe. The presence of diverse cyclostomes in the same beds shows that the low diversity recorded for the first 50 Ma of cheilostome history is not simply a preservational artefact.

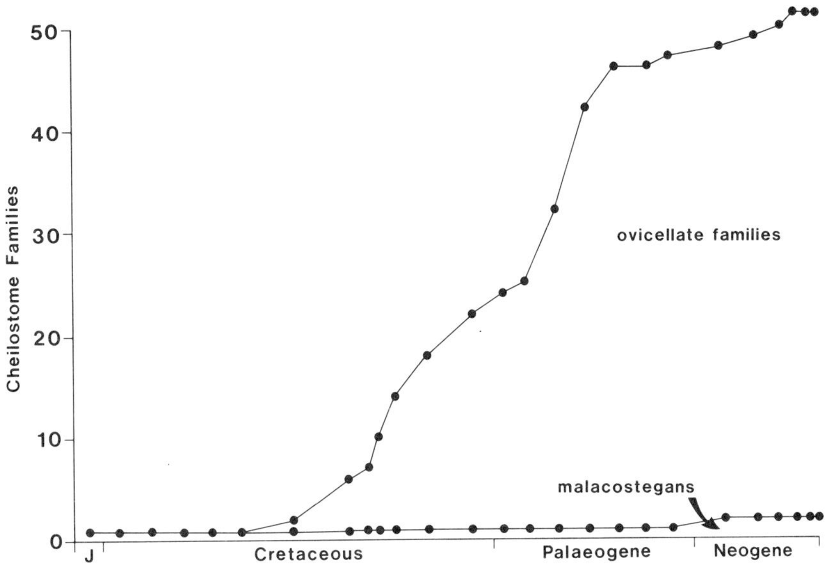

Fig. 10.4. Familial diversification of cheilostome bryozoans (after Taylor 1988*a*).

Phylogenetic relationships

1. *General features*

Identification of evolutionary innovations acquired at times of major radiation demands a knowledge of character transformations and phylogenetic relationships between the taxonomic groups involved. Unfortunately, few studies of bryozoan phylogeny have been published, and relationships between groups are poorly understood. Whereas early attempts using phenetic methods yielded conflicting and doubtful results (see Cuffey 1973; Boardman *et al.* 1983), cladistic methods are only now beginning to be applied (Carle and Ruppert 1983; d'Hondt 1986; Blake and Cuffey, in press). Therefore, the cladogram shown in Fig. 10.5 is tentative and incomplete; a firmer hypothesis of relationships demands a more comprehensive and rigorous analysis of character distribution across a greater range of taxonomic groups.

Phylactolaemates, a freshwater class of uncertain affinity (Jebram 1973; Dzik 1981; Mundy *et al.* 1981) and negligible fossil record, have been omitted from this analysis. Following Farmer (1977) and others, phoronids are shown as the sister-group of bryozoans, although

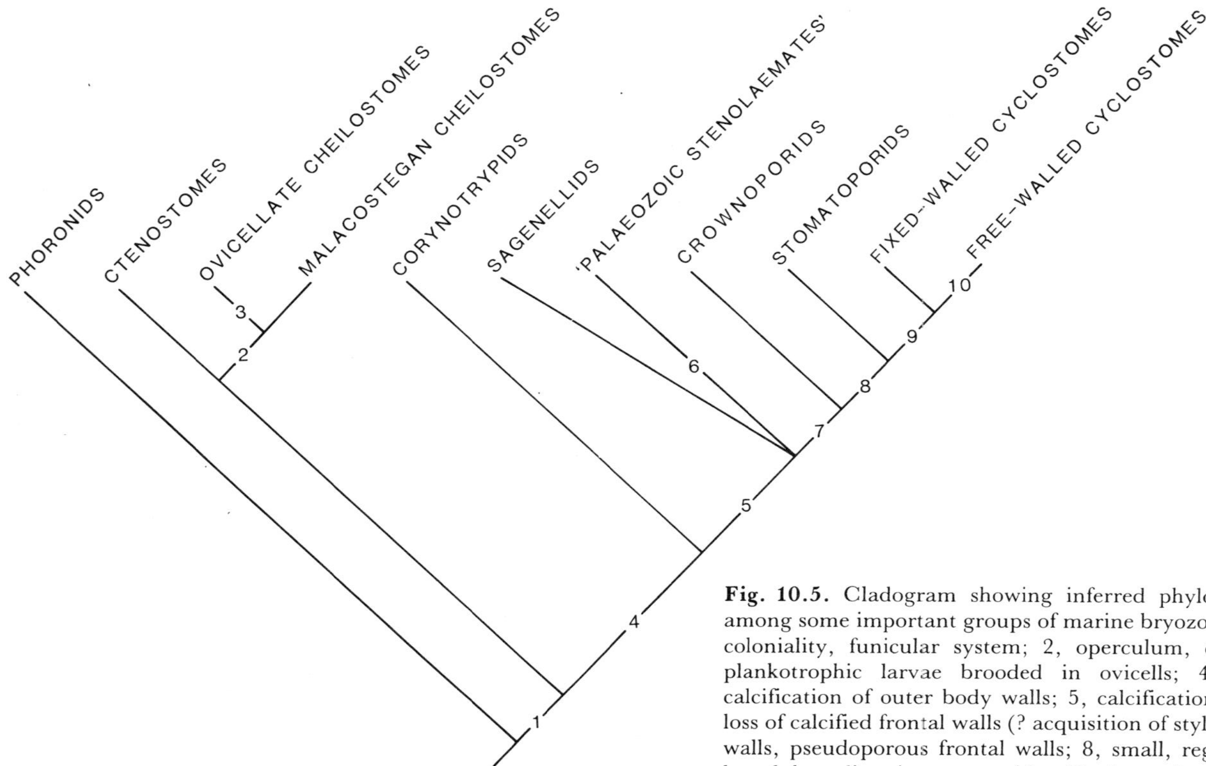

Fig. 10.5. Cladogram showing inferred phylogenetic relationships among some important groups of marine bryozoans. Apomorphies: 1, coloniality, funicular system; 2, operculum, calcification; 3, non-plankotrophic larvae brooded in ovicells; 4, membranous sac, calcification of outer body walls; 5, calcification of interior walls; 6, loss of calcified frontal walls (? acquisition of styles); 7, porous interior walls, pseudoporous frontal walls; 8, small, regular pseudopores; 9, larval brooding in gonozooids; 10, loss of calcified frontal walls. Groups lacking autapomorphies are presumed to be paraphyletic.

entoprocts are believed by some to be more closely related (Nielsen 1971). This ambiguity of outgroup does not affect the polarities of characters used to determine relationships within the Bryozoa.

Two principal groups of marine bryozoans are recognized by most bryozoan taxonomists: gymnolaemates and stenolaemates. An apomorphy for the gymnolaemates appears to be lacking and the group is probably paraphyletic. Gymnolaemates comprise two principal subgroups: ctenostomes and cheilostomes. Whereas cheilostomes are likely to be monophyletic, ctenostomes are regarded as a paraphyletic grouping of primitive marine bryozoans. Stenolaemates are monophyletic: important stenolaemate apomorphies include the membranous sac and protoecium, the latter a skeletal feature recognizable in fossils. Phylogenetic relationships within the stenolaemates have yet to be resolved adequately. Cyclostomes, as currently defined, appear to be a paraphyletic group; Boardman (1984) has even argued that they are polyphyletic. The view taken here is that corynotrypid 'cyclostomes' represent the primitive sister-group of all other stenolaemates. Relationships of the small number of remaining 'cyclostomes' found in the Palaeozoic are uncertain, except for the crownoporids (kukersellids) which are thought to be the primitive sister-group of post-Palaeozoic cyclostomes (Taylor 1985). The bulk of Palaeozoic stenolaemates belong to four 'free-walled' orders (see p. 219): trepostomes, cystoporates, cryptostomes, and fenestrates. The interrelationships of these orders have yet to be resolved, although Blake (in Boardman *et al.* 1983) has argued for a close relationship between cryptostomes and fenestrates. As there is as yet no good evidence to indicate that Palaeozoic free-walled stenolaemates are polyphyletic, the four orders are here grouped together and referred to as 'Palaeozoic stenolaemates' (in the absence of a formal taxonomic name). For the purposes of this paper, the post-Palaeozoic cyclostomes, which seem to be monophyletic (Brood 1976; cf. Boardman 1984), are divided into stomatoporids (apparently lacking brooding polymorphs), and advanced fixed-walled and free-walled cyclostomes both possessing brooding polymorphs. Stomatoporids and advanced fixed-walled cyclostomes are regarded as paraphyletic, whereas free-walled cyclostomes may be either monophyletic or, more probably, polyphyletic.

2. *Sister-group comparisons*

A useful approach towards identifying causal factors responsible for major evolutionary radiations is sister-group comparison. Here, the characters of a group which underwent radiation (and is therefore diverse) are compared with those of a sister-group which failed to radiate. This method has been used recently by Mitter *et al.* (1988) to show that the evolution of phytophagy promoted diversification in insects—clades with phytophagy are more diverse at the present day than their sister-groups which lack phytophagy.

Ideally, sister-group comparisons of diversity should be made between two monophyletic sister-groups (clades) and these, by definition, share a common ancestor and are therefore of equivalent age. However, comparisons between an advanced monophyletic clade of high diversity and its primitive paraphyletic sister-group of low diversity can also be made without biasing the result in favour of the hypothesis being tested. This is because the advanced, high-diversity clade must be younger than the primitive, low-diversity clade. Hence the difference in diversity cannot be explained by a greater longevity of the advanced clade.

Interpretations of the three radiations

1. *Ordovician stenolaemate radiation*

Ctenostomes are regarded as the primitive sister-group of stenolaemates (Fig. 10.5). Innovations pertinent to the origin and early radiation of stenolaemates from a 'ctenostome' stem-group are:

1. *Membranous sac and calcified skeleton*. The principal synapomorphy of the stenolaemates is the membranous sac which is interpreted as the detached mesoderm of the cystid wall (Nielsen and Pedersen 1979). A membranous sac is present in all investigated modern stenolaemates, and the remains of a possible membranous sac have been described in a Carboniferous stenolaemate (McKinney 1969). An associated synapomorphy is the calcified skeleton, also ubiquitous in modern stenolaemates.

2. *Calcified interior walls*. Corynotrypids are primitive among stenolaemates in having calcification of only exterior walls (walls secreted from the inside by an epithelium and covered on the outside by organic cuticle, see Boardman *et al.* 1983). These calcified exterior walls are of two kinds; basal walls ('basal laminae') which are normally adpressed to the substratum, and frontal walls located on the upper surface of the colony (Fig. 10.6). Stenolaemates more advanced than corynotrypids have the apomorphy of calcified interior walls (walls secreted from both sides by an enveloping epithelium and devoid of a cuticle). Forms in which the basal, frontal, and interior walls are all calcified have a fixed-walled skeletal organization in the terminology of Boardman *et al.* (1983) (= single-walled or stictocystic of previous authors). Fixed-walled organization is found in sagenellid and

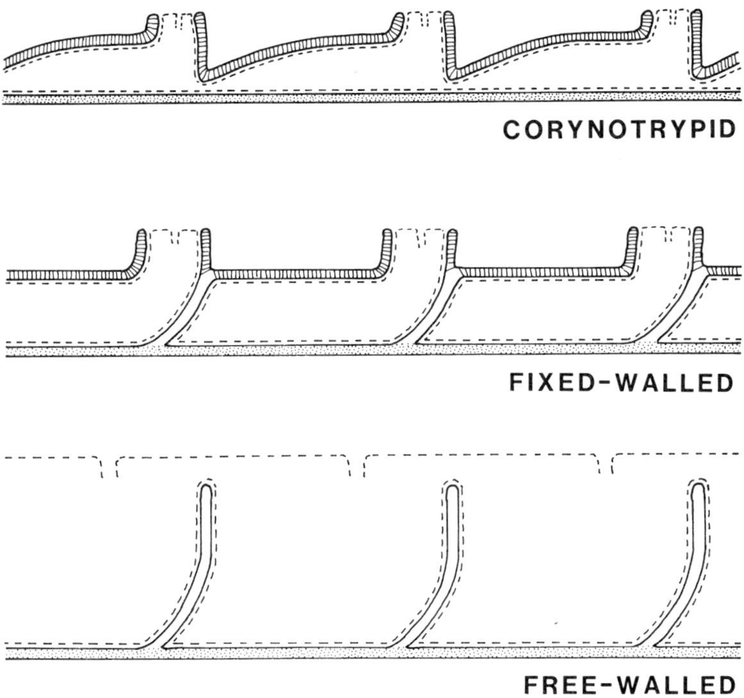

Fig. 10.6. The three main types of skeletal organization found in Palaeozoic stenolaemates depicted in diagrammatic longitudinal sections. Polypides are omitted; membranes are shown as dashed lines. Basal exterior walls are stippled, frontal exterior walls hatched, and interior walls unshaded.

crownoporid cyclostomes. Calcified interior walls provide a barrier between contiguous zooids. In most Palaeozoic stenolaemates, this barrier is complete, but in crownoporid cyclostomes (Buttler 1989) and some ceramoporid cystoporates (Utgaard in Boardman *et al.* 1983) interzooidal pores breach the calcified interior walls.

3. *Loss of calcified frontal walls.* Trepostomes, cystoporates, cryptostomes, and fenestrates, here grouped together as 'Palaeozoic stenolaemates', lack frontal exterior walls (Fig. 10.6) and have a free-walled organization (= double-walled or coelocystic). Loss of frontal walls is regarded as the principal synapomorphy of 'Palaeozoic stenolaemates'. This loss can be interpreted as zooidal paedomorphosis (Larwood and Taylor 1979; cf. Anstey 1987), frontal wall formation being deleted during phylogeny from the late ontogeny of the zooidal

skeleton. Also widely distributed within the 'Palaeozoic stenolaemates', but absent or rare (Brood 1973) in cyclostomes, are skeletal rods or styles. Styles may be an additional synapomorphy of the 'Palaeozoic stenolaemates'.

On functional morphological grounds there are good reasons to believe that the membranous sac and calcified skeleton were major innovations acquired more or less simultaneously during the origin of the stenolaemates. Their appearance is related to a change in the mode of tentacle protrusion. All bryozoans employ a hydrostatic system for everting the tentacle sheath which bears the tentacles (see Taylor 1981). Contraction of muscles attached to a membrane enclosing fluid in a body cavity (coelom and/or pseudocoelom) forces fluid into the introverted tentacle sheath causing it to evert. In soft-bodied ctenostomes these muscles act on a typically large area of the outer body wall which is flexible and lacks calcification. In stenolaemates the muscles act mostly on the membranous sac and body walls can be made rigid by calcification. Indeed, the system will only be mechanically efficient if these walls are rigidly calcified.

The acquisition of calcification in association with this novel mode of tentacle eversion extends beyond its consequence for the preservation of bryozoans in the fossil record. Because the outer body walls of ctenostome zooids deform during tentacle eversion, only limited contiguity is possible between adjacent zooids. Therefore, most ctenostome colonies have uniserial or biserial arrangements of zooids. With the exception of some genera with gelatinous colonies (e.g. *Alcyonidium*), ctenostomes have delicate colonies with encrusting runner-like or erect vine-like growth forms. In contrast, the possibility of extensive contiguity of zooids combined with the strength provided by calcification means that stenolaemates can develop a much greater variety of colony-forms. Therefore, the membranous sac can be viewed as an important structural innovation which released previous constraints on bryozoan colony-form (Larwood and Taylor 1979).

Nevertheless, the low taxonomic diversity and morphological variety of corynotrypids (Fig. 10.2), the first bryozoans presumed to have possessed a membranous sac, implies that significant radiation demanded further evolutionary innovations. The next apomorphy developed crownward from the corynotrypids was the acquisition of calcified interior walls (i.e. fixed-walled organization *s.s.*) which characterizes Palaeozoic cyclostomes minus corynotrypids (i.e. sagenellids and crownoporids, see Taylor 1985). Like the corynotrypids these achieved only a very limited diversity during the Ordovician (Fig. 10.2) and throughout the rest of the Palaeozoic (Brood 1973, 1975). The advanced sister-group of the fixed-walled cyclostomes are the 'Palaeozoic stenolaemates' comprising the trepostomes, cystoporates, cryptostomes, and fenestrates with a free-walled organization. As all four of these orders contributed significantly to the Ordovician radiation, especially trepostomes and cryptostomes, it is reasonable to suspect that free-walled organization was a key innovation. The functional significance of free-walled organization is twofold: it permits soft tissue continuity between zooids via epithelia and hypostegal pseudocoel which pass over the distal ends of the interior walls; and it allows prolonged distal growth of interior walls which are not halted by the secretion of a frontal wall. While such soft tissue continuity may have been important in the interzooidal translocation of nutrients and of substances controlling aspects of development and metabolism, prolonged distal growth of interior walls permitted the formation of long zooids and robust colonies. Therefore, free-walled organization further released constraints on colony growth, opening new morphospace.

The possibility of a switch from planktotrophic to non-planktotrophic larvae having a role in the Ordovician radiation deserves serious consideration in the light of events during cheilostome radiation (p. 225). Whereas outgroup comparison with phoronids suggests that planktotrophy is the primitive condition for bryozoans, all modern stenolaemates which have been studied have non-planktotrophic larvae (e.g. Nielsen 1970). Skeletal structures interpreted as larval brood chambers, and presumably signifying non-planktotrophy, have been described in various free-walled Palaeozoic stenolaemates. The earliest may be present in rhinoporid cystoporates which date from the late Ordovician (Buttler, in press), but only in Upper Palaeozoic fenestrates do these inferred brood chambers become reasonably common (Bancroft 1986; Morozova 1988). Unfortunately, with no good analogues among Recent stenolaemates, it is not known whether the absence of such structures in the bulk of Palaeozoic stenolaemates indicates that they were planktotrophs or, alternatively, non-planktotrophs without skeletal brood chambers. Assuming that non-planktotrophy evolved only once in Palaeozoic stenolaemates, the wide taxonomic distribution of inferred brood chambers implies that this occurred early in their history, before divergence of the major groups. Study of the ancestrulae of Palaeozoic stenolaemates offers some hope for establishing criteria for distinguishing between planktotrophs and non-planktotrophs.

Environmental aspects of the Ordovician stenolaemate radiation are poorly understood. If the hypothesis is correct that major innovations commonly originate in nearshore settings (see Jablonski and Bottjer, Chapter 2, this volume), the first stenolaemates may be expected to have inhabited such environments. Unfortunately, nearshore environments tend to be relatively underrepresented in the fossil record, and early Palaeozoic examples often comprise decalcified, clastic sediments in which bryozoans, especially inconspicuous encrusting colonies, are rarely searched for and are easily overlooked (Taylor and Cope 1987).

2. *Mid-Mesozoic cyclostome radiation*

Crownoporids are regarded as the primitive sister-group of all post-Palaeozoic cyclostomes (Brood 1976; Taylor 1985; Anstey 1986), on the basis of the synapomorphies of porous interior walls and pseudoporous exterior walls. Unfortunately, very little is known about crownoporids. Neither crownoporids nor their advanced sister-group, stomatoporids, have been recorded from the late Palaeozoic, but it must be assumed that one (or both) of these taxa survived from the Palaeozoic into the Mesozoic and gave rise to the post-Palaeozoic cyclostomes (Taylor 1985). Starting from a crownoporid ancestor, evolutionary innovations associated with the mid-Mesozoic cyclostome radiation are:

1. *Pseudopore size reduction and regularity*. The most primitive post-Palaeozoic cyclostomes, stomatoporids, apparently differ from crownoporids only in having pseudopores which are smaller, more uniform in size and more regularly distributed than those of crownoporids (see Buttler 1989).

2. *Gonozooids for larval brooding*. Like crownoporids, true stomatoporids appear to lack polymorphic zooids (gonozooids or gynozooids) for larval brooding. However, the description in a recent uniserial cyclostome assigned to *Stomatopora* of gonozooids budded high on the peristomes of the autozooids (Harmelin 1974) raises the possibility that these sturctures may have been present in fossil stomatoporids but not preserved due to loss of the delicate peristomes. An argument against this idea is that peristomial budding of gonozooids would appear to be an advanced feature; primitive gonozooids are expected to substitute for autozooids in the normal budding sequence, as in the early Jurassic genus *Microeciella* (Taylor and Seqeuiros 1982). Therefore, the presumption is made that the lack of gonozooids in fossil stomatoporids is a primary feature. As with Palaeozoic stenolaemates (p. 221), it is uncertain whether absence of recognizable brooding structures indicates lack of brooding (i.e. planktotrophy) or, alternatively, brooding without skeletal expression. The functional morphology of stomatoporids is compatible with (but does not prove) an hypothesis of non-brooding (Taylor 1979; see also Jablonski and Lutz 1983). The earliest known cyclostome with gonozooids is

Reptomultisparsa hybensis (Prantl), an encrusting tubuloporine from the late Triassic of Czechoslovakia (Taylor and Michalik, in press).

3. *Loss of calcified frontal walls*. This innovation closely parallels that seen at the origin of 'Palaeozoic stenolaemates'. However, accumulating evidence suggests that a free-walled organization evolved more than once in post-Palaeozoic cyclostomes; for example, melicerititids, a good monophyletic clade distinguished by their operculate zooids, consist primitively of fixed-walled genera but include two free-walled genera, *Foricula* and *Biforicula*. Free-walled organization is first recorded in the post-Palaeozoic in the apparent cerioporine *Heteropora tipperi* from the Toarcian of British Columbia (Henderson and Perry 1981).

Stomatoporids diversified very little during the mid-Mesozoic (Fig. 10.3). Most species were runner-like encrusters showing slight variations in zooidal budding pattern (Illies 1973, 1976). A few erect species, typically referred to the genus *Entalophora s.l.*, may also be stomatoporids based on the absence of gonozooids in large samples of branch fragments. The mid-Mesozoic radiation occurred almost entirely among 'advanced' cyclostomes with gonozooids (Fig. 10.3). Although some of the diversity of these advanced cyclostomes is due to taxonomic division made possible by gonozooidal characters, they undoubtedly evolved a considerably greater variety of colony-forms than is known among stomatoporids. Sheet-like encrusters and dendroid and foliaceous erect forms were all present by Middle Jurassic times (Walter 1970). Among these advanced cyclostomes, fixed-walled tubuloporines were the principal group to radiate (Fig. 10.3). Free-walled cyclostomes, initially represented only by cerioporines, made only a minor contribution to the Jurassic part of the mid-Mesozoic radiation, but diversified rather more in the early Cretaceous with the origin of the cancellates. In contrast to the Ordovician stenolaemate radiation, therefore, acquisition of a free-walled organization is not strongly implicated in the mid-Mesozoic cyclostome radiation. Possibly the presence of porous interior walls in post-Palaeozoic cyclostomes negated the advantages of a free-walled organization with respect to interzooidal communication. Also of note is the fact that some Mesozoic fixed-walled cyclostomes (e.g. *Terebellaria*, see Taylor 1978) were able to use multilamellar overgrowth to construct colonies equal in thickness to free-walled forms, thus bypassing constraints imposed by limited distal growth of interior walls.

The key innovation during the mid-Mesozoic cyclostome radiation may have been the evolution of gonozooids for larval brooding. It seems possible that the appearance of gonozooids signalled a switch from planktotrophic to non-planktotrophic larvae, with the same implication for speciation rates as discussed (p. 225) for cheilostomes. However, this hypothesis must remain speculative until reliable criteria have been established for inferring larval type in fossil cyclostomes.

3. *Late Mesozoic cheilostome radiation*

Ctenostomes are regarded as the primitive sister-group of cheilostomes (e.g. Banta 1975; Cheetham and Cook in Boardman *et al.* 1983; Taylor 1990; but see Dzik 1975 for an alternative hypothesis). Bioimmured Jurassic ctenostomes are of the general morphological type predicted for the ancestors of cheilostomes and include a genus, *Cardoarachnidium*, which is interpreted as a stem-group cheilostome because it possesses one of the two synapomorphies (operculate zooids) of crown-group cheilostomes (Taylor 1990). Innovations relevant to the late Mesozoic cheilostome radiation are:

1. *Calcification*. All cheilostomes, by definition, have calcification of the zooid body walls. In the oldest fossil cheilostomes the vertical outer body walls of the zooids are thickly calcified. Thin calcification, as developed in *Membranipora* and *Flustra*, appears to be a secondary feature.

2. *Larval brooding in ovicells*. Primitive cheilostomes ('malacostegans') possess planktotrophic cyphonautes larvae which are not brooded by the parent colony. Advanced cheilostomes have non-planktotrophic coronate larvae which are brooded by the parent colony. Larval brooding usually occurs in ovicells, hood-like skeletal brood chambers located distal of the orifice. A few other types of larval brooding are found in cheilostomes but are likely to be secondarily derived from ovicellular brooding (Taylor 1988*a*). Whereas the oldest cheilostomes known are from the late Jurassic, species with ovicells are not recorded until the mid-Cretaceous, notably *Wilbertopora mutabilis* from the late Albian–Cenomanian of Texas (Cheetham 1954, 1975). All pre-late Albian cheilostomes are morphologically similar to modern species of planktotrophic malacostegans (Taylor 1986, 1987*a*). Therefore, it seems likely that the mid-Cretaceous appearance of ovicells indicated the origin of non-planktotrophy in cheilostomes. Firm evidence for a polyphyletic origin of non-planktotrophy is currently lacking in cheilostomes (cf. molluscan groups), although the aberrant brooding method and unusual morphology of aeteids, a small group of lightly mineralized species with a poor fossil record, may point to an independent origin of non-planktotrophy from the bulk of the cheilostomes.

Significantly, the origin of calcification did not lead to an immediate radiation of cheilostomes. Radiation appears to have been delayed until the origin of non-planktotrophy as inferred by ovicells. Only non-planktotrophs diversified; even today, planktotrophs remain species-poor and of limited morphological variety (Taylor 1987*a*). There are likely to be important differences in population structure between planktotrophic and non-planktotrophic species of cheilostomes because of differences in dispersal. The larvae of planktotrophs are long-lived (4 weeks has been estimated for *Membranipora membranacea* larvae by Yoshioka 1982), whereas those of non-planktotrophs are short-lived (generally less than 24 hours, e.g. Cook 1985). Therefore, larval dispersal and gene flow both within and between populations are likely to be poorer in non-planktotrophs than in planktotrophs, as occurs in other benthic invertebrates (see Jablonski 1986). Some supportive evidence for these inferences is available for cheilostomes (see Taylor 1988*a*), but more data is needed on comparative geographical genetic variation in the two groups. Non-planktotrophic clades should have higher rates of speciation as a result of poorer gene flow which promotes various modes of allopatric and quasi-sympatric speciation (Jablonski 1986). On the basis of this prediction and the evolutionary pattern shown by cheilostomes, Taylor (1988*a*) proposed that cheilostome radiation may have been triggered by the origin of non-planktotrophy.

Bottjer and Jablonski (1988) have traced the environmental spread of early cheilostomes starting from the nearshore Tithonian species *Pyriporopsis portlandensis*. Significantly, 'malacostegan' cheilostomes apparently spread into offshore environments during the early Cretaceous, i.e. prior to the major radiation of non-planktotrophs in the mid- to late Cretaceous. At the present day, 'malacostegans' have a fairly wide environmental distribution as a group, although the majority of species are found in nearshore environments—many are algal epiphytes in intertidal habitats, while others inhabit brackish water.

Parallel evolution in colony-form

1. *Primitive colony-forms*

Primitive taxa at the base of the three major bryozoan radiations have strikingly similar colony-forms. In all cases, their colonies are encrusting with runner-like morphologies—zooids are arranged in uniserial or oligoserial branches which ramify across the surface of the substratum. This colony-form is shared by the Ordovician primitive stenolaemate *Corynotrypa* (see Bassler 1911), the mid-Mesozoic primitive cyclostome *Stomatopora* (see Illies 1973, 1976) and late Mesozoic primitive cheilostomes such as *Pyriporopsis* (see Pohowsky 1973; Taylor

1987*a*) and *Herpetopora* (see Taylor 1988*b*). Functional analysis of runner-like colonies indicates a fugitive growth strategy (Buss 1979; Jackson 1979; Bishop 1989). Runners 'disperse' their zooids widely over the substratum, thus increasing the probability that some part of the colony will locate a spatial refuge where chances of survival are enhanced. However, runners are highly vulnerable to destruction by predators, overgrowth by competitors for substratum space, and other sources of mortality. More so than in other colony-forms, long-term colony survival therefore depends upon the chance location of spatial refuges.

In general, runner-like colonies should be favoured over other colony-forms at high levels of disturbance because of their ability to encounter spatial refuges from such disturbance (McKinney and Jackson 1989). At least some runners may also be opportunists well-equipped to exploit temporarily unstable habitats through rapid recruitment by larvae and early sexual reproduction (although these attributes occur in other colony-forms as well, see Bishop 1989). Jackson (1979, p. 543) explains the fugitive strategy characterizing early representatives of new bryozoan groups as a result of the strong effect of competition with previously established groups. It may also be that the probability of radiation is greater starting from such less derived taxa than from highly derived forms.

2. *Advanced colony-forms*

Starting from runner-like ancestors, each of the three major radiations generated in parallel many advanced colony-forms. Closely homeomorphic colonies belonging to the three radiating clades can be distinguished from one another using characters at the zooidal level and below. To give just one example or parallelism, a recurrent morphology has been that of a narrow-branched tree with branches linked to form a planar, reticulate net-like colony. Feeding zooids open only on one side of the net, and the gaps between branches are termed fenestrules. This has been termed 'fenestrate' or 'reteporiform', and is included within the broader unilaminate category of McKinney (1986*a*,*b*). Homeomorphy of fenestrate colonies is well-known (e.g. Bigey 1981). Many genera of the Palaeozoic Order Fenestrata possess this colony-form, as do a few species of Mesozoic tubuloporine cyclostomes (see Taylor 1987*b*), several species of the Cainozoic cyclostome *Hornera* (see Mongereau 1972), and representatives of the Cainozoic cheilostome Family Sertellidae (see Hass 1948). The fenestrate colony-form reflects an efficient pattern of feeding in which there is a one-way flow of water through the fenestrules (Cowen and Rider 1972; Cook 1977; McKinney 1986*a*,*b*; McKinney and Jackson 1989). An increase in the relative proportion of fenestrate and other unilaminate colonies in bryozoan assemblages through both the Palaeozoic and the post-Palaeozoic has been shown by McKinney (1986*a*,*b*) and is interpreted as a trend resulting from progressive adaptive evolution (see McKinney and Jackson 1989).

Conclusions

Three major radiations of monophyletic clades can be recognized in the Bryozoa: an Ordovician radiation of stenolaemates; a mid-Mesozoic radiation of cyclostomes; and a late Mesozoic radiation of cheilostomes. Morphological divergence accompanied an increase in taxonomic diversity during these radiations. Each radiation began with primitive taxa having runner-like colonies, and generated an array of more complex and advanced colony-forms in parallel. Ecologically, these changes undoubtedly reflect shifts from purely fugitive species to species having a wide range of strategies.

Various internal factors—evolutionary innovations—which might have played a role in the three radiations can be suggested. Arguments in favour of specific innovations become stronger if the innovations are found in more than one radiation. Common to more than one radiation were the acquisition of a calcified skeleton (Ordovician stenolaemates and late Mesozoic cheilostomes), and of free-walled organization (Ordovician stenolaemates and mid-Mesozoic cyclostomes). Both of these innovations would have released constraints on colony-form, thereby allowing expansion into new morphospace. However, the origin of calcification in cheilostomes did not lead to an immediate radiation, and cyclostomes with a free-walled organization failed to diversify to the same extent as their fixed-walled primitive sister-group.

The origin of non-planktotrophic larvae is a third innovation which might possibly have characterized all three radiations. There is a good temporal correlation between the appearance of non-planktotrophy, inferred from the fossil history of brooding ovicells, and the late Mesozoic cheilostome radiation, but empirical evidence for the acquisition of non-planktotrophy during the other two radiations is poor or lacking. Nevertheless, a larval trigger for bryozoan radiations deserves serious consideration and has particular appeal because it is likely to have had a direct effect on speciation rate.

A scenario for major bryozoan radiations can be constructed as follows:

(1) origin (or survival following mass extinction) of a planktotrophic group with runner-like, encrusting colonies existing as fugitive strategists mainly in highly disturbed nearshore environments;

(2) origin of non-planktotropthic larvae resulting in increased speciation rates, and leading to the development of diverse, non-fugitive colony forms which spread into more offshore environments.

We acknowledge that bryozoan evolution has not been played out in an environmental vacuum; bryozoans must surely have been responsive to physical and biotic changes which were going on around them. However, relating these changes convincingly to radiation patterns is difficult at present. Resolving the relative roles of internal versus external processes in bryozoan radiation represents an important challenge for future research.

References

Anstey, R. L. (1986). Bryozoan provinces and patterns of generic evolution and extinction in the late Ordovician of North America. *Lethaia*, **14**, 33–51.

Anstey, R. L. (1987). Astogeny and phylogeny: evolutionary heterochrony in Paleozoic bryozoans. *Paleobiology*, **13**, 20–43.

Astrova, G. G. (1978). The history of development, systematics and phylogeny of the bryozoan order Trepostomata. *Trudy Paleontologicheskogo Instituta*, **169**, 1–240. (In Russian.)

Bancroft, A. J. (1986). Ovicells in the Palaeozoic bryozoan Order Fenestrata. *Palaeontology*, **29**, 155–64.

Banta, W. C. (1975). Origin and early evolution of cheilostome Bryozoa. *Documents des Laboratoires de Géologie de la Faculte des Sciences de Lyon*, hors série **3**, 565–82.

Bassler, R. S. (1911). *Corynotrypa*, a new genus of tubuliporoid Bryozoa. *Proceedings of the United States National Museum*, **39**, 497–527.

Bigey, F. (1981). Fenestellides Paleozoiques et Reteporides Actuels (Bryozoaires). Convergence morphologique et analogie ecologique. *Annales de Paleontologie (Invertebres)*, **7**(2), 59–80.

Bishop, J. D. (1989). Colony form and the exploitation of spatial refuges by encrusting Bryozoa. *Biological Reviews of the Cambridge Philosophical Society*, **64**, 197–218.

Bizzarini, F. and Braga, G. (1981). Prima segnelazione del genere *Stomatopora* (Bryozoa Cyclostomata) nel Trias superiore delle Dolomiti orientali (Italia). *Lavori Societa Veneziana di Scienze Naturali*, **6**, 135–44.

Bizzarini, F. and Braga, G. (1985). *Braiesopora voigti* n.gen. n.sp. (cyclostome bryozoan) in the S. Cassiano Formation in the Eastern Alps (Italy). In *Bryozoa: Ordovician to Recent* (ed. C. Nielsen and G. P. Larwood), pp. 25–33. Olsen and Olsen, Fredensborg.

Blake, D. B. and Cuffey, R. J. (in press). Cladistic analysis of the phylum Bryozoa. *Mémoire hors série Societé des Sciences Naturelles de l'Ouest de la France.*

Boardman, R. S. (1984). Origin of the post-Triassic Stenolaemata (Bryozoa): a taxonomic oversight. *Journal of Paleontology*, **58**, 19–39.

Boardman, R.S., Cheetham, A.H., Blake, D.B., Utgaard, J., Karklins, O.L., Cook, P.L., Sandberg, P.A., Lutaud, G. and Wood, T.S. (1983). Bryozoa (Revised). Volume 1. In *Treatise on invertebrate paleontology* (ed. R.A. Robison), pp. GI-G625. Geological Society of America, Boulder, Colorado and University of Kansas Press, Kansas.

Bottjer, D.J. and Jablonski, D. (1988). Paleoenvironmental patterns in the evolution of post-Paleozoic benthic marine invertebrates. *Palaios*, **3**, 540-60.

Brood, K. (1973). Palaeozoic Cyclostomata (a preliminary report). In *Living and fossil Bryozoa* (ed. G.P. Larwood), pp. 247-56. Academic Press, London.

Brood, K. (1975). Cyclostomatous Bryozoa from the Silurian of Gotland. *Stockholm Contributions in Geology*, **28**, 45-119.

Brood, K. (1976). Wall structure and evolution in cyclostome Bryozoa. *Lethaia*, **9**, 377-89.

Buss, L.W. (1979). Habitat selection, directional growth, and spatial refuges: why colonial animals have more hiding places. In *Biology and systematics of colonial organisms* (ed. G.P. Larwood and B.R. Rosen), pp. 459-97. Academic Press, London.

Buttler, C.J. (1989). New information on the morphology and skeletal ultrastructure of the Ordovician cyclostome bryozoan *Kukersella* Toots, 1952. *Paläontologisches Zeitschrift*, **63**, 215-27.

Buttler, C.J. (in press). Possible brooding structures in early Palaezoic cystoporate bryozoans. *Mémoire hors série Societé des Sciences Naturelles de l'Ouest de la France.*

Carle, K.J. and Ruppert, E.E. (1983). Comparative ultrastructure of the bryozoan funiculus: A blood vessel homologue. *Zeitschrift für Zoologische Systematik und Evolutionsforschung*, **21**(3), 181-93.

Cheetham, A.H. (1954). A new early Cretaceous cheilostome bryozoan from Texas. *Journal of Paleontology*, **28**, 177-84.

Cheetham, A.H. (1971). Functional morphology and biofacies distribution of cheilostome Bryozoa in the Danian Stage (Paleocene) of southern Scandinavia. *Smithsonian Contributions to Paleobiology*, **6**, 1-87.

Cheetham, A.H. (1975). Taxonomic significance of autozooid size and shape in some early multiserial cheilostomes from the Gulf Coast of the USA. *Documents des Laboratories de Géologie de la Faculté des Sciences de Lyon*, hors série **3**, 547-64.

Cook, P.L. (1977). Colony-wide water currents in living Bryozoa. *Cahiers de Biologie Marine*, **18**, 31-47.

Cook, P.L. (1985). Bryozoa from Ghana. *Zoologische Wetenschappen Musee Royal l'Afrique centrale Tervuren, Belgique*, **238**, 1-315.

Cowen, R. and Rider, J. (1972). Functional analysis of fenestellid bryozoan colonies. *Lethaia*, **5**, 145-64.

Cuffey, R.J. (1973). An improved classification, based upon numerical-taxonomic analyses, for the higher taxa of entoproct and ectoproct bryozoans. In *Living and fossil Bryozoa* (ed. G.P. Larwood), pp. 549-64. Academic Press, London.

d'Hondt, J.-L. (1986). Etat de connaissances sur la position phylogénétique et l'évolution des bryozoaires. *Bollettino di Zoologia*, **53**, 247-69.

Dzik, J. (1975). The origin and early phylogeny of the cheilostomatous Bryozoa. *Acta Paleontologica Polonica*, **20**, 395–423.

Dzik, J. (1981). Evolutionary relationships of the early Palaezoic 'cyclostomatous' Bryozoa. *Palaeontology*, **24**, 825–61.

Engeser, T.S. and Taylor, P.D. (1989). Supposed Triassic bryozoans in the Klipstein Collection from the Italian Dolomites redescribed as calcified demosponges. *Bulletin of the British Museum (Natural History), Geology Series*, **45**, 39–55.

Farmer, J.D. (1977). An adaptive model for the evolution of the ectoproct life cycle. In *Biology of bryozoans* (ed. R.M. Woollacott and R.L. Zimmer), pp. 487–517. Academic Press, New York.

Goryunova, R.V. (1985). Morphology, systematics and phylogeny of Bryozoa (Order Rhabdomesida). *Trudy Paleontologicheskogo Instituta*, **208**, 1–152. (In Russian.)

Goryunova, R.V. (1986). Early Ordovician Bryozoa: their systematics and distribution. *Abstracts of the Sixth All-Union Colloqium on fossil and Recent bryozoans*, pp. 15–6. Palaeontological Institute of the Academy of Sciences of the USSR, Moscow. (In Russian.)

Harmelin, J.-G. (1974). A propos d'une forme stomatoporienne typique, *Stomatopora gingrina* Jullien, 1882 (Bryozoaires Cyclostomes), et de son gonozoide. *Journal of Natural History*, **8**, 1–9.

Hass, H. (1948). Beitrag zur Kenntis der Reteporiden. *Zoologica. Stuttgart*, **101**, 1–138.

Henderson, C.M. and Perry, D.G. (1981). A Lower Jurassic heteroporid bryozoan and associated biota, Turnagain Lake, British Columbia. *Canadian Journal of Earth Sciences*, **18**, 457–68.

Hu Zhoa-Xun and Spjeldnaes, N. (in press). Early Ordovician bryozoans from China. *Mémoire hors série Societé des Sciences Naturelles de l'Ouest de la France*.

Illies, G. (1973). Different budding patterns in the genus *Stomatopora* (Bryozoa, Cyclostomata). In *Living and fossil Bryozoa* (ed. G.P. Larwood), pp. 307–15. Academic Press, London.

Illies, G. (1976). Budding and branching patterns in the genera *Stomatopora* Bronn, 1825 and *Voigtopora* Bassler, 1952 (Bryozoa Cyclostomata). *Oberrheinische Geologische Abhandlungen*, **25**, 97–110.

Jablonski, D. (1986). Larval ecology and macroevolution in marine invertebrates. *Bulletin of Marine Science*, **39**, 565–87.

Jablonski, D. and Lutz, R.A. (1983). Larval ecology of marine benthic invertebrates: paleobiological implications. *Biological Reviews*, **58**, 21–89.

Jackson, J.B.C. (1979). Morphological strategies of sessile animals. In *Biology and systematics of colonial organisms* (ed. G.P. Larwood and B.R. Rosen), pp. 499–555. Academic Press, London.

Jebram, D. (1973). The importance of different growth directions in the Phylactolaemata and Gymnolaemata for reconstructing the phylogeny of the Bryozoa. In *Living and fossil Bryozoa* (ed. G.P. Larwood), pp. 565–76. Academic Press, London.

Kobluk, D.R. (1984). *Archaeotrypa* Fritz 1947 (Cambrian, Problematica) reinterpreted. *Canadian Journal of Earth Sciences*, **21**, 1343–8.

Larwood, G.P. and Taylor, P.D. (1979). Early structural and ecological diversification in the Bryozoa. In *The origin of major invertebrate groups*, Systematics Association Special Volume No. 12 (ed. M.R. House), pp. 209–34. Academic Press, London.

Lavrenteva, V.D. (1985). Bryozoans of the suborder Phylloporina. *Trudy Paleontologicheskogo Instituta*, **214**, 1–98. (In Russian.)

McKinney, F.K. (1969). Organic structures in a late Mississippian trepostomatous ectoproct (bryozoan). *Journal of Paleontology*, **43**, 285–8.

McKinney, F.K. (1986*a*). Evolution of erect marine bryozoan faunas: repeated success of unilaminate species. *American Naturalist*, **128**, 795–809.

McKinney, F.K. (1986*b*). Historical record of erect bryozoan growth forms. *Philosophical Transactions of the Royal Society of London*, **B228**, 133–48.

McKinney, F.K. and Jackson, J.B.C. (1989). *Bryozoan evolution*, Unwin Hyman, London.

Mitter, C., Farrell, B. and Wiegmann, B. (1988). The phylogenetic study of adaptive zones: has phytophagy promoted insect diversification? *American Naturalist*, **132**, 107–28.

Mongereau, N. (1972). Le genre *Hornera* LAMOUROUX, 1821, en Europe. *Annalen des Naturhistorischen Museums. Wien*, **76**, 311–73.

Morozova, I.P. (1988). The morphology, systematics and colonial integration of bryozoans of the Order Fenestrida. *Trudy Palaeontologicheskogo Instituta*, **222**, 70–88. (In Russian.)

Morozova, I.P. and Viskova, L.A. (1977). Historical development of marine Bryozoa (Ectoprocta). *Paleontological Journal*, **1977**, 393–408.

Mundy, S.P., Taylor, P.D. and Thorpe, J.P. (1981). A reinterpretation of phylactolaemae phylogeny. In *Recent and fossil Bryozoa* (ed. G.P. Larwood and C.P. Nielsen), pp. 185–90. Olsen and Olsen, Fredensborg.

Nielsen, C. (1970). On metamorphosis and ancestrula formation in cyclostomatous bryozoans. *Ophelia*, **7**, 217–56.

Nielsen, C. (1971). Entoproct life-cycles and the entroproct/ectoproct relationship. *Ophelia*, **9**, 209–341.

Nielsen, C. and Pedersen, K.J. (1979). Cystid structure and protrusion of the polypide in *Crisia* (Bryozoa, Cyclostomata). *Acta Zoologica. Stockholm*, **60**, 65–88.

Pitt, L.J. and Taylor, P.D. (1990). Cretaceous Bryozoa from the Faringdon Sponge Gravel (Aptian) of Oxfordshire. *Bulletin of the British Museum (Natural History), Geology Series*, **46**, 61–152.

Pohowsky, R.A. (1973). A Jurassic cheilostome from England. In *Living and fossil Bryozoa* (ed. G.P. Larwood), pp. 447–61. Academic Press, London.

Pohowsky, R.A. (1978). The boring ctenostomate Bryozoa: taxonomy and paleobiology based on cavities in calcareous substrate. *Bulletins of American Paleontology*, **73**, 1–192.

Ross, J.R.P. (1985). Biogeography of Ordovician ectoproct (bryozoan) faunas. In *Bryozoa: Ordovician to Recent* (ed. C. Nielsen and G.P. Larwood), pp. 265–71. Olsen and Olsen, Fredensborg.

Schopf, T.J.M. (1977). Patterns and themes of evolution among the Bryozoa.

In *Patterns of evolution* (ed. A. Hallam), pp. 159–207. Elsevier. Amsterdam.
Sepkoski, J.J. (1982). A compendium of fossil marine families. *Milwaukee Public Museum Contributions to Biology and Geology*, **51**, 1–125.
Sepkoski, J.J. (1989). Periodicity in extinction and the problem of catastrophism in the history of life. *Journal of the Geological Society, London*, **146**, 7–19.
Smith, A.B. and Patterson, C. (1988). The influence of taxonomic method on the perception of patterns of evolution. *Evolutionary Biology*, **23**, 127–206.
Taylor, P.D. (1978). The spiral bryozoan *Terebellaria* from the Jurassic of southern England and Normandy. *Palaeontology*, **21**, 357–91.
Taylor, P.D. (1979). Functional significance of contrasting colony form in two Mesozoic encrusting bryozoans. *Palaeogeography, Palaeoclimatolgy, Palaeoecology*, **26**, 151–8.
Taylor, P.D. (1981). Functional morphology and evolutionary significance of differing modes of tentacle eversion in marine bryozoans. In *Recent and fossil Bryozoa* (ed. G.P. Larwood and C. Nielsen), pp. 235–47. Olsen and Olsen, Fredensborg.
Taylor, P.D. (1984). *Marcusodictyon* Bassler from the Lower Ordovician of Estonia: not the earliest bryozoan but a phosphatic problematicum. *Alcheringa*, **8**, 177–86.
Taylor, P.D. (1985). Carboniferous and Permian species of the cyclostome bryozoan *Corynotrypa* Bassler, 1911 and their clonal propagation. *Bulletin of the British Museum (Natural History), Geology Series*, **38**, 359–72.
Taylor, P.D. (1986). *Charixa* Lang and *Spinicharixa* gen. nov., cheilostome bryozoans from the Lower Cretaceous. *Bulletin of the British Museum (Natural History), Geology Series*, **40**, 197–222.
Taylor, P.D. (1987*a*). Skeletal morphology of malacostegan grade cheilostome Bryozoa. In *Bryozoa: present and past* (ed. J.R.P. Ross), pp. 269–76. Western Washington University, Bellingham.
Taylor, P.D. (1987*b*). Fenestrate colony-form in a new melicerititid bryozoan from the U. Cretaceous of Germany. *Mesozoic Research*, **1**, 71–7.
Taylor, P.D. (1988*a*). Major radiation of cheilostome bryozoans: triggered by the evolution of a new larval type? *Historical Biology*, **1**, 45–64.
Taylor, P.D. (1988*b*). Colony growth pattern and astogenetic gradients in the Cretaceous cheilostome bryozoan *Herpetopora*. *Palaeontology*, **31**, 519–49.
Taylor, P.D. (1990). Bioimmured ctenostomes from the Jurassic and the origin of the cheilostome Bryozoa. *Palaeontology*, **33**, 19–34.
Taylor, P.D. and Cope, J.C.W. (1987). A trepostome bryozoan from the Lower Arenig of South Wales: implications of the oldest described bryozoan. *Geological Magazine*, **24**, 367–71.
Taylor, P.D. and Curry, G.B. (1985). The earliest known fenestrate bryozoan, with a short review of Lower Ordovician Bryozoa. *Palaeontology*, **28**, 147–58.
Taylor, P.D. and Larwood, G.P. (1988). Mass extinctions and the pattern of bryozoan evolution. In *Extinction and survival in the fossil record*, Systematics Association Special Publication, no. 34 (ed. G.P. Larwood), pp. 99–119. Clarendon Press, Oxford.

Taylor, P. D. and Michalik, J. (In press). Late Triassic (Rhaetian) cyclostome bryozoans from the West Carpathians, Czechoslovakia. *Neues Jahrbuch für Geologie und Paläontologie, Abhandlungen.*

Taylor, P. D. and Sequeiros, L. (1982). Toarcian bryozoans from Belchite in north-east Spain. *Bulletin of the British Museum (Natural History), Geology Series*, **37**, 117–29.

Walter, B. (1970). Les bryozoaires jurassiques en France. *Documents des Laboratoires de Géologie de la Faculté des Sciences de Lyon*, **35** (for 1969), 1–327.

Yoshioka, P. M. (1982). Role of planktonic and benthic factors in the population dynamics of the bryozoan *Membranipora membranacea*. *Ecology*, **63**, 457–68.

11. Directional macroevolution of the diplograptacean graptolites: a product of astogenetic heterochrony and directed speciation

CHARLES E. MITCHELL

Department of Geology, State University of New York at Buffalo, Buffalo New York, USA

Abstract

Graptolite macroevolutionary history comprised a succession of rapid evolutionary radiations initiated by structural innovations in colony design. The Diplograptacea arose abruptly during the late Arenig, probably as the result of a major rearrangement of a four-branched proximal structure such as that of *Phyllograptus*. Subsequently, the stem-group differentiated along several paths through simplification of the primitive diplograptacean's complex pattern of colonial development (astogeny). By the late Llanvirn this radiation had established the subclades that came to dominate Ordovician graptoloid faunas. Rapid expansion of the Diplograptidae and Monograptidae during the late Arenig and Llandeilo suggests that diplograptaceans initially radiated at non-equilibrium rates. The Orthograptidae and Dicranograptidae radiated more slowly, achieving a peak generic diversity in the Ashgill that exceeded the diversity of the earlier radiation. The Monograptidae maintained very low generic diversity throughout the Ordovician but included the only diplograptaceans that survived the late Ordovician extinction. These survivors radiated explosively in the latest Ashgill and early Llandovery, and re-established most of the previous modes of colony design. They also evolved yet further simplified astogenies culminating in the appearance of uniserial scandent colonies (monograptids proper), which retained no specialized developmental stages other than the sicula itself.

Major Evolutionary Radiations (ed. P. D. Taylor and G. P. Larwood), Systematics Association Special Volume No. 42, pp. 235–264, Clarendon Press, Oxford, 1990.

Transitions from one astogenetic pattern to another during the evolution of the Diplograptacea invariably proceeded from a complex pattern to a less complex pattern. Thirteen such transitions are known and took place as a consequence of abrupt heterochronic rearrangements of colonial development. Clades characterized by colonies with complex astogenies waned while, in some cases, possessors of the newly simplified astogenies proliferated. By the Ashgill graptoloid faunas were almost exclusively composed of colonies with a highly simplified astogeny. Accordingly, the directional character of diplograptacean macroevolution was dominantly a consequence of directed speciation. There is little evidence that anagenetic progress was a significant agent in the evolution of diplograptacean colonial organization.

Introduction

Graptolites are unique among fossil invertebrates. The Graptoloidea are the only planktonic colonial creatures that possess a noteworthy fossil record—a record that is both intriguing and frustrating. We are certain, for instance, that graptoloid colonies drifted within the currents of the temperate and tropical oceans of the Ordovician to early Devonian (Rickards 1975), but it is not clear whether species occupied different depths in the water column or were arrayed along an onshore: offshore gradient, or both (compare, for example, Cisne and Chandlee 1982; Finney 1984). Despite a paucity of detailed information about graptoloid phylogeny (Bulman 1963*b*), an outline of their history has been clear since before the turn of the century (see recent summaries in Rickards 1977; Fortey and Cooper 1986). Early planktonic graptolites appear essentially at the base of the Ordovician System and the dominant forms of the Tremadoc were large, spreading or conical, multibranched colonies comprising dimorphic dwelling tubes (thecae). By early Arenig two-and four-branched species that had lost both thecal dimorphism and preservable stolon (which the earlier 'anisograptids' had retained from their benthic dendroid ancestors) in addition to having a simplified colony architecture became the dominant graptoloids. In relay-fashion, these graptoloids in turn were replaced by colonies with two branches that grew upward from their origin in contact with one another, and finally, in the Silurian, by graptoloids with but a single branch. Suggested causes for this strongly directional macroevolutionary history have varied widely. Nicholson and Marr (1895) and Elles (1922) suggested that passage through this series of grades involved the transformation of a multitude of independent lineages, whereas Bulman (1963*b*) and Rickards (1977) concluded that this history reflected successive radiation from one or a few innovative lineages. Initially, Bulman (1933*a*,*b*) considered the mechanism responsible for innovation to have been orthogenesis, but later, following Huxley (1958, who drew from Rensch 1954), he substituted the concepts of anagenesis (within-lineage transformation advancing from one grade of organization to another) and cladogenesis (divergent speciation, leading to elaborations on the themes established by anagenesis). All of these interpretations, however, share a commitment to a progressive view of the overall trend (see also Kirk 1979).

Like the Graptoloidea as a whole, diplograptaceans display a strongly directional macroevolutionary history. In contrast to gradual anagenetic progress, however, it appears that over most of their 80-Ma history, evolution iterated a fairly limited suite of homoplastic thecal and colony forms. Episodic directed speciation may have established the macroevolutionary vector.

Astogeny and phylogeny

Most graptoloid colonies were constructed of tubular thecae, the design of which was simply repeated (with minor changes in size and proportion) to form the colonial edifice (Figs 11.1 and 11.2). However, the conical sicula exhibits several unique structural specializations that reflect its formation by the colony's founder individual (Fig. 11.1), and these were retained in largely unmodified form during nearly the entire history of the Graptoloidea (Bulman 1970, p. V59; Fortey and Cooper 1986). Between the sicula and the iterated distal thecae there occur from one to five thecae that exhibit specialized ontogenies (primordial thecae, Mitchell 1987). These specializations serve to establish the foundation of the colony. They determine the number of initial branches, their growth direction, and the resulting symmetry of the rhabdosome (Cooper and Fortey 1983). The ontogeny of these specialized thecae and the resulting early astogeny also were evolutionarily conservative. Cooper and Fortey (1982, 1983), Fortey and Cooper (1986), Legrand (1987), and Mitchell (1986, 1987) have shown through studies of early growth patterns among a broad range of graptoloids that early astogeny in these organisms shows a close parallel with von Baer's Law (see Gould 1977). Fortey and Cooper (1986) and Mitchell (1987) employed this relationship to propose a cladistic interpretation of graptoloid phylogeny. This cladogenetic history, in turn, provides a substantially improved starting point from which to examine some of the evolutionary problems posed by graptoloid directional macroevolution.

Origin and initial radiation of the Diplograptacea

The Diplograptacea were the most diverse members of the Virgellina Fortey and Cooper, 1986. Primitively, diplograptaceans possessed two stipes arranged in an erect, back-to-back (biserial scandent) design. This architecture was determined from the outset of colony growth by a complex early astogeny that included five primordial thecae (not four as Mitchell 1987 and previous authors have believed; see Fig. 11.1). Diplograptaceans debut in the late Arenig (*Didymograptus hirundo* Zone or perhaps *Isograptus gibberulus* Zone, see Jenkins 1980; Fortey and Owens 1987). Even the earliest species, such as *Undulograptus austrodentatus* (Fig. 11.1A–C) exhibit a number of thecal characteristics (e.g., highly elaborate, double-sigmoidal thecae with inward-facing apertures) and features of their early astogeny (see Mitchell 1987, pp. 354–7) that unite the group and make them strikingly different from any preceding graptolite or likely sister group. Indeed, the origin of the clade has been one of the most perplexing problems in phylogenetic studies of graptoloids (Bulman 1963*a*,*b*). Several authors have attempted to account for the origins of the Diplograptacea (see Jenkins 1980 for a review and his own suggestion), but most have focused exclusively on the gross organization of the branches and the form of the thecae. Although Bulman (1936, 1945–47) repeatedly emphasized the need to account for the origin of their unique early astogeny, none of the previous attempts to derive diplograptaceans have succeeded. New data from three-dimensionally preserved specimens of *Undulograptus austrodentatus* (Figs 11.1A–C and 11.2A), pyritized material of '*Hallograptus*' sp. aff. '*H.*' *inutilis* (Hall) from the Lower *Didymograptus* Shale, Scania Sweden (Fig. 11.2B, see also Blake 1962), together with observations from Chen and Han's (1988) re-examination of *U. formosus* Mu and Lee, suggest a possible path by which these distinctive graptolites may have been derived from a quadriserial, *Phyllograptus*-like virgellinid.

Figure 11.1 illustrates the similarity in arrangement and attitude of the early thecae between *Undulograptus austrodentatus* (Fig. 11.1A,B) and *Phyllograptus typus* (Fig. 11.1E,F). (Following the standard convention, thecae are numbered according to their position in the thecal sequence and the side of the rhabdosome on which they occur—thus, in reverse view, $Th1^1$ is the first theca on the right (side 1), and $Th1^2$ is the first theca on the left (side 2).) As Cooper and Fortey (1983) pointed out, the early thecae of both graptolites exhibit a similarly recurved shape. On the obverse side of these colonies (Fig. 11.1B,F), the sicula is flanked by the first four thecae and is enclosed by the third pair of thecae. $Th2^1$ occupies the crotch formed by the U-shaped $Th1^1$. On the reverse side (Fig. 11.1A,E) the sicula of each is overgrown by the second and succeeding thecae. The second thecal pair ($Th2^1$ and $Th2^2$) in *P. typus* grows strongly toward the obverse and is nearly hidden in reverse view by the growth of latter thecae. In *U. austrodentatus* the early growth of

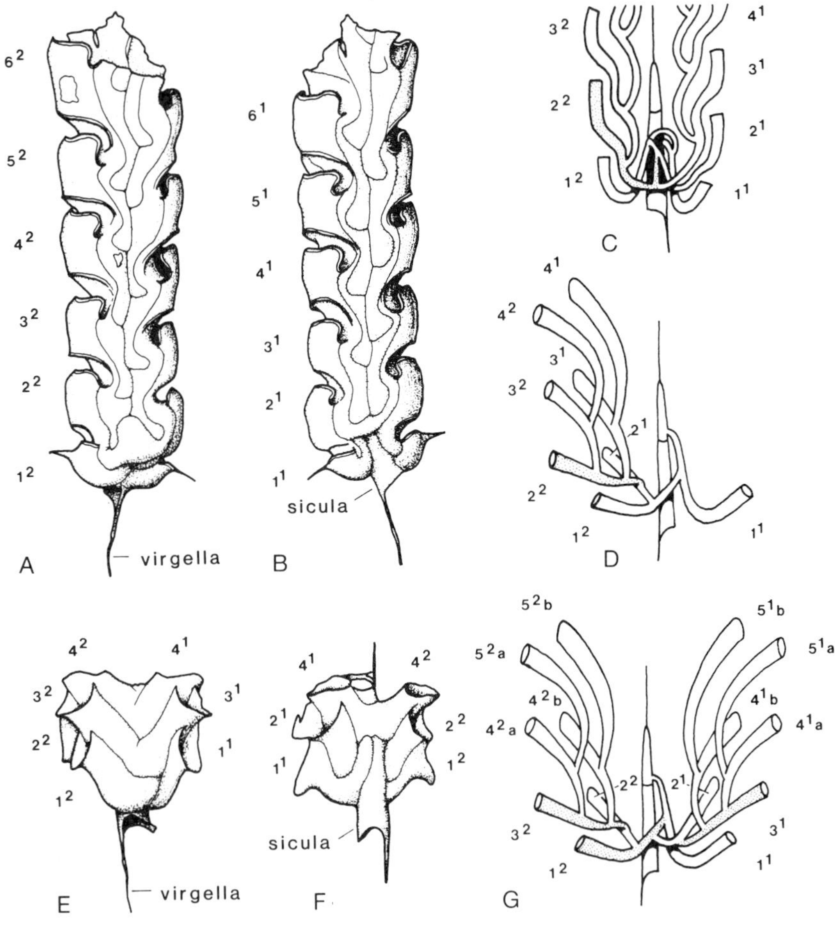

Fig. 11.1. Comparison of *Undulograptus austrodentatus* (Harris and Keble) and *Phyllograptus typus* Hall. A, B, reverse and obverse views, respectively, of isolated, three-dimensionally preserved *U. austrodentatus*, Cow Head Group, Western Newfoundland, sample SPS28B of Williams and Stevens (1988), × 15; C, thecal diagram showing budding pattern of early thecae in *U. austrodentatus*, dicalycal theca stippled; D, thecal diagram corresponding to budding pattern of *P. typus* but with first dichotomy suppressed, see text for discussion; E, F, reverse and obverse views, respectively, of isolated, three-dimensionally preserved growth stages of *P. typus*, Valhallfonna Formation, Spitsbergen (after Cooper and Fortey 1982), × 15; G, thecal diagram showing budding pattern of *P. typus*, dicalycal thecae stippled.

$Th2^1$ is also nearly hidden by later thecae, but there is no correspondingly obscure theca on the opposite side. Instead $Th2^2$ forms a pair with its daughter theca, $Th3^1$, and gives rise to a second theca as well—$Th3^2$. Thus, $Th2^2$ (which is known as a dicalycal theca because of this double budding) initiates a dichotomy; subsequent thecae form two branches united along their dorsal surface. This sequence of budding relationships for *U. austrodentatus* is shown diagrammatically in Fig. 11.1C with the dicalycal $Th2^2$ stippled. Figure 11.1G is the corresponding diagram for the astogeny of *P. typus*. Since the *Phyllograptus* colonies comprise four branches, this astogeny includes three dicalycal thecae. Figure 11.1D shows the same pattern, but with the first dichotomy suppressed such that $Th1^2$ produces only one descendant theca. In this case, the only remaining dicalycal theca in the reduced and renumbered sequence is the fourth theca—$Th2^2$, just as in the primitive diplograptacean astogeny. If such a mutation occurred in a *Phyllograptus*-like virgellinid, rearrangement of this new astogeny to re-establish bilateral symmetry and restore the geometrical relationships between the early thecae and the sicula might well have resulted in a growth pattern similar to that of the early diplograptaceans. However, important incongruities in both early astogeny and thecal architecture that do not appear to be directly connected in any obvious way to the changes described above remain between *Phyllograptus* and *Undulograptus*. Thus, *Phyllograptus* itself is not a good candidate for the ancestor of diplograptacean graptoloids.

Bulman (1936) also suggested asymmetric stipe reduction as a possible mechanism for the cryptogenetic origin of this clade, but later (Bulman 1945–47) abandoned the idea upon discovering that *Dicranograptus nicholsoni* possessed an astogeny identical to that of many archaic diplograptaceans. Because of their distinctly different colony form (Y-and V-shaped colonies with introverted, isolated thecal apertures), Bulman believed that dicranograptids had an ancestry independent of the Diplograptacea. Accordingly, the similar astogenies of these graptoloids could not have both been a reflection of a unique historical event such as asymmetric stipe reduction. Bulman (1945–47), therefore, turned to an explanation that emphasized constructional and functional constraints arising as a consequence of a few alterations in the form of the first two thecae, thus permitting what would otherwise be an implausibly precise convergence. Conversely, Fortey and Cooper (1986) and Mitchell (1987) have argued that this astogenetic similarity is homologous and reflects the derivation of dicranograptids from a diplograptacean ancestor.

Following their appearance in the late Arenig, diplograptaceans underwent a rapid series of speciation events during which several

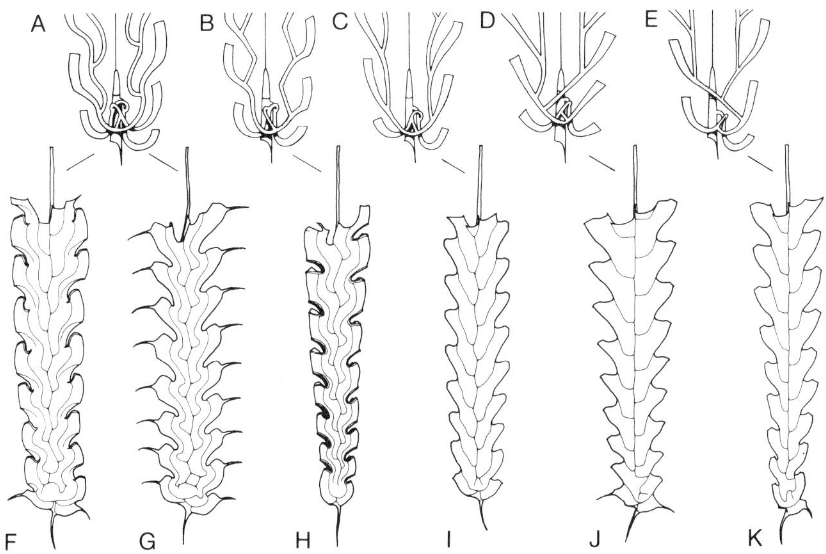

Fig. 11.2. A to E, thecal diagrams illustrating budding patterns and growth direction of early thecae in a range of astogenetic patterns corresponding to those of the species illustrated in F–K, dicalycal thecae stippled; A, Pattern U; B, Pattern C; C, 'archaic' Pattern A; D, 'advanced' Pattern A with delayed dicalycal theca; E, Pattern B; F to K, diagrammatic reconstructions of several late Arenig to early Llanvirn diplograptaceans; F, *Undulograptus austrodentatus* (Harris and Keble); G, *'Hallograptus'* aff. *'H.' inutilis* (Hall); H, *Pseudoclimacograptus (Archiclimacograptus) camptochilus* Skevington; I, *Oelandograptus oelandicus* (Bulman); J, *Hustedograptus* cf. *H. teretiusculus* (Hisinger); K, *Eoglyptograptus dentatus* (Brongniart); all figures approximately × 6.

further reorganizations of their early astogeny and overall colony design took place. Figure 11.2 depicts reconstructed rhabdosomes for several species from the *Didymograptus hirundo* and *D. artus* Zones (latest Arenig and early Llanvirn), including representatives of the stem-group diplograptaceans (Fig. 11.2A,F,G), the Diplograptidae (Fig. 11.2B,H), Orthograptidae (Fig. 11.2C,D,I,J), and the Monograptidae (Fig. 11.2E,K). The Monograptidae is here considered to include the entire clade of which the uniserial scandent monograptids are a terminal branch (see Mitchell (1987) for a discussion of the redefinition of the diplograptacean families).

A cladistic interpretation of the phylogenetic relationships among these and other early diplograptacean species for which reliable data are available on astogeny and rhabdosome architecture is shown in

Fig. 11.3. *Phyllograptus* forms the outgroup and, if the location of the dicalycal theca at Th2² is taken as synapomorphic with the outgroup, as argued above, then the shared astogenetic features of *Undulograptus formosus* and *U. austrodentatus* are primitive for the group. This polarity accords with the stratigraphic occurrence of these graptolites as well. The *Undulograptus* stem-group is the sister taxon to the rest of the Diplograptacea, which share a derived astogeny in which the location of the dicalycal theca is accelerated to Th2¹. Only the first four thecae of diplograptaceans in this group retain specialized ontogenies and the arrangement of buds is altered to produce a less densely populated proximal end. This derived group likewise falls into two large divisions: (1) the Diplograptidae; and (2) the Orthograptidae plus Dicranograptidae and Monograptidae. A paraphyletic stem-group

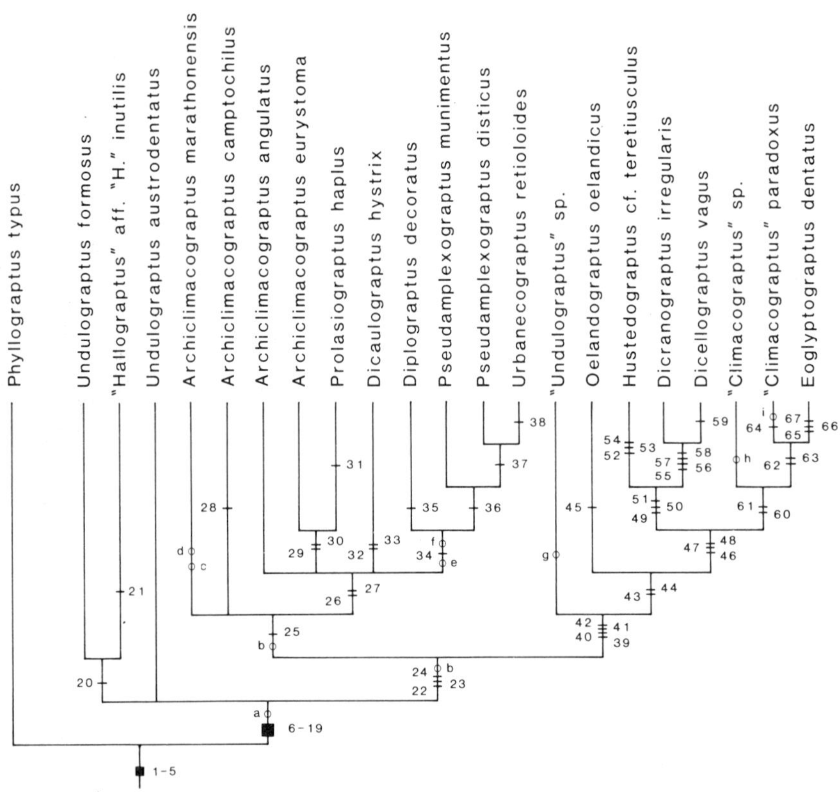

Fig. 11.3. Cladogram, constructed manually, illustrating a possible interpretation of the phylogenetic relationships among a suite of late Arenig to Llandeilo diplograptacean graptolites. *Phyllograptus typus* constitutes the outgroup. *'Hallograptus* aff. *'H.' inutilis* (Fig. 11.2G, see also Blake 1962;

Williams and Stevens 1988, plate 34, figs. 11–13), *'Undulograptus'* sp (*Didymograptus* Shale, Scania Sweden, which combines a Pattern A astogeny and undulograptid thecae), and *'Climacograptus'* sp., (Williams and Stevens 1988, plate 32, figs 4, 6, 8, 11 and plate 33, fig. 1, as *U. austrodentatus*) are undescribed species. Apomorphies: 1, upturned proximal thecae; 2, blunt, symmetrical proximal end; 3, $Th1^1$ and $Th1^2$ descend to near sicular aperture before turning outward; 4, sicula enclosed by third obverse thecal pair; 5, platycalycal, scandent colony structure: 6, ventral metasicular origin of $Th1^1$; 7, $Th1^2$ with left-handed origin from $Th1^1$; 8, dichotomy d1 suppressed; 9, $Th1^2$ with pronounced streptoblastic curvature; 10, $Th2^1$ with right-handed origin from $Th1^2$ and streptoblastic curvature; 11, $Th2^2$ with U-shaped crossing canal; 12, $Th3^1$ primordial and with J-shape; 13, prothecae extremely long and overlap 2/3; 14, prothecae with double sigmoidal curvature and pronounced prothecal fold; 15, metathecal shape geniculate sigmoidal (pseudoclimacograptid); 16, thecal aperture geniculate; 17, $Th1^1$ and $Th1^2$ with prominent apertural spines; 18, $Th2^1$ and later thecae lack spines; 19, median septum undulose; 20, non-geniculate sigmoidal metathecae; 21, mesial spines on all proximal thecae; 22, $Th2^1$ dicalycal with $Th2^1$ and $Th2^2$ forming as pair from fusion of $Th2^1$ crossing canal with flange; 23, $Th2^2$ L-shaped; 24, $Th3^1$ non-primordial; 25, $Th3^1$ arises from resorbed foramen; 26, sigmoidal thecae with reduced overlap; 27, zig-zag median septum; 28, rimmed thecal apertures; 29, $Th2^1$ flange arises on dorsal side of $Th1^2$ crossing canal; 30, asymmetrical sicula deflected to dorsal side; 31, 'lasiograptid' metathecae; 32, fenestrate apertural lobes; 33, dorsal spines on sicular aperture; 34, distal thecae with non-lobed, horizontal apertures; 35, distal metathecae orthograptid; 36, thecae amplexograptid; 37, $Th1^1$ and $Th1^2$ with short orthograptid metathecae; 38, gymnograptid metathecae with prominent lists and genicular spines; 39, J-shaped $Th2^1$ with left-handed origin from $Th1^2$; 40, rounded symmetrical proximal end; 41, $Th3^1$ arises from dorsal side of protheca; 42, prothecal fold absent; 43, long prothecae with 1/2 overlap; 44, sigmoidal thecae; 45, oelandograptid thecae with subtle geniculum near aperture; 46, short thecal overlap of less than 1/2; 47, cuspate thecal apertures; 48, asymmetrical rounded proximal end; 49, $Th2^1$ flange arises on $Th1^1$ metatheca; 50, $Th2^1$ crossing canal short; 51, elaborated sicular aperture with paired dorsal flanges; 52, glyptograptid metathecae; 53, straight median septum; 54, blunt, asymmetrical proximal end; 55, notched and lappette-bearing sicular aperture; 56, wholly to partly uniserial stipes; 57, isolated, introverted thecal apertures; 58, geniculate dicranograptid metathecae with prominent growth-line unconformity within prothecae; 59, reclined, two-stiped colonies; 60, $Th1^2$ metatheca strongly recurved producing narrow, asymmetrical proximal end; 61, $Th1^2$ without spine; 62, $Th2^2$ non-primordial; 63, $Th2^1$ flange suppressed; 64, strongly geniculate climacograptid metathecae; 65, glyptograptid metathecae; 66, straight median septum; 67, prosoblastic $Th1^2$. Plesiomorphies: a, $Th2^2$ dicalycal; b, right-handed, steptoblastic $Th2^1$; c, prothecal folds; d, 2/3 thecal overlap of double sigmoidal thecae; e, $Th3^1$ arises via resorbed foramen in $Th2^1$ metatheca; f, broad proximal end with introverted apertures in $Th1^1$ and $Th1^2$; g, 2/3 overlap of double sigmoidal thecae; h, geniculate metathecae; i, streptoblastic $Th1^2$.

represented by '*Undulograptus*' sp. (an undescribed species similar to *U. klabavensis* (Boucek)) and *Oelandograptus oelandicus* (Bulman) are included here among the Orthograptidae on the basis of their similar thecal forms for the purpose of the analysis of clade diversity discussed below. The Diplograptidae possess a Pattern C astogeny (in the terminology of Mitchell 1987, see Fig. 11.2B,H) that I formerly suggested was derived from the Pattern A astogeny (Fig. 11.2C,I) of a stem-group orthograptid (Mitchell 1987). However, it now appears more likely that most of the features of the astogeny and the colony form (such as sharply sigmoidal thecae with inward-facing apertures and zig-zag median septum) present in primitive members of the Diplograptidae are plesiomorphic. The Pattern A astogeny, synapomorphic to the group including '*Undulograptus*' sp. in Fig. 11.3 and the species to the right of it, is interpreted as derived in the reconstruction given here. Not only is the dicalycal theca accelerated to $Th2^1$, but this theca has a simpler, left-handed origin from $Th1^2$ in colonies exhibiting a Pattern A astogeny (compare Fig. 11.2B,C). This pattern is retained largely unmodified through most of the cladogenetic events that produced the species in this subclade during the late Arenig and Llanvirn, with the exception of the *Eoglyptograptus dentatus* lineage. In this early monograptid the astogeny is further simplified. Only the first three thecae retain specialized ontogenetic features (Fig. 11.2E,K). Despite the substantially derived condition of *E. dentatus*, it appears in the *D. artus* Zone at the base of the Llanvirn, scarcely more than one zone above the first appearance of the earliest diplograptaceans.

Clade diversity of the Ordovician Diplograptacea

Figure 11.4 illustrates the pattern of generic diversity within families of the Diplograptacea during the Ordovician. The vertical axis is scaled to reflect the relative duration, in Ma, of the Ordovician series and graptolite biozones, modified somewhat from the time-scale of Ross *et al.* (1982*a*,*b*). This is necessarily a rough approximation to an absolute scale and doubtless includes many inaccuracies. In addition, the zonal units within which diversity was compiled are of markedly uneven length. At present it is not possible to compile ranges consistently on a shorter or more even scale. Therefore, generic ranges were treated as if they occupied the entire duration of the zones. The stratigraphic ranges of the genera were taken from compilations given in Mitchell (1987) and Melchin and Mitchell (1991); however, the full ranges undoubtedly are not known for many diplograptacean genera. These difficulties notwithstanding, the relative pattern of diversity among families and the general temporal history shown are probably not grossly in error. I

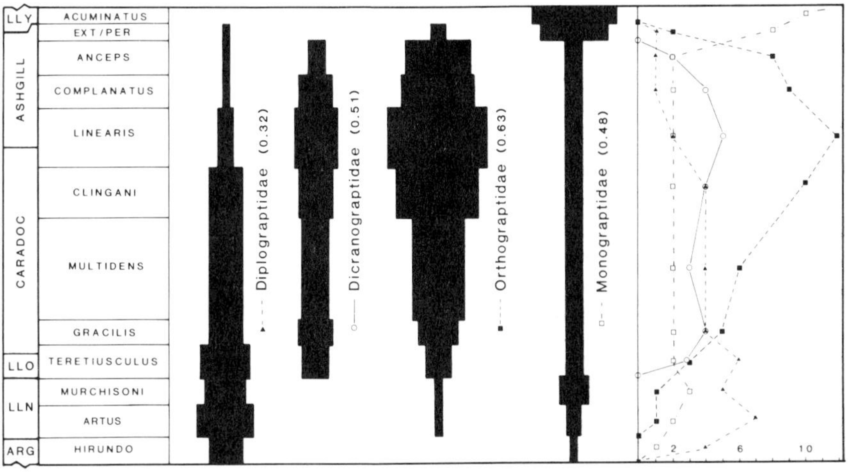

Fig. 11.4. Generic diversity within families of the Diplograptacea by graptolite zones plotted against the British series from the late Arenig to early Llandovery illustrated by spindle diagrams (corresponding centres of gravity in brackets) and smoothed histograms. Data from Mitchell (1987) with revisions from Melchin and Mitchell (1990). ARG, Arenig; LLN, Llanvrin; LLO, Llandeilo; LLY, Llandovery.

calculated the centre of gravity for each family's clade diversity diagram following the method of Gould *et al.* (1987) with unequal sampling intervals.

The stem-group diplograptaceans (the *Undulograptus austrodentatus* and '*Hallograptus*' aff. '*H.*' *inutilis* group in Fig. 11.3) shares a substantial set of plesiomorphic similarities in colony form with the Diplograptidae and, consequently, I grouped these graptolites together in Fig. 11.4. Similarly, the stem-group orthograptids such as *Oelandograptus oelandicus* were included with the Orthograptidae for this analysis. I excluded the late Ashgill–Llandovery secondary radiation of the Monograptidae from the calculation of the centre of gravity for this clade since this expansion was a response to the mass extinction of the other diplograptaceans near the end of Ordovician (see Rickards *et al.* 1977; Koren' 1991; Melchin and Mitchell 1991) and was not a reflection of the dynamics of the early Ordovician radiation of the superfamily.

The radiation of the Diplograptacea occurred in two phases. During the first phase the Diplograptidae and the Monograptidae radiated rapidly and reached their peak generic diversity in the Llanvirn. The centres of gravity of these clades are 0.32 and 0.48, respectively. The Orthograptidae and Dicranograptidae arose later, and their slower diversification with its peak in the late Caradoc (centres of gravity 0.63 and 0.51, respectively) comprise the second phase. This pattern of early arising, rapidly radiating clades with centre of gravity below 0.5 and later arising clades with centre of gravity of greater than or equal to 0.5 and slower rates of diversification recalls the distribution of clade diversity histories in the early Palaeozoic radiation of metazoans (Gould *et al.* 1977; Sepkoski 1979) or the post-mass extinction radiation of mammals (Gould *et al.* 1987) and benthic molluscs (Hansen 1988) in the early Tertiary. This similarity suggests that the phase 1 radiation of the Diplograptacea may represent an initial non-equilibrium radiation —in this case, a rapid filling of ecological opportunities opened up by the structural reorganization that attended their origin. Despite the fact that a biserial scandent colony design had already been achieved independently by several other lineages (i.e. *Glossograptus*, *Cardiograptus*, and *Apiograptus*; see Cooper and McLaurin 1974; Finney 1978; Fortey and Cooper 1986), diplograptaceans appear to have discovered a previously unexploited adaptive zone.

During phase 2, more derived taxa radiated while less derived diplograptaceans steadily waned. Several large-scale features of diplograptacean history suggest that this two-phased radiation may reflect some form of ecological interaction among these basic colony designs.

1. The phase 2 radiation produced a substantially higher generic diversity at its peak (16 genera) than did the phase 1 radiation (9 genera). This may indicate that the species of the Orthograptidae and Dicranograptidae shared some feature or features that permitted evolution of an array of more refined or adaptively specialized colony designs compared to a somewhat adaptively generalized or less versatile pattern in the Diplograptidae and Monograptidae.

2. As noted above, the Monograptidae included the only lineage that survived the late Ashgill mass extinction and they subsequently experienced a rapid radiation following the extirpation of the dominant orthograptids and dicranograptids (see Rickards 1977; Rickards *et al.* 1977; Melchin and Mitchell 1991). During the course of the monograptid radiation in the Llandovery (early Silurian), many species evolved colony designs and thecal shapes that were strikingly convergent on the forms previously developed by Ordovician lineages (Fig. 11.5). Many other examples of iterated colony design exist among the Ordovician Diplograptacea (e.g. *Dicaulograptus* and *Brevigraptus*; Mitchell 1988) and within the Graptoloidea in general (see Cooper and Fortey 1983, pp. 165–8; Rickards *et al.* 1977, pp. 83–97).

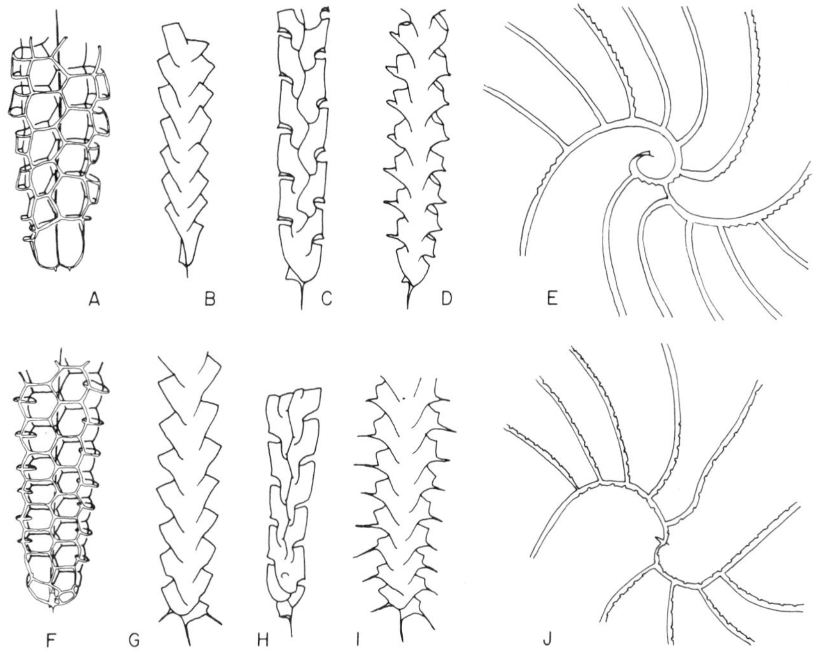

Fig. 11.5. Comparison of convergent Silurian (A–E) and Ordovician (F–J) colony forms. A, *Plectograptus macilentus* (after Bulman 1970), × 5; B, *Pseudorthograptus cyperoides* (after Rickards *et al.* 1977), × 5; C, *Metaclimacograptus hughesi* (after Bulman and Rickards 1968), × 10; D, *Paraclimacograptus innotatus* (after Riva 1974), × 5; E, *Averianowograptus magnificus* (after Bulman 1970), × 0.8; F, *Reteograptus geinitzianus* (after Finney 1980), × 7; G, *Orthograptus abbreviatus* (after Williams 1982), × 7; H, *Pseudoclimacograptus haddingi* (new, L02410t, Paleontological Institute, Lund), × 10; I, *Paraorthograptus pacificus kimi* (after Koren' 1979), × 5; J, *Nemagraptus gracilis* (after Finney 1985), × 2.5.

3. Experimental studies of rhabdosome architecture (Fortey and Bell 1987; Rigby and Rickards 1989) suggest this homoplasy is the consequence of adaptive convergence on a few colony designs constrained by functional limitations imposed by a planktonic, suspension-feeding existence and constructional factors reflecting the graptoloid Bauplan.

Figure 11.6 illustrates the stratigraphic range and sequence of derivation (inferred from the cladogenetic history given in Mitchell 1987) of the 14 astogenetic patterns exhibited by diplograptaceans in the Ordovician and earliest Silurian. During the course of their evolutionary diversification, each of the four diplograptacean clades

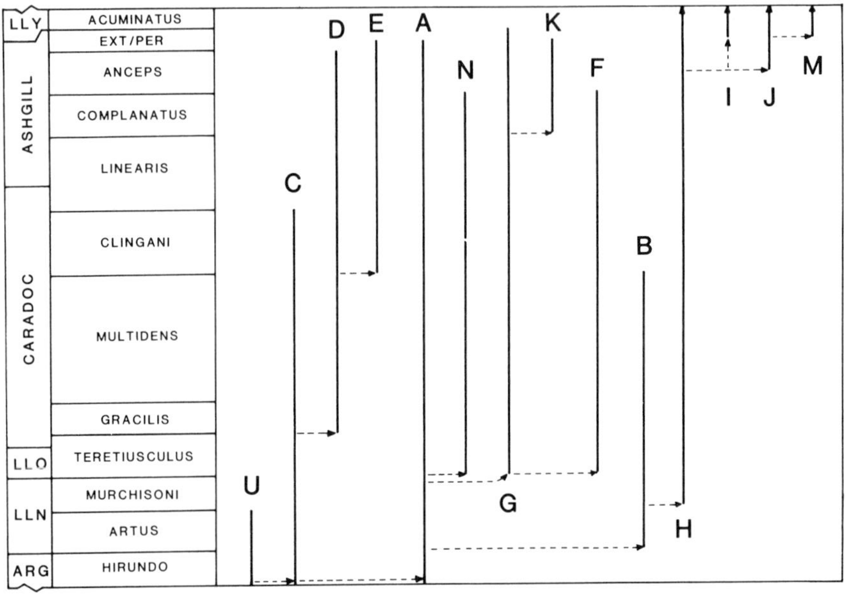

Fig. 11.6. Summed stratigraphic ranges of species exhibiting each of the several diplograptacean astogenetic patterns and inferred sequence of descent (from phylogenetic analysis of diplograptacean genera, Mitchell 1987). Time scale and abbreviations as in Fig. 11.4.

produced lineages with a succession of increasingly simplified astogenies (Pattern C to Pattern E in the Diplograptidae, A to K and F in the Orthograptidae, A to N in the Dicranograptidae, and B to M in the Monograptidae). Within the Diplograptidae, Orthograptidae, and Monograptidae, taxa with more simple astogenies increased in diversity while those with more complex astogenies waned (Fig. 11.7). By mid-Ashgill time graptolite faunas were dominated almost exclusively by diplograptaceans with highly simplified astogenies in which only two or three thecae retained ontogenetic specializations, all thecae but $Th1^1$ grew upward from their origin, and the proximal end was markedly less crowded (compare Fig. 11.9B,E, and H). If one excludes the Dicranograptidae and considers only the fully biserial diplograptaceans, this alteration of relative diversity is even more graphic.

In several cases the appearance of a simplified astogeny coincided with renewed diversification. For instance, the principle radiation of the orthograptids coincides with the origin of the Pattern G astogeny and a broad array of different colony architectures were based on this astogenetic pattern (Fig. 11.8). Similarly, a large segment of the

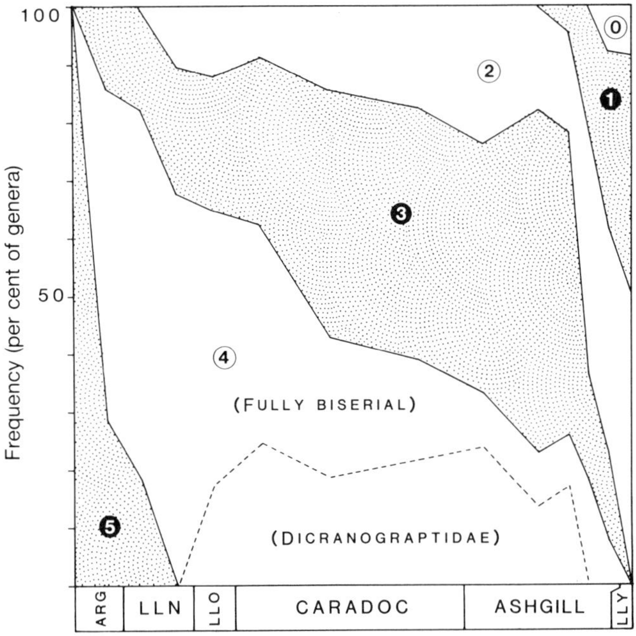

Fig. 11.7. Changing relative proportion of diplograptacean genera exhibiting astogenetic patterns of different degrees of complexity during the interval of late Arenig to early Llandoverty (time-scale and abbreviations as in Fig. 11.4). Genera were combined based on the number of specialized thecae present in their astogenetic pattern, from 5 (most complex) to 0 (least complex). Field labelled '5' reflects the relative abundance of genera with a Pattern U astogeny (five specialized thecae); '4': patterns A, C, D, and N; '3': patterns B, G, and K; '2': patterns, E, F, and H; '1': patterns I and J; and '0': Pattern M. Dashed region within field of genera with four specialized thecae corresponds to the diversity of the Dicranograptidae.

diversity generated during monograptid radiation in the latest Ashgill and early Llandovery was based upon the highly simplified Pattern I and Pattern M astogenies. However, not all lineages with simplified astogenies diversified (see below).

Among the Dicranograptidae this directional evolution was less well developed. *Nemagraptus*, and perhaps the other nemagraptinids, does appear to have a slightly simpler developmental pattern than the Pattern A astogeny exhibited by *Dicellograptus* (Finney 1985), and the nemagraptinids may be derived from the dicranograptinids (Mitchell 1987), but this burst of diversification was short-lived. In general, the

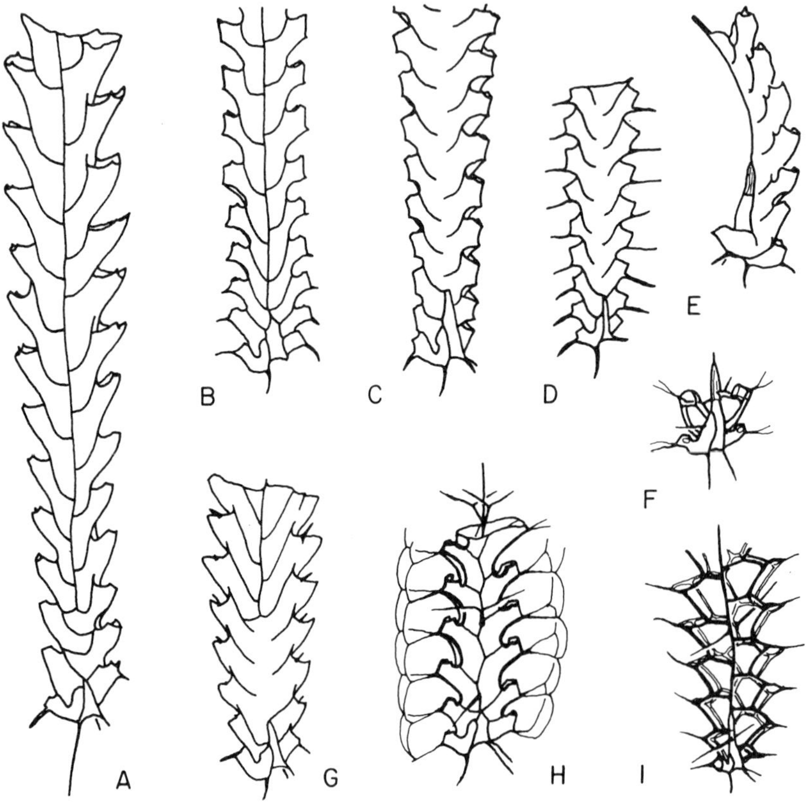

Fig. 11.8. Representatives of the Orthograptidae, illustrating range of colony architectures with a Pattern G astogeny. All figures approximately × 8. After Mitchell (1987), except where noted otherwise. A, *Hustedograptus uplandicus*, a possible sister species to the derived orthograptids that retains a Pattern A astogeny; B, *Amplexograptus bekkeri*, note strong gradient in thecal form and reduction of spines in distal thecae; C, *Amplexograptus preatypicalis* (reconstructed from Bergström and Mitchell 1986) colony with uniform geniculate thecae and no median septum; D, *Paraorthograptus pacificus kimi* (after Koren' 1979) exhibiting paired genicular spines on all thecae; E, *Peiragraptus fallax* (after Strachan 1954), note largely uniserial scandent architecture; F, *Brevigraptus quadrithecatus* (after Mitchell 1988) progenetic retiolitid comprising only a sicula and four thecae; G, *Orthograptus quadrimucronatus* with delayed median septum and paired apertural spines; H, *Lasiograptus harknessi* (reconstructed from Mitchell 1988 and unpublished data) illustrating branched nema and development of peridermal mesh (lacinia) exterior to thecae; I, *Orthoretiolites hami* (new) retiolitid with simple orthograptid thecae constructed of peridermal rods (clathria).

family behaved quite conservatively *vis-à-vis* astogenetic evolution. Even the latest representatives appear to have retained a Pattern A astogeny. Their conservatism coincided with a similarly unusual pattern of evolution in colony architecture: dicranograptids repeatedly evolved more complex, multibranched rhabdosomes (see Mu 1963).

Directed speciation and astogenetic heterochrony

At least thirteen transitions from one astogenetic pattern to another took place during the evolution of the Diplograptacea (assuming no undetected parallelisms). Based on the phylogenetic reconstructions given here (Figs. 11.3, 11.6) and in Mitchell (1987), all of these transitions were from a more complex astogeny to a less complex astogeny. This polarity also accords with the order of first appearance of the complex and more simple astogenies during the history of each family (see Fig. 11.6). As Bulman (1963*b*) noted, changes in thecal form, in contrast, were dominantly non-directional. Similar thecal types arose repeatedly, suggesting evolution of thecal form was limited to numerous variations on a few basic themes. Bulman considered the directional character of graptoloid evolution to have been the result of more-or-less ongoing anagenesis whereas cladogenesis yielded only non-progressive change: 'The processes of thecal differentation appear to me to be essentially cladogenetic, not anagenetic, and the primary relationship between the thecal groups is lateral not vertical' (Bulman 1963*b*, p. 413). The principal differences between early and late representatives of the Diplograptacea arose from changes in the early astogeny of the colonies—changes that owe their directionality to the preferential origin of simplfied colonial architectures.

In an attempt to understand the cause of this striking birth bias, it is instructive to examine the processes by which the structural reorganizations themselves took place. Figure 11.9 depicts early growth stages of several species of the Orthograptidae and illustrates the mode of formation of the derived Pattern G astogeny. Stem-group orthograptids (represented here by *Oelandograptus oelandicus*) have a compact proximal end in which the thecae develop with substantial overlap in the period of their ontogeny (Figs 11.2C,I and 11.9H). Six thecae are present by the time the third theca ($Th2^1$) reaches maturity. $Th1^2$ has a pronounced S-shape and develops in tandem with $Th1^1$ (Fig. 11.9F). Early in its ontogeny $Th1^2$ bifurcates to form the crossing canal of $Th2^1$ (Fig. 11.9G). $Th2^1$ then grows down toward the sicula where it fuses with an upward-growing flange. This occurs contemporaneously with the maturation of $Th1^2$, and yields a pair of foramina from which the continuation of $Th2^1$ and the beginning of $Th2^2$ arise (fig. 11.9H).

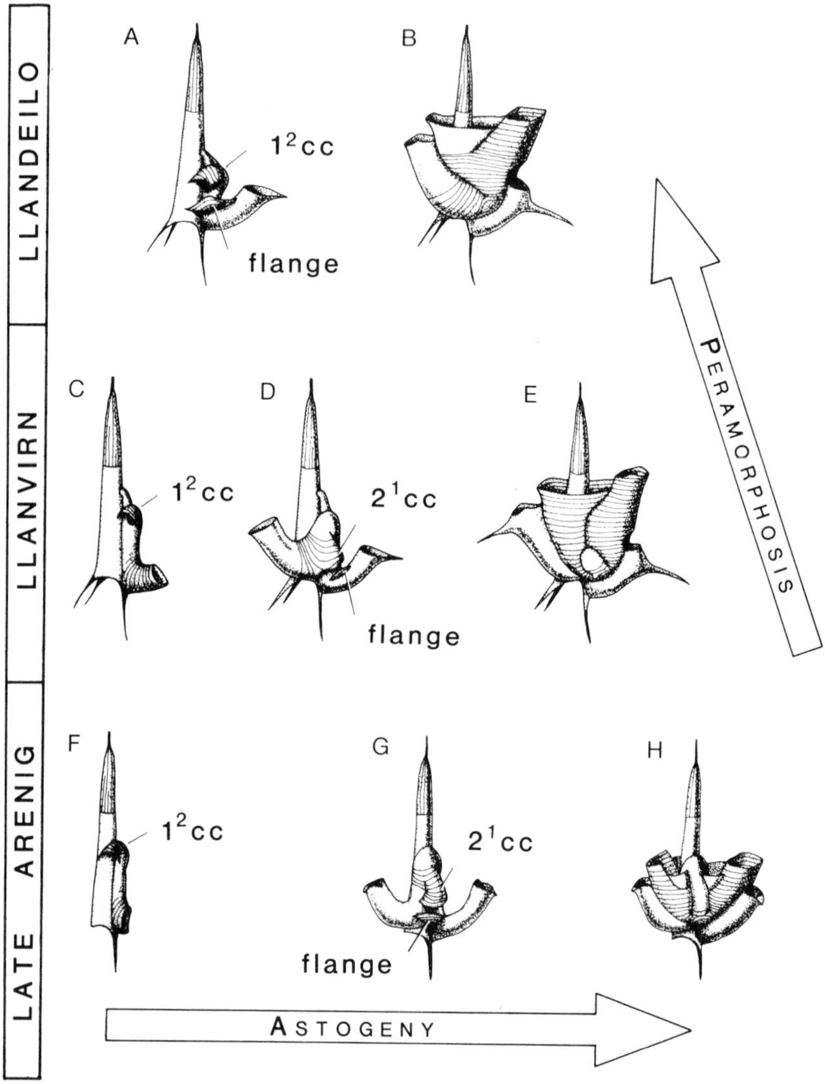

Fig. 11.9. Diagrammatic sketches of immature colonies at comparable growth stages for species with a Pattern G astogeny (A, B; *Amplexograptus maxwelli*), a derived Pattern A astogeny (C–E; *Hustedograptus uplandicus*), and a primitive Pattern A astogeny (F–H; *Oelandograptus oelandicus*), illustrating a possible mode of transition from the Pattern A to the Pattern G astogeny. Location of examples on time-scale reflects likely timing of origin of species with these astogenetic characters rather than that of the species chosen (by virtue of being well understood) to illustrate them. CC: crossing canal. See text for further discussion.

Thus, Th2[1] is dicalycal and each of the first four thecae in this pattern exhibit ontogenetic specializations. The fifth and subsequent thecae develop via a common, iterated ontogeny.

In more derived Pattern A-bearing orthograptids such as *Hustedograptus uplandicus* (Figs 11.8A and 11.9C–E), the thecae bud in the same manner as above but the proximal end is less compact and more asymmetrical. Th1[2] now has a less strongly convoluted form and its development is slightly delayed relative to the maturation of Th1[1] (Fig. 11.9C). The descending crossing canal of Th2[1] is shorter and the appearance of the upward growing flange is accelerated relative to the growth of Th1[2] (Fig. 11.9D). These changes result in less temporal overlap in the ontogenies of the proximal thecae. However, there are still four primordial thecae and this degree of restructuring probably represents the limit to which simplification can proceed within the constraints imposed by this developmental pattern.

The derived orthograptid astogeny, Pattern G, was achieved by a species of *Orthograptus* that probably was descended from a *Hustedograptus uplandicus*-like ancestor (compare Fig. 11.8A and G). Pattern G (Fig. 11.9A,B) is substantially less complex than even the derived Pattern A of *H. uplandicus*. Only the first three thecae retain specialized ontogenetic features. Th2[2] has a thecal form and mode of origin from Th2[1] that is identical to the iterated distal thecal ontogeny (Fig. 11.9B). The descending crossing canal of Th2[1], which was prominent in the primitive Pattern A and reduced in the derived Pattern A, is here entirely suppressed. The upward-growing flange, which formerly fused with this Th2[1] crossing canal, is accelerated into the ontogeny of preceding theca, Th1[2], where it fuses instead with the descending Th1[2] crossing canal (Fig. 11.8A). As in Pattern A, this fusion produces two foramina and two thecae develop from them, but they are not the third and fourth thecae. The continuation of Th1[2] develops from the larger of the two, as Th2[1] had done previously, and Th2[1] arises from the smaller in a manner like that of Th2[2] in the ancestral astogeny.

Thus, species with a Pattern G astogeny exhibit a peramorphic colony from (*sensu* Alberch *et al.* 1979, but see also Pandolfi 1988) relative to that of their Pattern A-bearing ancestor: at a common astogenetic stage (such as the stage at which Th2[2] develops) the derived orthograptid colony exhibits a morphology like that of more astogenetically mature stages of their ancestors. This morphological result appears to be largely a consequence of astogenetic predisplacement.

Astogenetic peramorphosis by predisplacement also appears to be responsible for the sequence of astogenetic changes that occurred within the Diplograptidae and the Monograptidae. Figure 11.10 depicts a set of monograptid growth stages and illustrates a history of proximal-ward

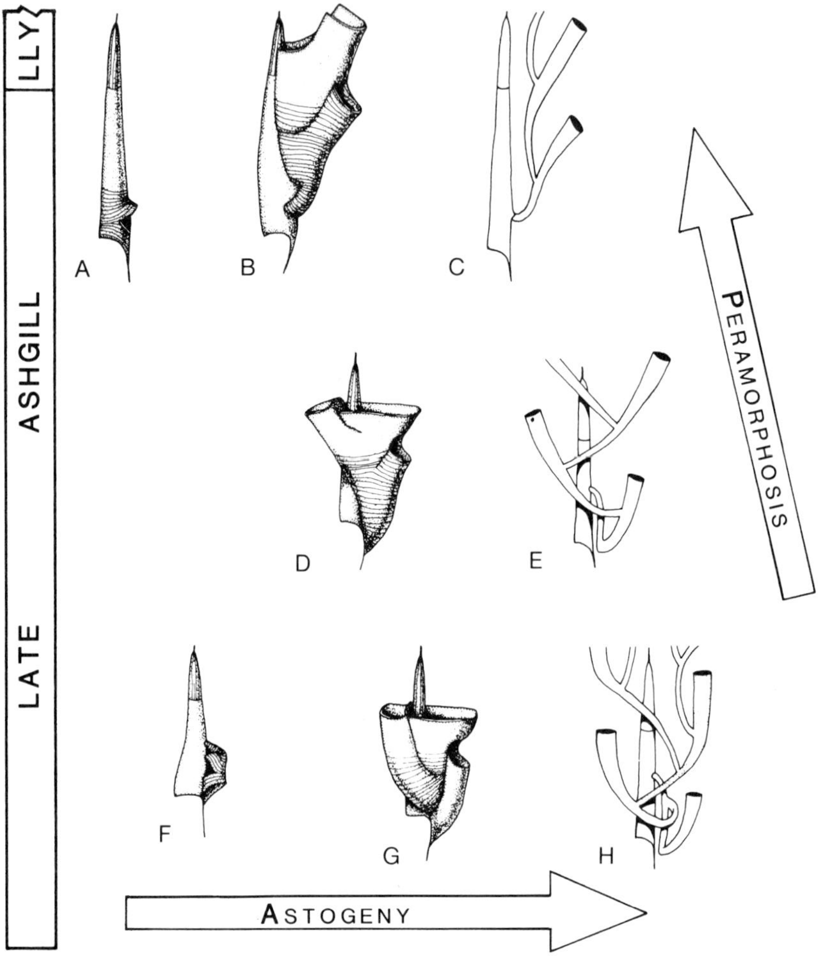

Fig. 11.10. Sketches of immature colonies at comparable growth stages together with thecal diagrams for species with a Pattern M astogeny (A–C; *Pristograptus* sp), a Pattern I astogeny (D, E; *Petalograptus* sp.), and a Pattern H astogeny (F–H; *Normalograptus brevis*) showing successive predisplacement of simple distal thecal ontogenesis and loss of ontogenetic specialization in early thecae. Location of examples on time-scale reflects likely timing of origin of species with these astogenetic characters rather than that of the species chosen (by virtue of being well understood) to illustrate them.

predisplacement of the iterated distal thecal ontogeny and a corresponding reduction in the number of primordial thecae. Consequently, species with a Pattern I astogeny (Fig. 11.10D,E) possess only one specialized theca ($Th1^2$ arises by differentiation from the distal part of $Th1^1$ in the manner of all astogenetically mature thecae), and those with Pattern M possess none (Fig. 11.10A–C). The few remaining ontogenetic peculiarities of the first theca have been accelerated into the otherwise refractory ontogeny of the sicula itself. All previous graptoloids produced a foramen in the wall of the sicula (to allow development of the first theca) by resorption. In the monograptinids this foramen is constructed as an integral part of the ontogeny of the sicula via a pattern of fabrication that is remarkably similar to that employed by earlier monograptids to produce the $Th1^2$ foramen (compare Fig. 11.10A and F).

The macroevolutionary history arising from this intrepretation is a step-like pattern in which the long-term trend arose episodically from a small number of heterochronic transformations that exhibit a striking directional bias (Fig. 11.11). The discontinuous character of the changes excludes the possibility of origin through anagenetic transformation or a succession of speciation events. New astogenetic patterns must have arisen abruptly and spread through a polymorphic population, probably during allopatric speciation. Although most of the tenure of the superfamily was occupied by modification of rhabdosome design and thecal shape, this aspect of colonial evolution contributed little to the direction of the group's long-term history. That is, change in thecal characters was non-directional and was not intimately related to alteration of the conservative early astogeny of diplograptacean graptolites. Nearly all thecal forms evolved more than once and several, such as the climacograptid or glyptograptid theca, occur in all major diplograptacean lineages as well as in both early and late members of the superfamily, be they biserial or uniserial.

I do not mean to imply, however, that the processes involved in the rare events that established the macroevolutionary trend of the Diplograptacea were fundamentally different than the processes that contributed to speciation at other times during their history. Heterochronic changes probably were also common and important in many more modest changes associated with the origin of new species (see for example Urbanek 1973; Rickards 1977; Finney 1986; Mitchell 1986; Pandolfi 1988).

Although directed speciation of the consistency suggested here, combined with extinction at random with respect to astogenetic pattern, would have been sufficient to produce the macroevolutionary trends exhibited by the Diplograptacea, it is also possible that some form of

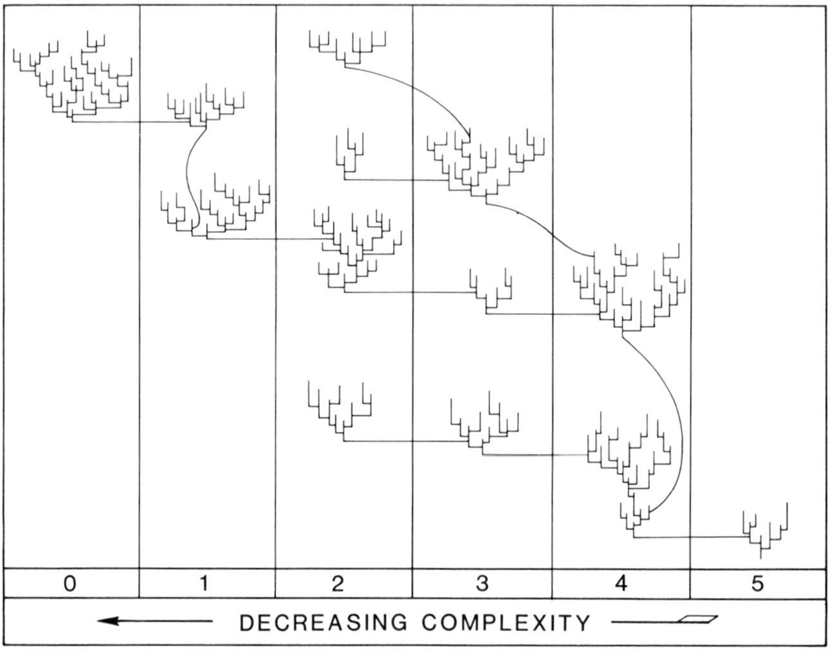

Fig. 11.11. Conceptual diagram illustrating episodic pattern of declining astogenetic complexity during diplograptacean macroevolution. Horizontal axis is the number of early thecae with ontogenetic specializations; vertical axis depicts descent but is not uniform with respect to time. Vertical line segments symbolize species; lengths of horizontal or curved lines correspond to morphological differentiation during speciation. Large displacements represent heterochronic reorganization of astogeny and are contrasted with the more modest offsets usual in the group.

species-level sorting may have contributed as well. Among colonial organisms for instance, Lidgard and Jackson (1989) have demonstrated that changes in colonial growth patterns, and the resulting improved ability to capture and hold substrate, has fuelled a major change in relative diversity among encrusting cheilostome bryozoans during the Tertiary (effect macroevolution *sensu* Vrba 1980; Vrba and Gould 1986), and Taylor (1988; see also Taylor and Larwood, Chapter 10, this volume) provided evidence of an acceleration of evolutionary radiation in the Cheilostomata as a consequence of populational changes emerging from a shift to non-planktotrophic larvae (species selection). Unfortunately neither the data on species abundance and longevity, nor the ecological consequences of the observed astogenetic changes are currently well enough known in the case of the Diplograptacea to evaluate the contribution of species-level processes to their macroevolution.

Directional macroevolution and progress

In addition to gradual, within-lineage transformation, the other major conceptual component of anagenesis is progressive change—not mere maintenance of function but true improvement. Kirk (1979) has made one of the most detailed claims in this mode. However, Kirk's progressivist interpretation depends upon a number of assumptions and inferences about graptolite zooid anatomy and colonial ecology that are not widely accepted (see Crowther 1981; Urbanek 1987; Rigby and Rickards 1989). More generally, I am not convinced that the character of diplograptacean macroevolution either requires or supports a progressive interpretation. Patterns of thecal evolution are already excluded as a source of progressive trends: thecal evolution was highly repetitive and non-directional. Evolution did produce a suite of highly directional trends in rhabdosome structure through changes in astogeny, but astogenetic simplification was not a guarantee of subsequent evolutionary success. The *Climacograptus* (*Diplacanthograptus*), *Geniculograptus*, and *Normalograptus* lineages, for instance, all achieved highly simplified astogenies but underwent no accelerated speciation or expansion in diversity following the innovations. Indeed, the *Normalograptus* lineage was probably the first to achieve an astogeny in which the number of specialized thecae was reduced to only two, but it nevertheless persisted as a species-poor lineage from its origin in the Llandeilo until the latest Ashgill. Were it not for the mass extinction of all other diplograptaceans, the great Silurian expansion of the monograptids probably would not have taken place. Overall, seven of the 13 lineages founded by this process did exhibit evolutionary radiation (at the generic level) following their origin (see Mitchell 1987, fig. 17). If we assume the probability of radiation following astogenetic reorganization = probability of not radiating = 0.5, then, based on the binomial distribution, the probability that exactly seven of 13 clades radiate is 0.2095 and the cumulative probability that seven or more radiate is 0.4953. Thus, by this measure of success the evolutionary history of the diplograptacean clades does not differ significantly from the null hypothesis that reorganization to simpler astogeny produces no increased likelihood of evolutionary radiation over random expectations. Astogenetic simplification, like the other changes in diplograptacean colonial design, was apparently a consequence of locally adaptive changes within individual lineages. These changes occasionally opened the way to further evolutionary radiation, but just as often did not. Furthermore, any attempt to interpret this history as progressive must also admit that it was a most unusual form of progress. Rather than the steady elaboration of structure and differentiation of function typical of Spencerian progressivism (Bowler 1984; Richards 1989), the outcome of this trend was an episodic reduction of complexity and loss of functional specialization.

If not a general arrow of progress, what was the underlying cause of the directional bias in colonial evolution among the Diplograptacea? Astogenetic peramorphosis appears to have acted as an evolutionary ratchet. This character arises for several reasons.

1. Astogeny, like ontogeny, is an inherently directional process. As McNamara (1982) noted, heterochronic evolution is constrained to parallel this directionality.
2. Specialized ontogenetic structures of the early thecae once lost cannot be regained.
3. The character of diplograptacean astogeny itself disallowed increases in complexity. Individual thecal ontogenies in the proximal astogeny were specialized to produce a specific result: the establishment of the biserial arrangement of thecae. Repetition of one or more of these ontogenies would not produce a coherent, biserial colony. Thus, paedomorphic alterations of early astogeny were limited for the most part to postdisplacement of the double-budding, dicalycal theca and a corresponding delay in formation of the median septum. This change occurred many times, in a number of different lineages (Bulman 1963*b*, p. 410), independently of other alterations of the early astogeny (Mitchell 1987, p. 357).

The issue of what immediate selective value astogenetic simplification may have had to a diplograptacean graptolite remains to be solved. One possible answer is suggested by the fact that the reorganizations were achieved by astogenetic predisplacement. Onset of sexual reproduction in colonial organisms is commonly associated with attainment of a particular colony size or stage of astogenetic development (e.g. Silén 1966; van Duyl *et al.* 1981; Wahle 1983). Predisplacement among the diplograptaceans accelerated astogenetically mature thecae into earlier stages of colony growth. In addition, the less compact, more upward-directed growth may have resulted in a more rapidly elongating colony. Hence, accelerated astogenetic maturation may have been associated with a similar acceleration of sexual reproduction, resulting in higher rates of population increase. This hypothesis must be tested against models of graptolite ecology, but it raises the possibility that diplograptacean macroevolution was driven not by progressive improvements in colony engineering so much as by increases in population growth rates and consequent enhancements in their ability effectively to exploit ephemeral plankton blooms.

In this connection, it is interesting to recall that it was precisely those diplograptaceans with colony forms least likely to have been adapted to opportunistic life-styles—the Dicranograptidae with their elaborate multibranched rhabdosomes—that also were least affected by the tendency to develop simplified early astogenies.

Conclusion

According to the synthetic theory of evolution articulated by Dobzhansky (1937), Mayr (1942), Huxley (1942), and Simpson (1944), major alterations of adaptive type occur not by some special mechanism operating only at this level, but through the same processes of adaptation by natural selection that operate within populations (for recent summaries see Schaeffer and Hecht 1965; Dobzhansky *et al.* 1977; Grant 1985; Levinton 1988). Change within the Diplograptacea is consistent with this view. In part as a reaction to saltational and macromutational theories of the origin of higher taxa, advocates of the modern synthesis have also generally emphasized that the evolutionary differentiation of related higher taxa takes place gradually. It develops in additive fashion through a combination of speciation and divergent anagenesis operating over the duration of a succession of species. Such differentiation may occur at an accelerated rate as in Simpson's (1944, 1953) quantum evolution. This model is inconsistent with the evolutionary history of the Diplograptacea. The new astogenetic patterns that established the separate diplograptacean lineages did not arise through a series of successive approximations toward a new adaptive zone; rather, they occurred during the course of single speciation events (Fig. 11.11). Thereafter, the new astogenetic patterns were retained in largely unmodified form while evolutionary alteration of thecal shape and colony architecture iterated the derived astogeny through a diversity of new species—species that in many cases were convergent upon the morphology of members of earlier lineages. Thus, the origin of the subclades of the Diplograptacea exhibits a decidedly non-gradual character (Fig. 11.11).

The macroevolutionary trend exhibited by the Diplograptacea cannot be understood from the perspective of anagenetic progress. Instead, it appears that the directional history of the clade originated largely from a small number of directed speciation events driven by astogenetic peramorphosis.

Acknowledgements

S. Henry Williams generously supplied me with limestone blocks bearing *Undulograptus austrodentatus*, which have yielded specimens critical to this work. Stig M. Bergström and an anonymous reviewer provided comments on the manuscript that have helped to improve it. This research was supported by grant BSR 8806586, from the Division of Biotic Systems and Resources, National Science Foundation.

References

Alberch, P., Gould, S.J., Oster, G. and Wake, D. B. (1979). Size and shape in ontogeny and phylogeny. *Paleobiology*, **5**, 296–317.

Bergström, S. M. and Mitchell, C. E. (1986). The graptolite correlation of the North American Upper Ordovician standard. *Lethaia*, **19**, 247–66.

Blake, D.H. (1962). A new Lower Ordovician graptolite fauna from the Trondheim region. *Norsk Geologisk Tidsskrift*, **42**, 223–38, plates 1, 2.

Bowler, P.J. (1984), *Evolution, the history of an idea*. University of California Press, Berkeley.

Bulman, O. M.B. (1933*a*) Notes on the evolution and morphology of certain Graptoloidea. *Arkiv för Zoologi*, **24A**, no. 13 (for 1932).

Bulman, O.M.B. (1933*b*). Program-evolution in the graptolites. *Biological Reviews*, **8**, 311–34.

Bulman, O.M.B. (1936). On the graptolites prepared by Holm, Pt. 7. The graptolite fauna of the *Orthoceras* Limestone of Hälludden, Öland, and its bearing on the evolution of the lower Ordovician graptolites. *Arkiv for Zoologi*, **28A**, no. 17.

Bulman, O.M.B. (1945–47). Monograph of the Caradoc (Balclatchie) graptolites from the limestones in Laggan Burn, Ayrshire. *Palaeontographical Society [Monographs]*. no. 1 (1945), pp. 1–42, plates 1–3; no. 2 (1946), pp. 43–58, plates 4–6; no. 3 (1947), pp. i–xi, 59–78, plates 7–10.

Bulman, O.M.B. (1963*a*). On *Glyptograptus dentatus* (Brongniart) and some allied species. *Palaeontology*, **6**, 655–89.

Bulman, O. M. B (1963*b*). The evolution and classification of the Graptoloidea. *Quarterly Journal of the Geological Society of London*, **119**, 401–18.

Bulman, O.M.B. (1970). Graptolithina with sections on Enteropneusta and Pterobranchia. In *Treatise on invertebrate paleontology*, Pt. V. (2nd edn), (ed. C. Teichurt). Geological Society of America and University of Kansas Press, Bolder, Colorado and Lawrence, Kansas.

Bulman, O.M.B. and Rickards, R.B. (1968). Some new diplograptids from the Llandovery of Britain and Scandanavia. *Palaeontology*, **11**, 1–15.

Chen, Xu and Han, Nai-ren (1988). SEM examination on *Pseudoclimacograptus (Undulograptus) formosus* Mu et Lee. *Acta Palaeontologica Sinica*, **27**, 141–9, plate 1.

Cisne, J.L. and Chandlee, G.O. (1982). Taconic Foreland Basin graptolites: age zonation, depth zonation, and use in ecostratigraphic correlation. *Lethaia*, **15**, 343–63.

Cooper, R.A. and Fortey , R.A. (1982). The Ordovician graptolites of

Spitsbergen. *Bulletin of the British Museum (Natural History), Geology Series*, **36**, 157–302.

Cooper, R.A. and Fortey, R.A. (1983). Development of the graptolite rhabdosome. *Alcheringa*, **7**, 201–21.

Cooper, R.A. and McLaurin, A.N. (1974). *Apiograptus* gen. nov. and the origin of the biserial graptolite rhabdosome. *Special Papers in Palaeontology*, **13**, 75–86.

Crowther, P.R. (1981). The fine structure of graptolite periderm. *Special Papers in Palaeontology*, **26**, 1–119.

Dobzhansky, T. (1937). *Genetics and the origin of species* (1st edn). Colombia University Press, New York.

Dobzhansky, T., Ayala, F.J., Stebbins, G.L. and Valentine, J.W. (1977). *Evolution*. W.H. Freeman and Co, San Francisco.

Elles, G.L. (1922). The graptolite faunas of the British Isles. A study in evolution. *Proceedings of the Geologists' Association*, **33**, 168–200.

Finney, S.C. (1978). Affinities of *Isograptus, Glossograptus, Cryptograptus, Corynoides*, and allied graptolites. *Acta Palaeontologica Polonica*, **23**, 481–95.

Finney, S.C. (1980). Thamnograptid, dicograptid and abrograptid graptolites from the middle Ordovician Athens Shale of Alabama. *Journal of Paleontology*, **54**, 1184–208.

Finney, S.C. (1984). Biogeography of Ordovician graptolites in the southern Appalachians. In *Aspects of the Ordovician system*, Palaeontological Contributions of the University of Oslo, no. 295 (ed. D.L. Bruton), pp. 167–76. Universitets forlaget, Oslo.

Finney, S.C. (1985). Nemagraptid graptolites from the Middle Ordovician Athens Shale, Alabama. *Journal of Paleontology*, **59**, 1100–37.

Finney, S.C. (1986). Heterochrony, punctuated equilibrium, and graptolite zonal boundaries. In *Palaeoecology and biostratigraphy of graptolites*, Geological Society Special Publication, no. 20 (ed. C.P. Hughes and R.B. Rickards), pp. 103–13. Blackwell, Oxford.

Fortey, R.A. and Bell, A. (1987). Branching geometry and function of multiramous graptoloids. *Paleobiology*, **13**, 1–19.

Fortey, R.A. and Cooper, R.A. (1986). A phylogenetic classification of the graptoloids. *Palaeontology*, **29**, 631–54.

Fortey, R.A. and Owens, R.M. (1987). The Arenig Series in south Wales: Stratigraphy and palaeontology. *Bulletin of the British Museum (Natural History), Geology Series*, **43**, 69–364.

Gould, S.J. (1977). *Ontogeny and phylogeny*. Belknap Press, Cambridge, Massachusetts.

Gould, S.J., Gilinsky, N.L. and German, R.Z. (1987). Asymmetry of lineages and the direction of evolutionary time. *Science*, **236**, 1437–41.

Gould, S.J., Raup, D.M., Sepkoski, J.J. Jr, Schopf, T.J.M. and Simberloff, D.S. (1977). The shape of evolution: a comparison of real and random clades. *Paleobiology*, **3**, 23–40.

Grant, V. (1985). *The evolutionary process: a critical review of evolutionary theory*. Columbia University Press, New York.

Hansen, T.A. (1988). Early Tertiary radiation of marine molluscs and the long-term effects of the Cretaceous–Tertiary extinction. *Paleobiology*, **14**, 37–51.

Huxley, J. (1942). *Evolution: the modern synthesis*. Harper, New York.
Huxley, J. (1958). Evolutionary processes and taxonomy with special reference to grades. *Uppsala University Årsskrift*, **6**, 21–39.
Jenkins, C. (1980). *Maeandrograptus schmalenseei* and its bearing on the origin of the diplograptids. *Lethaia*, **13**, 289–302.
Kirk, N.H. (1979). Thoughts on coloniality in the Graptolithina. *Biology and systematics of colonial organisms*, Systematics Association Special Volume, no. 11 (ed. G. Larwood and B.R. Rosen), pp. 411–32. Academic Press, London.
Koren', T.N. (1979). *Pacificograptus*, a new late Ordovician diplograptid genus. *Paleontological Journal*, **1979**, 65–70.
Koren', T.N. (1991). Evolutionary crisis of Ashgill graptolites. In *Proceedings of the Fifth International Symposium on the Ordovician System*, Geological Survey of Canada, Special Publication (ed. S.H. Williams and C.R. Barnes). (In press.)
Legrand, P. (1987). Modo de desarrollo del suborden Diplograptina (Graptolithina) en el Ordovícico Superior y en el Silúrico. *Revista Española de Paleontología*, **2**, 59–64.
Levinton, J. (1988). *Genetics, paleontology, and macroevolution*. Cambridge University Press, Cambridge.
Lidgard, S. and Jackson, J.B.C. (1989). Growth in encrusting cheilostome bryozoans: I. Evolutionary trends. *Paleobiology*, **15**, 255–82.
McNamara, K.J. (1982). Heterochrony and phylogenetic trends. *Paleobiology*, **8**, 130–42.
Mayr, E. (1942). *Systematics and the origin of species*. Columbia University Press, New York.
Melchin, M.J. and Mitchell, C.E. (1991). Late Ordovician extinction among the Graptoloidea. In *Proceedings of the Fifth International Symposium on the Ordovician System*, Geological Survey of Canada, Special Publication (ed. S.H. Williams and C.R. Barnes). (In press.)
Mitchell, C.E. (1986). Morphometric studies of *Climacograptus* Hall and the phylogenetic significance of astogeny. In *Palaeoecology and biostratigraphy of graptolites*, Geological Society Special Publication, no. 20 (ed. C.P. Hughes and R.B. Rickards), pp. 119–29. Blackwell, Oxford.
Mitchell, C.E. (1987). Evolution and phylogenetic classification of the Diplograptacea. *Palaeontology*, **30**, 353–405.
Mitchell, C.E. (1988). The morphology and ultrastructure of *Brevigraptus quadrithecatus* n. gen., n. sp. (Diplograptacea), and its convergence upon *Dicaulograptus hystrix* (Bulman). *Journal of Paleontology*, **62**, 448–63.
Mu, En-zhi (1963). On the complication of the graptolite rhabdosome. *Acta Palaeontologica Sinica*, **11**, 346–77. (In Chinese with an English summary.)
Nicholson, H.A. and Marr, J.E. (1895). Notes on the phylogeny of the graptolites. *Geological Magazine*, **42**, 529–39.
Pandolfi, J.M. (1988). Heterochrony in colonial marine animals. In *Heterochrony in evolution: a multidisciplinary approach*, Topics in Geobiology, no. 7 (ed. M.L. McKinney), pp. 135–58. Plenum Press, New York.
Rensch, B. (1954). *Neuere Probleme der Abstammungslehre*, (2nd edn). Ferdinand Enke, Stuttgart.

Richards, R.J. (1989). The moral foundation of the idea of evolutionary progress: Darwin, Spencer, and neo-darwinians. In *Evolutionary progress* (ed. M.H. Nitecki), pp. 129–49. University of Chicago Press, Chicago.

Rickards, R.B. (1975). Palaeoecology of the Graptolithina, an extinct class of the Phylum Hemichordata. *Biological Reviews*, **50**, 397–436.

Rickards, R.B. (1977). Patterns of evolution in the graptolites. In *Patterns of evolution, as illustrated by the fossil record* (ed. A. Hallam), pp. 333–58. Elsevier, Amsterdam.

Rickards, R.B., Hutt, J.E. and Berry, W.B..N. (1977). Evolution of the Silurian and Devonian graptoloids. *Bulletin of the British Museum (Natural History), Geology Series*, **28**, 1–120.

Rigby, S. and Rickards, R.B. (1989). New evidence for the life habit of graptoloids from physical modelling. *Paleobiology*, **15**, 402–13.

Riva, J. (1974). Graptolites with multiple genicular spines from the Upper Ordovician of western North America. *Canadian Journal of Earth Sciences*, **11**, 1455–60.

Ross, R.J. Jr *et al.* (1982*a*). Fission track dating of British Ordovician and Silurian stratotypes. *Geological Magazine*, **119**, 135–53.

Ross, R.J. Jr *et al.* (1982*b*). *The Ordovician System in the United States, correlation chart and explanatory notes.* International Union of Geological Sciences, Publication no. 12. International Union of Geological Sciences, Ottawa.

Schaeffer, B. and Hecht, M.K. (1965). Symposium: the origin of higher levels of organization. Introduction and historical resume. *Systematic Zoology*, **14**, 245–8.

Sepkoski, J.J. Jr (1979). A kinematic model of Phanerozoic taxonomic diversity. II. Early Phanerozoic families and multiple equilibria. *Paleobiology*, **7**, 222–51.

Silén, L. (1966). On the fertilization problem in the gymnolaematous Bryozoa. *Ophelia*, **3**, 113–40.

Simpson, G.G. (1944). *Tempo and mode in evolution.* Columbia University Press, New York.

Simpson. G.G. (1953). *The major features of evolution.* Columbia University Press, New York.

Strachan, I. (1954). The structure and development of *Peiragraptus fallax*, gen. et sp. nov. *Geological Magazine*, **91**, 509–13.

Taylor, P.D. (1988). Major radiation of cheilostome bryozoans: triggered by the evolution of a new larval type? *Historical Biology*, **1**, 45–64.

Urbanek, A. (1973). Organization and evolution of graptolite colonies. In *Animal colonies: development and function through time*, (ed. R.S. Boardman, A.H. Cheetham, and W.A. Oliver, Jr), pp. 441–514. Dowden, Hutchinson and Ross, Stroudsburg, Pennsylvania.

Urbanek, A. (1987). The phylogenetic debate on the affinities of graptolites. An attempt at a methodological analysis. *Bulletin of the Geological Society of Denmark*, **35**, 231–44.

van Duyl, F.C., Bak, R.P.M. and Sybesma, J. (1981). The ecology of the tropical compound ascidian *Trididemuum solidum*. I. Reproductive strategy and larval behavior. *Marine Ecology: Progress Series*, **6**, 35–42.

Vrba, E. S. (1980). Evolution, species and fossils: how does life evolve? *South African Journal of Science*, **76**, 61–84.

Vrba, E. S. and Gould, S. J. (1986). The hierarchical expansion of sorting and selection: sorting and selection cannot be equated. *Paleobiology*, **12**, 217–28.

Wahle, C. M. (1983). The role of age, size, and injury in sexual reproduction among Jamaican gorgonians. *American Zoologist*, **23**, 961.

Williams, S. H. (1982). The Late Ordovician graptolite fauna of the Anceps Bands at Dob's Linn, southern Scotland. *Geologica et Palaeontologica*, **16**, 29–47.

Williams, S.H. and Stevens, R.K. (1988). Early Ordovician (Arenig) graptolites of the Cow Head Group, western Newfoundland, Canada. *Palaeontographica Canadiana*, **5**, 1–167.

12. Evolutionary diversification of echinoderms during the early Palaeozoic

ANDREW B. SMITH
Department of Palaeontology, Natural History Museum, London, UK

Abstract

The greatest radiation in the history of echinoderms is supposed to have taken place during the Cambrian when many taxa of low diversity and high categorial rank suddenly appeared in the fossil record. This stems, however, from our poor knowledge of the Cambrian fauna and an arbitrary system of taxonomy. Both morphological and ecological diversification at this period are, in reality, much less spectacular. A phylogenetic tree, constructed from cladistic analysis and stratigraphical occurrence data, provides the basis for an evolutionary interpretation of events during this period. By considering the scheme of synapomorphies from a functional point of view a picture of the ecological diversification of echinoderms is built up.

The earliest radiate echinoderms were passive low-level suspension feeders that were totally reliant on environmental currents and lacked any form of attachment device. This presumably restricted them to environmental regimes of low current strength. The Cambrian diversification of echinoderms to a large extent represents the group's evolution to enhance its feeding capability by (1) improving the design of the food-capturing surface and (2) developing attachment structures to cope with life in stronger current regimes. Specific adaptational changes related to gaseous exchange or defence are virtually absent until the Ordovician. Echinoderms continued to diversify after the Cambrian and became active suspension feeders, deposit feeders, predators, and grazers.

Major Evolutionary Radiations (ed. P.D. Taylor and G.P. Larwood) Systematics Association Special Volume No. 42, pp. 265–286, Clarendon Press, Oxford, 1990.

Introduction

Over the past 15 years there have been several reviews of the early history of echinoderms (Ubaghs 1975; Paul 1977, 1979, 1988*b*; Sprinkle 1980*a*, *b*, 1981, 1983; Paul and Smith 1984; Campbell and Marshall 1987; Smith 1988). Most of these have argued that echinoderms underwent an explosive radiation during the Cambrian giving rise to many poorly adapted low-diversity groups and that subsequently many of these groups were weeded out through competition as the Palaeozoic marine community became more crowded (see Paul 1979; Sprinkle 1983). Furthermore, Sprinkle (1980*b*), Campbell and Marshall (1987), and Paul (1988*b*) have all argued that evolutionary diversification occurred in two phases, one in the Cambrian and a second beginning in the Ordovician, as suggested by Sepkoski (1979) for marine metazoans in general. The evidence presented in support of these models comes largely from analysis of taxonomic data and in particular from first appearances of major taxonomic categories such as classes. Thus virtually all echinoderm classes, as recognized in the *Treatise on invertebrate paleontology*, had evolved by the end of the Ordovician (Sprinkle 1983).

This view of the early evolution of echinoderms has been challenged by me (Smith 1988). Specifically, the traditional model is based on inappropriate assumptions about the correlation between taxonomic rank and morphological or ecological diversity. Biases are also introduced because Cambrian echinoderms really are quite poorly known.

1. *Poor knowledge of echinoderm faunas during 'major radiations'*

Many echinoderms lack a firmly articulated skeleton and are highly prone to disarticulate into individual elements upon death. Special sedimentological conditions must be met to preserve such species intact—creating the so-called 'starfish-bed' Lagerstätten. These are rather rare and so many echinoderms are apparently restricted in both time and space simply because they have only ever been found in these unique Lagerstätten. For example, only some 6 per cent of Cambrian and early Ordovician species have ever been recorded from more than one restricted locality (Table 12.2). Smith (1988) showed that there was a strong correlation between the number of such Lagerstätten within a geological period and the number of taxa apparently originating at that time. This suggests that the observed pattern of species diversity in the Cambrian might simply be a reflection of the distribution and number of fossiliferous localities that have been discovered.

There is an extremely strong geographical bias to our record of

Table 12.1. Number of species of echinoderms based on test/thecal material in the Cambrian and early Ordovician and number of species of echinoids based on test material in the Triassic (from Smith 1990), tabulated according to provenance.

	Number of species (per cent)	
	Cambrian ($N = 101$)	Triassic ($N = 49$)
Europe	32 (32)	44 (90)
North America	49 (48)	2 (4)
Australia/New Zealand	12 (12)	1 (2)
Africa	7 (7)	—
USSR	1 (1)	—
India	—	1 (2)
Middle East	—	1 (2)
Asia	—	—
South America	—	—

Cambrian echinoderms. Eighty per cent of recorded species come from North America and western Europe (Table 12.1). Australia has a relatively large echinoderm fauna due almost entirely to recent exploratory work by Jell *et al.* (1985).

Nor is the fauna particularly well documented. Approximately half of the named species are inadequately known; that is to say they are based on incomplete material in which basic anatomical details remain unknown (Table 12.2).

Thus the Cambrian echinoderm fauna is relatively poorly known and strongly biased. A very similar situation exists for echinoids in the Triassic (Table 12.2) which represents another period of supposed 'major adaptive radiation'.

2. *Taxa, clades, and ecological guilds*

Higher taxonomic categories of echinoderms have been created for a variety of reasons (see Smith 1988) and only a proportion represent monophyletic groupings. This is because of the arbitrary way in which categorial rank has been assigned in the past and the different approaches that workers have adopted in creating taxonomies. There is a belief, most clearly stated by Sprinkle (1976, p. 87) for echinoderms, that taxonomic rank reflects or correlates with morphological distinctiveness and consequently gives a measure of ecological differentiation. The problem is that evolutionary systematists have created higher

Table 12.2. Comparison of Cambrian to early Ordovician and Triassic records of echinoderm 'adaptive radiations'. Early Palaeozoic data taken from Smith (1988) with additions from Smith and Jell (1990). Triassic echinoid data taken from Smith (1989).

	Cambrian	Triassic
Time interval (Ma)	*c.* 45	*c.* 35
Total number of species described	101	49
Taxonomic level of originations	Class/order	Subclass/order
% of species fully known	24	6
% of species known from more than one general locality	6	8
Geographical bias	Strong	Very strong
Correlation between number of species described and number of fossiliferous localities in a given time interval	Strong	Strong

taxonomic categories for such a variety of reasons that they have become meaningless for analysis. If evolutionary patterns are to be deduced from the stratigraphical distribution of higher taxa then we must at least make sure that the units we use are of a similar nature; otherwise the patterns will reflect taxonomic practices rather than any real biological pattern (Smith and Patterson 1988).

There are two approaches to studying evolutionary patterns. First, we can take a topological approach and analyse the shape of the evolutionary tree and its branching pattern. This requires the branches of the tree to be defined in a consistent and non-arbitrary way and thus requires the recognition of only monophyletic groups. Second, we can fit species into ecologically defined groupings or guilds in which we are interested and track their history. Unfortunately, under evolutionary systematics, higher taxa have tried to fulfil both roles and have consequently failed to satisfy the criteria of either, falling between two stools. They are neither monophyletic groups nor ecological guilds but a mixture of these together with other forms of arbitrary groupings. As such, few meaningful results can be expected from their analysis.

Phylogenetic relationships and an evolutionary tree

Here I shall be dealing only with radiate echinoderms, and specifically exclude carpoids and their relatives. Previously (Smith 1988), I have given a cladistic analysis of relationships amongst Cambrian and Lower Ordovician echinoderms and a more recent analysis of Lower Palaeozoic eleutherozoan taxa is given in Smith and Jell (1990) (Fig. 12.1). A detailed cladistic treatment of Lower Palaeozoic cystoid groups is given by Paul (1988*a*) and of primitive crinoids by Donovan (1988). These papers establish the phylogenetic relationships of many of the early Cambrian echinoderms. Although the major groupings and the majority of relationships are agreed upon, there remain a few areas of difference between the phylogenies proposed by Smith (1988) and Paul (1988*a*). Neither I nor Paul subjected our data to a numerical analysis and, whereas I subjectively gave greater weight to the evolution of ambulacra extending free from the body wall, Paul placed emphasis on the evolution of a holomeric stalk. In order to clarify these differences a numerical analysis has been undertaken for the major pelmatozoan clades using the combined data of Smith (1988) and Paul (1988*a*), using the Hennig86 program (Farris 1988). This produced nine equally parsimonious solutions which can be simplified to

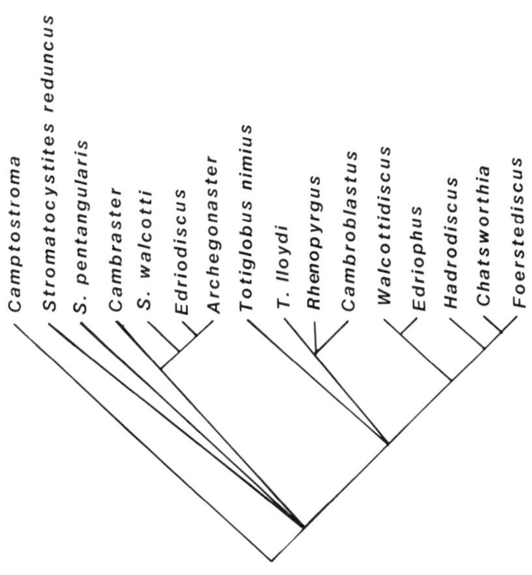

Fig. 12.1. Phylogenetic relationships of early Palaeozoic eleutherozoans established using the numerical cladistic program Hennig86 (Farris 1988). (From Smith and Jell 1990.)

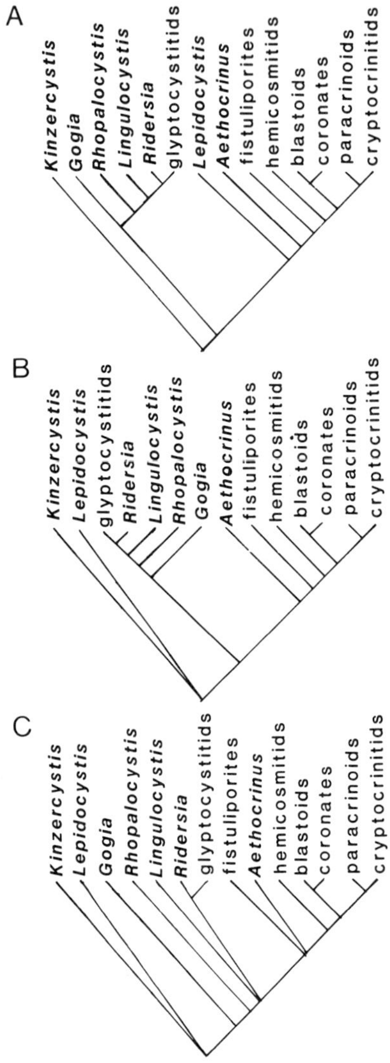

Fig. 12.2. Phylogenetic relationships of selected early Palaeozoic pelmatozoan groups. Table 12.3 gives the data matrix on which this is based. Multistate characters are treated as unordered; other characters are polarized using *Camptostroma* as outgroup. Nine equally parsimonious trees were found using the 'ie' option and these simplify to the three topologies shown. Solution A is favoured here.

three distinct topologies (Fig. 12.2), one of which is consistent with the cladogram of Smith (1988). Furthermore, in all nine trees pelmatozoans with free arms are grouped together, with the exception of *Lepidocystis*, and all indicate a holomeric column evolved two times independently. However, the tree given by Smith (1988, fig. 9) needs correcting in that it fails to treat arm-bearing pelmatozoans as monophyletic and this is corrected in Fig. 12.3. The analysis presented here suggests that crinoids may be sister-group to arm-bearing cystoids.

The cladistic analysis of relationships taken with stratigraphical data produces our best estimate of the evolutionary tree. Stratigraphical ranges for most taxa are given in Smith (1988) and additions are found in Smith and Jell (1990).

Ecological diversification during the early Palaeozoic

In order to examine ecological aspects of the early Palaeozoic diversification of echinoderms, the functional significance of apomorphic characters used to establish their phylogenetic relationships has been explored. These characters have been subdivided into those affecting different 'vital functions' (as defined by Paul 1977), namely feeding, defence, respiration, posture/attachment, excretion, growth, and reproduction. Because the cladistic analyses try to look at all morphological changes that have taken place during the evolution of echinoderms in the Cambrian, it is valid to ask what proportion can be ascribed to each vital function. This should indicate which factors were important to echinoderms during their early diversification. Table 12.3 lists the proportion of characters (taken from Smith 1988, and Smith and Jell 1990) that can be related to specific vital functions. For example, the evolution of arms as free appendages is clearly related specifically to modification of the food-capturing surface, while the evolution of an attachment disc is related to posture. A certain percentage of characters refer to changes in skeletal organization that have no obvious functional significance and are left in an open category. Many of these reflect increasing organization in skeletal arrangement and may be related to simplification of developmental pathways or structural strengthening.

Table 12.4 shows that, for Cambrian and early Ordovician echinoderms, the majority of skeletal changes with obvious functional significance relate to feeding, and a smaller proportion to posture and respiration. Interestingly, Cambrian echinoderms show no adaptational changes for gaseous exchange, the only changes being, in fact, loss of epispires. It was not until the start of the Ordovician that the evolution of respiratory features became significant.

Fig. 12.3. Evolutionary diversification of early Palaeozoic echinoderms. Stipple, unconsolidated substratum; diagonal hatching, hard substratum.

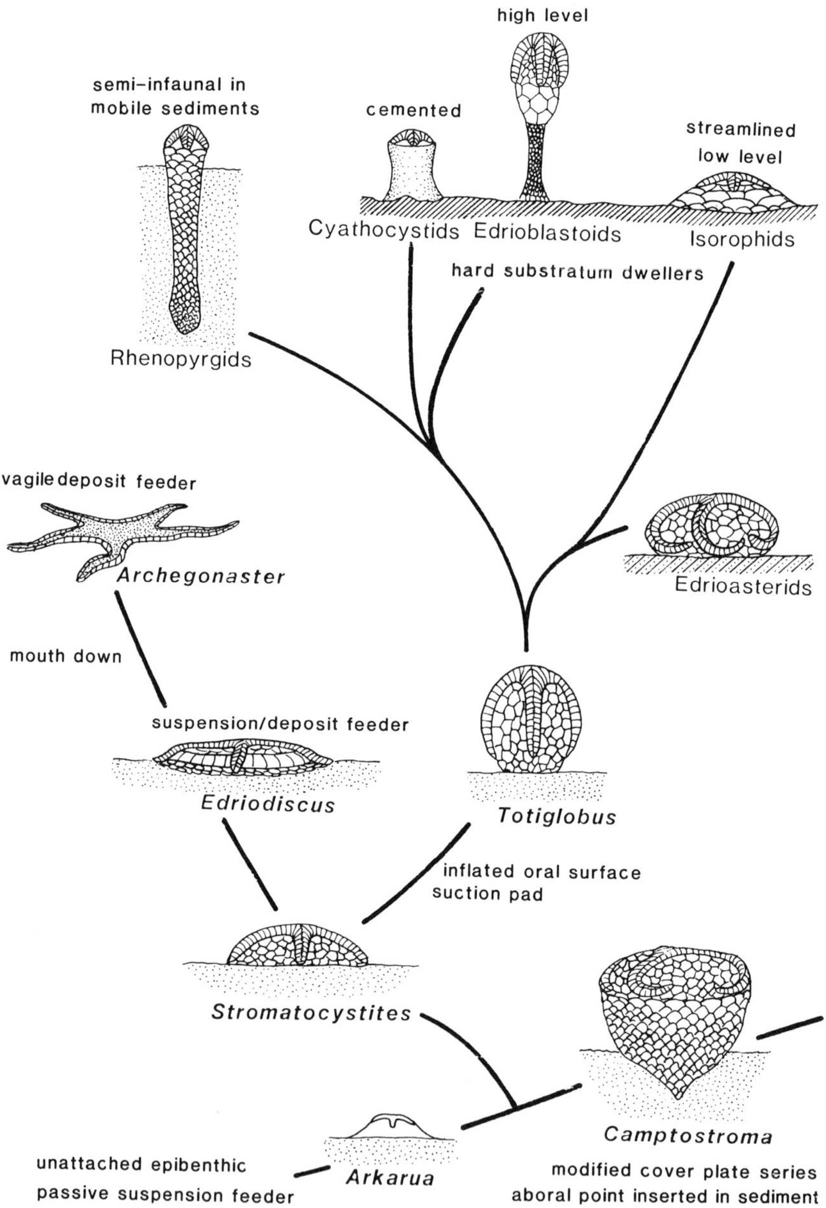

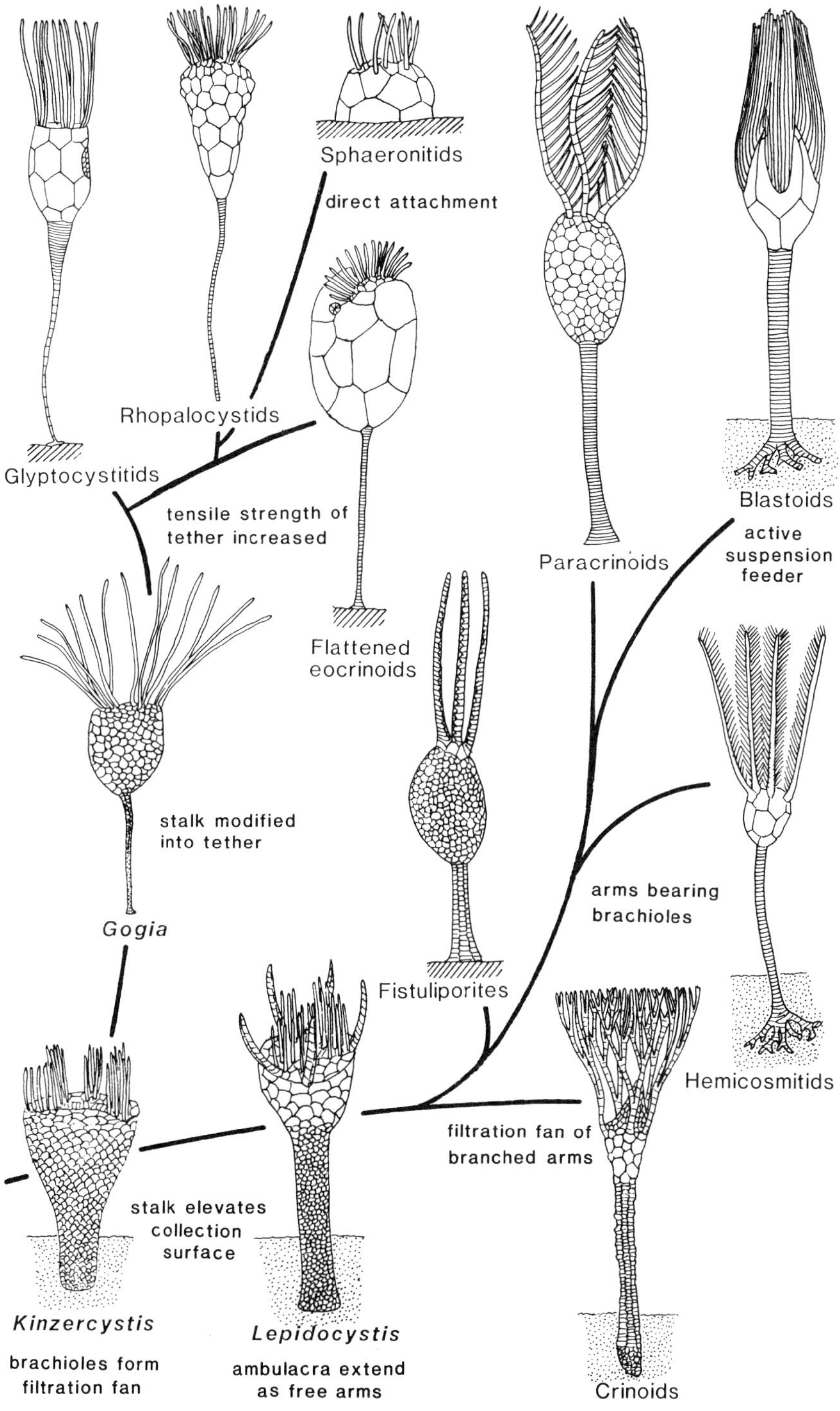
Sphaeronitids
direct attachment
Rhopalocystids
Glyptocystitids
tensile strength of tether increased
Flattened eocrinoids
Paracrinoids
Blastoids
active suspension feeder
Gogia
stalk modified into tether
Fistuliporites
arms bearing brachioles
Hemicosmitids
filtration fan of branched arms
stalk elevates collection surface
Kinzercystis
brachioles form filtration fan
Lepidocystis
ambulacra extend as free arms
Crinoids

Table 12.3. Data matrix for cladistic analysis of pelmatozoan groups presented in Fig. 12.2. All multistate characters were entered unordered; others were polarized using *Camptostroma* as outgroup

Derived states for characters 1–18

1, Aboral surface elongated to form a stalk.
2, Brachioles developed from the ambulacra.
3, Skeleton composed of tesselate plates.
4, Ambulacra (arms) extend free of the theca (1); arms secondarily recumbent (2).
5, Brachioles arise from a single ambulacral plate, not from the suture between two adjacent ambulacrals.
6, Arms (ambulacra) uniserial.
7, Column composed of holomeric columnals (1); pleurimeric, composed of columns of brick-like ossicles (2).
8, Stem xenomorphic, with a proximal expanded zone.
9, Lumen of stem triangular in section.
10, Oral area raised as a spout-like platform.
11, Epispires lost.
12, Respiratory rhombic passage systems developed; pectinirhombs (1); fistulipores (2).
13, Basals differentiated: 5 basals (1); four basals (2); three basals (3); one fused basal (4).
14, Basals unequal, smallest basal opposite the periproct.
15, Arms extend free of the theca immediately adjacent to the peristome rather than around the thecal perimeter.
16, Stalk slender, acting as a tether (true of only some *Gogia*).
17, Deltoid crests developed.
18, Thecal plating highly organized in discrete circlets: hemicosmitid pattern (1); glyptocystitid pattern (2); blastoid pattern (3).

Data matrix

Kinzercystis	110000000000000000
Lepidocystis	110100000000000000
Fistuliporites	101100200112001000
Aethocrinus	101101200010100000
Hemicosmitids	111100100111201001
Coronates	111100100010311013
Blastoids	111200100010311013
Paracrinoids	111111101110311001
Cryptocrinitids	111110101110311001
Gogia	111000000000000100
Ridersia	111000110010200102
Glyptocystitids	111000110011200102
Lingulocystis	111010100010200100
Rhopalocystis	111010100000000100

Table 12.4. The distribution of phylogenetically informative characters in relation to their functional significance for Cambrian to Lower Ordovician echinoderms. Characters used in the construction of cladograms are listed according to the vital function that they primarily affect. Characters for eleutherozoans are taken from Smith and Jell (1990), those for pelmatozoans from Smith (1988). Figures in brackets give percentages of total.

Vital function	Eleutherozoans (*N* = 16)	Pelmatozoans (*N* = 29)	All echinoderms (*N* = 45)
Feeding	7 (44)	10 (35)	17 (38)
Posture	4 (25)	7 (24)	11 (24)
Respiration	1 (6)	5 (17)	6 (13)
Excretion	1 (6)	2 (7)	3 (7)
Defence	—	—	—
Reproduction	—	—	—
Skeletal organization	3 (19)	5 (17)	8 (18)

The single most important characteristic of primitive echinoderms, and the one that provides the key to understanding their evolutionary diversification, is that they were passive suspension feeders. This means that, unlike suspension feeders that create their own currents via ciliary activity, they must have been totally reliant on environmentally produced water currents to carry suspended food particles to the ambulacra where they could be captured. As the earliest radiate echinoderms had no apparent means of attachment, they were presumably restricted to areas of relatively weak (and presumably, therefore, less predictable) current activity, to avoid washout.

In this light the predominance of morphological changes directly affecting the geometry of the food-capturing surfaces and mode of attachment become clear. Echinoderms in the Cambrian were evolving to improve their feeding capability by developing: (1) more efficient geometrical arrangements of the food-capturing surface; and (2) methods of fixation for life in stronger current regimes. The evolution of a tesselate theca occurred in several groups independently and strengthened the theca against impact forces.

The earliest radiate echinoderm

It is important to establish the mode of life of the very earliest radiate echinoderm prior to discussing the ecological diversification of echinoderms during the Cambrian. There are two animals that must be considered: *Arkarua*, which represents a possible late Precambrian pentaradiate echinoderm, and *Helicoplacus*, a Lower Cambrian tri-radiate echinoderm that has been identified as primitive sister-group to pentaradiate echinoderms (Paul and Smith 1984).

Arkarua, described from the late Precambrian Pound Sandstone Group of Australia by Gehling (1987), may be the oldest known echinoderm. Unfortunately, preservation is generally poor and many details of its anatomy remain unknown. However, it is a disc-shaped animal with five rays (?ambulacra) radiating from a central depression (mouth) on its upper surface. It was clearly a sedentary epibenthic passive suspension feeder and apparently simply rested on the sea floor. It is not known whether it had a blind-ending gut or an anus.

Helicoplacus was originally interpreted as a free-living deposit feeder (Durham 1967) but was later reinterpreted as a sessile, erect suspension feeder (Derstler 1981; Paul and Smith 1984). It has three rather short spiral ambulacra leading to a laterally-positioned mouth. It is not known whether an anus was present. The tapering lower part of the body was presumably inserted into the sediment but lacks specific attachment specializations. It was a suspension feeder, using its ambulacra as food-collecting surfaces. By analogy with modern echinoderms, small suspended particles are likely to have been captured by mucus-rich tube feet and passed to the mouth along the ambulacral grooves. Ambulacra are spiral to give all round coverage and extend total food gathering capacity.

Both *Arkarua* and *Helicoplacus* indicate that the earliest radiate echinoderms were passive suspension feeders using specialized parts of their body wall (the ambulacra) as collecting surfaces. They were sessile and either epifaunal or with their base loosely inserted into sediments of the sea floor. Respiration occurred across the body wall without obviously specialized exchange surfaces.

Feeding

1. *Modifications to the design of the collecting surface*

For passive suspension feeders the efficiency with which food particles can be collected from the surrounding water is greatly dependent upon the geometry of the surface that is used to capture these particles. In echinoderms food particles are captured by the tube feet which run along the ambulacra. It is not surprising, therefore, to find that during the Cambrian there were significant changes to the design of ambulacra. There are four obvious ways in which ambulacral arrangement was modified for more efficient food capture.

1. *Curvature of ambulacra.* In eleutherozoans at least three lines independently evolved curved ambulacra. In addition to the obvious advantage of simply increasing the surface area available for capturing particles, there is a second possible advantage: curved ambulacra provide a method for extending the ambulacra around virtually the entire periphery. Because current-induced drag resuspends bottom material downstream, the peripheral ambulacra are ideally positioned for capturing this resuspended material. This is especially true for edrioasterids where the ambulacra virtually rest on the sea floor. Curved ambulacra had already evolved in one group by the late Lower Cambrian and both isorophids and edrioasterids developed curved ambulacra independently.

2. *Inflation of the oral surface.* The relative area of ambulacra can be increased by inflating the oral surface relative to the aboral surface. This happened in the eleutherozoan *Totiglobus* and its descendants. Although this directly increases the relative proportion of body surface involved in food harvesting, it may also serve to elevate much of the collection surface above the sediment/water interface to take advantage of natural currents. Both would directly benefit feeding efficiency.

3. *Extension of ambulacra as arms.* This occurred at least once and possibly twice in pelmatozoans. Having ambulacra that project free of the body serves to increase the food-gathering surface, extends this collection surface well above the boundary layer into 'free-stream' flow, and also offers the potential for developing a filtration fan. Initially, particles were presumably captured by tube feet from the thecal surface. Extending the structures that carried the tube feet away from the body surface as arms must have improved current flow around the capture surface and marks a significant improvement in design. In open-network structures, such as those found in early Palaeozoic pelmatozoans, aerosol filtration (Rubenstein and Koehl 1977) rather than sieving would have been the only possible mechanism of particle capture.

4. *Extending and calcifying tube-feet as brachioles.* Brachioles appear to have arisen by extension and calcification of tube feet from the ambulacra (Paul and Smith 1984). In *Camptostroma* the cover-plate

series shows an early stage in this development. Brachioles, like arms, increase the food-gathering surface and elevate much of it above the boundary layer of the theca. In some groups, such as blastoids and parablastoids, the brachioles formed a dense filtration fan, but in many cases the brachioles were relatively sparse. According to Meyer (1982) more open filtration networks in crinoids are found in species inhabiting lower velocity flow regimes. Brachioles evolved once and were apparently lost by crinoids and fistuliporite cystoids, which may be sister-groups.

2. *Elevation of the feeding surface above the sediment/water interface*

Because water flow is reduced by boundary effects more efficient use of natural currents can be made by elevating the food-capturing surface above the sediment/water interface. Elevation of the ambulacra occurred in two different lineages. Pelmatozoans evolved a stem to raise the oral surface above the sea floor during the Cambrian and this was retained by almost all subsequent descendants. Where the stem was lost, as, for example, in sphaeronitid cystoids, this was usually associated with firm attachment to a hard substratum, presumably an adaptation for life in stronger current regimes. One group of eleutherozoans also evolved a stem and its more derived members, such as *Cambroblastus* and *Astrocystites*, closely mimic pelmatozoans (Smith and Jell 1990).

It must be noted, however, that not all stems acted to elevate the body physically and in some cases they became modified as a tether for life in strong-current regimes (see below).

3. *Other modes of feeding*

All Cambrian radiate echinoderms were passive suspension feeders. Later echinoderms diversified to become active suspension feeders, deposit feeders, grazers, and predators.

No living echinoderm is an active suspension feeder but it is possible that both blastoids and their sister-group, the coronates, adopted active suspension feeding. Previously, blastoids have been interpreted as passive suspension feeders (Breimer and Macurda 1972), but their well developed hydrospire system must clearly have created ciliary-driven currents passed the brachioles. By drawing water in at the base of the brachioles and expelling it upwards out through the centre of the filtration fan, a continuous influx of nutrients would be brought past the ambulacral system. In coronates the large deltoid crests that border the ambulacral system have been interpreted as respiratory adaptations (Brett *et al*. 1983). Although I agree that these surfaces were preadapted for development into thin-walled surfaces for gaseous exchange as seen in blastoids, I believe that they initially evolved as a ciliary pump. Later ridging and crenulation of this surface in fissiculate blastoids increased the surface area of the ciliated surface and improved water flow past the brachioles. It was only at this stage that radial folds became thin-walled and secondarily took on a respiratory function in addition to their role in creating feeding currents.

Parablastoids have a similar arrangement of brachioles and may also have been active suspension feeders, but here water currents were drawn in peripherally and expelled immediately beside the base of the brachioles (Paul and Cope 1982); this arrangement would seem to be less efficient. Active suspension feeding did not evolve until the Lower Ordovician. The oldest member of the coronate/blastoid clade is Llanvirn (Brett *et al.* 1983), and the oldest parablastoid Arenig (Paul and Cope 1982). Blastoids later achieved considerable evolutionary diversity with some 90 genera currently recognized (Waters 1988).

The earliest true deposit feeders are the early Ordovician somasteroids (see Smith and Jell 1990 for a detailed description of the most primitive of these, *Archegonaster*). These gave rise to ophiuroids and asteroids. However, the closest Cambrian relative to *Archegonaster*, the edrioasteroid *Edriodiscus*, has a very flattened body and ambulacra that extend over the peripheral rim right to the margin of the body (Jell *et al.* 1985; Smith and Jell 1990). These arm tips thus extended to the sediment-water interface and it seems likely that *Edriodiscus* collected detrital material both from the rain of suspended particles and from the surrounding sediment, either directly or by resuspension. However, the adoption of deposit feeding involved a major anatomical inversion in orientation with the mouth facing downwards. Interestingly, *Edriodiscus* and its close relatives are the only Cambrian eleutherozoans to lack any form of aboral attachment.

Some asteroids may have become predators by the late Ordovician, when specimens of some species are preserved with their oral cavity filled with small shelly fossils. This mode of wholesale ingestion of prey presumably developed from deposit feeding. The more specialized predatory asteroids that we generally think of today did not evolve until the Mesozoic (Gale 1987).

Grazing on algal mats may also have been a late Ordovician innovation, since it is then that the first echinoids with properly developed lanterns appear. Once again, however, these early lantern feeders may simply have used their lanterns as scoops for deposit feeding, and the earliest undoubted evidence of intensive grazing on hard substrata comes from Triassic echinoids.

Posture and locomotion

The early radiate echinoderms have little in the way of attachment structure and forms such as *Stromatocystites* and *Arkarua* presumably rested on the sea floor. These early echinoderms were faced with a problem—they needed environmental currents to bring food particles within reach, but lacked a means of attachment to resist being washed away by these currents. In order to inhabit environments with stronger current regimes and gain access to more predictable food supply, echinoderms developed a variety of specialized methods of attachment. Brett (1981) has documented the various types of attachment structures found in echinoderms. Only a limited range of such structures had evolved during the Cambrian and Lower Ordovician.

1. *Modes of attachment*

1. *Insertion into unconsolidated sediment.* A certain amount of stability is achieved by inserting the aboral pole within the sediment. This occurred, for example, in *Camptostroma*, helicoplacoids, and primitive pelmatozoans such as *Kinzercystis, Lepidocystis*, and *Aethocrinus*. Later pelmatozoan groups evolved a system of branching rootlets for greater anchorage within sediments.

One group of edrioasteroids, the rhenopyrgids, evolved for life within mobile sediments by developing a long retractile peduncle. Rhenopyrgids probably lived in a similar manner to lingulid brachiopods, using their peduncle to withdraw at times of strong current activity. There is a small, plated scoriaceous attachment pad at the base of a long, retractile peduncle and this disc presumably used some sort of mucus secretion for adhesion. Rhenopyrgids are commonly preserved with the majority of the peduncle fully articulated but with the uppermost part of the theca either collapsed or missing (personal observation). This supports the hypothesis that rhenopyrgids lived largely infaunally with only the apical surface above the sea floor.

2. *Basal attachment pad.* Modification of the aboral surface into an attachment pad occurred early on in the evolution of eleutherozoans. Several of the species of *Stromatocystites* had developed a complex internal system of ridging that has been interpreted as a suction pad (Smith and Jell 1990). This suction pad was developed and enlarged in the lines that gave rise to isorophids and edrioasterids, the former becoming relatively streamlined to resist strong currents. The earliest isorophids are found in relatively shallow-water trough-and cross-bedded calcarenites (Smith and Jell 1990), where adhesion would have been essential.

Only one pelmatozoan group displays a comparable attachment pad, namely *Marjumicystis* and its descendants, the aristocystitids. They have an aboral disc which strongly resembles that of *Stromatocystites* in external appearance (see Ubaghs and Robison 1985, fig. 16.3; Kesling 1967, fig. 148.1g). However, this appears to have been rigid and is often massive, so cannot have been a suction pad. Attachment here was presumably by epithelial adhesion, as in sphaeronitids.

Echmatocrinus is preserved attached directly to a worm tube (Sprinkle 1973) but it is not known how this attachment was achieved.

Sphaeronitid cystoids developed direct attachment, moulding their aboral surface to the contours of the substratum. However, they were apparently not firmly cemented, as they are almost always encountered unattached in lag deposits, and Paul and Bockelie (1983) have argued that they used some form of epithelial adhesion. They also clearly lived in relatively strong current regimes in current-orientated sediments (Paul and Bockelie 1983).

3. *Cementation to the substratum.* The distal ends of columns in several lines of pelmatozoan appear to have been cemented to hard substrata. This is true of rhipidocystids (Lewis *et al.* 1987), paracrinoids (Parsley and Mintz 1975), glyptocystitids (Kesling 1967), and echinosphaeritids. Edrioblastoids may also have had a discoidal holdfast cemented directly to the sea floor, although details of the very base of the column are missing (Smith and Jell 1990). Direct cementation without a stalk took place in cyathocystitids. These are the most specialized of attached primitive echinoderms and have a cylindrical calcite turret composed of a single element (Bockelie and Paul 1983). They lived in shallow-water areas, possibly even intertidally (Bell 1980).

2. *Modification of the stem into a tether*

The pelmatozoan stem, as seen in Lower Cambrian forms such as *Kinzercystis* and *Lepidocystis*, was originally stout and sturdy and clearly served to elevate the food-capturing surfaces. This role continued in groups such as the fistuliporites, blastoids, and early crinoids, for example. However, certain groups reduced their stem to such an extent that it can hardly have served to support the theca, as noted by previous workers (Ubaghs 1960; Paul 1968). In such cases the stem probably functioned as a tether, securing the animal to the substratum and letting the theca and food-gathering surfaces move with the current.

Tethers need only a strong tensile strength and this is best achieved through the use of cable-like structures (Wainwright *et al.* 1976). Reducing the stem to a narrow, collagen-bound series of ossicles fulfils this requirement. The stereom ultrastructure of Middle Cambrian holomeric columnals from Sweden (Berg-Madsen 1986) and Australia (unpublished data) shows that bundles of long penetrative collagen fibres passed through these columnals. This arrangement marks a tremendous improvement in tensile strength over the unorganized peduncle of early stalked echinoderms. Flattened eocrinoids such as *Lingulocystis* are found in shallow-water, high-energy environments and their thecal shape may be streamlined like a kite for lift (Ubaghs 1960). *Protocrinites* and *Glyptosphaerites* have a stem with a much reduced diameter that is unlikely to have had a supportive function. Similarly, *Ridersia* and many glyptocystitids have a xenomorphic stem which has a broad proximal segment with a large lumen and a long, narrow, tapering distal portion whose termination is never seen. Paul (1968) noted that the earliest glyptocystitid *Macrocystella* must have been top-heavy and suggested that it might have lived unattached. This may indeed be the case but basal holdfasts are known in a number of younger glyptocystitids (viz. *Lepadocystis*, Kesling and Mintz 1967) and current-aligned *Ridersia* occur in abundance in channel-sand bodies in the Upper Cambrian of Australia, demonstrating that this species had an affiliation for strong currents. In such cases the absence of a basal holdfast could be due to storm detachment.

3. *Free-living*

One group of eleutherozoans evolved a very different approach to collecting food and became deposit feeders. As discussed above, the adoption of deposit feeding required a major morphological inversion in life habit to bring the mouth face-down on to the substratum. As yet the way in which this came about is not clear, although very few modifications are required to transform a suspension-feeding form such as *Edriodiscus* into the deposit feeder *Archegonaster*. *Archegonaster* has modified or lost its ambulacral cover plates and lost its anus, but has an otherwise basically identical body plan (Smith and Jell 1990).

Other vital functions

1. *Respiration*

Surprisingly little in the way of evolutionary innovation that can be related to respiratory adaptations occurred during the Cambrian. Indeed the most significant changes during this period were not specializations but the loss of epispires, which occurred at least twice independently. It was not until the start of the Ordovician that certain groups began to develop more complex systems for gaseous exchange. As argued above, the development of a hydrospire system for gaseous exchange in blastoids was probably a secondary derivation from ciliary pump suspension feeding.

2. *Protection*

No morphological changes in Cambrian echinoderms can be recognized that relate directly and unambiguously to defence against predators. However, the general strengthening of the theca that took place through structural changes in plate arrangement might be considered as promoting survival to reproductive age.

3. *Reproduction*

Virtually nothing is known about reproductive strategies in primitive echinoderms and there are no obvious morphological changes that can be associated directly with reproduction or dispersal. However, the gonopore is found at the summit of the theca and elevation of the theca above the boundary layer may, in a limited way, have acted to improve dispersal capability.

Species diversity

Assuming that sister-groups have retained approximately similar preservational potential, it may be instructive to compare species diversity achieved within sister-groups where the two branches have evolved different modes of life. For example, if we take the dichotomy between eleutherozoan and pelmatozoan taxa in the early Cambrian it is clear that very many more species of pelmatozoan than eleutherozoan had evolved by the end of the Cambrian. The diversities of the two sister groups are significantly different ($p < 0.01$) using a binomial statistic as estimated by total number of species or total generic diversity up to the Arenig and by species diversity in the Arenig. However, the form that the filtration fan took appears to have had very little influence on species diversity. Comparing species diversity for the monophyletic group derived from *Kinzercystis*, which bore brachioles only, with its sister-group derived from *Lepidocystis*, which had freely extending arms, shows no statistically significant difference in diversity by the end of the Cambrian for both total generic diversity up to the Arenig ($p = 0.44$) and species diversity in the Arenig ($p = 0.43$).

It is tempting to speculate that this success might be due to the development of an improved feeding technique in pelmatozoans. Whereas eleutherozoans retained the original form of thecal surface capture, pelmatozoans developed a projecting filtration fan. Furthermore, although the development of a filtration fan may have made pelmatozoans more successful in terms of species diversity than eleutherozoans, the precise form that the filtration fan took during this initial period had little obvious influence. However, there seems to be no logic for linking filtration efficiency with speciation processes and other scenarios are equally plausible and just as untestable.

The earliest radiate echinoderms were passive low-level suspension feeders that were totally reliant on environmental currents and lacked any form of attachment device. As such they were presumably restricted to environmental regimes of low current strength. The Cambrian record of echinoderms to a large extent documents the group's evolutionary diversification towards more efficient feeding techniques and the ability to live in stronger current regimes. This may therefore represent an example of an offshore-to-onshore radiation which is contrary to the majority of cases in post-Palaeozoic echinoderms (Jablonski and Bottjer 1988; Chapter 2, this volume).

References

Bell, B.M. (1980). Edrioasteroids. *University of Tennessee Department of Geological Sciences Studies in Geology*, **3**, 158–74.

Berg-Madsen, V. (1986). Middle Cambrian cystoid (*sensu lato*) stem columnals from Bornholm, Denmark. *Lethaia*, **19**, 67–80.

Bockelie, J.F. and Paul, C.R.C. (1983). *Cyathotheca suecica* and its bearing on the evolution of the Edrioasteroidea. *Lethaia*, **16**, 257–64.

Breimer, A. and Macurda, D.B. (1972). The phylogeny of the fissiculate blastoids. *Verhandelingen der Koninklijke Nederlandse Akademie van Wetenschappen, afd. Natuurkunde*, **26**, 1–390.

Brett, C.E. (1981). Terminology and functional morphology of attachment structures in pelmatozoan echinoderms. *Lethaia*, **14**, 343–70.

Brett, C.E., Frest, T.J., Sprinkle, J., and Clement, C.R. (1983). Coronoidea: a new class of blastozoan echinoderms based on taxonomic reevaluation of *Stephanocrinus*. *Journal of Paleontology*, **57**, 627–51.

Campbell, K.S.W. and Marshall, C.R. (1987). Rates of evolution among Palaeozoic echinoderms. In *Rates of evolution* (ed. K.S.W. Campbell and M.F. Day), pp. 61–100. Allen and Unwin, London.

Derstler, K.L. (1981). Morphological diversity of early Cambrian echinoderms. *Short Papers for the second international symposium on the Cambrian System. US Geological Survey Open File report*, **81–743**, 71–5.

Donovan, S.K. (1988). The early evolution of the Crinoidea. In *Phylogeny and evolutionary biology of echinoderms* (ed. C.R.C. Paul and A.B. Smith), pp. 235–44, Oxford University Press, Oxford.

Durham, J.W. (1967). Notes on the Helicoplacoidea and early echinoderms. *Journal of Paleontology*, **41**, 97–102.

Farris, J.S. (1988). *Hennig86, version 1.5*. Computer program.

Gale, A.S. (1987). Phylogeny and classification of the Asteroidea (Echinodermata). *Zoological Journal of the Linnean Society*, **89**, 107–32.

Gehling, J.G. (1987). Earliest known echinoderm—a new Ediacaran fossil from the Pound Subgroup of South Australia. *Alcheringa*, **11**, 337–45.

Jablonski, D. and Bottjer, D.J. (1988). Onshore-offshore evolutionary patterns in post-Paleozoic echinoderms: a preliminary analysis. In *Echinoderm biology* (ed. R. D. Burke, P. V. Mladenov, P. Lambert, and R. L. Parsley), pp. 81–90. Balkema, Rotterdam.

Jell, P. A., Burrett, C. F., and Banks, M. R. (1985). Cambrian and Ordovician echinoderms from eastern Australia. *Alcheringa*, **9**, 183–208.

Kesling, R. V. (1967). Cystoids. In *Treatise on invertebrate paleontology*. Part S. *Echinodermata 1* (ed. R. C. Moore), pp. 85–267. Geological Society of America and University of Kansas Press, Boulder, Colorado and Lawrence, Kansas.

Kesling, R. V. and Mintz, LW. (1961). Notes on *Lepadocystis moorei* (Meek) an Upper Ordovician callocystitid cystoid. *Contributions from the Museum of Paleontology of the University of Michigan*, **17**, 123–48.

Lewis, R. D., Sprinkle, J., Bailey, B., Moffit, J., and Parsley, R. L. (1987). *Mandalacystis*, a new rhipidocystid eocrinoid from the Whiterockian Stage (Ordovician) in Oklahoma and Nevada. *Journal of Paleontology*, **61**, 1222–35.

Meyer, D. L. (1982). Food and feeding mechanisms: Crinozoa. In *Echinoderm nutrition* (ed. M. Jangoux and J. M. Lawrence), pp. 25–42. Balkema, Rotterdam.

Parsley, R. L. and Mintz, L. W. (1975). North American Paracrinoidea (Ordovician: Paracrinozoa, New Echinodermata). *Bulletins of American Paleontology*, **68**, 5–115.

Paul, C. R. C. (1968). *Macrocystella* Callaway, the earliest glyptocystitid cystoid. *Palaeontology*, **113**, 580–600.

Paul, C. R. C. (1977). Evolution of primitive echinoderms. In *Patterns of evolution* (ed. A. Hallam) pp. 123–58. Elsevier, Amsterdam.

Paul, C. R. C. (1979). Early echinoderm radiation. In *The origin of major invertebrate groups*, Systematics Association Special Publication, no. 21 (ed. M. R. House), pp. 415–34. Academic Press, London.

Paul, C. R. C. (1988*a*). The phylogeny of cystoids. In *Echinoderm phylogeny and evolutionary biology* (ed. C. R. C. Paul and A. B. Smith), pp. 199–214, Oxford University Press, Oxford.

Paul, C. R. C. (1988*b*). *Extinction and survival in the echinoderms*. In *Extinction and survival in the fossil record*, Systematics Association Special Publication, no. 34 (ed. G. P. Larwood), pp. 155–70. Clarendon Press, Oxford.

Paul, C. R. C. and Bockelie, J. F. (1983). Evolution and functional morphology of the cystoid *Sphaeronites* in Britain and Scandinavia. *Palaeontology*, **26**, 687–734.

Paul, C. R. C. and Cope, J. C. W. (1982). A parablastoid from the Arenig of South Wales. *Palaeontology*, **25**, 499–507.

Paul, C. R. C. and Smith, A. B. (1984). The early radiation and phylogeny of echinoderms. *Biological Reviews*, **59**, 443–81.

Rubenstein, D. I. and Koehl, M. A. R. (1977). The mechanisms of filter feeding: some theoretical considerations. *American Naturalist*, **111**, 981–94.

Sepkoski, J. J. (1979). A kinetic model of Phanerozoic taxonomic diversity, II, early Phanerozoic families and multiple equilibria. *Paleobiology*, **5**, 222–51.

Smith, A.B. (1988). Patterns of diversification and extinction in early Palaeozoic echinoderms. *Palaeontology*, **32**, 799–828.
Smith, A.B. (1990). Echinoid evolution from the Triassic to Lower Jurassic. *Cahiers de l'Institut Catholique de Lyon, series des Sciences*, **3**, 79–117.
Smith, A.B. and Jell, P.A. (1990). New and poorly known Cambrian edrioasteroids from Australia and a redescription of the early Ordovician starfish *Archegonaster*. *Memoirs of the Queensland Museum*. (In press.)
Smith, A.B. and Patterson, C. (1988). The influence of taxonomic method on the perception of patterns of evolution. *Evolutionary Biology*, **23**, 127–216.
Sprinkle, J. (1973). *Morphology and evolution of blastozoan echinoderms*. Museum of Comparative Zoology, Harvard University, Cambridge, Massachusetts.
Sprinkle, J. (1976). Classification and phylogeny of 'pelmatozoan' echinoderms. *Systematic Zoology*, **25**, 83–91.
Sprinkle, J. (1980*a*). An overview of the fossil record. In *Echinoderms: notes for a short course* (ed. T.W. Broadhead and J.A. Waters). University of Tennessee Department of Geological Sciences Studies in Geology, no. 3, pp. 15–26.
Sprinkle, J. (1980*b*). Early diversification. In *Echinoderms: notes for a short course* (ed. T.W. Broadhead and J.A. Waters). University of Tennessee Department of Geological Sciences Studies in Geology, no. 3, pp. 86–93.
Sprinkle, J. (1981). Diversity and evolutionary patterns of Cambrian echinoderms. *Short Papers for the second international symposium on the Cambrian System. US Geological Survey Open File report*, **81–743**, 219–21.
Sprinkle, J. (1983). Patterns and problems in echinoderm evolution. *Echinoderm Studies*, **1**, 1–18.
Ubaghs, G. (1960). Le genre *Lingulocystis* Thoral. *Annales de Paleontologie*, **46**, 81–115.
Ubaghs, G. (1975). Early Palaeozoic echinoderms. *Reviews on Earth and Planetary Science*, **3**, 79–98.
Ubaghs, G. and Robison, R.A. (1985). A new homoiostelean and a new eocrinoid from the Middle Cambrian of Utah. *Paleontological Contributions of the University of Kansas*. no. 115, 1–24. University of Kansas Press. Lawrence, Kansas.
Wainwright, S.A., Briggs, W.D., Currey, J.D., and Gosline, J.M. (1976). *Mechanical design in organisms*. Edward Arnold, London.
Waters, J.A. (1988). The evolutionary palaeoecology of the Blastoidea. In *Echinoderm phylogeny and evolutionary biology* (ed. C.R.C. Paul and A.B. Smith), pp. 215–34. Oxford University Press, Oxford.

13. The radiation of post-Palaeozoic echinoderms

MICHAEL J. SIMMS
Department of Geology, Trinity College, Dublin, Ireland

Abstract

Of the five extant echinoderm classes at least two, crinoids and echinoids, experienced a decline in diversity in the late Palaeozoic, with only a handful of taxa surviving into the early Triassic, whilst cladistic analysis of the asteroids suggests that they too passed through an evolutionary bottle-neck at this time. Only among echinoids and crinoids is the fossil record sufficiently well documented and representative for their subsequent post-Palaeozoic history to be interpreted. From very low diversity in the early Triassic, both groups diversified rapidly in the mid-Triassic to an early Carnian peak. Several important Triassic clades became extinct in the mid-Carnian, followed by a second phase of diversification in the Norian and early Jurassic when the fundamental character of post-Triassic echinoderm faunas was established. Throughout their post-Palaeozoic history, diversity has increased through two evolutionary strategies: expansion into juvenile morphotypes through paedomorphosis, and the optimization of existing adult morphology. By comparison with their Palaeozoic antecedents, post-Palaeozoic crinoids show few, if any, morphological innovations related to feeding and largely occupy niches similar to those inhabited by various Palaeozoic crinoids. Echinoids have experienced a number of major adaptive breakthroughs in feeding strategy following the first appearance of irregular echinoids. The initial breakthrough was the development of tube-foot feeding and later the appearance of pencillate tube feet in atelostomates and numerous tiny tube feet in clypeasteroids. It is doubtful that any of these strategies had evolved in Palaeozoic echinoids, yet irregular echinoids now account for almost 50 per cent of extant echinoid diversity.

Major Evolutionary Radiations (ed. P.D. Taylor and G.P. Larwood) Systematics Association Special Volume No. 42, pp. 287–304, Clarendon Press, Oxford, 1990.

Introduction

Throughout much of the late Palaeozoic echinoderms were an abundant and diverse group, yet towards the end of the Permian they suffered a substantial decline in which many monophyletic taxa became extinct (Paul 1988). By Permian times at least six classes of echinoderm remained, of which five survived into the Triassic and are extant. Blastoids remained reasonably diverse into the late Permian (Waters 1988) but failed to cross the Permo-Triassic boundary. However, considering the magnitude of the overall decline in diversity among those groups for which there is adequate documentation, it is perhaps surprising that blastoids were the only echinoderm class to disappear at the end of the Permian.

Documentation of the post-Palaeozoic history of each of the five extant classes varies enormously. Present knowledge of ophiuroids during this interval is obscured by an outdated taxonomic treatment and a poorly documented fossil record so that nothing can be said concerning their diversity through time. For holothurians the fossil record is extremely poor. Isolated spicules are often abundant but intact material is known from very few localities or horizons. Recent work shows many late Palaeozoic sclerite morphotypes passing through into the Triassic, followed by a large increase in morphological diversity within the Triassic (P. Gilliland, personal communication) but little else can be said about their history. The fossil record of asteroids is very poor during the Triassic but cladistic analyses indicate that post-Palaeozoic, asteroids are a monophyletic group derived from a common ancestor in the late Permian or early Triassic (Blake 1987; Gale 1987). Although reaching the same general conclusion, these two accounts otherwise differ very considerably in their opinion of the polarity of characters and the relationships of major groups.

For the remaining two groups, the crinoids and echinoids, cladistic analyses also indicate that the taxonomic diversity of both groups was very low in the earliest Triassic (Smith 1984, in press; Simms 1988*a*). These conclusions are borne out by evidence from the fossil record which in both groups is reasonably well documented. The current diversity of each of the five classes lies between 600 and 2000 extant species so in all cases it is obvious that there has been a considerable increase in diversity since the early Triassic. In the following account I shall attempt to interpret when and how this diversification occurred in those two groups which have the best documented fossil record —crinoids and echinoids.

The diversification of crinoids

Documentation of crinoids in the Triassic and early Jurassic is better than for most of the remaining Mesozoic and Tertiary. The fossil record of this group is rather poor through much of the Ladinian and Norian but otherwise is felt to be reasonably representative of true diversity levels between the late Scythian and the Pliensbachian, judging from the relatively low incidence of Lazarus taxa at generic and higher level during this interval (unpublished data). However, for the Toarcian and most of the remainder of the Jurassic the record appears less representative, as well as being more poorly documented, so that real diversity changes are difficult to detect with any confidence.

In the early Triassic only a small number of species of *Holocrinus* appear to have been present. There was a fairly rapid diversification in the Middle Triassic, notably with the appearance of the encrinids, reaching a peak in the early Carnian (Fig. 13.1). There then followed a major decline in the mid-Carnian, though this might initially be attributed to sampling error associated with the poor fossil record in the Norian. However, the complete absence of encrinids, a diverse and highly distinctive clade, beyond the early Carnian indicates that this was a real extinction event. A significant episode of biotic turnover around the Lower–Middle Carnian boundary has also been docu-

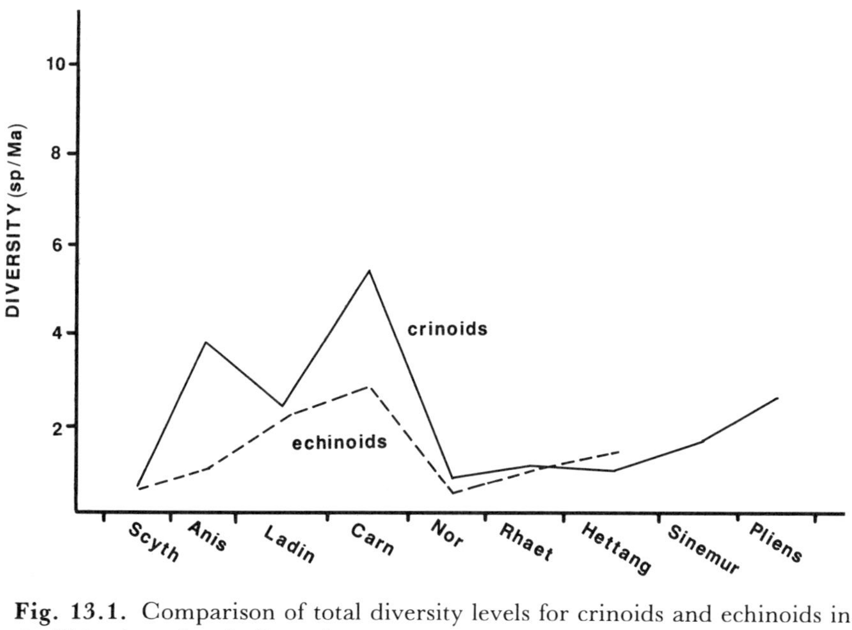

Fig. 13.1. Comparison of total diversity levels for crinoids and echinoids in the Triassic and early Jurassic.

mented for a variety of other marine invertebrate groups (Benton 1986; Johnson and Simms 1989), suggesting that diversity changes across this interval were controlled by extrinsic factors, probably associated with a major climatic change in the Middle and Upper Carnian (Simms and Ruffell 1989, 1990). Although the Norian record is poor, several important post-Triassic groups are first encountered in the late Norian and Hettangian, among them the pseudoplanktonic pentacrinitids, the paracomatulids, and the true comatulids (Kristan-Tollmann 1989). The subsequent history of crinoids in the early Jurassic is one of gradual expansion through the Sinemurian and Pliensbachian before the record becomes too unreliable in the Toarcian to interpret meaningfully.

The diversification of crinoids, both before and after the mid-Carnian decline, occurred through the extension of habitat range. This was achieved by two main strategies: paedomorphosis and the optimization of existing adult morphology, the latter often occurring through processes generally interpreted as heterochronous but operating on particular structures rather than the entire organism.

1. *Paedomorphosis*

The existence of large adult crinoids necessarily implies that the much smaller juvenile morphology is equally viable. Diversification into juvenile morphotypes through heterochrony therefore seems almost inevitable. This may occur via a number of different pathways, summarized by McNamara (1986). In progenesis (precocious sexual maturation) the descendant taxon may most closely approach the ancestral juvenile in its morphology. In contrast, forms arising through neoteny (reduction in the rate of morphological development) or post-displacement (delayed onset of growth) may have the ancestral juvenile morphology modified to a greater or lesser extent by functional or constructional limitations imposed by differences in size between the descendant adult and the ancestral juvenile. Thus paedomorphosis does not represent a simple pathway into juvenile morphologies but represents an evolutionary gateway into a broader range of morphotypes.

There are many examples of important articulate crinoid clades which have arisen through paedomorphosis. The Cretaceous to Recent bourgueticrinids represent an offshoot of the unstalked comatulids which have returned to a stalked mode of life through the neotenous retention of the stem found in larval comatulids (Simms 1988*a*). Similarly, an important component of the early Jurassic diversification of the Isocrinidae occurred through progenesis, with the appearance of the genus *Balanocrinus* (Simms 1988*b*, *c*), and this has probably occurred on many other occasions yet to be documented. Possibly the most striking example of paedomorphosis is found in the three Mesozoic groups of supposedly planktonic microcrinoids: the Triassic Somphocrinidae, the late Jurassic to late Cretaceous Saccocomidae, and the Cretaceous Roveacrinidae. The small size of these crinoids is probably due to paedomorphism which has been so extreme that it is virtually impossible to establish the relationships of these groups to other articulate groups. However, unlike the Palaeozoic cladid and disparid crinoids, articulates seem never to have attained the diversity or minute size of the benthic microcrinoids found in the Palaeozoic. The reason for this remains enigmatic.

2. *Optimization of existing adult morphology*

Optimization of existing adult morphology can operate on various aspects of the crinoid's morphology but is usually most evident in feeding structures. An improvement in feeding efficiency will allow a greater proportion of the total energy budget to be invested in reproduction, thereby increasing the organism's fitness. Selection pressure for improved efficiency of the crinoid filtration fan is reflected in several distinct 'trends' (these are not trends in the strict sense but represent the extremes within a general increase in variance, as discussed by Gould 1988). One such 'trend' documented among Triassic crinoids is seen in the development of biserial arms in the encrinids. In all articulates the pinnules arise from only one side of each brachial; hence a biserial arrangement of brachials in the arms allows a greater number of pinnules than in a uniserial arm of comparable length (Fig. 13.2). Encrinids were the largest single clade in the Triassic, diversifying

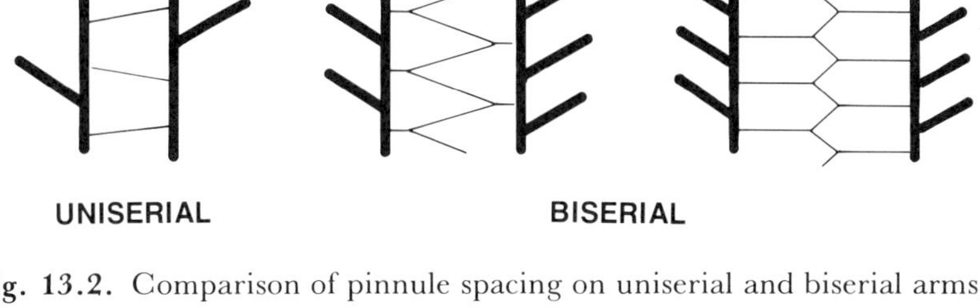

Fig. 13.2. Comparison of pinnule spacing on uniserial and biserial arms.

rapidly from their first appearance in the early Anisian. This might be interpreted as evidence that the development of biseriality in this group was a major adaptive breakthrough. However, throughout their history encrinid diversity was almost exactly matched by diversity levels among the remaining, uniserially-armed crinoids (Fig. 13.3). The implication of this is that biseriality in itself did not confer a major adaptive advantage over uniseriality. Biseriality failed to re-evolve following the demise of the encrinids in the mid-Carnian, further evidence that this strategy for increasing pinnule frequency was not the panacea that it might at first seem. One possible reason for this is perhaps indicated by a bizarre encrinid, *Vostacovacrinus boreus*, described from the USSR (Eltycheva and Poliarnaya 1986). In this species high pinnule frequency was achieved by reducing the height of the brachials to less than half that seen in the strongly biserial arms of its sister species, *Traumatocrinus caudex*. Such a strategy involves no extra expenditure of material but results in much greater flexibility of the arms.

A more widespread strategy to optimize filtration fan efficiency, which has occurred independently on many occasions, is achieved through an increase in the number of divisions in the arms thereby also

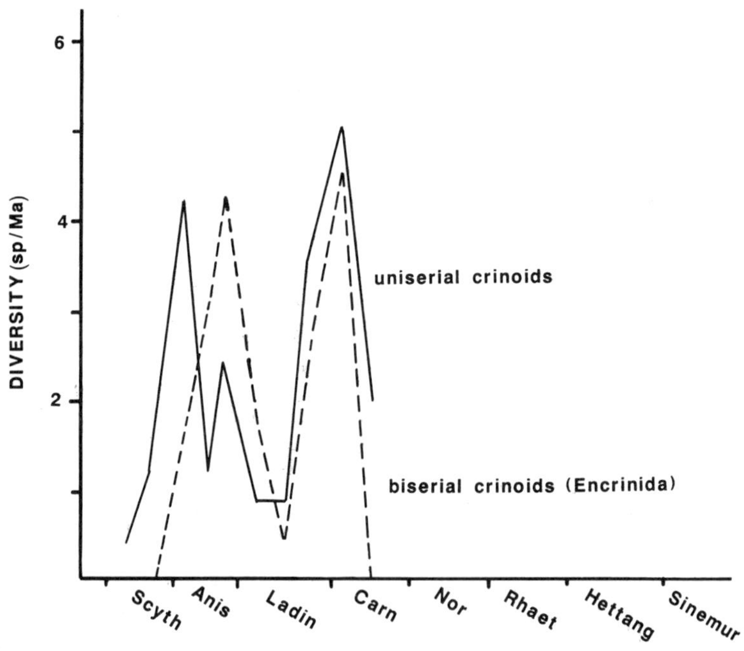

Fig. 13.3. Comparison of diversity levels in biserial- and uniserial-armed crinoid taxa through the Triassic.

increasing the total number of pinnules and hence tube feet for a crinoid of a given size. Such 'trends', in the sense of Gould (1988), towards increased arm branching are seen in the encrinids too. The earliest forms had only 10 arms but they were soon joined by species with 20 arms. The 'trend' was carried to its ultimate conclusion in *Traumatocrinus* which had a complex endotomous pattern of arm branching (Mu 1949). Endotomy has been considered to represent the optimum arm branching arrangement for an efficient filtration network necessitating the minimum expenditure of energy and material in its construction. An analogy has been drawn with the ideal arrangement of roads on a banana plantation (Cowen 1981). An endotomous filtration fan has significantly fewer gaps between pinnules than in the more usual isotomous arrangement of arm branching. Other examples of endotomy have been observed in *'Encrinus' varians*, among the Jurassic Pentacrinitidae (Simms 1986), and in the extant genus *Teliocrinus*, but again, like biseriality, this strategy is much less common than might be expected considering the obvious advantages it confers. The reason for this is unclear. It may be related to the relative clumsiness of endotomous arms when not actually being held in a parabolic filtration fan, the sheer number of arm branches preventing them being closed together as a protective screen.

Although the development of biserial arms among the encrinids was clearly a significant breakthrough for Triassic articulates, it was by no means a morphological innovation for crinoids as a whole, having been developed independently by a variety of Palaeozoic crinoid groups. Indeed, in terms of feeding strategies the articulate crinoids do not appear to have experienced any major adaptive breakthrough since, with the possible exception of some encrinids which may have been active filter feeders (Jefferies 1989), all post-Palaeozoic crinoids adopted essentially the same basic feeding method of passive filter feeding. In terms of their methods of attachment and locomotion, and their position in the water column, three distinct groups of articulates appear to have exploited niches which were significantly different from those occupied by any Palaeozoic crinoids.

In late Triassic and early Jurassic times the Pentacrinitidae adopted an obligate pseudoplanktonic life-style (Simms 1986). With the possible exception of some late Devonian melocrinitid camerates (Wells 1941; McIntosh 1978) that may have been facultative pseudoplankton (Wignall and Simms 1990), the pentacrinitids were unique among crinoids. However, although this placed them in a relatively stable environment with a rich food, supply, their diversity remained low, with never more than a single species of each of the two genera existing at any one time. This may reflect the severe functional and biological constraints associated with attachment to floating driftwood. Hence, this episode in crinoid evolution may perhaps be regarded as a significant adaptive breakthrough without a subsequent adaptive radiation.

At least three distinct groups of microcrinoid are known from the Mesozoic: the late Triassic Somphocrinidae, the late Jurassic to late Cretaceous Saccocomidae, and the Cretaceous Roveacrinidae. Their small size, light construction, and cosmopolitan distribution have been taken as evidence for a planktonic mode of life. Assuming this to be correct then they are unparalleled by any Palaeozoic group. Among Palaeozoic crinoids only the large camerate genus *Scyphocrinites* has been interpreted as planktonic and it adopted a quite different strategy to keep afloat, with the development of a large buoyant 'lobolith' (Haude 1972) at the end of the stem. The Mesozoic planktonic microcrinoids have often been grouped together in the Order Roveacrinida, though their relationship to each other and to other crinoid groups remains unclear. Although they were the only articulate crinoids to adopt a planktonic life-style (a planktonic mode of life is now considered unlikely for the Cretaceous genera *Uintacrinus* and *Marsupites* (unpublished data)) and were often extraordinarily abundant, they never achieved particularly high diversity levels and only one group, the saccocomids, survived for more than a few stages. Somewhat enigmatically, saccocomids were the least diverse of the three despite having the longest range, from the Oxfordian to the Maastrichtian (J.W.M. Jagt, personal communication), an interval of some 90 Ma. The reason for their apparent lack of success, at least in terms of taxonomic diversity if not in terms of actual abundance at any one time, remains unknown.

The evolution of comatulid crinoids is often cited as an example of adaptive radiation (Meyer and Macurda 1977). Certainly it is true that their present diversity is considerably greater than that of all other crinoids put together, and it is also probably true that many extant comatulid taxa have much better developed locomotory capabilities than were found in any Palaeozoic forms. However, attainment of this highly motile morphology in comatulids did not occur in any 'explosive radiation' but developed over a considerable time interval during which diversity levels remained comparatively low. The initial evolution of the basic comatulid morphotype appears to have been fairly rapid. The earliest known pentacrinitids, sister-group to the Comatulidina (Simms 1988*a*), occur in the Upper Norian and hence are virtually contemporaneous with the oldest paracomatulid, currently being described (Hans Hagdorn, personal communication), in which the very short stem is effectively a 'segmented centrodorsal'. The earliest true comatulid, in which the centrodorsal comprises only a single ossicle, is found only slightly later, in the succeeding Hettangian (Kristan-Tollmann 1989).

However, despite this apparently rapid transition between stalked crinoids and true comatulids, comatulid diversity appears to have remained low until well into the Jurassic. Between the late Triassic and the end of the early Jurassic only five species of comatulid or paracomatulacean are known (Simms 1988*a*, *d*, 1989), with only a further 13 species recorded from the Middle Jurassic (Biese 1935–37), a total interval of some 50 Ma (Harland *et al*. 1982). It is clear from their disjunct stratigraphic distribution that paracomatulaceans are grossly underrepresented in the fossil record. This is widely held to be the case for true comatulids too. Their apparently poor fossil record during the Jurassic is often attributed to the rapid disarticulation of the skeleton following death (Liddell 1975; Meyer and Meyer 1986) and they are often held to have a lower preservation potential than other articulate crinoids (Bambach 1985), though there are no documented comparisons between different articulates to support this. Admittedly, the intact comatulid skeleton has a very low preservation potential, though probably no worse than for stalked articulates, but isolated centrodorsals are comparatively robust and should have a preservation potential at least as good as that for pluricolumnals of stalked crinoids. The possibility that comatulids inhabited high-energy environments, thereby reducing their preservation potential, seems insufficient to account for their rarity in the Middle Jurassic where such facies commonly yield abundant remains of millericrinids. Selective removal of calyces by predators, as suggested for recent comatulids by Meyer and Meyer (1986), also seems unlikely since the teleost fish which now prey on comatulids did not appear until late Cretaceous times and there is minimal evidence for predation on crinoids during the Triassic and Jurassic (Schneider 1988). Considered together these lines of evidence indicate that comatulid diversity remained comparatively low until at least the start of the Upper Jurassic and that, although they are now the most successful single crinoid group, their diversification was far from an 'explosive radiation'. The eventual success of comatulids is probably attributable to the greater ecological flexibility attained by using the arms and cirri for attachment to the substrate and for locomotion. The highly muscular arms of articulates, a character otherwise found only in a few advanced cladids of the late Palaeozoic, represented a pre-adaptation to their use by comatulids for anchorage and locomotion. Despite the apparent advantages of the comatulid morphotype, it is clearly not superior in all respects to the stemmed morphotype since at least one comatulid offshoot, the bourgueticrinids, have reverted to a stalked mode of life through neotenous retention of the larval comatulid stem.

Echinoids

The Triassic and early Jurassic diversification of echinoids shows similar patterns to that of the crinoids, with two distinct episodes of expansion interrupted by a mid-Carnian decline (Fig. 13.1; Smith, in press). Echinoids have also adopted essentially the same strategies for expansion: paedomorphosis and optimization.

The small size of many Triassic and early Jurassic echinoids indicates that paedomorphosis has played at least some part in their evolution. In particular, Smith (1981) has claimed an important role for paedomorphosis in the origin of the early Jurassic *Eodiadema*, and hence for all irregular echinoids. Somewhat earlier in time, the early Carnian *Tiarechinus* has been so profoundly modified by paedomorphosis that, like the planktonic microcrinoids already discussed, its sister-group relationships are unclear (Kier 1977).

A major component of the diversification of post-Palaeozoic echinoids has occurred through optimization of their feeding strategies as algal grazers or general scavengers. Both cidaroids and euechinoids independently evolved stronger teeth with stronger and lighter lanterns operated by more powerful and efficient lantern muscles attached to a perignathic girdle. Typical cidaroid tooth and lantern support structures, the latter developed from interambulacral processes, were already present in Permian times (Kier 1984; Smith and Hollingworth 1990). A second group of 'miocidarids' which lacked any type of perignathic girdle also crossed from the Permian into the Triassic where, in late Triassic times, they gave rise to the euechinoids with lantern supports developed in the ambulacra. The lantern mechanism was further improved by an increase in the rigidity of the test, a necessary development for the efficient operation of the lantern muscles. However, despite these improvements regular echinoids experienced a relatively modest radiation. In terms of numbers of genera, cidaroids have diversified since the Triassic but only to levels comparable with those lost by the extinction of the exclusively Palaeozoic orders (Bambach 1985). The two main regular euechinoid clades, the Diadematacea and the Echinacea, diversified in the late Triassic and Jurassic but failed to expand further and still retain their mid-Mesozoic diversity levels (Bambach 1985). Like cidaroids, diademataceans are primarily herbivorous epifaunal forms but were much more lightly constructed and more mobile. They also had a more fully developed perignathic girdle for attachment of the lantern muscles. Compared with the diademataceans, the echinaceans have a superior lantern design and have developed carnivory as well as herbivory. Some even developed an ability to bore into rocky substrate. Thus the echinaceans were able to exploit a more diverse range of niches than diademataceans and this is reflected in the greater generic diversity of echinaceans, which is almost four times that of the former group (Bambach 1985). Even within the Echinacea there has been further optimization of lantern structure and the progressive displacement of stirodont echinoids by camarodont echinoids, which have a mechanically superior lantern, is a superb example of this strategy of diversification and replacement through optimization of an existing feeding strategy.

In early Jurassic times the first irregular echinoids appeared. The small, early Jurassic regular echinoid *Eodiadema* probably lies close to the common ancestry of all irregular echinoids since it shares with them the synapomorphy of diamond-shaped teeth. In the Sinemurian '*Plesiechinus*' *hawkinsi* the test is flattened to provide greater stability on loose sediment and the typical diadematoid arrangement of one large tubercle and spine per plate is replaced by three to five smaller tubercles and spines, anticipating the development of a uniform spine canopy in later irregulars, though paralleling the arrangement found in some late Triassic species of *Diademopsis* and '*Hemipedina*'. In terms of defining the irregular echinoid clade, the most significant morphological development in these early forms was the migration of the periproct away from the apex of the test in response to the adoption of a sediment-rich diet. The position of the periproct is only slightly eccentric in '*Plesiechinus*' *hawkinsi* and it almost certainly still fed using the lantern to collect detritus from the sediment rather than swallowing large quantities of sediment.

By Toarcian times two distinct groups of irregular echinoids had developed. The Holectypoida refined the feeding strategy of *Plesiechinus* though still using the lantern to collect sediment. The further migration of the periproct indicates the ingestion of much larger quantities of sediment but holectypoids do not appear to have been efficient burrowers able to plough through sediment, but merely buried themselves during the day as protection against predators (Smith 1978). Despite the adoption of a sediment-rich diet, holectypoids were morphologically conservative and never attained particularly high diversity levels (Fig. 13.4; Rose and Olver 1985). Only a few species survive today.

The second group of irregulars in the Toarcian were the Galeropygidae, represented first by *Eogaleropygus*. This group developed a fairly uniform aboral spine canopy and, somewhat later in time (Middle Jurassic), an anal sulcus to channel faeces away from the periproct. Both features are consistent with a fully infaunal mode of life in coarse sediment. More significantly, however, the tube feet had started to become differentiated, with the aboral tube feet becoming thin-walled for gaseous exchange while suckered peri-oral tube feet were

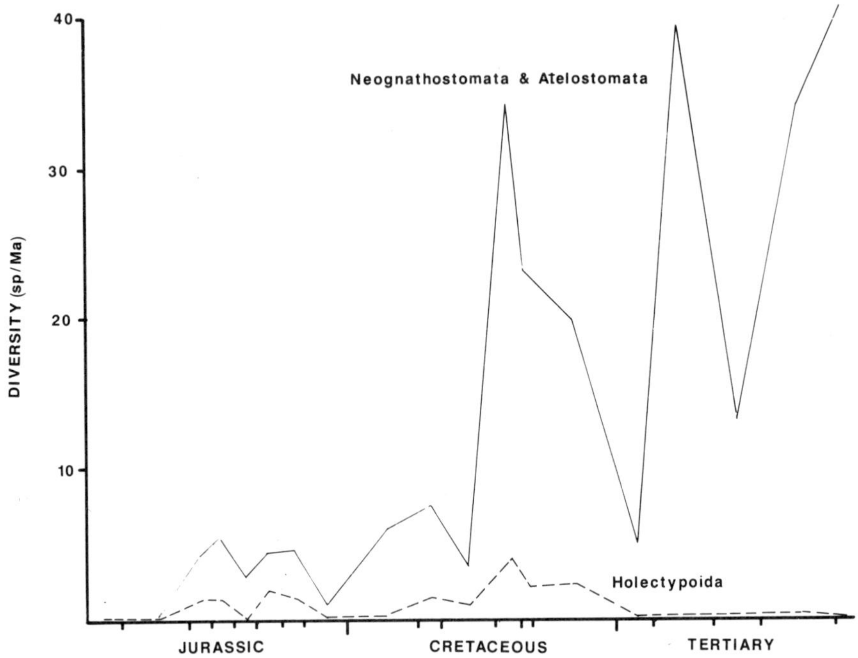

Fig. 13.4. Comparison of diversity levels between irregular echinoids with tube-foot feeding (atelostomates and neognathostomates) and those lacking tube-foot feeding (holectypoids) (based on Kier 1974).

used to collect material and transfer it to the mouth. The lantern was lost entirely in the adult.

Galeropygids and their derivatives diversified rapidly from Middle Jurassic times onwards and are now broadly divided into two groups: the Atelostomata, comprising the holasteroids and spatangoids; and the Neognathostomata, comprising the cassiduloids and clypeasteroids. In total these two groups comprise almost half of all extant echinoids. They are clearly very much more successful than the holectypoids (Fig. 13.4) despite their slightly later appearance. The obvious implication of this is that the migration of the periproct in association with a sediment-rich diet was not, in itself, a major adaptive breakthrough whereas the adoption of peri-oral tube foot feeding clearly was and opened up a major new trophic resource for echinoids to exploit.

Although much of the Jurassic diversification of neognathostomates and atelostomates can be attributed to the optimization of this basic feeding strategy, at least part of their overall success can be ascribed to two further adaptive breakthroughs. Cassiduloids became specialized for swallowing large quantities of sand-sized particles and so were able to exploit sediments of relatively low organic content. The Middle and Upper Jurassic disasteroids were the first important offshoot from the cassiduloids. They diversified slowly until the Jurassic when pencillate tube feet probably appeared for the first time (Smith 1980, 1984). The evolution of pencillate tube feet enabled them to collect finer particles, by mucous adhesion, than was possible with suckered oral tube feet. This development enabled them to feed in finer muds than other echinoids and probably accounted for the major radiation of holasteroids and spatangoids in the Cretaceous (Fig. 13.5). The other major offshoot from the cassiduloids was the clypeasteroids. This group diversified rapidly from their first appearance in the Palaeocene genus *Togocyamus*. The adaptive breakthrough here was in the development, through heterochrony, of large numbers of tiny suckered tube feet which enabled them to manipulate finer sediments than could the cassiduloids. A second innovation, whose function, however, is contested by Telford *et al.* (1985), was the development of spine-free food grooves. Ciliary currents associated with the respiratory tube feet are thought to bring fine particles, sieved from the overlying sediment through the aboral spine canopy, along these grooves towards the mouth. The optimum limits of this feeding strategy were rapidly

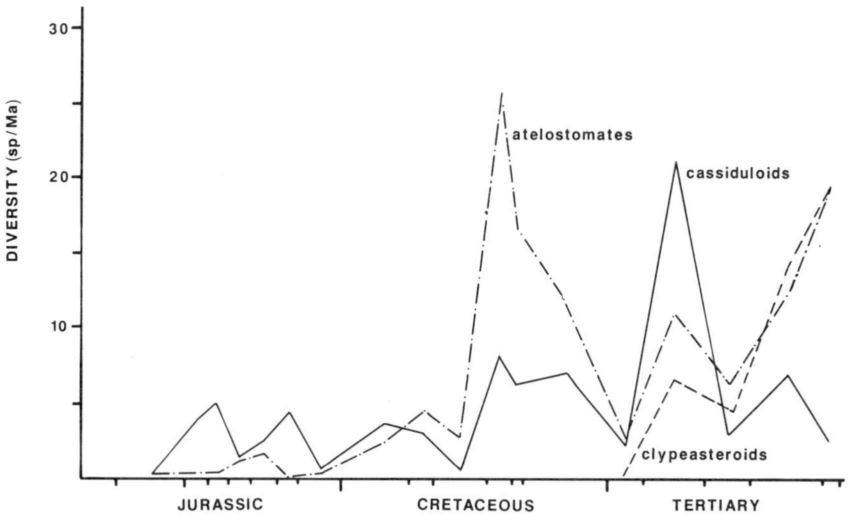

Fig. 13.5. Comparison of diversity levels of cassiduloids (solid line), atelostomates (dashed and dotted line) and clypeasteroids (dashed line) since the early Jurassic showing the relative magnitudes of the adaptive breakthroughs in the latter two groups (based on Kier 1974).

approached, following the initial breakthrough, by flattening of the test so that only the top layer of sediment was sieved (Kier 1982), and by branching of the food grooves to collect from the entire margin of the test. This was carried still further by the development, again through heterochrony (McNamara 1988), of notches and lunules which increased the effective length of the margin. The same overall trends evolved independently in several clypeasteroid groups (Smith 1984), testifying to their adaptive significance, while the very rapid diversification of the entire clypeasteroid clade through the Tertiary (Fig. 13.5) demonstrates the importance of this newly opened trophic resource.

Conclusions

The patterns of diversification of both crinoids and echinoids since the early Triassic show a number of similarities, particularly until early Jurassic times. From very low diversity, both groups diversified rapidly through heterochronous 'migration' into juvenile morphotypes (paedomorphosis) and the optimization of existing adult morphology, particularly relating to feeding strategies. Diversity in both groups was reset to low levels by a Carnian 'extinction event' before increasing again into the Jurassic (Fig. 13.1). The post-Triassic history of crinoids largely follows the same pattern of clade expansion through paedomorphosis and optimization of existing strategies. Adaptive breakthroughs did occur, with the development of planktonic and pseudoplanktonic taxa at various times during the Mesozoic, but neither strategy proved especially successful in terms of clade diversity or longevity. Although comatulids are now the dominant crinoid group, they show a very significant 'macroevolutionary lag' (Jablonski and Bottjer, Chapter 2, this volume) in the early part of their history. Nonetheless, attainment of the comatulid morphology and the high degree of motility that eventually developed from this was a significant adaptive breakthrough.

In contrast, echinoids experienced a major adaptive breakthrough in the adoption of a sediment-rich diet. The much greater success of galeropygids and their derivatives compared with the more primitive holectypoids (Fig. 13.4) demonstrates that the development of tube-feet feeding was of much greater adaptive significance than the migration of the periproct away from the apex, a character used to define all irregular echinoids. Irregular echinoids now represent almost half of all extant echinoid species (Kier 1974) and, of these, all but one genus are derived from the galeropygids, testifying to the importance of this adaptive breakthrough and the scale of the trophic resource which it opened up.

Comparison of present diversity levels in crinoids, with about 600 species, against echinoids, with about 900 species, is rather meaningless as a measure of their relative success since their ecologies and the tropic resources which are available to them are so different. However, comparison with Palaeozoic crinoids and echinoids respectively may perhaps provide a better indication of their success. Both groups were reduced to extremely low diversity levels in the early Triassic; hence neither has any significant advantage over the other in terms of initial diversity. The morphological diversity of crinoids from early Triassic to Recent times has been significantly less than for much of the Palaeozoic, when crinoids were by far the most diverse echinoderm group (Paul and Smith 1984). In contrast, echinoid diversity through much of the Mesozoic and up to the present has been considerably greater than at any time during the Palaeozoic, even after the relatively poor fossil record of Palaeozoic echinoids is taken into account. This suggests that echinoids have experienced a real post-Palaeozoic adaptive radiation, in the sense that they have developed the ability to exploit trophic resources not accessible to their Palaeozoic predecessors and hence have no parallels among Palaeozoic taxa. In contrast, the morphology of post-Palaeozoic crinoids has remained relatively conservative by comparison with Palaeozoic forms, with only the comatulids lacking any obvious parallel among Palaeozoic crinoids. Whether this difference in success between the two groups reflects fundamental changes in biotic or abiotic factors between Palaeozoic and post-Palaeozoic times, or is merely due to chance, remains open to speculation.

Acknowledgements

I thank Andrew B. Smith for comments made on the original manuscript and Dave Jablonski, George Sevastopulo, and Hans Hagdorn for useful discussion.

References

Bambach, R. K. (1985). Classes and adaptive variety: The ecology of diversification in marine faunas through the Phanerozoic. In *Phanerozoic diversity patterns: profiles in macroevolution* (ed. J. W. Valentine), pp. 191–253. Princeton University Press, Princeton, New Jersey.

Benton, M. J. (1986). More than one event in the late Triassic extinction. *Nature*, **321**, 857–9.

Biese, W. (1935–37). Crinoidea jurassica. In *Fossilium Catalogus, I. Animalia* (ed. W. Quenstedt), pt. 70, 73, 76. W. Junk (s'Gravenhage).

Blake, D. B. (1987). A classification and phylogeny of post-Palaeozoic sea stars (Asteroidea: Echinodermata). *Journal of Natural History*, **21**, 481–528.

Cowen, R. (1981). Crinoid arms and banana plantations: an economic harvesting analogy. *Paleobiology*, **7**, 332–43.

Eltycheva, R.S. and Poliarnaya, Z.A. (1986). Traumatocrinid findings from the Triassic sediments of Novosibirsk Islands. In *Parastratigraphical groups of fauna and flora of the Triassic*, Proceedings of the All Union Order of Lenin A.P. Karpenski Scientific-Technical Institute of Geology, new series, vol. 334. (ed. A.N. Oleinekov and A.E. Khamoel), pp. 112–16. (In Russian.)

Gale, A.S. (1987). Phylogeny and classification of the Asteroidea (Echinodermata). *Zoological Journal of the Linnaean Society*, **89**, 107–32.

Gould, S.J. (1988). Trends as changes in variance: A new slant on progress and directionality in evolution. *Journal of Paleontology*, **62**, 319–29.

Harland, W.B., Cox, A.V., Lewellyn, P.G., Pickton, C.A.G., Smith, A.G. and Walters, R. (1982) *Geological time scale*. Cambridge University Press, Cambridge.

Haude, R. (1972). Bau und funktion der *Scyphocrinites*-lobolithen. *Lethaia*, **5**, 95–125.

Jefferies, R.P.S. (1989). The arm structure and mode of feeding of the Triassic crinoid *Encrinus liliiformis*. *Palaeontology*, **32**, 483–97.

Johnson, A.L. and Simms, M.J. (1989). The timing and cause of late Triassic marine invertebrate extinctions: evidence from scallops and crinoids. In *Mass extinctions: processes and evidence* (ed. S.K. Donovan), pp. 174–94. Belhaven Press, London.

Kier, P.M. (1974). Evolutionary trends and their functional significance in the post-Palaeozoic echinoids. *Paleontological Society Memoir*, **5**, 1–95.

Kier, P.M. (1977). Triassic echinoids. *Smithsonian Contributions to Paleobiology*, **30**, 1–88.

Kier, P.M. (1982). Rapid evolution in echinoids. *Palaeontology*, **25**, 1–9.

Kier, P.M. (1984). Echinoids from the Triassic (St. Cassian) of Italy, their lantern supports, and a revised phylogeny of Triassic echinoids. *Smithsonian Contributions to Paleobiology*, **56**, 1–41.

Kristan-Tollmann, E. (1989). *Palaeocomaster styriacus* n. sp., eine Schwebcrinoide (Comatulida) aus dem alpinen Lias des Salzkammergutes, Osterreich. *Geologica et Palaeontologica*, **22**, 81–7.

Liddell, W.D. (1975). Recent crinoid biostratinomy. *Geological Society of America, Abstracts with Programs*, **7** (7), 116.

McIntosh, G.C. (1978). Pseudoplanktonic crinoid colonies attached to Upper Devonian logs. *Geological Society of America, Abstracts with Programs,* **10** (7), 453.

McNamara, K.J. (1986). A guide to the nomenclature of heterochrony. *Journal of Paleontology*, **60**, 4–13.

McNamara, K.J. (1988). Heterochrony and the evolution of echinoids. In *Echinoderm phylogeny and evolutionary biology* (ed. C.R.C. Paul and A.B. Smith), pp. 149–63. Oxford University Press, Oxford.

Meyer, D.L. and Macurda, D.B. (1977). Adaptive radiation of the comatulid crinoids. *Paleobiology*, **3**, 74–82.

Meyer, D.L. and Meyer, K.B. (1986). Biostratinomy of Recent crinoids (Echinodermata) at Lizard Island, Great Barrier Reef, Australia. *Palaios*, **1**, 294–302.

Mu, A.T. (1949). On the discovery of the crown of *Traumatocrinus*. *Bulletin of the Geological Society of China*, **29**, 85–92.

Paul, C.R.C. (1988). Extinction and survival in the echinoderms. In *Extinction and survival in the fossil record*, Systematics Association Special Volume, no. 34 (ed G.P. Larwood), pp. 155–70. Clarendon Press, Oxford.

Paul, C.R.C. and Smith, A.B. (1984). The early radiation and phylogeny of echinoderms. *Biological Review*, **59**, 443–81.

Rose, E.P.F. and Olver, J.B.S. (1985). Slow evolution in the Holectypidae, a family of primitive irregular echinoids. In *Proceedings of the 5th International Echinoderm Conference, Galway* (ed. B.F. Keegan and B.D.S. O'Conner), pp. 81–9. Balkema, Rotterdam.

Schneider, J.A. (1988). Frequency of arm regeneration of comatulids in relation to life habit. In *Proceedings of the 6th International Echinoderm Conference, Victoria, British Columbia* (ed. R.D. Burke, P.V. Mladenov, P. Lambert, and R.L. Parsley), pp. 531–8. Balkema, Rotterdam.

Simms, M.J. (1986). Contrasting lifestyles in Lower Jurassic crinoids: A comparison of benthic and pseudopelagic Isocrinida. *Palaeontology*, **29**, 475–93.

Simms, M.J. (1988*a*). The phylogeny of post-Palaeozoic crinoids. In *Echinoderm phylogeny and evolutionary biology* (ed. C.R.C. Paul and A.B. Smith), pp. 269–84. Oxford University Press, Oxford.

Simms, M.J. (1988*b*). The role of heterochrony in the evolution of post-Palaeozoic crinoids. In *Proceedings of the 6th International Echinoderm Conference, Victoria, British Columbia* (ed. R.D. Burke, P.V. Mladenov, P. Lambert, and R.L. Parsley), pp. 97–102. Balkema, Rotterdam.

Simms, M.J. (1988*c*). Patterns of evolution among Lower Jurassic crinoids. *Historical Biology*, **1**, 17–44.

Simms, M.J. (1988*d*). An intact comatulid crinoid from the Toarcian of southern Germany. *Stuttgarter Beitrage zur Naturkunde, Serie B*, **140**, 7 pp.

Simms, M.J. (1989). *British Lower Jurassic crinoids*. Palaeontographical Society Monograph, **142**, 103 pp., 15 pts.

Simms, M.J. and Ruffell, A.H. (1989). Synchroneity of climatic change and extinctions in the late Triassic. *Geology*, **17**, 265–8.

Simms, M.J. and Ruffell, A.H. (1990). Climatic and biotic change in the late Triassic. *Journal of the Geological Society of London*, **147**, 321–7.

Smith, A.B. (1978). A comparative study of the life style of two Jurassic irregular echinoids. *Lethaia*, **11**, 57–66.

Smith, A.B. (1980). The structure, function and evolution of tube feet and ambulacral pores in irregular echinoids. *Palaeontology*, **23**, 39–84.

Smith, A.B. (1981). Implications of lantern morphology for the phylogeny of post-Palaeozoic echinoids. *Palaeontology*, **24**, 779–801.

Smith, A.B. (1984). *Echinoid palaeobiology*. George Allen and Unwin, London.

Smith, A.B. (in press). Echinoid evolution from the Triassic to Lower Jurassic.

Cashiers Scientifiques de l'Institut Catholique de Lyon, Triassic–Jurassic Boundary Special Volume.

Smith, A. B. and Hollingworth, N. J. (1990). Tooth structure and phylogeny of the Upper Permian echinoid *Miocidaris keyserlingi. Proceedings of the Yorkshire Geological Society*, **48**, 47–60.

Telford, M., Mooi, R. and Ellers, O. (1985). A new model of podial deposit feeding in the sand dollar, *Mellita quinquiesperforata* (Leske): the sieve hypothesis challenged. *Biological Bulletin*, **169**, 431–48.

Waters, J. (1988). The evolutionary palaeoecology of the Blastoidea. In *Echinoderm phylogeny and evolutionary biology* (ed. C. R. C. Paul and A. B. Smith), pp. 215–33. Oxford University Press, Oxford.

Wells, J. W. (1941). Crinoids and *Callixylon. American Journal of Science*, **239**, 454–6.

Wignall, P. B. and Simms, M. J. (1990). Pseudoplankton. *Palaeontology*, **33**, 359–78.

14. Placoderm fishes: diversity through time

B.G. GARDINER
Biosphere Sciences, King's College, London, UK

Abstract

The placoderms, with 250 genera, constitute a small but much diversified group of bony fishes. They are sometimes classified as stem-group osteichthyans and, together with the acanthodians, are the earliest true jawed fishes. They are found in the mid-Silurian, but flourished in the Devonian reaching their greatest diversity in the late Devonian (Frasnian). Although 13 genera made their first appearance in the Famennian, no placoderm survived into the Carboniferous. The Arthrodira (150 genera) is the largest group, while the coccosteoids are the most successful clade within the arthrodires. The last major diversification of the arthrodires occurred in the late Famennian with the appearance of the largest of all placoderms. Stratigraphical distributions of placoderm genera are reviewed, and phylogenetic relationships are inferred between the principal groups.

Introduction

Placoderm fishes are often regarded as stem-group osteichthyans (Forey 1980; Gardiner 1984*a*, *b*; Maisey 1986; Gardiner and Schaeffer 1989) and to be either the stem-group of the osteichthyans or the sister-group of the acanthodians + osteichthyans. They are also believed to be the sister-group of the chondrichthyans (Goujet 1982), while Young (1986) considers them to be the most plesiomorphic gnathostomes.

Major Evolutionary Radiations (ed. P.D. Taylor and G.P. Larwood) Systematics Association Special Volume No. 42, pp. 305–319, Clarendon Press, Oxford, 1990.

The fossil record

The placoderms flourished in the Devonian and are almost entirely restricted to this period. However, antiarch and actinolepid remains have recently been recorded from the marine Upper Silurian of the Guandi Formation in the Qujing District, eastern Yunnan Province, and in western Hunan Province, China (Pan and Dineley 1988), while more primitive forms related to palaeacahthaspids have been recovered from

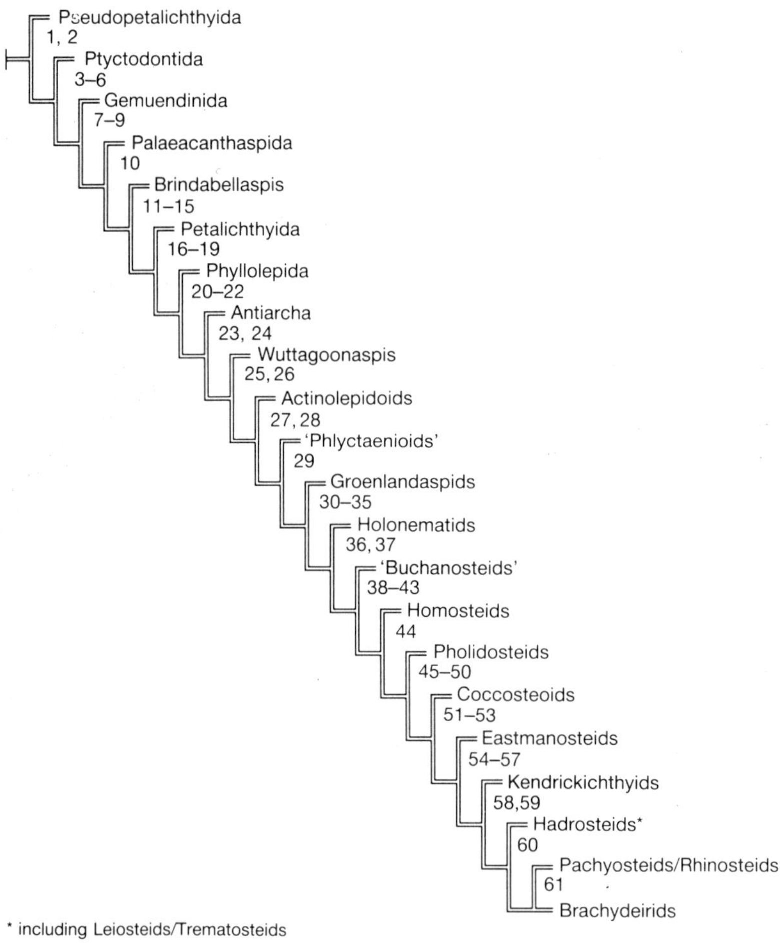

Fig. 14.1. Character phylogeny of some of the better known placoderm groups. Characters are listed in Appendix 1.

somewhat lower marine middle Silurian sediments of the Yulongssu Formation, Qujing (Chang Mee-Mann, personal communication).

Since the antiarchs and arthrodires (which include the actinolepidoids) are the two most cladistically derived groups (Fig. 14.1), their occurrence in the Upper Silurian points to the incompleteness of the placoderm fossil record, but this incompleteness does not necessarily imply inadequacy (Paul 1985) as I hope to demonstrate in this chapter. Interestingly, at their earliest occurrence both groups (antiarchs, arthrodires) are marine, whereas the Phyllolepida and the stem-group arthrodire *Wuttagoonaspis* occur in freshwater deposits, as do the Gedinnian actinolepidoids and the majority of the antiarchs. However, the most primitive placoderm groups are almost exclusively marine including the Pseudopetalichthyida, Gemuendinida, Palaeacanthaspida, *Brindabellaspis*, and Petalichthyida, whereas the Ptyctodontida are marine except for the genus *Rhamphodopsis*.

So far some 246 genera of placoderms have been described (see Denison 1978), less than 60 of which contain more than one species. Of this total at least 17 are synonyms (congeneric with earlier described genera), a further 19 are only attributable to Arthrodira incertae sedis, and another 7 to Placodermi incertae sedis. *Stensioella* is not included in this total since it cannot be shown to be a placoderm.

Diversity

The Arthrodira with 150 genera (including 19 incertae sedis not listed in Appendix 2) is the largest group, containing 61 per cent of all known placoderm genera and, when combined with its sister-group the Antiarchi (27 genera), this figure becomes 70 per cent.

Of the 203 genera of known age and affinity (Appendix 2), 14 make their first appearance in the Gedinnian, 32 in the Siegenian, 32 in the Emsian, 27 in the Eifelian, 24 in the Givetian, 61 in the Frasnian, and 13 in the Famennian. Even allowing for the incompleteness of the fossil record (referred to above) and bearing in mind the numbers of endemic genera (China 36; Australia 35; Germany 32; USA 22 (which includes 5 Arthrodira incertae sedii), USSR 11) which suggest possible restrictions in distribution and/or sampling bias (including the inflating effect of Lagerstätten) there is little doubt that the placoderms reached their maximum diversity in the Frasnian when 6 of the 8 major groups (20 of the 33 clades listed in Appendix 2) are represented. Thus the prevailing picture of diversification of placoderms during the Devonian given by Miles (1969) of the replacement of the smaller groups by the presumed competitively superior arthrodires is only partly true since several of these smaller groups continue into the Upper Devonian (Ptyctodontida five genera; Gemuendinida one genus; Petalichthyida one genus; Phyllolepida four genera). On the other hand his scenario (Miles 1969) of the four successive levels of organization that he used to describe the arthrodires (Actinolepid level; Phlyctaenaspid level, Coccosteomorph level, and Pachyosteomorph level), each of which was defined by characters relating to adaptations in the locomotor and feeding mechanisms, is still broadly applicable.

The diversity of the arthrodires is correlated with the acquisition of two superognathals (character 28), and successive modifications of the tooth plates including the co-ossification of the blade and biting regions of the inferognathal (character 38) and the addition of posterior processes to the superognathals (character 47) which brought about the changes (improvements) in feeding mechanisms visualized by Miles (1969). This diversity as well as that of the Antiarchi (28 genera) may further be correlated with the strong attachment of the palatoquadrate to both the suborbital and postsuborbital and the resulting change in the position of insertion of the adductor musclature.

Arthrodira

The antinolepidoids with some 18 genera are the least cladistically derived of the arthrodires. Although they first appeared in the Upper Silurian (see above), they reached their maximum diversity in the Siegenian (15 genera).

The 'phlyctaenoids' are paraphyletic with at least two separate lineages (*Dicksonosteus* and *Phlyctaenius*, Gardiner and Miles 1990). Nevertheless, they all possess an important acquisition over the actinolepidoids—a para-articular process on the paranuchal and a subglenoid process on the inner surface of the obstantic process of the anterior dorsolateral (character 29*b*) which allowed the head to rotate through some 15°. This character is therefore correlated with the feeding mechanism, as is the ventral keel on the median dorsal plate (character 29*a*) which presumably served for the anchorage of muscles and ligaments concerned with raising the head. These two synapomorphies define the 'phlyctaenoids' or the second successive level of arthrodiran organization postulated by Miles (1969). The 'phlyctaenoids' first occur in the Siegenian and continue into the Frasnian, reaching their greatest diversity in the Emsian. Nominally included in this group are some 20 genera.

Interspersed between the 'phlyctaenoids' and Miles's (1969) third or coccosteomorph level are several small groups. In ascending order these are: the groenlandaspids (three genera); holonematids (seven genera);

'buchanosteids' (seven general); heterosteids (three genera); homosteids (seven genera); and pholidosteids (three genera). As we progress stepwise through these six groups we see the acquisition of a series of synapomorphies, many of which are related either to the feeding mechanism or to locomotion. Thus at the groenlandaspid level there are four characters intimately linked to the movement of the head: the posterior expansion of the nuchal plate (character 31), the paired pits on its visceral surface (character 35), the appearance of a postnuchal process on the paranuchal (character 32), and a nuchal thickening on the visceral surface of the skull roof (character 34). At the holonematid level synapomorphies include the enlargement of the pectoral fenestra and exposure of the scapulocoracoid which presumably gave improved fin control (character 37), while the 'buchanosteids' and above share the enlargement of the inferognathals in which the blade and biting divisions are co-ossified (character 38), thereby providing an insertion for the adductor mandibulae muscles and increasing the effectiveness of the bite. The pholidosteids share with the coccosteoids and above the acquisition of stout dorsal processes on the anterior and posterior superognathals (character 47) which contact the ethmoid region of the neurocranium and the autopalatine respectively. These processes brace the dentition (Miles and Dennis 1979) and are correlated with the formation of the suborbital lamina (character 45) and the separation of the quadrate and autopalatine bones (character 46). They may also be correlated with the formation of a detent process on the quadrate and the supraglenoid on the articular (character 48) which presumably contributed to the accuracy of the shearing bite.

The coccosteoids show further modifications concerned with locomotion, including the enlargement of the pectoral fenestrae (character 53) and the reduction of the spinal plate (character 52). They also show one modification concerned with both feeding and respiration in the hinging of the cheek to the neurocranium via the submarginal and hyomandibula. This may possibly be correlated with the further consolidation of the skull roof (character 51). The coccosteoids comprise some 23 genera and as such are the largest arthrodiran group. The earliest members are Eifelian and maximum diversity was achieved in the Frasnian (12 genera); nevertheless, these genera are closely related and distinguished by small differences in the shape and proportion of the dermal bones, the closure of the spiracular notch, the length of the snout, and variations in the size of the pectoral fenestra and spinal plate.

The remaining arthrodires are usually referred to as the Pachyosteina (Miles's level 4) and are typified by the pectoral embayments (character 54) that are widely open behind, and by the reduction of the posterior extent of the armour of the flank (characters 55, 57, 60). These changes related to increased efficiency in swimming (e.g. the broad based pectoral fins were efficient hydrofoils) brought about a diversification in the Upper Devonian with the addition of 36 genera belonging to at least eight (Fig. 14.1) and more, probably 13 (Appendix 2), separate clades.

The last major diversification of these 'pachyosteomorph' arthrodires occurred in the Upper Famennian with the appearance of the largest of all placoderms. These belong to two separate lineages, the eastmanosteids and the titanichthyids. The eastmanosteids include *Dunkleosteus*, *Dinichthys*, and *Gorgonichthys*, with the latter reaching a length of over 3 metres (skull length 70 cm). *Titanichthys*, on the other hand, was comparable in size with *Dunkleosteus* (skull length 50 cm).

Summary

There are at least 16 successive levels of organization within the arthrodires with the coccosteoids having the greatest number of genera. The most cladistically derived arthrodires (hadrosteids, leiosteids, trematosteids, pachyosteids, rhinosteids, and brachydeirids) are confined to Germany, France, and the USA, with the exception of the trematosteid *Simosteus* which occurs in the Frasnian of Australia.

Conclusions

The placoderms reached their greatest diversity in the Upper Devonian (Frasnian) with some 70 genera belonging to 21 clades. However, only seven of these genera survived into the Famennian, although within this period a further 13 genera representing eight clades made their first appearance. A few fragmentary remains have been described from the base of the Carboniferous (Miles 1967) but these may well have been derived from earlier sediments.

References

Denison, R. H. (1978). Placodermi. In *Handbook of Paleoichthyology*, Vol. 2 (ed. H.-P. Schultze), vii + 128 pp. Fischer Verlag, Stuttgart.

Forey, P. L. (1980). *Latimeria*: a paradoxical fish. *Proceedings of the Royal Society of London*, **B208**, 369–84.

Forey, P. L. and Gardiner, B. G. (1986). Observations on *Ctenurella* (Ptyctodontida) and the classification of placoderm fishes. *Zoological Journal of the Linnean Society*, **86**, 43–74.

Gardiners B. G. (1984*a*). The relationships of placoderms. *Journal of Vertebrate Paleontology*, **4**, 379–95.

Gardiner, B. G. (1984*b*). The relationships of palaeoniscid fishes, a review based on new specimens of *Mimia* and *Moythomasia* from the Upper

Devonian of Western Australia. *Bulletin of the British Museum (Natural History), Geology Series*, **37**, 173–428.
Gardiner, B.G. and Miles, R.S. (1990). Further primitive eubrachythoracid arthrodires from Gogo, Western Australia. *Zoological Journal of the Linnean Society*, **99**, 159–204.
Gardiner, B.G. and Schaeffer, B. (1989). Interrelationships of lower actinopterygian fishes. *Zoological Journal of the Linnean Society*, **97**, 135–87.
Goujet, D.F. (1981). Les affinités des Placodermes, une revue des hypothèses actuelles. *Géobios, Mémoire Speciale*, **6**, 27–38.
Maisey, J. (1986). Heads and tails: a chordate phylogeny. *Cladistics*, **2**, 201–256.
Miles, R.S. (1967). 'Class Placodermi'. In *The fossil record* (ed. W.B. Harland *et al.*), p. 640. Geological Society of London.
Miles, R.S. (1969). Features of placoderm diversification and the evolution of the arthrodire feeding mechanism. *Transactions of the Royal Society of Edinburgh*, **68**, 123–70.
Miles, R.S. and Dennis, K.D. (1979). A primitive eubrachythoracid arthrodire from Gogo, Western Australia. *Zoological Journal of the Linnean Society*, **66**, 31–62.
Pan, S.J. and Dineley, D.L. (1988). A review of early (Silurian and Devonian) vertebrate biogeography and biostratigraphy of China. *Proceedings of the Royal Society of London*, **B235**, 20–61.
Paul, C.R.C. (1985). The adequacy of the fossil record reconsidered. *Special Papers in Palaeontology*, **33**, 7–15.
Young, G.C. (1986). The relationships of placoderm fishes. *Zoological Journal of the Linnean Society*, **88**, 1–57.

Appendix 1

Characters used to construct the placoderm phylogeny shown in Fig. 14.1

1. Anterior lateral plate present.
2. Main cephalic sensory canals running parallel to the margin of the skull roof; synarcual articulates with occiput.
3. Median dorsal plate present.
4. Anterior dorsolateral plate present.
5. Spinal plate present; body armour forming complete ring.
6. Central plate present.
7. Prepectoral process developed on scapulocoracoid.
8. Dorsal nares.
9. Sensory canals in open grooves.
10. Posterior dorsolateral plate present; palatoquadrate firmly attached to suborbital.
11. Ginglymoid neck joint.
12. Eye completely enclosed within inflexible bones.
13. Centrals excluded from posterior skull margin.

14. Two paranuchals; nuchal elongated.
15. Postmarginal plate present.
16. Posterior ventrolateral plate present.
17. Anterior median ventral plate present.
18. Anterior ventrolaterals meeting in midline suture.
19. Posterior ventrolaterals meeting in midline suture.
20. Anterior ventrolateral and separate interolateral plates present.
21. Posterior median ventral plate present.
22. Great width of skull-roof across posterolateral angles.
23. Posterior lateral plate present.
24. Closed pectoral fenestra.
25. Sliding dermal neck joint.
26. Central plate polygonal with posterior and middle pit lines; nuchal plate pointed anteriorly.
27. Anterior median ventral completely separating anterior ventrolaterals and meeting posterior median ventral.
28. Two superognathals.
29. *a* Median dorsal plate long and narrow and with a ventral keel.
    b Para-articular process on the paranuchal and a subglenoid process on the inner surface of the obstantic process of the anterior dorsolateral.
30. Suture lines distinct; main plates with extensive overlaps.
31. Nuchal plate posteriorly expanded.
32. Paranuchal plate with postnuchal process.
33. Supraorbital sensory grooves extend on to centrals.
34. Nuchal thickening on visceral surface.
35. Paired pits on visceral surface of nuchal plate.
36. Pre-endolymphatic thickening.
37. Scapulocoracoid widely exposed in enlarged, laterally facing pectoral fenestra.
38. Inferognathal with blade and biting regions complete and co-ossified.
39. Skull-roof with lateral consolidated region.
40. Submarginal plate small and closely incorporated in the cheek unit.
41. Absence of supravagal process and pravagal fossa.
42. Skull-roof with well developed supraorbital vault.
43. Median dorsal plate with carinal process.
44. Preorbital plate with dermal preorbital lamina.
45. *a* Suborbital plate with slender suborbital lamina.
    b Suborbital plate with internal lamina.
46. Palatoquadrate with separate autopalatine and quadrate bones.
47. Superognathals with posterior processes.

48. Detent process on quadrate; supraglenoid on articular.
49. Nuchal pits in trough partly divided by a stout septum and with a transverse posterior ridge.
50. Suborbital plate with strongly developed linguiform process.
51. Well developed postocular process, often of bifurcate form.
52. Spinal reduced.
53. Pectoral fenestrae enlarged.
54. Pectoral incision.
55. Pseudospinal short and squat without a spinal pit.
56. Anterior lateral plate elbowed posteriorly at level of the ventral edge of the posterior lateral plate.
57. Interolateral plate with reduced ventral lamina; produced anteriorly into a prominent postbranchial lamina.
58. Marginal-paranuchal suture in transverse plane of posterodorsal angles of the skull-roof.
59. Parasphenoid with strongly developed posterolateral processes and a hypophysial depression with both anterior and posterior divisions.
60. Loss of spinal plate.
61. Marginal plate forming posterior margin of enlarged orbital fenestra.

(For an extended explanation of these characters see Forey and Gardiner 1986; Gardiner and Miles 1990.)

Appendix 2

Stratigraphical distributions of placoderm genera arranged systematically. (Standard geological stage abbreviations are used.)

Pseudopetalichthyida	
Pseudopetalichthys	SIG
Ptyctodontida	
Campbellodus	FRS
Chelyophorus	FAM
Ctenurella	FRS
Goniosteus	EIF-GIV
Palaeomylus	EIF-GIV
Ptyctodopsis	EIF
Ptyctodus	EIF-FRS
Rhamphodopsis	EIF-GIV
Rhynchodus	EIF-FRS
Tollodus	SIG

Gemuendinida	
Asterosteus	EIF
Bolivosteus	EIF
Gemuendina	EMS
Jagorina	FRS
Palaeacanthaspida	
Breizosteus	SIG
Dobrowlania	GED
Kimaspis	GED
Kolymaspis	EMS
Kosoraspis	GED
Murrindalaspis	EMS
Palaeacanthaspis	GED
Radotina	GED-SIG
Romundina	GED
Weejasperaspis	EMS
Brindabellaspis	EMS
Petalichthyida	
Diandongpetalichthys	GED
Ellopetalichthys	GIV
Epipetalichthys	FRS
Latipetalichthys	EIF
Lunaspis	SIG-EMS
Macropetalichthys	EIF-GIV
Neopetalichthys	EMS
Notopetalichthys	EMS
Quasipetalichthys	EIF
Shearsbyaspis	EMS
Sinopetalichthys	SIG
Wijdeaspis	SIG
Xinanpetalichthys	GED
Phyllolepida	
Antarctaspis	FRS
Austrophyllolepis	FRS
Phyllolepis	FAM
Placolepis	FRS
Antiarcha	
Asterolepis	EIF-FRS

Bothriolepis	GIV–FAM
Byssacanthus	EIF–FRS
Dianolepis	GIV
Gerdalepis	EIF–FRS
Grossaspis	GIV
Grossilepis	FRS
Hillsaspis	FRS
Hohsienolepis	GIV
Hunanolepis	GIV
Lepadolepis	FRS
Liujiangolepis	SIG
Microbrachius	GIV
Monarolepis	GIV
Pambulaspis	FRS
Phymolepis	SIG
Procondylolepis	GED
Pterichthyodes	EMS–GIV
Qujinolepis	GED
Remigolepis	FRS–FAM
Sherbonaspis	EIF–GIV
Sinolepis	FRS–FAM
Stegolepis	FRS
Wudinolepis	GIV
Xichonolepis	GIV
Yunnanolepis	GED
Zhanjilepis	GED
Wuttagoonaspida	
Wuttagoonaspis	EIF
Arthrodira	
Actinolepidoids	
Actinolepis	EMS–GIV
Aethaspis	SIG
Ailuracantha	SIG
Anarthraspis	SIG
Asiacanthus	SIG
Baringaspis	GED
Bollandaspis	SIG
Bryantolepis	SIG
Kujdanowiaspis	GED–SIG
Kunmingolepis	SIG

Lehmanosteus	SIG
Mediaspis	SIG
Murmur	SIG
Proaethaspis	SIG
Sigaspis	SIG
Simblaspis	SIG
Svalbardaspis	SIG-EMS
Szelepis	SIG
'Phlyctaenioids'	
Aggeraspis	SIG
Arctaspis	SIG
Arctolepis	EMS
Barrydalaspis	GIV
Changyanophyton	GIV
Denisonosteus	GIV
Diadsomaspis	EMS
Dicksonosteus	SIG
Elegantaspis	SIG
Exutaspis	GIV
Gaspeaspis	EMS
Heintzosteus	SIG
Heterogaspis	EMS
Hunginaspis	EIF
Kolpaspis	EMS
Kueichowlepis	SIG
Neophlyctaenius	FRS
Pageauaspis	EMS
Phlyctaenius	EMS
Stuertzaspis	EMS
Yangaspis	GIV
Williamsaspis	EMS
Groenlandaspids	
Groenlandaspis	FRS-FAM
Prosphymaspis	EMS
Tiaraspis	SIG
Holonematids	
Aspidichthys	FRS
Deirosteus	GIV-FRS
Deveonema	FRS

Gyroplacosteus	FRS
Holonema	EIF-FRS
Megaloplax	FRS
Rhenonema	GIV
'Buchanosteids'	
Arenipiscis	EMS
Buchanosteus	EMS
Burrinjucosteus	EMS
Errolosteus	EMS
Goodradigbeon	EMS
Kweichowlepis	SIG
Toombsosteus	EMS
Gemuendenaspis	EMS
Heterosteids	
Herasmius	EIF
Heterosteus	EIF-GIV
Yinosteus	GIV
Homosteids	
Antineosteus	EMS
Atlantidosteus	EMS-EIF
Euleptaspis	SIG
Homosteus	EMS-GIV
Luetkeichthys	EIF
Taemasosteus	EMS
Tityosteus	EMS
Pholidosteids	
Malerosteus	FRS
Pholidosteus	FRS
Tapinosteus	FRS
Coccosteoids	
Ardennosteus	FAM
Belgiosteus	GIV
Camuropiscis	FRS
Clarkeosteus	EIF
Coccosteus	EIF-GIV
Dickosteus	EIF-GIV
Eldenosteus	FRS

Fallacosteus	FRS
Harrytoombsia	FRS
Incisoscutum	FRS
Jiuchengia	GIV
Kiangyosteus	GIV
Kimberleyichthys	FRS
Latocamurus	FRS
Livosteus	EIF
Millerosteus	EIF–GIV
Panxiosteus	FRS
Plourdosteus	FRS
Protitanichthys	EIF–GIV
Rolfosteus	FRS
Torosteus	FRS
Tubonasus	FRS
Watsonosteus	GIV
Eastmanosteids	
Dinichthys	FRS–FAM
Dunkleosteus	FRS–FAM
Eastmanosteus	GIV–FAM
Golshanichthys	FRS
Gorgonichthys	FAM
Heintzichthys	FAM
Hussakofia	FAM
Westralichthys	FRS
Kendrickichthyids	
Bullerichthys	FRS
Bruntonichthys	FRS
Kendrickichthys	FRS
Rachiosteus	FRS
Titanichthyids	
?Tafilalichthys	FAM
Titanichthys	FAM
Dinomylostoma	FRS
Mylostoma	FAM
Trematosteids	
Belosteus	FRS

Brachyosteus	FRS
Braunosteus	FRS
Bungartius	FAM
Parabelosteus	FRS
Paramylostoma	FAM
Simosteus	FRS
Trematosteus	FRS
Leiosteids	
Erromenosteus	FRS
Hadrosteids	
Hadrosteus	FRS
Gymnotrachelus	FAM
Selenosteus	FAM
Pachyosteids	
Enseosteus	FRS
Microsteus	FRS
Pachyosteus	FRS
Rhinosteids	
Melanosteus	FRS
Rhinosteus	FRS
Brachydeirids	
Brachydeirus	FRS
Oxyosteus	FRS
Synauchenia	FRS
Leptosteus	FRS

15. The radiations of temnospondyl amphibians

ANDREW R. MILNER
Department of Biology, Birkbeck College, London, UK

Abstract

The temnospondyls are the largest group of archaic amphibians, currently placed in 40 families and 160 genera. They first appear in the Visean, are represented by a wide range of forms from the Westphalian to the Spathian, and then decline in diversity, the latest specimen being of early Cretaceous age. A provisional phylogenetic analysis of temnospondyl families combined with their chronological distribution suggests that the patterns of diversity and evolution of the group can be modelled in terms of two major radiations, here termed the Carboniferous radiation and the Griesbachian radiation. The Carboniferous radiation commenced with the appearance of temnospondyls in the Visean and full diversification by the late Westphalian largely, if not entirely, within Euramerica. This was followed by relative stability through the Permian culminating in extinction of most families, probably at the end-Permian event. This extinction may have been centred in the northern region of Pangaea with differential survival of southern Pangaea temnospondyls through the Permo-Triassic boundary. The Griesbachian radiation of the stereospondyls may have been triggered by the ecological opportunities left by the end-Permian extinctions. The stereospondyl diversification was largely complete by the end of the Dienerian and only one family-level branching event may have occurred later. There is little evidence of regional endemism during this phase of temnospondyl evolution. The Griesbachian stereospondyl radiation was probably accompanied by a comparable radiation of the lissamphibians (temnospondyl descendants), although the fossil record shows little direct evidence of this.

Non-neotenous temnospondyls had different larval and adult morphologies which seem to have provided a fruitful basis for heterochrony

Major Evolutionary Radiations (ed. P.D. Taylor and G.P. Larwood) Systematics Association Special Volume No. 42, pp. 321–349, Clarendon Press, Oxford, 1990.

to act on, in both radiations. Neotenous forms with a single morphology, have lost this potential and, in the Griesbachian, the non-neotenous stereospondyls radiated while the neotenous families remained at low diversity although the latter survived unchanged for longer than the stereospondyls.

Introduction

The temnospondyls are the largest group of archaic amphibians and are represented in the fossil record from the Visean to the Lower Cretaceous. Comprising about 40 families and 160 genera, they are more numerous than all the other early amphibian groups combined, and their representatives occur in most late Palaeozoic and Triassic assemblages of continental tetrapods, sometimes in abundance. Until recently, the temnospondyls have generally been lumped with other groups in the paraphyletic grade-group 'Labyrinthodontia' and, even when considered separately, their pattern of diversification and extinction has been simply depicted as a spindle-shaped rise and fall in diversity, with no indication of whether this represented one cycle or several merged together (e.g. Romer and Parsons 1986, fig. 39). The family-level diversity of temnospondyls (including colosteids) through time has been graphed more precisely by DeFauw (1987, fig. 1; 1989, fig. 1) and shows a gradual increase from Visean to Sakmarian, followed by a decline to the Tatarian. This is succeeded by a sudden increase in diversity in the Induan followed by a gradual decline in diversity to the Bajocian, with a slight Middle Triassic dip. In the absence of cladistic analysis of the families involved, it is not clear what patterns of turnover occur within this general pattern or indeed whether the families involved represent monophyletic units. One of the few arguments for temnospondyl turnover as part of higher biotic phenomena was that of Pitrat (1973) who used the temnospondyl suborders Rhachitomi and Stereospondyli as groups supporting a faunal turnover at the Permian–Triassic boundary. However, the Rhachitomi was simply a basal grade of temnospondyls while the Stereospondyli, as used by Pitrat, was a polyphyletic grade, some members of which occurred in the Upper Permian.

The following account is intended to be an initial attempt to superimpose a cladogram of monophyletic temnospondyl families on a stratigraphical time-scale in order to examine the chronology of adaptive radiation within the group, together with phases of stability and extinction. The cladogram is also briefly examined for geographical, ecological, and heterochronic/morphological patterns.

Temnospondyl amphibians—defining a paraphyletic group. The temnospondyl amphibians are widely perceived as a grade of organization between primitive tetrapods and some or all lissamphibians, but their precise taxonomy has been relatively neglected in recent years. Because they represent a grade of organization it is probable that their systematic status will change over the next decade. This could happen in one of two ways. Following the methodology of Gauthier (1986), the term Temnospondyli could be made fully inclusive of all descendant groups and the Lissamphibia could be placed within a monophyletic–holophyletic Temnospondyli. This would result in a confusing change in use of well-known terminology and is not used here. Alternatively, if the methodology of Patterson and Rosen (1977) were followed, then eventually the Temnospondyli would be replaced by a string of plesions ranked with respect to the Lissamphibia and the entire group could be considered to be stem-lissamphibians, just as the dinosaurs are stem-birds. The formal taxonomic term might be lost but it is likely that the group would retain utility as an object of palaeobiological study and should be capable of definition as a monophyletic–paraphyletic grade for palaeobiological purposes. It is treated as such in this account.

The base of the Temnospondyli has been variously defined. Romer (1947) included the ichthyostegids, loxommatids, and colosteids in the Temnospondyli which was thus defined entirely on primitive tetrapod features. There is a general but not unanimous consensus that ichthyostegids are the most primitive stem-tetrapods and have no specific relationship to any one other tetrapod subgroup (Gaffney 1979; Gardiner 1983; Godfrey 1989). Smithson (1985) and Panchen and Smithson (1988) have argued that ichthyostegids fall within a 'batrachomorph' clade with other amphibian groups but this does not entail an immediate relationship to the temnospondyls either. Loxommatids also share only primitive tetrapod characters with temnospondyls and were first removed from that group by Panchen (1980). Gardiner (1983) suggested that they were stem-tetrapods, but Panchen and Smithson (1988) have argued that they are stem-amniotes. Neither proposal provides support for the view that they are temnospondyls. The systematic position of the Colosteidae is more controversial. Most authors still consider them to be basal temnospondyls (e.g. Godfrey *et al.* 1987), but their temnospondyl features are all either common to microsaurs as well, or are highly homoplasious. I believe that they fall outside the temnospondyl–microsaur group (Milner *et al.* 1986; Panchen and Smithson 1988) and they are here excluded from the temnospondyls. The characters in Node **A1** of

Fig. 15.1 are those used here to define the base of the temnospondyls with *Caerorhachis* as their most primitive representative.

The boundary between the temnospondyls and their crown-group the lissamphibians, although arbitrary, is represented by a large gap in the fossil record and so presents little controversy. The peliontid *Doleserpeton* has been argued to be the closest approach to a lissamphibian (Bolt 1969). Both peliontids (see Appendix for generic content) and branchiosaurids share a few (different) characters with the Lissamphibia but lack many more (Milner 1988) and either might represent the sister-group of the lissamphibians. This is reflected in the terminal trichotomy in Fig. 15.1. There are no stem-lissamphibians higher than members of these two families, all known fossil lissamphibians being clearly part of the crown-group with the possible exception of the Albanerpetontidae (Milner 1988). The resulting content of the Temnospondyli is that compiled in the Appendix to this chapter, in which all families and component valid genera are listed.

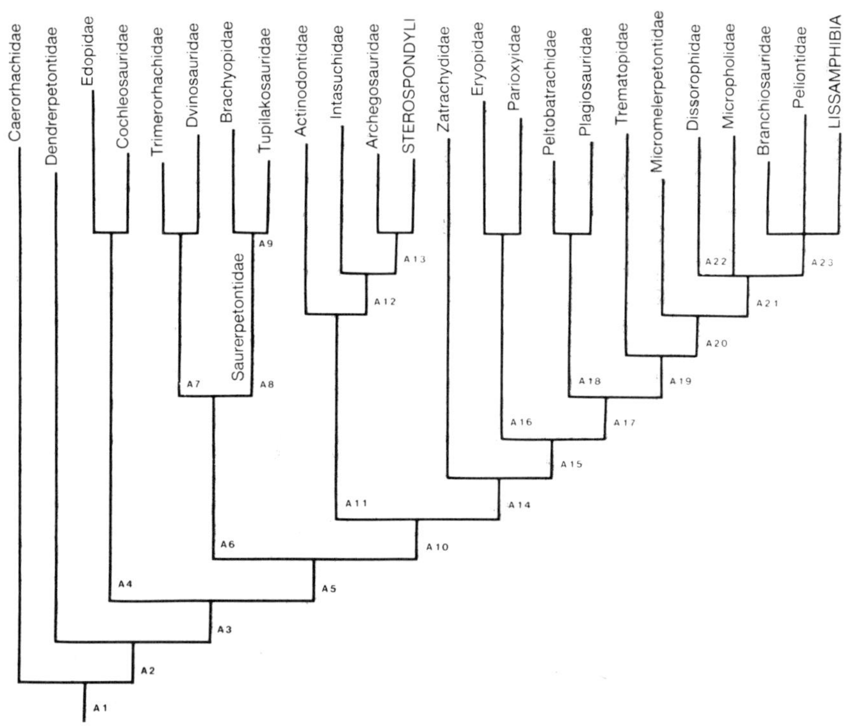

Fig. 15.1. Cladogram of relationships of temnospondyl families other than Stereospondyli. Defining apomorphies as listed in Table 15.1.

Table 15.1. Apomorphies defining nodes in Fig. 15.1.

A1. Distinct interpterygoid vacuities bordered by triradiate pterygoids; strut-like parasphenoid with long cultriform process overlapping or suturing with vomers and with a broad flat basal plate; broad flat vomers; ribs straight or slightly curved in large forms.

A2. Premaxillary with narrow marginal ramus and dorsal alary process bordering relatively large naris; semicircular tympanic notch not reaching quadrate region; weakly developed posterolateral flange on pterygoid; dorsolaterally orientated stapes; presacral column reduced to 25 vertebrae.

A3. Stapes with elongate, rod-like, anteroposteriorly compressed shaft; finely striated triangular septomaxillary wedged between nasal and lacrimal or maxillary posterior to external naris; humerus without entepicondylar foramen; ilium with rudimentary caudal process.

A4. Enlarged premaxillaries with a long common medial suture and extending well back along the jaw margin behind the level of the common suture; small inset external naris set well back along snout; jugal-prefrontal contact excluding lacrimal from orbit margin.

A5. Vomers entering margin of interpterygoid vacuities.

A6. Anterior palatal fenestrae in palate; pterygoid withdrawn partway down sides of interpterygoid vacuity; broad cultriform process, tympanic notch shallow to absent; enlarged retroarticular process, medial exposure of angular in mandible, limbs reduced in size and incompletely ossified.

A7. Vomer–pterygoid contact produced by posterior extension of the vomer; tooth-rows on vomer and ectopterygoid; presacral column increased to 31–5.

A8. L-shaped postparietals; coronoid process on the mandible; quadrate anterior to level of occiput; cheek deep with quadrate below level of parasphenoid; stapes small and compact, not rod-like.

A9. Loss of intertemporals; loss of separate lacrimals; pterygoids broadly sutured to parasphenoid.

A10. Loss of intertemporals; exoccipitals enlarged to give bilobed occipital condyle; reduced basioccipital still participating in condyles.

A11. Short vomerine tooth-row/clump medial to choana; palatine with fang and short (2–4) tooth row; ectopterygoid with 2–10 teeth behind fang; paired anterior palatal fossae (not vacuities); lateral-line sulci generally present in adults.

A12. Skull narrower with midline length: width ratio of 1.55 or more.

A13. Palatine tooth-row of fang plus 7–8 teeth; palatine with short exposure in margin of interpterygoid vacuity; ectopterygoid with tooth row but no fang; pterygoid flange lost; 1–2 anterior palatal vacuities present; elongate interclavicle—length at least twice width.

Table 15.1. *continued*

A14. Short broad skull table with abbreviated postorbital not wedged between supratemporal and squamosal; pterygoid flange enlarged and downturned; semicircular interpterygoid vacuities, together wider than long; lateral-line sulci usually absent; small rounded interclavicle and narrow-bladed clavicles.

A15. Septomaxillary not an ornamented bone in skull roof but a detached ossification inside the external naris.

A16. Large anterior expansion of jugal, contacting prefrontal and excluding lacrimal from orbit margin.

A17. Relatively large interpterygoid vacuities bordered by slender lateral palatal series.

A18. Dermal bones usually bearing 'pustular' ornament (not *Plagiosternum*); projecting quadratojugal horns lateral to jaw hinge; backwardly sloping occiput; holospondylous trunk vertebrae with massive notochordal centra; extensive dorsal armour of dermal scutes, either in multiple rows or a mosaic.

A19. Anteriorly reduced jugal extending only to midlevel of orbit; maxilla and (sometimes) palatine in orbit margin; tympanum suspended from dermal skull elements forming an otic notch, and from dorsal process of the quadrate.

A20. Palatine broadly entering margin of interpterygoid vacuity; humerus without supinator process.

A21. Humerus an elongate and slender bone.

A22. Distinct supratympanic flange with a semilunar component from the squamosal. One or two rows of large dermal scutes above or fused to the neural spines of the anterior trunk vertebrae.

A23. Extreme abbreviation of the skull table; ectopterygoid reduced to a strip of bone; ribs extremely reduced, interclavicle reduced to a poorly ossified disc of bone.

Family interrelationships

This chapter is based on a cladistic study of temnospondyl relationships prepared during compilation by the author of a volume in the 'Handbuch der Palaeoherpetologie' series (Milner, in preparation). It is iterative but encompasses 39 families, some minor monotypic families having been omitted (Kourerpetontidae, Sclerothoracidae).

The cladogram is split between Figs 15.1 and 15.2, the latter depicting the clade Stereospondyli. The characters used to support this set of relationships are listed in Tables 15.1 and 15.2, respectively. For

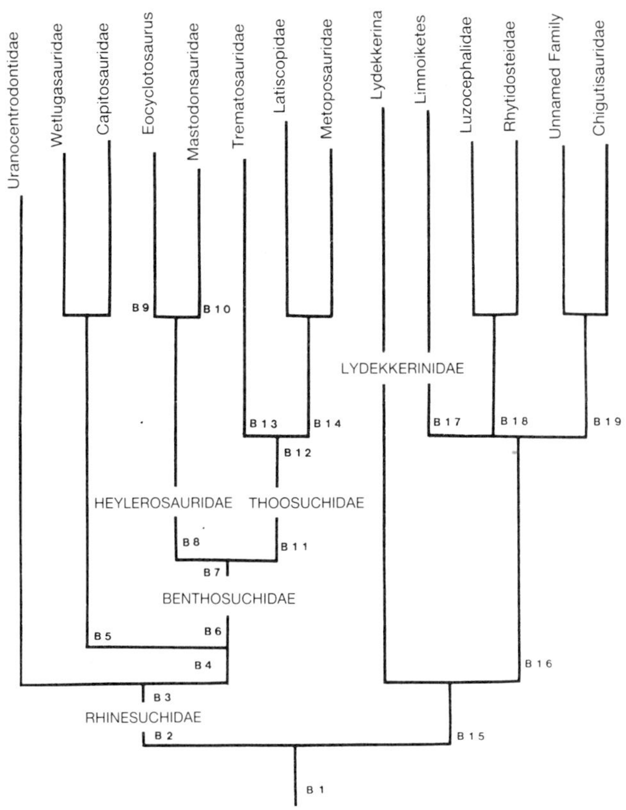

Fig. 15.2. Cladogram of relationships of stereospondyl families. Defining apomorphies as listed in Table 15.2.

Table 15.2. Apomorphies defining nodes in Fig. 15.2

B1. Braincase and pterygoids broadly sutured; single anterior palatal vacuity; curved line of teeth between choanae; palatine ramus of pterygoid reduced giving long palatine exposure in margin of interpterygoid vacuity; no arterial grooves or foramina on ventral surface of basal plate of parasphenoid; vertebrae with reduced or unossified pleurocentra.

B2. Oblique ridge present on pterygoid (lost in Trematosauridae); ectopterygoid fangs generally absent; at least 16 teeth in row behind palatine fang.

B3. Opisthotic not forming part of paroccipital bar.

B4. Posterior coronoid bearing tooth-row instead of denticles; tiny basioccipital taking no part in occipital condyles.

Table 15.2. *continued*

B5. Extreme elongation of anterior region of jugal; enlarged posterolaterally directed tabulars.

B6. Bilobed anterior palatal vacuity.

B7. Paired anterior palatal vacuities.

B8. Laterally directed tabular horns.

B9. Tabulars enlarged to completely enclose otic notch.

B10. Paired premaxillary fenestrae to accommodate symphyseal fangs; extreme flattening of skull; orbits proportionately enlarged with frontal entering orbit margin; at least anterior trunk vertebrae stereospondylous.

B11. Elongation of skull table with distinct zone of rapid growth; small laterally situated orbits; small tabulars; palatine ramus of pterygoid short and ectopterygoid entering margin of interpterygoid vacuity; parasphenoid with deep narrow cultriform process with dorsal notch and concave sides (broadened in metoposaurids); weakly developed oblique ridge on pterygoid; large posterior Meckelian fenestra (at least 0.8 of adductor fossa length).

B12. Common exoccipital–pterygoid suture.

B13. Exoccipitals underplated by the parasphenoid; elongate pterygoid–parasphenoid suture; loss of oblique ridge on pterygoid.

B14. Large, closely set terminal nares; palatine ramus of pterygoid very short with posteromedial ramus of ectopterygoid forming part of strut.

B15. Reduced allometric growth of skull and elongation of skull table giving orbits about halfway along skull; width across quadratojugals almost as great as skull midline length; upper part of palatoquadrate fissure present (refilled in chigutisaurids); squamosal–quadratojugal trough; reduced basioccipital with paired occipital condyles formed by exoccipitals; 'pockets' on posteroventral surface of parasphenoid very reduced or absent.

B16. Paroccipital process not including opisthotic; small tooth row medial to each palatine fang; palatine fang almost as large as choana.

B17. Interpterygoid vacuities semicircular, not elongate.

B18. Anterior palatal vacuity rounded, not wider than long; lateral skull margins straight, not curved.

B19. Snout abbreviated and skull rounded; lacrimal absent; tabulars large and posterolaterally directed; otic notch shallow; wide pterygoid–parasphenoid plate; ventrolateral borders of pterygoids at least slightly downturned; quadrate condyles below level of occipital condyles; palatal tooth rows across vomers between vomerine fangs and along medial margin of choanae back to palatine fangs; posttemporal fossae wider than deep; supracoracoid foramen open ventrally.

reasons of space, it is a simplified list and omits various reversals and variations of character states that occur within families. I have emphasized the characters linking two or more monophyletic families, rather than defined the families themselves, which is beyond the scope of this chapter. Higher-resolution studies, defining families or examining relationships within families, include Coldiron (1978) on saurerpetontids, Ingavat and Janvier (1981) on some capitosaurids, Warren and Hutchinson (1983) on brachyopids and chigutisaurids, Hammer (1987) on trematosaurids, Boy (1989) on zatrachydids, Berman *et al.* (1987) on trematopids, and Werneburg (1989) on micromelerpetontids and branchiosaurids.

One result of this study that requires some comment is the recognition that several temnospondyl families bracketing the Permo-Triassic boundary are paraphyletic and their 'extinctions' are merely taxonomic. I have reinstated several little-used families and subfamilies by breaking up the Rhinesuchidae *s.l.* (splitting off the Uranocentrodontidae), the Lydekkerinidae *s.l.* (splitting off the Luzocephalidae), and the Benthosuchidae *s.l.* (splitting off the Thoosuchidae and Heylerosauridae). These splits permit a slightly more precise analysis of relationships but many of the resultant subunits are still not demonstrably monophyletic and may ultimately be resolved as strings of plesion-genera. Several earlier families are also paraphyletic as defined, including the Dendrerpetontidae, Saurerpetontidae, Actinodontidae, and Peltobatrachidae (see Appendix for details). Although depicted as a monophyletic family in Figs 15.1 and 15.3, the Dendrerpetontidae may yet prove to be a post-caerorhachid grade with the potential to incorporate basal forms of several later taxa. If the Peliontidae are ultimately shown to be the stem-group of the Lissamphibia then they too will be a paraphyletic family.

Temnospondyl families through time

In Fig. 15.3, the cladogram of temnospondyl family relationships has been transcribed on to a stratigraphical stage chart. The distribution of each family is indicated as a broad line spanning the first and last appearances in the fossil record (data in Appendix). These represent the latest possible date of appearance, and the earliest possible date of extinction of each family. Thin lines represent earlier stages when a given family is inferred to have been present because of the presence of its sister-group. The temnospondyl record is poor for many families as evidenced by regular new discoveries of specimens which extend the chronological range of families. For example, in the last decade the upper stratigraphic ranges of the following families have been extended:

Fig. 15.3. Cladogram of temnospondyl families superimposed on a stratigraphic stage chart to show the Carboniferous and Griesbachian radiations. All di/trichotomies are depicted at the latest possible stage of occurrence, i.e. the first appearance of one of the daughter groups. The monotypic Sclerothoracidae is omitted, and the paraphyletic Benthosuchidae, Thoosuchidae, and Lydekkerinidae are not labelled.

Kungurian Trematopidae (Martens 1988) and Branchiosauridae (Maňourová 1981); Ufimian Eryopidae (Gubin 1983); Carnian Trematosauridae (Hellrung 1987); Pliensbachian Chigutisauridae (Warren and Hutchinson 1983); Bajocian/Bathonian Brachyopidae (Dong 1985) and Valanginian/Albian Plagiosauridae (Jupp and Warren 1986; Milner 1989). Any conclusions drawn from this study must inevitably be seen as provisional while such discoveries are occurring on a nearly annual basis, and the terms 'first appearance' and 'last appearance' are more accurate than 'origin' and 'extinction' in describing the fossil record of a temnospondyl family.

Examination of Fig. 15.3 reveals that the Temnospondyli shows two cycles of diversification, stability, and last appearances. The two phases of diversification are here referred to as the Carboniferous radiation and the Griesbachian radiation and are discussed sequentially. Dates in Ma are taken from Harland *et al.* (1982).

The Carboniferous radiation

1. *Chronology*

The Carboniferous radiation had commenced by the Brigantian Stage of the Visean (about 338 Ma) when the first temnospondyls appear in the East Kirkton Limestone of Scotland (Milner *et al.* 1986). The subsequent diversification is poorly documented because the Visean/Namurian/lower Westphalian tetrapod record is deficient. However, the radiation was almost complete by the late Westphalian D (about 295 Ma). The seven families—Cochleosauridae, Saurerpetontidae, Trematopidae, Dissorophidae, Peliontidae, Micromelerpetontidae, and Branchiosauridae—were tangibly present, but the pattern of sister-group relationships suggests that another eight lineages—the Edopidae, stem-trimerorhachoids, stem-Stereospondyli, Zatrachydidae, Eryopidae, Parioxyidae, Peltobatrachidae, and Micropholidae—should also have been present (see Fig. 15.3). Their absence is almost certainly due to taphonomic overlay. The Westphalian D tetrapod assemblages (Linton, Mazon Creek, Florence and Nýřany), though rich, are ecologically limited (Milner 1987) and the absence of many of the latter groups is not surprising, as they are all associated with flood-plain or lacustrine deposits when they appear in the record.

Thus by the late Westphalian, most of the Palaeozoic temnospondyl families were present, either tangibly or by phylogenetic inference. Clearly 43 Ma is a long period of time and it is impossible to assess whether this radiation occurred gradually through this span or whether it took place rapidly at some stage within the 43 Ma. A few early

Westphalian temnospondyls (notably the stem-saurerpetontid *Eugyrinus*) demonstrate that it was not entirely a late Westphalian radiation, but only the discovery of further Visean–Westphalian material will permit this radiation to be chronologically characterized. The new temnospondyl material from the Visean of East Kirkton in Scotland may prove to be valuable in this respect. The only Palaeozoic families which may not have originated until the Stephano-Permian are the Trimerorhachidae and Dvinosauridae (first appearance of former in Stephanian), and the stem-stereospondyl families, Actinodontidae, Intasuchidae, Archegosauridae, and Rhinesuchidae. The latter families were all either tangibly or inferentially present by the Artinskian.

2. *Geography*

In the Upper Palaeozoic, Euramerica and Gondwanaland had coalesced to form the greater part of Pangaea, although Angara and Cathaysia were still separate platelets (Parrish *et al.* 1986, fig. 1). Almost all temnospondyls from the Visean to the Ufimian have been found between New Mexico and the Urals along a belt representing equatorial Pangaea. There are no Carboniferous exceptions and the only anomaly is an assemblage of uncertain Lower Permian age from Kashmir. If this distribution is a genuine phenomenon, it demonstrates that the Carboniferous phase of this radiation took place entirely within the Euramerican region of Pangaea, close to the equator. It matches part of the distribution of coals in the Carboniferous and may reflect the restriction of temnospondyls to hot humid environments. The alternative explanation would be that the distribution is merely an artefact of non-preservation of Carboniferous–Early Permian temnospondyls in Gondwanan, Angaran, and Cathaysian regions (the latter two regions having coal deposits at this time). If the latter were so, one might expect some late Permian temnospondyls in these areas to be anomalously distinct from the Euramerican forms. Most Upper Permian temnospondyls from the Gondwanan region have Lower Permian sister-families in the Euramerican region and are consistent with a Lower Permian range extension of temnospondyls following the withdrawal of the Gondwanan glaciation in the Lower Permian. However, the Peltobatrachidae and Micropholidae are anomalous in their appearance and have no obvious close relatives in the Laurasian faunas. If their position in the cladogram is correct, they have long unknown histories and may represent descendants of a Carboniferous range-extension of temnospondyls into the Gondwanan region. It is also possible that they are highly derived descendants of Euramerican Lower Permian families with some character reversals which have confused the systematic analysis.

Many of the Palaeozoic temnospondyl families occur throughout the known Euramerican range and geographical endemism was exceptional if it occurred at all. One possible example is the early stem-stereospondyl radiation (Actinodontidae, Intasuchidae, and Archegosauridae), which is based entirely on a complex suite of forms from the Lower Permian of Europe and Kashmir, suggesting a Lower Permian radiation genuinely restricted to eastern Euramerica extending into Gondwana. The Micromelerpetontidae are also currently known only from the Permo-Carboniferous of Europe.

Upper Permian temnospondyls are known from Europe, Africa, India, Australia, and South America and may be assumed to have been global in distribution. This was probably the result of a Lower Permian range extension following the contraction of the Gondwanan glaciation. It represents a distinct phenomenon from the Carboniferous radiation itself. There is no evidence for any major Upper Permian cladogenesis as a result of this range extension but the Gondwana temnospondyls do show some evidence of morphological innovation, discussed below under 'Stability and extinctions'.

3. *Morphology and heterochrony*

We can only speculate on the initial basis for the successful diversification of the temnospondyls. Among the few skeletal innovations that we can recognize near the base of the Temnospondyli are a tympanic ear and modified stapes capable of transducing high-frequency air-borne sound, and a modified palate suggesting an efficient buccal pump associated with air-breathing. These suggest that the earliest temnospondyls were relatively terrestrial as adults and the earliest known forms from East Kirkton appear to have been terrestrial (lateral-line sulci absent; ossified carpals and tarsals present). Like modern frogs, many non-neotenous temnospondyls had gilled larvae and non-gilled adults, and probably radically remodelled the skull at metamorphosis to accommodate the large tympanum, only of use in the adult. This structural dichotomy, between a larva with a short broad gape, large hyobranchial skeleton, and low-frequency ear, and an adult with a deep gape, reduced hyobranchial skeleton, and high-frequency ear, probably characterized the earliest temnospondyls and may have been a fruitful basis for heterochronic processes to operate on. Other early tetrapods do not appear to have had such a radical dichotomy of larval and adult morphologies and the heterochronic potential of this may be a factor in the great diversity of temnospondyls in comparison with most other Permo-Carboniferous tetrapod groups.

Three forms of heterochronic process (using the terminology of McNamara 1986) can be readily recognized in the Carboniferous radiation, namely, local acceleration, progenesis, and neoteny. The most primitive temnospondyls were short-snouted and local acceleration of growth of the snout region has produced crocodile-like piscivores on one (Node **A3** reversing at **A6** and **A17**) or two (Nodes **A4** and **A10** reversing at **A17**) occasions. Progenetic dwarfing, progressive reduction of adult size but without neoteny, seems to be operating through the Dissorophoidea between **A19** and **A23**. Neoteny apparently operated at **A6** and also to give the families Micromelerpetontidae and Branchiosauridae.

4. *Ecology*

Although temnospondyls seem to have been initially terrestrial, heterochronic retention of larval characters has permitted the temnospondyls to have radiated into a range of water-associated niches. Most were probably associated with freshwater bodies but the neotenous trimerorhachids and saurerpetontids (Node **A6**) regularly occur in lowland-deltaic environments and may even have been euryhaline (Milner 1987). The edopoid (**A4**) and the stem-stereospondyl (**A11–A13**) parts of the radiation were apparently based on diversification of feeding techniques in semi-aquatic crocodile-like groups, the latter restricted to lacustrine habitats. Another part of the Carboniferous radiation became more specialized in terrestrial niches (**A15**) and includes the eryopids, parioxyids, peltobatrachids, trematopids, dissorophids, and peliontids, most of which are found in association with reptiles rather than fishes. Within this clade there are at least three neotenous subclades, two of which (Micromelerpetontidae and Branchiosauridae) were small lacustrine carnivores, associated with fish-poor water-bodies.

5. *Stability and extinctions*

In the 47 Ma from the late Westphalian (295 Ma) to the end of the Permian (248 Ma), little family-level cladogenesis appears to have occurred apart from the stem-stereospondyls noted above. The Upper Permian temnospondyls of the Laurasian region (effectively the assemblages from pre-Uralian Russia) all resemble their Lower Permian relatives so closely as to be seen as confamilial (e.g. Actinodontidae, Intasuchidae, Archegosauridae, Eryopidae, Dissorophidae). Only the Dvinosauridae are distinctive and they may be simply highly derived trimerorhachids. Some morphological innovation did take place however, apparently in lineages which colonized the Gondwana region.

The Upper Permian temnospondyls of the Gondwana region however are sufficiently highly derived that they have always merited distinct family status (e.g. Brachyopidae *contra* Lower Permian Saurerpetontidae; Rhinesuchidae *contra* other stem-stereospondyls; Peltobatrachidae with no obvious Lower Permian relatives; and Micropholidae (Lower Triassic but with inferred Upper Permian relatives) likewise). Only the Archegosauridae appear as a Gondwana family with close resemblances to Lower Permian forms (Barberena and Daemon 1974). Thus, although the Permian was apparently a time of little family-level cladogenesis, there seems to have been a certain amount of morphological innovation following the colonization of the Gondwana region.

Although this review is concerned with radiations, it is necessary to discuss the end-Permian extinctions in order to place the Griesbachian radiation in context. No monophyletic temnospondyl families terminate prior to the Permian. Carboniferous 'extinctions' are the Caerorhachidae (monotypic with a single first and last appearance) and the Dendrerpetontidae (possibly paraphyletic). In the Permian, a series of final appearances of holophyletic families runs through all stages namely Asselian (Cochleosauridae), Sakmarian (Edopidae), Artinskian (Trimerorhachidae, Zatrachydidae, Parioxyidae, Micromelerpetontidae, and Peliontidae), Kungurian (Trematopidae and Branchiosauridae), Ufimian (Intasuchidae, Eryopidae), Kazanian (Dissorophidae), and Tatarian (Dvinosauridae, Archegosauridae). The Saurerpetontidae, Actinodontidae, Rhinesuchidae, and Peltobatrachidae are taken to be paraphyletic with respect to later families. The Upper Permian amphibian record is obviously depauperate (Carroll 1977; DeFauw 1987, 1989) and this pattern of cessation of appearance of families is probably a taphonomic artefact. As noted above, it is still changing from year to year as new discoveries are made, particularly in the Kungurian and Ufimian of eastern Europe. The direct evidence is not sufficient to indicate whether the Carboniferous radiation gradually declined in diversity or whether it was partly truncated by the end-Permian extinction event. Two indirect lines of evidence suggest the latter. One is the regional pattern of clade-extinction versus clade-survival. As can be seen in Fig. 15.3, the immediate relatives of the Triassic temnospondyls are the earliest Brachyopidae (Upper Permian of Australia), Rhinesuchidae and Uranocentrodontidae (Southern Africa and India), and Peltobatrachidae (East Africa). The Lower Triassic micropholids (South Africa and Australia) and lissamphibian *Triadobatrachus* (Madagascar) have no Upper Permian relatives but are themselves Gondwanan forms. In other words all the post-Permian temnospondyls and descendants can be argued to have been of Gondwanan origin. The clade-families which make their last appearances during the Upper Permian (Ufimian–Tatarian) are all exclusively Laurasian apart from the widespread Archegosauridae. This suggests a regional phenomenon, with extinction of the conservative Laurasian/equatorial temnospondyls and survival into the Triassic of the Gondwanan temnospondyls. A Laurasia-based mass extinction at the end of the Upper Permian might explain this phenomenon. The second line of evidence supporting an end-Permian mass extinction, is the explosive diversification in the Griesbachian as outlined below. If the Permian decline were genuinely gradual, one might expect an equally gradual appearance of Triassic families, even in a depauperate fauna. However, all recent additions to our knowledge of Upper Permian temnospondyls are late extensions of Lower Permian forms, not earlier occurrences of Triassic forms.

The end-Permian extinction is now recognized as having had a major effect on the Upper Permian amniote assemblages (Benton 1988) with 21 families making their last appearance in the Tatarian. It may well also have had a similar effect on the temnospondyl amphibians. This is in interesting contrast to the end-Cretaceous event which had no apparent effect on the amphibians of that time. There are no lissamphibian families with last appearances in the Upper Cretaceous (Milner, submitted).

6. *Post-Permian non-stereospondyl survivors*

Much of the post-Permian temnospondyl record is comprised of the stereospondyl clade, but four other families survived the end-Permian event even though they did not diversify greatly. They were the Micropholidae, Tupilakosauridae, Brachyopidae, and Plagiosauridae. The first of these families comprise enigmatic small peliontid-like forms while the latter three families comprise specialized neotenous forms.

The Micropholidae were a progenetic dwarf amphibious family of unknown Permian origin and described from South Africa and Australia (Warren and Hutchinson 1990). They make their first and last appearance in the Griesbachian and little can be said about them.

The Brachyopidae and Tupilakosauridae are sister-families of Permian origin. The Tupilakosauridae have been described only from a few Lower Triassic sites in Greenland and Russia but tupilakosaurid-like vertebrae have been collected from the Upper Permian of South Africa and are currently being described by G. King and the author. The Brachyopidae extend from the Upper Permian to the Bajocian/Bathonian as a structurally conservative group occurring in low diversity (only one genus in any assemblage).

The Plagiosauridae are a neotenous family of Griesbachian first appearance and probably descended directly from the Upper Permian terrestrial *Peltobatrachus*. The transition from terrestrial peltobatrachid to neotenous plagiosaurid may have occurred around the Permo-Triassic and be in some way connected with their survival of the end-Permian extinction. Like the brachyopids, the plagiosaurs were structurally conservative, once established, and occur in low diversity (probably two lineages) throughout the Triassic, persisting to the Valanginian/Albian in Australia.

Apart from the Rewan Formation of Australia, brachyopids and plagiosaurs have never been found together, although both were geographically widespread, and there may have been some ecological partitioning between the two families. Both families may have been morphologically conservative because their neotenous members lacked heterochronic potential.

The Griesbachian radiation

1. *Chronology*

Apart from the Lissamphibia and the post-Permian survivors noted above, about 16 families of post-Permian temnospondyls are recognized although one (Sclerothoracidae) is monotypic and others (Uranocentrodontidae, Benthosuchidae, Heylerosauridae, Thoosuchidae, Lydekkerinidae, and Luzocephalidae) are probably paraphyletic. All are part of the Stereospondyli, a clade with its stem in the Permian. Of the 16 families, only the Uranocentrodontidae spans the Permo-Triassic boundary. Of the other families, the Capitosauridae, Trematosauridae, Lydekkerinidae, Luzocephalidae, Rhytidosteidae, an unnamed family, and Chigutisauridae all first appear in the Griesbachian or early Dienerian of the Southern Hemisphere (248–245 Ma). There are no Griesbachian temnospondyls in the northern hemisphere (Cosgriff 1984) but in the Dienerian of Russia, the Wetlugasauridae, Benthosuchidae, Thoosuchidae first appear, and in the Spathian of Germany, the Heylerosauridae and Mastodonsauridae make their earliest appearances. The pattern of dichotomies superimposed on the stratigraphical scale suggests most must have been present in the Griesbachian or earliest Dienerian. The only family-level dichotomy that need be post-Dienerian is the latiscopid—metoposaurid dichotomy which may be as late as Ladinian when the first metoposaurid is known (Morales 1988). The stem-group of the metoposaurid—latiscopid clade must have been present in the Griesbachian, however, as it is the sister-group to the Trematosauridae.

Thus, during the Griesbachian, the Stereospondyli underwent a vast adaptive radiation of families and paraphyletic families, only one of which (Uranocentrodontidae) is known prior to the Triassic but almost all of which were in existence within 3 Ma of the Permo-Triassic event. The large amount of early Triassic cladogenesis has resulted in the presence in the record of many stem-groups linking terminal families and several of these have themselves achieved family status. Most of them do not persist beyond the Lower Triassic although the Heylerosauridae (mainly stem-mastodonsaurs) does extend into the Middle Triassic.

2. *Geography*

In the early Triassic, almost all of the continental masses were still aggregated as Pangaea (Parrish *et al.* 1986, fig. 14), and several temnospondyl families (Capitosauridae, Trematosauridae) were so widespread as to be probably global in distribution. There were no major barriers to movement and only a few patterns hint at the possibility that the Griesbachian radiation may have had regional components followed by later range extensions.

1. The rhytidostean radiation (**B15** upwards) of about 20 genera is almost entirely from the Gondwana region apart from *Luzocephalus*, *Boreopelta*, and *Peltostega* (the latter two probably immediately related) and may represent a Gondwana radiation with two range extensions.
2. The basal forms of the mastodonsauroid–trematosauroid radiation (**B6** upwards) are largely from the Laurasian region. Only *Kestrosaurus*, and some members of the families Trematosauridae and Metoposauridae represent range extensions.

3. *Morphology and heterochrony*

The Griesbachian radiation differs from the Carboniferous radiation in that it derived from a large long-snouted amphibious morphology, rather than a smaller 'salamandriform' terrestrial morphology. However, there is some evidence for distinct larval and adult morphologies in these forms (Warren and Hutchinson 1988) and the same basis for heterochrony to operate. Again, local acceleration, progenesis, and neoteny can be identified within the radiation. Most of the radiation comprises amphibious carnivores and large neotenous forms.

The wetlugasaurid–capitosaurid (**B5**) clade was the most conservative and both families are difficult to diagnose on unambiguous characters which are not subject to homoplasy. The benthosuchid-based clade (**B6**) seems to have been based on partial neoteny to give more nearly aquatic piscivores but retaining variations of the elongate snout and the high-frequency ear. Within this clade, the Metoposauridae may represent a shift to more complete neoteny, and were homeomorphic with the earlier trimerorhachids.

If the stem-stereospondyls are assumed to have been rhinesuchid-like then the move to a *Lydekkerina*-like morphology (**B15**) involved progenetic dwarfing, as the lydekkerinid grade is superficially juvenile-like. This phase of dwarfing may have permitted the remodelling which, when enlarged, gave rise to the relatively aberrant rhytidosteids and relatives. Neoteny subsequently operated within this clade (**B19**) to give the brachyopid-convergent chigutisaurids and relatives.

4. *Ecology*

The inferred ecology of the stereospondyls resulting from the Griesbachian radiation has been discussed by DeFauw (1987, 1989) and the following is an interpretation of her observations in the context of the phylogeny and systematics used here. The basal members of the radiation (Rhinesuchidae, Uranocentrodontidae, Sclerothoracidae, Lydekkerinidae), the capitosauroid (**B5**) clade (Wetlugasauridae and Capitosauridae), and some of the rhytidostean clade (Chigutisauridae) are all identified by DeFauw as semi-aquatic freshwater carnivores, similar to modern crocodiles and alligators. It would appear that this was the basal niche of the stereospondyls and the capitosaurids continued to fill it throughout the Triassic, remaining structurally the most conservative of the stereospondyls. The benthosuchids (including heylerosaurids and thoosuchids) were recognized by DeFauw (1987) as similar freshwater forms but probably completely aquatic suggesting a greater degree of aquatic specialization at the **B6** node. Aquatic specialization in the benthosuchid complex accords with their position at the base of the **B10** and **B12** clades. Three clades of stereospondyls appear to have shifted from freshwater to euryhaline existence and most are structurally derived swimming specialists. The **B12** clade comprises three families (Trematosauridae, Latiscopidae, and Metoposauridae) which DeFauw identifies as fully aquatic euryhaline forms and I suspect that the Mastodonsauridae may also fall into this category rather than semi-aquatic euryhaline as DeFauw classifies them. Some trematosaurids became marine and are found in deep-water deposits in Spitsbergen. The **B18** (Luzocephalidae + Rhytidosteidae) clade paralleled the **B12** clade in also producing euryhaline and marine aquatic forms.

The Stereospondyli had little success at radiating into large terrestrial niches (the aberrant rhytidosteid *Laidleria* from the Lower Triassic of South Africa is the only known example) probably because of more diverse amniote competition (DeFauw 1987).

The stem-stereospondyl lineage is inferred to have originated in the Upper Carboniferous and have acquired the basic stereospondyl characters by the Upper Permian but this radiation did not take place until the basal Triassic. The explosive nature of the Griesbachian radiation following the end-Permian extinction strongly suggests that this radiation was triggered by ecological opportunism following the extinction of Permian temnospondyls, rather than a specific morphological innovation.

5. *Stability and extinctions*

The Scythian/Lower Triassic (248–243 Ma) was apparently a period of rapid radiation but subsequently there is little evidence of family-level cladogenesis. The Heylerosauridae and early Mastodonsauridae represent a complex taxonomic problem. The Heylerosauridae may be monophyletic or paraphyletic with respect to the Mastodonsauridae but, if the former, the dichotomy must have taken place by the end of the Lower Triassic. As noted above, only the Latiscopidae and Metoposauridae may have diverged later. Some families (Capitosauridae, Mastodonsauridae) remained conservative in their appearance throughout their post-Lower Triassic existence.

The Griesbachian radiation shows a gradual pattern of last appearances, complicated by a slightly depauperate Middle Triassic record. Ignoring the monotypic Sclerothoracidae and the pseudoextinctions of paraphyletic families (which cease to appear earlier or at the same time as their descendants), last appearances are Smithian (unnamed family: see Appendix), Spathian (Wetlugasauridae, Rhytidosteidae), Carnian (Mastodonsauridae, Trematosauridae), Norian (definite Capitosauridae, Latiscopidae, definite Metoposauridae), Hettangian (possible for Metoposauridae, Murry 1987), Pliensbachian/Toarcian (Chigutisauridae), and Callovian (possible Capitosauridae, Nessov 1988).

Many families have ceased to appear by the Norian but the ecologically and geographically patchy nature of the Middle Triassic record makes recognition of extinction events difficult. An end-Spathian event has been argued for amniotes (e.g. Benton 1988) and it may be noted that, apart from the tupilakosaurids, wetlugasaurids, and rhytidosteids, the Trematosauridae would have been included until the recent recognition of a Carnian genus (Hellrung 1987).

6. *Lissamphibia*

The Lissamphibia (frogs, salamanders, and caecilians) can be considered as a highly derived subgroup of temnospondyls which persist in great diversity to the present day. Cladistic analysis suggests that they are relatives/descendants of either the peliontids or the branchiosaurids and that the Salientia (frogs and stem-frogs) and Caudata (salamanders) are immediately related with the Gymnophiona (caecilians) more distantly related (Milner 1988). By the Jurassic, all three groups are tangibly present in the fossil record but the only Triassic fossils are *Triassurus*, a doubtful salamander from the Upper Triassic of the Kirgiz SSR (Ivakhnenko 1978); and *Triadobatrachus*, a primitive salientian from the Lower Triassic of Madagascar (Rage and Roček 1989). Given the relationships of the modern groups as described immediately above, the presence of a primitive frog implies that primitive salamanders and caecilians were also present and that the lissamphibian radiation was under way in the Lower Triassic. It is possible that the end-Permian extinction event also wiped out archaic small non-temnospondyl amphibians (nectrideans, aistopods, microsaurs) and left vacant niche space for a group of highly modified small temnospondyls to exploit, thus initiating a Griesbachian radiation of lissamphibians.

Conclusions

This provisional account of temnospondyl phylogeny suggests that the history of the group can be modelled as two cycles of radiation, stability, and extinction. The initial radiation of the group occurred in the Euramerican/equatorial region of Pangaea, commencing in the Visean and being substantially complete by the late Westphalian. It may have been triggered by morphological innovation based on heterochronic processes acting on a group characterized by distinct larval and adult morphologies. From the Stephanian to the Tatarian there was family-level stability in the Euramerican region except for some diversification of the stereospondyl stem in the Lower Permian of Europe. From the Kazanian onwards, morphologically innovative temnospondyls appeared in the Gondwana region. Some are immediately related to Euramerican forms and probably represent derived vicariant groups evolving after an early Permian range extension of temnospondyls into the Gondwanan region following the withdrawal of the Palaeozoic southern glaciation. Temnospondyls appear to have been subject to a drastic extinction event at the end of the Permian, the effects of which may have been less extreme in Gondwana. This was immediately followed by the stereospondyl radiation in the Griesbachian, which was probably triggered by ecological opportunism rather than morphological innovation. Through the Triassic, a period of family-level stability ensued. The last appearances of the second cycle of temnospondyl families are spread between the Spathian and the Lower Cretaceous and do not, by themselves, suggest any one mass extinction. The post-Triassic survivors are largely known from east Asia and Australia and appear to have been large neotenous forms.

Acknowledgements

I thank Sandy Sequeira for drafting the illustrations for this chapter, and Dr Jennifer Clack for constructive criticism of the typescript.

References

Barberena, M.C. and Daemon, R.F. (1974). A primeira ocorrencia de Amphibia (Labyrinthodonta) na Formaçao Rio de Rasto; implicacoes geocronologicas e estratigraficas. *Congresso Brasileiro de Geologia, Anais*, **28**, 251–61.

Benton, M.J. (1988). Mass extinctions in the fossil record of reptiles: paraphyly, patchiness and periodicity (?). In *Extinction and survival in the fossil record* Systematics Association (ed. G.P. Larwood), pp. 269–94. Clarendon Press, Oxford.

Berman, D.S., Reisz, R.R. and Eberth, D.A. (1987). A new genus and species of trematopid amphibian from the late Pennsylvanian of north-central New Mexico. *Journal of Vertebrate Paleontology*, **7**, 252–69.

Bolt, J.R. (1969). Lissamphibian origins: possible protolissamphibian from the Lower Permian of Oklahoma. *Science*, **166**, 888–91.

Boy, J.A. (1989). Über einige vertreter der Eryopoidea (Amphibia: Temnospondyli) aus dem europäischen Rotliegend (?hochstes Karbon–Perm) 2. *Acanthostomatops*. *Paläontologische Zeitschrift*, **63**, 133–51.

Carroll, R.L. (1977). Patterns of amphibian evolution: an extended example of the incompleteness of the fossil record. In *Patterns of evolution* (ed. A. Hallam), pp. 405–37, Elsevier, Amsterdam.

Coldiron, R.W. (1978). *Acroplous vorax* Hotton (Amphibia, Saurerpetontidae) restudied in light of new material. *American Museum of Natural History, Novitates*, **2662**, 1–27.

Cosgriff, J.W. (1984). The temnospondyl labyrinthodonts of the earliest Triassic. *Journal of Vertebrate Paleontology*, **4**, 30–46.

DeFauw, S.L. (1987). A new perspective on the last phases in the evolution of the Temnospondyli (Amphibia, Labyrinthodontia). 4th Symposium of Mesozoic Terrestrial Ecosystems, Short Papers (ed. P.J. Currie and E.H. Koster). *Occasional Papers of the Tyrrell Museum of Paleontology*, **3**, 61–9.

DeFauw, S.L. (1989). Temnospondyl amphibians: a new perspective on the last phases in the evolution of the Labyrinthodontia. *Michigan Academician*, **21**, 7–32.

Dong, Z. (1985). A Middle Jurassic labyrinthodont (*Sinobrachyops placenticephalus* gen. et sp. nov.) from Dashanpu, Zigong, Sichuan Province. *Gu Jizhui Dongwu Xuebao—Vertebrata PalAsiatica*, **23**, 301–6.
Gaffney, E. S. (1979). Tetrapod monophyly: a phylogenetic analysis. *Bulletin of the Carnegie Museum of Natural History*, **13**, 92–105.
Gardiner, B. G. (1983). Gnathostome vertebrae and the classification of the Amphibia. *Zoological Journal of the Linnean Society*, **79**, 1–59.
Gauthier, J. (1986). Saurischian monophyly and the origin of birds. In *The origin of birds and the evolution of flight*. (ed. K. Padian). *Memoirs of the California Academy of Sciences*, **8**, 1–55.
Godfrey, S.J. (1989). The postcranial skeletal anatomy of the Carboniferous tetrapod *Greererpeton burkemorani* Romer 1969. *Philosophical Transactions of the Royal Society of London*, **B323**, 75–133.
Godfrey, S.J., Fiorillo, A. R. and Carroll, R. L. (1987). A newly discovered skull of the temnospondyl amphibian *Dendrerpeton acadianum* Owen. *Canadian Journal of Earth Sciences*, **24**, 796–805.
Gubin, Y. M. (1983). The first eryopids from the Permian of the east European platform. *Paleontologicheskiy Zhurnal*, **1983** (4), 110–15 (In Russian.) *Paleontological Journal*, **17** (4), 105–10. (In English.)
Hammer, W. R. (1987). Paleoecology and phylogeny of the Trematosauridae. In *Gondwana six: stratigraphy, sedimentology and paleontology*, American Geophysical Union, Geophysical Monograph Series, no. 41 (ed. G. D. McKenzie), pp. 73–83. American Geophysical Union, Washington, DC.
Harland, W. B., Cox, A. V., Llewellyn, P. G., Pickton, C. A. G., Smith, A. G. and Walters, R. (1982). *A geologic time scale*. Cambridge University Press, Cambridge.
Hellrung, H. (1987). Revision von *Hyperokynodon keuperinus* Plieninger (Amphibia: Temnospondyli) aus dem Schilfsandstein von Heilbronn (Baden-Württemberg). *Stuttgarter Beiträge zur Naturkunde Serie B, Geologie and Paläontologie*, **136**, 1–28.
Ingavat, R. and Janvier, P. (1981). *Cyclotosaurus* cf. *posthumus* Fraas (Capitosauridae: Stereospondyli) from the Huai Hin Lat Formation (Upper Triassic), northeastern Thailand with a note on capitosaurid biogeography. *Geobios*, **14**, 711–25.
Ivakhnenko, M. F. (1978). Urodelans from the Triassic and Jurassic of Soviet Central Asia. *Paleontologicheskiy Zhurnal*, **1978** (3), 84–9. (In Russian.) *Paleontological Journal*, **12** (3), 362–8. (In English.)
Jupp, R. and Warren, A. A. (1986). The mandibles of the Triassic temnospondyl amphibians. *Alcheringa*, **10**, 99–124.
Maňourová, M. (1981). Nový nález Branchiosauru na lokalitě Horní Kalná u Vrchlabí. *Casopis Národního Muzeu v Praze, Řada Přírodovědná*, **150**, 169–72.
Martens, T. (1988). Die bedeutung der Rotsedimente für die analyse der lebewelt des Rotliegenden. *Zeitschrift für Geologische Wissenschaften, Berlin*, **16**, 933–8.
McNamara, K. J. (1986). A guide to the nomenclature of heterochrony. *Journal of Paleontology*, **60**, 4–13.

Milner, A.R. (1987). The Westphalian tetrapod fauna; some aspects of its geography and ecology. *Journal of the Geological Society of London*, **144,** 495–506.

Milner, A. R. (1988). The relationships and origin of living amphibians. In *The phylogeny and classification of the tetrapods*, Vol. 1 (ed. M.J. Benton), pp. 59–102. Clarendon Press, Oxford.

Milner, A. R. (1989). Late extinctions of amphibians. *Nature (London)*, **338,** 117.

Milner, A. R. (submitted). Amphibian-grade Tetrapoda. In *The fossil record* (ed. M.J. Benton and M. A. Whyte). Chapman and Hall, London and New York.

Milner, A.R. Smithson, T.R., Milner, A.C., Coates, M.I. and Rolfe, W.D.I. (1986). The search for early tetrapods. *Modern Geology*, **10,** 1–28.

Morales, M. (1988). New metoposaurid and capitosauroid labyrinthodonts from the Triassic of Germany and the Soviet Union. Abstract in *Journal of Vertebrate Paleontology*, **8** (3, Suppl.), 23A.

Murry, P. A. (1987). Notes on the stratigraphy and paleontology of the Upper Triassic Dockum Group. *Journal of the Arizona–Nevada Academy of Science*, **22,** 73–84.

Nessov, L. A. (1988). Late Mesozoic amphibians and lizards of Soviet Middle Asia. *Acta Zoologica Cracoviensis*, **31,** 475–86.

Panchen, A. L. (1980). The origin and relationships of the anthracosaur Amphibia from the late Palaeozoic. In *The terrestrial environment and the origin of land vertebrates* (ed. A. L. Panchen), pp. 319–50. Academic Press, London.

Panchen, A.L. and Smithson, T.R. (1988). The relationships of early tetrapods. In *The phylogeny and classification of the tetrapods*, Vol. 1 (ed. M.J. Benton) pp. 1–32. Clarendon Press, Oxford.

Parrish, J.M., Parrish, J.T. and Ziegler, A.M. (1986). Permian-Triassic paleogeography and paleoclimatology and implications for therapsid distribution. In *The ecology and biology of mammal-like reptiles* (ed. N. Hotton III, P.D. MacLean, J.J. Roth and E.C. Roth), pp. 109–131. Smithsonian Press, Washington, DC.

Patterson, C. and Rosen, D. E. (1977). Review of ichthyodectiform and other Mesozoic teleost fishes and the theory and practice of classifying fossils. *Bulletin of the American Museum of Natural History*, **158,** 81–172.

Pitrat, C.W. (1973). Vertebrates and the Permo-Triassic extinction. *Palaeogeography, Palaeoclimatology, Palaeoecology*, **14,** 249–64.

Rage, J.-C. and Roček, Z. (1989). Redescription of *Triadobatrachus massinoti* (Piveteau, 1936) an anuran amphibian from the early Triassic. *Palaeontographica Abt. A*, **206,** 1–16.

Romer, A. S. (1947). Review of the Labyrinthodontia. *Bulletin of the Museum of Comparative Zoology at Harvard College*, **99,** 1–138.

Romer, A. S. and Parsons, T. S. (1986). *The vertebrate body* (6th edn). Saunders, Philadelphia.

Smithson, T.R. (1985). The morphology and relationships of the Carboniferous amphibian *Eoherpeton watsoni* Panchen. *Zoological Journal of the Linnean Society*, **85,** 317–410.

Warren, A. A. and Hutchinson, M. N. (1983). The last labyrinthodont? A new brachyopoid (Amphibia, Temnospondyli) from the early Jurassic Evergreen Formation of Queensland, Australia. *Philosophical Transactions of the Royal Society of London*, **B303**, 1–62.

Warren, A. A. and Hutchinson, M. N. (1988). A new capitosaurid amphibian from the early Triassic of Queensland, and the ontogeny of the capitosaur skull. *Palaeontology*, **31**, 857–76.

Warren, A. A. and Hutchinson, M. N. (1990). *Lapillopsis*, a new genus of temnospondyl amphibians from the early Triassic of Queensland. *Alcheringa*, **14**, 149–58.

Werneburg, R. (1989). Labyrinthodontier (Amphibia) aus dem Oberkarbon und Unterperm Mitteleuropas—systematik, phylogenie und biostratigraphie. *Freiberger Forschungshefte Series C*, **436**, 7–57.

Appendix

Families of temnospondyl amphibians

The following listing of temnospondyl families includes stratigraphical range, generic content, and notes on phylogenetic status, and forms the basis of Fig. 15.3. The data were drawn from Milner (submitted) and Milner (in preparation).

Stratigraphical abbreviations are as follows: ALB, Albian; ANS Anisian; ART, Artinskian; ASS, Asselian; BAJ, Bajocian; BTH, Bathonian; CAR, Carnian; CLV, Callovian; DIE, Dienerian; GRI, Griesbachian; HET, Hettangian; Jur., Jurassic; K., Cretaceous; KAZ, Kazanian; KUN, Kungurian; Mis., Mississippian; NAM, Namurian; NOR, Norian; P., Permian; Pen., Pennsylvanian; PLI, Pliensbachian; SAK, Sakmarian; SMI, Smithian; SPA, Spathian; STE, Stephanian; TAT, Tatarian; TOA, Toarcian; Tr., Triassic; UFI, Ufimian; VIS, Visean; VLG, Valanginian; WES, Westphalian.

@, Family based on material from two or more localities and believed to represent a monophyletic clade; !, monotypic family based on a single species from one locality; *, monotypic family based on single specimen; +, family believed to be paraphyletic with respect to other families as listed.

'Stem-temnospondyls'

Caerorhachidae * ?Mis. (?NAM)
Caerorhachis.

Dendrerpetontidae + Mis. (VIS)–Pen. (WES.B)
Dendrerpeton and unnamed material from East Kirkton. (Family may be paraphyletic with respect to all other temnospondyls except Caerorhachidae.)

'Edopoidea'

Edopidae @ P. (ASS-SAK)

Edops.

Cochleosauridae @ Pen. (WES.B)-Pen./P. (STE/SAK)

Chenoprosopus, *Cochleosaurus*, unnamed specimen from Joggins, Nova Scotia.

'The trimerorhachoid-brachyopoid group'

Trimerorhachidae @ Pen. (STE)-P. (ART)

Dawsonerpeton, *Lafonius*, *Neldasaurus*, *Trimerorhachis.*

Dvinosauridae @ P. (TAT)

Dvinosaurus.

Saurerpetontidae + Pen. (WES.A)-P. (ART)

Acroplous, *Eugyrinus*, *Isodectes*, *Saurerpeton.*

(Family may be paraphyletic with respect to Tupilakosauridae and Brachyopidae, see Coldiron 1978.)

Tupilakosauridae @ Tr. (DIE-SPA)

Batrachosuchoides, *Tupilakosaurus.*

Brachyopidae @ P. (KAZ/TAT)-Jur. (BAJ/BTH)

Austrobrachyops, *Batrachosuchus*, *Blinasaurus*, *Bothriceps*, *Brachyops*, *Hadrokkosaurus*, *Notobrachyops*, *Platycepsion*, *Sinobrachyops*, *Xenobrachyops.*

Unplaced trimerorhachoid—brachyopoid genera

Erpetosaurus, *Kourerpeton*, *Nannospondylus*, *Perryella*, *Slaughenhopia*, *Tungussogyrinus.*

'Stem-stereospondyls'

Actinodontidae + P. (ASS-UFI)

Sclerocephalus (= *Actinodon*, *Lusor*), *Syndyodosuchus.* (Family may be paraphyletic with respect to all other stem-stereospondyls and stereospondyls.)

Intasuchidae @ P. (ART-UFI)

Cheliderpeton, *Intasuchus.*

Archegosauridae @ P. (ART-TAT)

Archegosaurus, *Bashkirosaurus*, *Collidosuchus*, *Melosaurus*, *Platyoposaurus*, *Prionosuchus.*

'Crown-group stereospondyls'

Rhinesuchidae + P. (KAZ-KAZ/TAT)

Rhinesuchoides, *Rhinesuchus.*

(Family may be paraphyletic with respect to those following from Uranocentrodontidae to Metoposauridae.)

Uranocentrodontidae + P. (TAT)-Tr. (GRI)
Rhineceps, *Uranocentrodon* (= *Laccocephalus*, *Laccosaurus*). (Family may be paraphyletic with respect to those following from Sclerothoracidae to Metoposauridae.)

Sclerothoracidae ! Tr. (SPA)
Sclerothorax.

Wetlugasauridae @ Tr. (DIE-SPA)
Wetlugasaurus.

Capitosauridae @ Tr. (GRI/DIE-NOR)-Jur. (CLV)?
Cyclotosaurus, *Eryosuchus*, *Parotosuchus* (= *Capitosaurus*), *Stenotosaurus*, *Wellesaurus*.
(Uncertain Callovian record is temnospondyl fragments reported as capitosaurid by Nessov 1988.)

Benthosuchidae + Tr. (DIE-SPA)
Benthosphenus, *Benthosuchus*.
(Family may be paraphyletic with respect to Heylerosauridae, Mastodonsauridae, Thoosuchidae, Trematosauridae, Latiscopidae, and Metoposauridae.)

Heylerosauridae + Tr. (SMI-SPA/ANS)
Eocyclotosaurus, *Kestrosaurus*, *Meyerosuchus*, *Odenwaldia*. (Family may be paraphyletic with respect to Mastodonsauridae.)

Mastodonsauridae @ Tr. (SPA/ANS-CAR)
Bukobaja, *Mastodonsaurus*.

Thoosuchidae + Tr. (DIE/SPA)
Thoosuchus, *Trematotegmen*.
(Family may be paraphyletic with respect to Trematosauridae, Latiscopidae, and Metoposauridae.)

Trematosauridae @ Tr. (GRI-CAR)
Aphaneramma, *Erythrobatrachus*, *Gonioglyptus* (= *Glyptognathus*, *Panchetosaurus*), *Halobatrachus*, *Hyperokynodon*, *Inflectosaurus*, *Lyrocephaliscus*, *Microposaurus*, *Platystega*, *Stoschiosaurus*, *Tertrema*, *Trematosaurus*, *Wantzosaurus*.

Latiscopidae @ Tr. (CAR/NOR)
Almasaurus, *Latiscopus*.

Metoposauridae @ Tr. (LAD)-Tr./Jur. (NOR/HET?)
Anaschisma, *Buettneria*, *Koskinonodon*, *Metoposaurus*. (Possible Hettangian record noted by Murry 1987.)

Lydekkerinidae + Tr. (GRI)
Limnoiketes, *Lydekkerina*.
(Family may be paraphyletic with respect to Luzocephalidae, Rhytidosteidae, unnamed family, and Chigutisauridae.)

Luzocephalidae + Tr. (GRI–SMI)
Chomatobatrachus, *Luzocephalus* and '*Lydekkerina*' *panchetensis*.
(Family may be paraphyletic with respect to Rhytidosteidae).
Rhytidosteidae @ Tr. (GRI–SPA)
Boreopelta, *Deltasaurus*, *Indobrachyops*, *Laidleria*, *Mahavisaurus*, *Peltostega*, *Pneumatostega*, *Rhytidosteus*.
Family to be named @ Tr. (GRI–SMI)
Arcadia, *Acerastea*, *Derwentia*, *Rewana*.
Chigutisauridae @ Tr. (GRI/DIE)–Jur. (PLI/TOA)
Keratobrachyops, *Pelorocephalus* (= *Chigutisaurus*), *Siderops*.
Unplaced Stereospondyl Genera
Boreosaurus, *Cryobatrachus*, *Deltacephalus*, *Gondwanosaurus*, *Moenkopisaurus*, *Pachygonia*, *Promastodonsaurus*, *Rhadalognathus*, *Yarengia*

'The eryopid–dissorophoid group'
Zatrachydidae @ P. (SAK–ART)
Acanthostomatops, *Dasyceps*, *Zatrachys*.
Eryopidae @ Pen. (STE)–P. (UFI)
Clamorosaurus, *Eryops*, *Onchiodon*, *Osteophorus*.
Parioxyidae @ P. (SAK–ART)
Parioxys.
Peltobatrachidae ! + P. (KAZ/TAT)
Peltobatrachus.
(This monotypic family is a metataxon (*sensu* Gauthier 1986) with respect to the Plagiosauridae.)
Plagiosauridae @ Tr. (GRI/DIE)–K. (VLG/ALB)
Gerrothorax, *Melanopelta*, *Plagiobatrachus*, *Plagiosaurus*, *Plagioscutum*, *Plagiosternum*, *Plagiosuchus*, and unnamed mandible from Lower Cretaceous Strzelecki Formation, Victoria, Australia (Jupp and Warren 1986).
Trematopidae @ Pen. (WES.D)–P. (KUN)
Acheloma (= *Trematops*), *Actiobates*, *Anconastes*, *Ecolsonia*, *Mordex*, and undescribed trematopid from the Tambacher Formation of East Germany (Martens 1988).
Micromelerpetontidae @ Pen. (WES.D)–P. (ART)
Branchierpeton, *Limnogyrinus*, *Micromelerpeton*.
Dissorophidae @ Pen. (WES.D)–P. (KAZ)
Alegeinosaurus, *Aspidosaurus*, *Astreptorhachis*, *Brevidorsum*, *Broiliellus*, *Cacops*, *Conjunctio*, *Dissorophus*, *Fayella*, *Iratusaurus*, *Kamacops*, *Longiscitula*, *Platyhistrix*, *Stegops*, *Zygosaurus*.
Micropholidae @ Tr. (GRI)
Lapillopsis, *Micropholis*.

Branchiosauridae @ Pen. (WES.D)- P. (KUN)
Apateon (= *Protriton*), *Branchiosaurus*, *Melanerpeton* (= *Leptorophus*), *Schoenfelderpeton*.

Peliontidae @ Pen. (WES.D)-P.(ART)
Amphibamus, *Doleserpeton*, *Tersomius*.
(One of the above two families may be paraphyletic with respect to the Lissamphibia.)

Unplaced Eryopid-Dissorophoid Genera
Arkanserpeton, *Hesperoherpeton*, *Macrerpeton*.

16. Interpretations of early land plant radiations: 'facile adaptationist guesswork' or reasoned speculation?

D. EDWARDS
Department of Geology, University of Wales College of Cardiff, Cardiff, UK
M.S. DAVIES
School of Pure and Applied Biology, University of Wales College of Cardiff, Cardiff, UK

Abstract

The diversification of higher plants following initial colonization of the land is analysed at the species level both qualitatively and quantitatively based on data from Laurussia. The overall pattern of diversification broadly parallels that described for marine invertebrates in that, following the appearance of pteridophyte-like plants (*Cooksonia*: rhyniophytoid) in the late Wenlock, numbers remained low until the end of the Silurian—early Devonian, when a proliferation of rhyniophytoids occurred. This was followed by a diversification of Zosterophyllophytina in the late Gedinnian—early Siegenian, and Trimerophytina and taxa of less certain affinity later in the Lower Devonian. In the absence of direct information on habitats, of the adaptive significance (if any) of variation in individual characters, and indeed the genetic status of the 'morphospecies' involved, interpretation must be tentative. The radiation of the rhyniophytoids is evaluated as that of the archetypal founder population with presumed limited gene pool and simplicity of organization with little variation. The succeeding zosterophylls show far greater diversity in growth form and reproductive capacity, largely as a result of release from the archetypal constraints of isotomous branching and terminal sporangia seen in the majority of rhyniophytoids. The record thus supports the hypothesis of competitive replacement of

Major Evolutionary Radiations (ed. P.D. Taylor and G.P. Larwood), Systematics Association Special Volume No. 42, pp. 351–376, Clarendon Press, Oxford, 1990.

rhyniophytoids by zosterophylls in the Gedinnian, but interrelationships of higher taxa later in the Lower Devonian are less clear-cut.

Introduction

The quantitative analysis of evolutionary patterns in early vascular plants undertaken by Knoll *et al.* (1984) was both comprehensive and courageous—the latter because the data base suffers from severe shortcomings both in quantity and quality (Edwards, in press) when compared with that of marine invertebrates. Nevertheless, those authors were able to demonstrate from megafossil data that patterns of diversification, origination, and extinction fitted broadly the kinetic model for marine invertebrates proposed by Sepkoski (1979, 1981). They anticipated modification of their findings as new data became available and here, half a decade on, we propose to reconsider megafossil data for the initial phases of land colonization by pteridophyte-like plants. We incorporate in particular new information from Britain and Belgium, generously provided by Drs U. Fanning, P. Kenrick, and P. Gerrienne (Fig. 16.1). Our analysis is deliberately restricted to a limited geographical area, viz. Laurussia (broadly equivalent to the Old Red Sandstone Continent), because it is only from this region that we have the sufficiently detailed stratigraphic, sedimentological, and taphonomic background data, plus confidence in identification and correlation, that are the prerequisites for sensible interpretation. This approach does not mean that we are convinced that the 'cradle' of the higher land plants was in Laurussia. Data from Australia already indicate greater diversity (e.g. in the record of *Baragwanathia* in the Silurian, Garratt *et al.* 1984), while less well documented assemblages from the Soviet Union and China, although sometimes controversial in identification of taxa and age (Edwards 1990), will undoubtedly lead to re-evaluation on a global scale.

1. *Sources of data and their temporal framework*

The most satisfactory taxon used in the diversity curves is the species (Fig. 16.2). Analysis at this level is possible because of the comparatively few plants recorded and has the advantage of avoiding troublesome questions of generic/species delimitation (Edwards and Edwards 1986; Niklas *et al.* 1980, 1985). Records were taken from original papers and from research currently being undertaken in Britain and Belgium. Correlation is based on the spore zonation scheme developed by Richardson and McGregor (1986) with added data from Steemans and Streel (e.g. Streel *et al.* 1987) for Belgium, and McGregor (e.g. McGregor and Camfield 1976) for north-east America. As correlation between continental/shallow marine and deeper water facies is still in its infancy, we continue to use Gedinnian and Siegenian rather than the now legitimate Lochkovian and Pragian. The choice of the absolute time-scale is subjective (see e.g. Holland 1985); we use that of McKerrow *et al.* (1985) largely because it emphasizes the relatively small and large amounts of time encompassed by Pridoli Series and Emsian Stage, respectively. The positioning of our points in relation to the absolute time-scale is also subjective. Where possible in the Lower Devonian we have divided each stage into 2–3 Ma intervals and plotted the value half-way through. In the case of the Upper Gedinnian and the Ludlow and Wenlock Series, data are recorded from the latter part, while in the Pridoli most records are early and come from just above the Ludlow Bone Bed or its equivalent.

2. *Affinities of taxa*

Of the three subdivisions erected by Banks (1968) from the 'psilophytes', the Zosterophyllophytina has best stood the tests of usage and additional information. It is clearly monophyletic, although its 'upper limits' *vis-à-vis* its most derived members are conjectural (Schweitzer 1983; Crane 1989). For the purposes of this analysis we exclude, as Banks did, the taxa *Baragwanathia*, *Drepanophycus*, and *Asteroxylon*. These are grouped with the Lycophytina, which is represented in the Lower Devonian of Laurussia by *Leclercqia*. The Trimerophytina with far fewer genera presents greater problems in that *Psilophyton* with a number of usually well-defined species is known both anatomically and morphologically, while *Trimerophyton* and the more complex *Pertica* lack anatomy. Particularly intriguing are the affinities of Emsian sterile plants known only from anatomy (e.g. Gensel 1984; Remy and Hass 1986*b*) which may well link trimerophytes and progymnosperms or iridopteridaleans. Such plants are here grouped under 'other' which also encompasses older taxa, e.g. *Estinnophyton*, *Krithodeophyton*, *Sartilmania*, *Sennicaulis*. Recent reviews of the Rhyniophytina (Edwards and Edwards 1986; Taylor 1988) emphasize just how few taxa assigned to that group are demonstrably vascular plants. This is particularly so for Silurian and earliest Devonian representatives, which are thus described as rhyniophytoid. Whether or not they form a natural taxon or are linked by convergence as a result of their morphological simplicity remains conjectural. The earliest representatives are compressions and thus lack evidence relating to the actual existence of conducting tissues or, if present, on their nature. Caution is advisable after the demonstration of moss-like conducting cells in *Aglaophyton* (*Rhynia*) *major* whose morphology typifies that of a rhyniophyte (D. S. Edwards 1986). Further, it is perhaps unreasonable

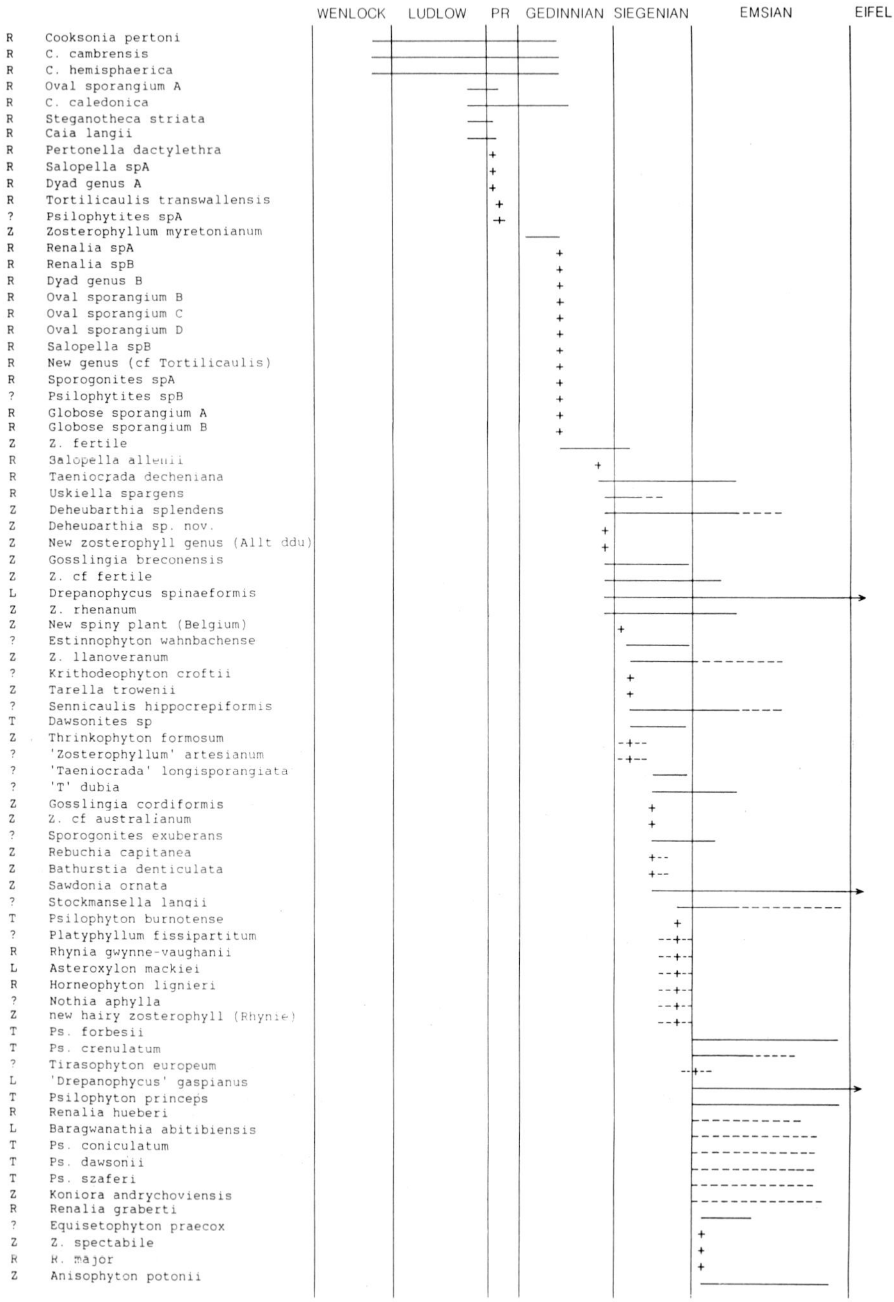
WENLOCK
LUDLOW
PR
GEDINNIAN
SIEGENIAN
EMSIAN
EIFEL
R Cooksonia pertoni
R C. cambrensis
R C. hemisphaerica
R Oval sporangium A
R C. caledonica
R Steganotheca striata
R Caia langii
R Pertonella dactylethra
R Salopella spA
R Dyad genus A
R Tortilicaulis transwallensis
? Psilophytites spA
Z Zosterophyllum myretonianum
R Renalia spA
R Renalia spB
R Dyad genus B
R Oval sporangium B
R Oval sporangium C
R Oval sporangium D
R Salopella spB
R New genus (cf Tortilicaulis)
R Sporogonites spA
? Psilophytites spB
R Globose sporangium A
R Globose sporangium B
Z Z. fertile
R Salopella allenii
R Taeniocrada decheniana
R Uskiella spargens
Z Deheubarthia splendens
Z Deheubarthia sp. nov.
Z New zosterophyll genus (Allt ddu)
Z Gosslingia breconensis
Z Z. cf fertile
L Drepanophycus spinaeformis
Z Z. rhenanum
Z New spiny plant (Belgium)
? Estinnophyton wahnbachense
Z Z. llanoveranum
? Krithodeophyton croftii
Z Tarella trowenii
? Sennicaulis hippocrepiformis
T Dawsonites sp
Z Thrinkophyton formosum
? 'Zosterophyllum' artesianum
? 'Taeniocrada' longisporangiata
? 'T' dubia
Z Gosslingia cordiformis
Z Z. cf australianum
? Sporogonites exuberans
Z Rebuchia capitanea
Z Bathurstia denticulata
Z Sawdonia ornata
? Stockmansella langii
T Psilophyton burnotense
? Platyphyllum fissipartitum
R Rhynia gwynne-vaughanii
L Asteroxylon mackiei
R Horneophyton lignieri
? Nothia aphylla
Z new hairy zosterophyll (Rhynie)
T Ps. forbesii
T Ps. crenulatum
? Tirasophyton europeum
L 'Drepanophycus' gaspianus
T Psilophyton princeps
R Renalia hueberi
L Baragwanathia abitibiensis
T Ps. coniculatum
T Ps. dawsonii
T Ps. szaferi
Z Koniora andrychoviensis
R Renalia graberti
? Equisetophyton praecox
Z Z. spectabile
R R. major
Z Anisophyton potonii

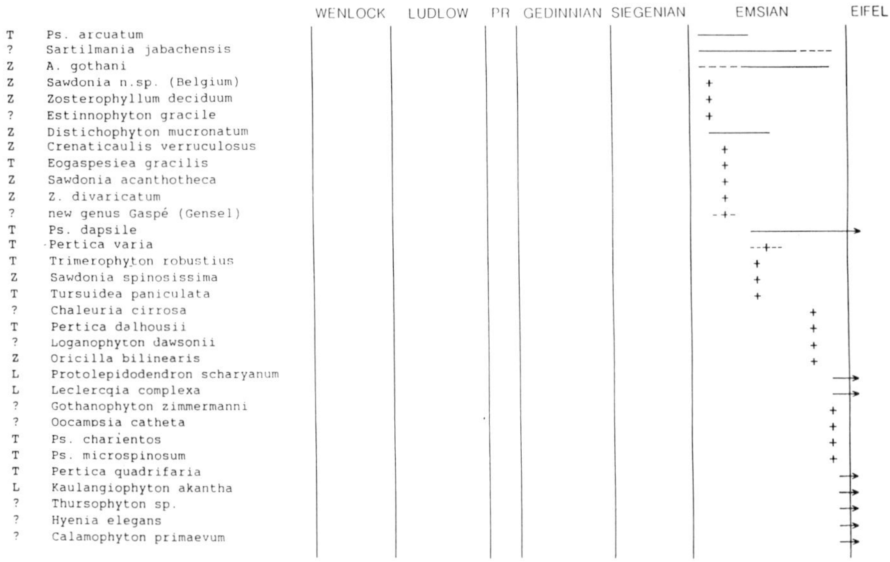

Fig. 16.1. Stratigraphic ranges of plants included in the analysis. Dashed line indicates that precise age is unknown; + represents a single record; R, Rhyniophytina/rhyniophytoid; Z, Zosterophyllophytina; T, Trimerophytina; L, Lycophytina and includes 'prelycophytes'.

to postulate that Silurian pteridophyte-like plants possessed *all* the anatomical attributes of land plants. Sequential acquisition is a possibility, beginning with 'sporopollenin'-impregnated spores, then cuticle and stomata, with evolution of tracheids following as the plants became taller (Edwards *et al.* 1986). Thus, although for this analysis we have grouped all those morphologically simple plants together, we may well be dealing with a radiation of a number of higher taxa, and whether or not these were monophyletic remains to be resolved.

3. *The missing data*

The megafossil record is dominated by fragments of aerial sporophytes while convincing examples of independent gametophytes, as anticipated from their homospory, are very few. Low fossilization potential, as deduced from the parenchymatous construction, and lack of homoiohydric attributes of extant representatives may explain this. However, the most acceptable records—*Lyonophyton* and an unnamed form (Remy 1982; Remy and Hass 1986*a*), both from the Rhynie Chert—possessed cuticles and conducting tissues, as did the more equivocal *Sciadophyton* (Schweitzer 1983). Whether or not these are typical awaits the recognition of further gametophytes in clastic sediments, but the present lack of direct evidence should not be used as an excuse to ignore the relevance of this generation to considerations on the ecology and spread of early land plants, nor to overlook the significance of its haploidy when discussing evolutionary processes.

In the haploid gametophyte recessive mutants, pertaining to that generation, would be expressed in the phenotype. A favourable recessive trait might thus become established in the population more quickly than in the diploid phase. However, selection would act immediately on a new mutant and the long-term advantage of a hidden pool of genetic resources maintained in the heterozygous state in the diploid would not occur. Further gene expression is clearly different in the two generations and phenotypic features, both anatomical and physiological, selected for in one generation might be disadvantaged or even eliminated in the next if linked, either genetically or physiologically, to a deleterious trait in that generation. Such contradictory selection in the two generations might delay considerably the fixation of advantageous characters compared to organisms not showing alternation of independent generations, and selection would favour the loosening or breaking of linkage between favourable traits in one generation and disadvantageous traits in the other. Similarly, any mechanism that would result in closer linkage between advantageous characters in the gametophytic and sporophytic generations respectively (e.g. chromosome inversions) might be favoured.

Although habitats (see p. 357) remain conjectural it seems reasonable to assume that the water requirements and water dependence of early pteridophyte gametophytes were similar to those of extant forms, and that they colonized damp habitats. Thus, attainment of reproductive maturity in drying environments would have been favoured by rapid germination under favourable conditions, physiological desiccation tolerance, and speed of reaching maturity, and there may well have been intense inter- and intraspecific competition for water, light, and scarce nutrients (Beerbower 1985) during vegetative growth. It now seems unrealistic to postulate that the pioneering pteridophytes colonized almost 'empty' land surfaces, and indirect evidence points to an existing ground cover of thalloid organisms (Gray 1985*a*, *b*; Selden and Edwards 1989; Strother 1988). Competition between this and the early pteridophyte gametophyte may well have been intense and initially successful sexual reproduction low. This in turn may give a partial explanation for the time-lag illustrated in Fig. 16.3 (see also Jablonski and Bottjer, Chapter 2, this volume).

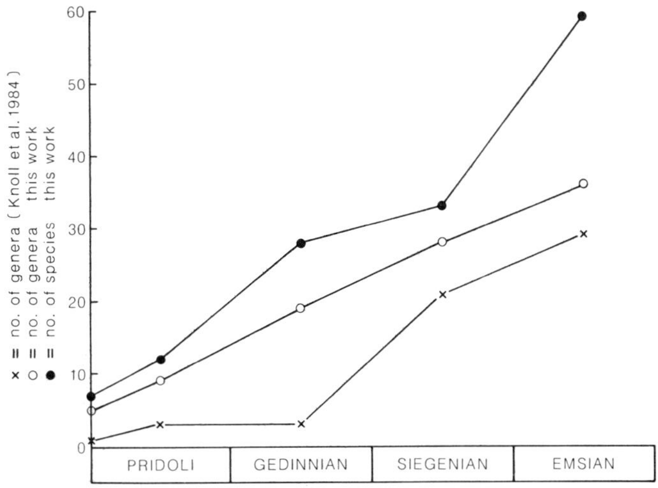

Fig. 16.2. Total numbers of late Silurian/early Devonian species and genera from this analysis plotted on the graph for generic richness produced by Knoll *et al.* (1984).

4. *Habitats*

All Silurian megafossils are recovered from marine sediments and give no clues to the habitats of the living plants. Most workers favour moist environments at the terrestrial–freshwater interface (Selden and Edwards 1989; but see Schweitzer 1983). Since latest Silurian records are in marginal marine and early Devonian records in fluvial sediments, coastal plain communities in a variety of habitats seem likely. Such habitats would include mudflats, dune slacks, and stream-, river-, and lake-sides and are all classified as variable in time and space. This environmental variability would act as a disruptive selective force with the net effect of increasing the range of genetic diversity within early land plants, in that different phenotypes would have optimal fitness in the different microenvironments. On the other hand extremely disturbed environments (e.g. those subjected to regular flooding) would be unlikely habitats in that there would have been no opportunity for the buildup of nutrient-and water-retentive soil, the preferred substrate for the establishment of higher plants (Beerbower 1985).

Presentation of data

1. Ranges of species are given in Fig. 16.1. A comprehensive bibliography is impossible in the space allocated, but will be included

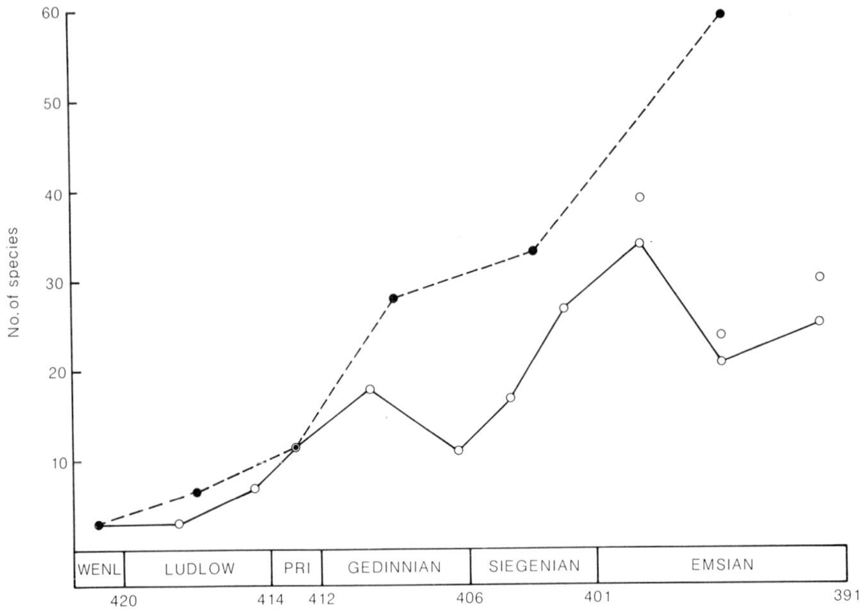

Fig. 16.3. Plot of numbers of species present in the upper Silurian and Lower Devonian. Interrupted lines join points (solid circles) for total number within the series or stage. Circles not connected by lines are possible maximum numbers of species within the Emsian.

in a paper being prepared with colleagues from Liège. This will give a more detailed stratigraphic column and information on localities.

2. Numbers of species and genera are expressed graphically in Figs 16.2 and 16.3. Note that when *totals* for series and stages are plotted (Fig. 16.2) the shape of the curve broadly parallels that produced by Knoll *et al.* (1984). A more accurate picture is obtained by subdivision into 2 or 3 Ma intervals. Origination and extinction rates have not been calculated because of the high percentage of single records (cf. Niklas *et al.* 1985). These are indicated in Fig. 16.1 by ' + ' symbols.

Radiation of rhyniophytoids

1. *Quantitative overview*

All fertile Laurussian Silurian records are of rhyniophytoids; the first zosterophyll is recorded in the early Gedinnian. The increase in number of species in the graph thus documents a radiation of rhyniophytoids,

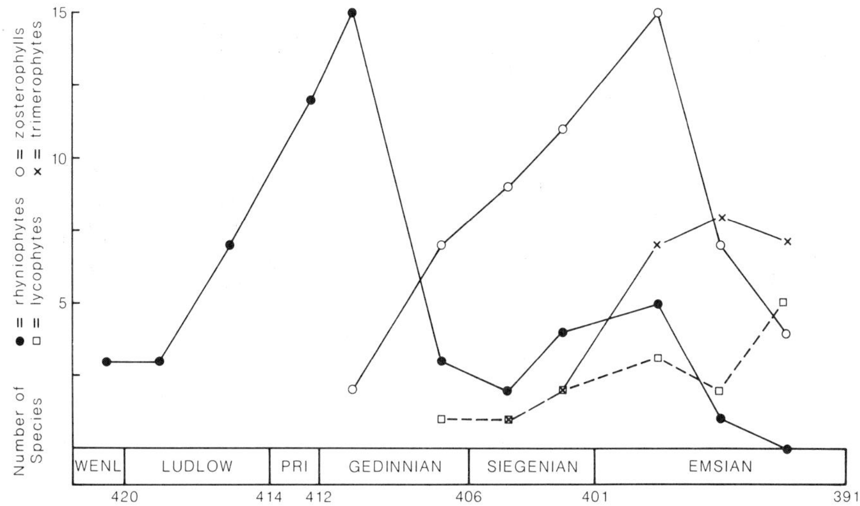

Fig. 16.4. Changes in diversity at the species level of subdivisions present in the late Silurian and early Devonian. Species of uncertain affinity are excluded.

and shows slow increase in Wenlock and Ludlow times followed by diversification in the Pridoli and early Gedinnian (Fig. 16.4). This parallels the pattern described by Knoll *et al.* (1984) except that the lag phase occurs earlier in the Silurian. Rhyniophytoids/rhyniophytes quite rapidly decline in species number in the Siegenian and latest Emsian, with minor recovery in between, although the affinities of younger examples are conjectural (see p. 363).

2. *Qualitative analysis of variation*

Rhyniophytoids are remarkably similar in overall morphology: isotomously branched, usually smooth, axes are terminated by solitary homosporous sporangia, their shape (globose, reniform, discoidal, fusiform) providing the major source of variation (see Fig. 16.5). On the basis of *in situ* spore studies branched sporangia originally thought to be new taxa (Edwards and Fanning 1985) are now known to be variants of oval or fusiform kinds (e.g. *Salopella*). Most of the sporangia are smooth, but two recently described genera bear prominent projections (Fanning *et al.* 1990, in press). Thickened distal sporangial margins (e.g. as in *Cooksonia caledonica*) evince predetermined dehiscence, but such anatomical modifications are not common in early rhyniophytoids. The isolation of spores from a large number of taxa (Edwards *et al.*, in press) has provided further characters for taxon delimitation and indicates the

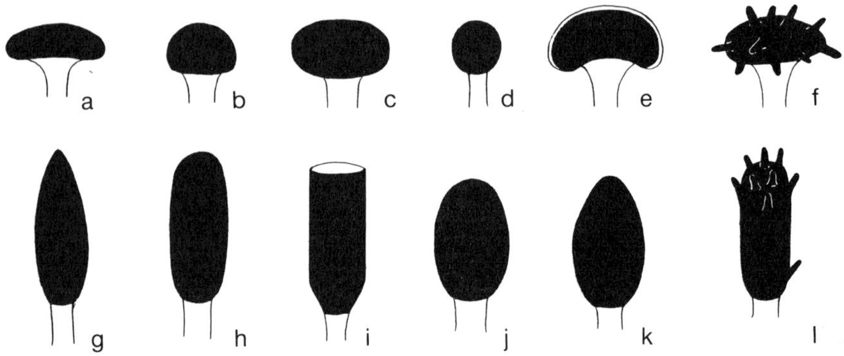

Fig. 16.5. Silhouettes of sporangia of late Silurian and basal Devonian rhyniophytoids to demonstrate morphological diversity. Drawings not to scale: a, *Cooksonia pertoni*; b, *C. hemisphaerica*, c, d, *C. cambrensis*; e, *C. calidonica*; f, new genus; g, *Salopella*; h, j, k, unnamed taxa; i, *Steganotheca*; l, *Caia*.

existence of more species than is apparent from morphology alone. We have incorporated this information into our data base (Fig. 16.1), although many of the taxa are not yet formally described.

Gedinnian rhyniophytoids, although slightly more numerous, show little morphological advancement on Silurian representatives. More complex branching is recorded, the most important type being a departure from isotomous branching by overtopping. Axes are usually smooth. Exceptions include sterile axes with triangular projections from the Pridoli of Dyfed and those with short spines from the Gedinnian of Targrove (Edwards and Fanning 1985, plate 1, fig. 14) but their affinities are obscure. Siegenian rhyniophytes are more complex; the most completely known come from the Rhynie Chert, e.g. *Rhynia gwynne-vaughanii* and possibly *Horneophyton*. Only in the latter is there a clear demonstration of centrarch xylem. The relatively undifferentiated sporangia in *R. gwynne-vaughanii* were abscissed from a branching system with more sophisticated growth patterns (D.S. Edwards 1980). In contrast, in the recently described *Uskiella*, branching is conservative but sporangia possessed anatomical modifications for predetermined dehiscence (Shute and Edwards 1989). The remaining putative rhyniophytes are known only from impressions or compressions. The most controversial Emsian example, *Renalia*, has a very well-defined pseudomonopodial branching system, and reniform sporangia with marginal dehiscence terminating lateral branch systems (Gensel 1976). A number of taxa here recorded as 'other' are sometimes considered to be possible intermediates between rhyniophytes and other taxa, but again anatomical evidence is lacking.

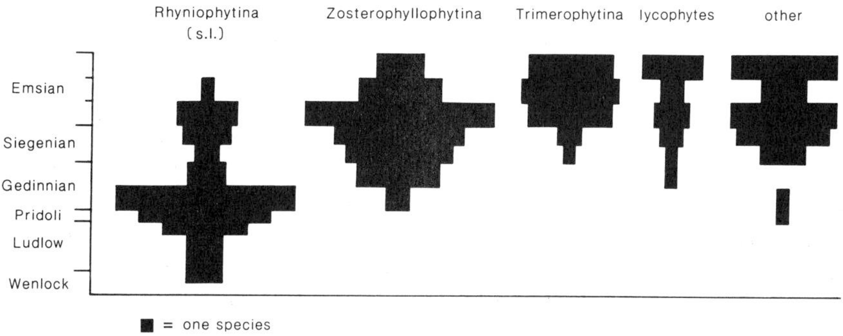

Fig. 16.6. Numbers of Silurian and Lower Devonian species plotted according to affinity.

3. *Analysis and explanation of pattern*

The nature of the megafossil record in the Silurian raises the possibility that the lag in diversification in the Ludlow results from a combination of taphonomic bias and ignorance. All records are in marine sediments (Richardson and Edwards 1989), but the older plants were transported farther and buried in deeper water, while Pridoli examples were preserved closer to the shore and thus perhaps better represent the composition of (coeval) terrestrial vegetation. Greater diversity is seen in Gedinnian assemblages preserved in fluvial sediments. Older records also contain fewer specimens. Thus facies bias undoubtedly contributes to the lag phase, but Ludlow and Wenlock records from more marginal facies are needed to assess its extent. The dispersed spore record which is more independent of facies shows a similar pattern, but its analysis is beyond the scope of this chapter (Richardson and Burgess, in progress).

Anatomical data are rare in pre-Pridoli fossils and taxon delimitation is based solely on morphology. We have shown already that sporangia of Silurian *Cooksonia pertoni* contain two kinds of spores (Fanning *et al.* 1988) and that this situation is paralleled in fusiform (*Salopella*-like) and oval sporangia (Edwards *et al.*, in press). We have taken this sort of information into account when compiling Pridoli and Gedinnian lists, but have no relevant data on pre-Pridoli rhyniophytoids. A further source of cryptic diversity, but not yet identified in the Silurian, relates to the nature of conducting cells—*Aglaophyton* and *Rhynia* might well be reunited if known only as compressions! We thus have little doubt that Silurian numbers are underestimates but also question that inadequacies of preservation and recognition are entirely responsible. We consider it possible that there was an initial period when the pioneering land plants, having attained or being in the process of evolving the anatomical and biochemical sporophytic attributes essential for homoiohydry and being subjected to intense competition of gametophytes (p. 356), were struggling to maintain a foothold in terrestrial ecosystems. Thus there would have been a period of increasing physiological fine-tuning, when numbers of individuals may well have been low. Using the terminology of modern population ecology, these were small isolated founder populations with very limited gene pools and perhaps lacking potential for major morphological diversification. This tendency would be amplified by archetypal constraints, in particular those relating to isotomous branching and determinate growth imposed by terminal sporangia. Indeed, the proliferation of taxa recorded at the end of the Silurian results overwhelmingly from variation in sporangial and spore characters and provides little evidence for release from these archetypal constraints.

It is tempting to describe the late Silurian increase as an adaptive radiation, but this presupposes knowledge of plant habitats and the functional morphology and hence adaptive significance of the innovations involved. Three examples are given to illustrate the type of reasoning involved.

1. *Exine ornamentation.* The significance of exine structure has been extensively debated elsewhere (e.g. Chaloner 1976) and has been related to buoyancy, defence, lodging, and timing of germination. The significance and relative merits of different kinds of ornament (coni, verrucae, spinae, etc.) is even more controversial. We have evidence of certain trends in different lineages (Fanning *et al.* 1988), suggesting convergence in response to similar selective pressures (e.g. those resulting from colonization of increasingly drier habitats), or perhaps to developmental constraints, but the fact remains that a wide range of similar morphologies is present in both the Silurian and early Devonian, although with increase in diversity in the latter.

2. *Anatomical modification for predetermined dehiscence.* A distinctive sporangial border characterizes a small number of late Silurian rhyniophytoids, and work in progress in Cardiff indicates cellular differentiation in this area. Such predetermined dehiscence ensures that the maximum number of spores are exposed on splitting into two equal valves and raises the possibility of control over timing of dehiscence. Similar modifications are much commoner in early Gedinnian isolated terminal sporangia and, of course, characterize the zosterophylls.

3. *Emergences on axes and sporangia.* Axial spines may have had photosynthetic or protective functions (e.g. Kevan *et al.* 1975). They are

rare in the Silurian and only slightly less so in the early Devonian. The recent demonstration of prominent sporangial projections in two Pridoli taxa prompted speculation on their functions, in nutrition, defence, and prevention of desiccation (Fanning *et al.* 1990, in press).

All these deliberations are based on small fragments of the aerial parts of plants and we are mindful of Professor J. Harper's condemnation as 'facile adaptationist guesswork' of the practice by ecologists working on extant vegetation of searching for interpretations on any isolated part of an organism and giving it an explanation as an optimal system (Harper 1982, p. 17). His comment is even more salutary for palaeobotanists, where there is no chance of evaluating the sets of compromises that are inevitable in the partitioning of resources between, for example, aerial vegetative growth, acquisition of mineral resources, and reproductive capacity. It has been suggested that such plants are an early example of *r* selection with allocation of assimilate to spore production and less to production of axes, although this 'strategy' would have placed them at a disadvantage for dispersal and light harvesting. It has also been argued that such continuously unproductive environments produce 'adversity selection' (Whittaker 1975) and favour individuals that make smaller demands on scarce environmental resources and thus are likely to be slower growing and have low reproductive effort (Grime and Hodgson 1987). Here, admittedly at the threshold, rather than the product of long-term selection pressures, we are dealing with plants showing reduction in both vegetative and reproductive vigour in response to putatively poor environmental conditions. Certainly in considering Siegenian to Emsian rhyniophytes growing in presumably ameliorating substrate there appears to be no marked increase (even in *Renalia*) in sporangial/spore production relative to 'length' of axis (although this needs quantifying). Instead, selective pressures produced increased height or sophistication of dehiscence. Increase in efficiency of capture and recycling of scarce mineral nutrient resources by many extant plants growing in unproductive habitats is achieved by mutualistic associations, e.g. lichens and mycorrhizal-root associations. It has been postulated that associations between algae and aquatic fungi were the forerunners of mutualistic associations between early land plants and fungi, conferring on these plants a greater resistance to desiccation and a greater efficiency of capturing scarce mineral nutrients (Pirozynski and Malloch 1975).

Finally, each isolated population represents a different genetic resource in terms of the mutants and recombinants it contains. The direction and magnitude of evolutionary change in response to a particular selective pressure will be constrained by the genetic variability available and thus may well differ between populations. In each population the evolutionary response would be 'along the line of least resistance' (Stebbins 1970, p. 320) and would not necessarily be the optimum response to that particular environmental change. Thus we cannot conclude that features we observe in organisms represent optimum responses and adaptations to past selective forces and environments. Indeed, it is possible that some diversity of characters may not have arisen as a result of adaptive responses to differing selection pressures but may be the result of association with unknown physiological or morphological adaptive features caused by pleiotropy or genetic linkage (linkage disequilibrium). The search for adaptationist explanations for each trait may also be confounded by oversimplified concepts of environmental factors. Natural selection forces are most often considered in terms of physical and chemical measures of the environment but powerful selection on plant populations may operate via biotic forces, including competition from neighbours, herbivory, and parasitism. Such factors may be extremely complex and vary over very short distances resulting in considerable effective heterogeneity in environments which are seemingly uniform in terms of physical and chemical parameters (e.g. Burdon 1980; Harper 1982). Such 'Sisyphean' forces may also result in coadaptation and coevolution between neighbours, both between and within species. Therefore we should be aware of the dangers of underestimating the great diversity of selective forces acting upon plant populations and of trying to produce 'adaptationist' explanations for every character variant.

Vegetation in the late Silurian and early Devonian

We can envisage late Silurian and early Devonian rhyniophytoid-dominated communities forming 'mixed turfs', tens of millimetres in height, exploiting a variety of microenvironments maintained by ever-changing physical and biological factors. As a result of inter- and intraspecific competition morphotypes may have gained temporary advantage, but such ephemeral dominance is impossible to identify in the geological record; our evidence comes from a very small number of horizons. In the Gedinnian there was a general increase in size and height with some overtopping, a release from isotomous branching constraints. In addition, the first records of *Zosterophyllum* indicate the capacity of clonal growth and hence local dominance, for which we have no direct evidence in the Silurian of Laurussia.

Radiation of Zosterophyllophytina

1. *Quantitative overview*

The Laurussian zosterophylls appear near the base of the Devonian, show major proliferation at the end of the Gedinnian and early Siegenian, and begin to decline in the mid-Emsian. The final record is in the upper Devonian. When compared with the rhyniophytoids there is no pronounced lag between origin and diversification. Examination of raw numbers of early and late Gedinnian species in Fig. 16.3 fails to reveal the striking change in affinity of the taxa involved (Figs 16.4 and 16.6). Early Gedinnian assemblages, where numbers are somewhat inflated by our current research on two Dittonian localities in the Welsh Borderland, are predominantly rhyniophytoids, whereas in the uppermost Gedinnian zosterophylls are more numerous.

2. *Qualitative analysis of variation*

The proliferation of zosterophylls in the late Gedinnian and early Siegenian produced almost all the variation known in the group. Later innovations include asymmetry in sporangial dehiscence and solitary sporangia. The uniting features of zosterophylls are a substantial/prominent exarch xylem strand and lateral sporangia in which height rarely exceeds breadth and which split in a predetermined way into two valves. Variation exists in the distribution of sporangia, sporangial orientation and symmetry of dehiscence, branching patterns, and presence or absence of emergences (see e.g. Kenrick and Edwards 1988; Edwards *et al.* 1989; Niklas and Banks 1990; Fig. 16.7). The earliest member, *Zosterophyllum*, possessed a terminal strobilus (or spike) of sporangia which are spirally inserted in *Z. myretonianum*, and borne in two rows in the slightly younger *Z. fertile*. Such organization clearly increases the sporing capacity of these plants when compared with rhyniophytoids, while basipetal maturation ensured spore production over a longer period (Edwards 1975), and anatomical differentiation related to sporangial dehiscence may have afforded additional control over the timing of dispersal. The K- and H-configurations recorded in vegetative branches of *Z. myretonianum* and a number of other species are believed to be concentrated in the basal regions of the plants and to have resulted in a tufted or creeping growth form, with a number of strobili produced by the same plant (Gerrienne 1988). The extent of such clonal behaviour remains unknown but, in that a single plant had the potential to cover a considerable area, it could have cut down inter- and intraspecific competition. Tiffney and Niklas (1985) postulated that such prolonged vegetative development would also have reduced reliance on more risky sexual reproduction.

a
b
c
d
e
f
g
h

Shortage of space precludes a comprehensive description of zosterophyll diversity; we concentrate on just four aspects and somewhat hesitatingly (see p. 363) speculate on any adaptive significance.

1. *Orientation of sporangia.* In the ancestral and commonest condition exemplified by *Zosterophyllum*, sporangia are held upright on short stalks and split so that one valve (the adaxial) is close to the axis. In the orientation that characterizes *Gosslingia*, the subtending stalks are at right angles to the strobilus axis and the sporangium splits into right and left valves which are capable of unrestricted opening when compared with the *Zosterophyllum*-type where movement of the adaxial valve could have been restricted by the strobilus axis. We know nothing about the ontogeny of *Gosslingia*-type except that it is frequently associated with circinate tips (Niklas and Banks 1990). However, the orientation of the *Zosterophyllum*-type is clearly an advantage during the development of a strobilus where older sporangia overlap and could protect younger sporangia.

2. *Distribution of sporangia.* This ranges from compact to lax strobili, to examples where sporangia are more widely distributed, perhaps occasionally forming more defined zones. In the latter there is sometimes quite a precise relationship to branching (e.g. *Gosslingia*, Edwards 1970; *Tarella*, Edwards and Kenrick 1986). As mentioned earlier, more extensive distribution of sporangia increases duration of spore production, an attribute which in the later Carboniferous lycophytes (DiMichele and Phillips 1985) has been related to colonization of disturbed habitats. It would also spread temporally the energy demand of sporangium and spore production. A terminal strobilus may well have imposed a considerable drain on the photosynthetic resources of the erect axis; the extent to which the basal ramifying axes were exposed is conjectural (Walton 1964). It would also have necessitated efficient conducting tissue, and the size of the zosterophyll strand, which is comparatively large when compared with that in rhyniophytes, may have initially evolved in response to the requirements of the strobilus sink. We have no evidence for amounts of phloem, but it would presumably have formed a correspondingly more extensive layer around the xylem. Such speculation may well (and justly) be cited as an

---

Fig. 16.7. Reconstructions of representative late Gedinnian/early Siegenian zosterophylls. a, *Thrinkophyton*; b, *Gosslingia*; c, *Zosterophyllum myretonianum*; d, *Z. fertile*; e, hypothetical reconstructed basal regions of a *Zosterophyllum*; f, *Deheubarthia* (spines omitted); g, part of vegetative axis of *Deheubarthia* with spines; h, *Tarella*.

excellent example of the oversimplistic approach (see p. 363) especially as we have, for example, no good comparative evidence for three-dimensional xylem from rhyniophytes outside the Rhynie Chert which may represent an atypical environment.

Staggering of sporangial development would thus have spread energy demands of spore production, and perhaps increased sporing capacity. Despite limitation of the data set it would be interesting to compare the number of sporangia per unit volume of axis in the more completely known zosterophylls. The most inefficient in terms of productivity would undoubtedly be Emsian *Koniora* where a solitary sporangium occurs below a branch point (Zdebska 1982). Here it is tempting to suggest that energy resources were diverted into production of numerous spines!

3. *Symmetry in branching*. One of the most remarkable features of a number of zosterophylls is the planar nature of the aerial, presumed upright, branching system. Bilateral symmetry in sporangial arrangement (viz. the two-rowed strobilus of *Z. fertile*) was already established early in zosterophyll evolution and, in the late Gedinnian, lateral branches were alternately produced on opposite sides of the axis such that the overall system was in one plane (*Gosslingia*, *Deheubarthia*). Similarly organized plants are also recorded in the Siegenian (*Thrinkophyton*, *Sawdonia*). In isolation such a flattened system would surely have been readily blown over but, in that we have some evidence for monospecific stands in, for example, *Gosslingia breconensis* and *Tarella trowenii* in the early Siegenian, it is possible that, when associated with clonal growth, such branch systems may have allowed closer packing with less mutual interference and hence more efficient light harvesting.

4. *Plasticity in branching*. Phenotypic plasticity in the vegetative phase would have been advantageous in plants growing in disturbed habitats, and this may have been achieved, at least in part, by the lateral projections present in certain zosterophylls, which are not part of the main branching system, but show varying degrees of development. Most common are prominent elevations occurring just below a branching point (axillary tubercles/subaxillary projections, e.g. Edwards *et al.* 1989) which may well be analagous to the rhizophores in extant *Selaginella* where they usually function as roots, but can develop into leafy shoots on damage to growing points. Thus in late Gedinnian–Siegenian *Deheubarthia splendens* downwardly directed spiny branches occur in the subaxillary position and in Emsian *Crenaticaulis verruculosus*, in which aerial axes possess two rows of tooth-like enations, similarly placed branches are smooth. In both taxa more work is needed

to determine the extent of development of such branches, and of their morphological variability. In early Siegenian *Tarella trowenii*, more randomly placed projections range from short-coiled or hooked structures, reminiscent of circinate tips, to more extended branched axes and it was suggested by Edwards and Kenrick (1986) that these were dormant 'buds' capable of developing following partial burial or 'uprooting' of the plants, during flooding. Finally, in *Sawdonia ornata* the small projection on the abaxial surface of a lateral branch close to its departure was interpreted as a possible dormant 'root-like' primordium in the planar aerial system (Rayner 1983).

Vegetational change in the Gedinnian–early Siegenian

The rhyniophyte-dominated coastal plain vegetation of southern Britain persisted into the Gedinnian but the 'turf' was taller and there were occasional stands of *Zosterophyllum*. In contrast, coeval assemblages preserved in the internal facies of the Old Red Sandstone of Scotland were dominated by *Z. myretonianum* and the almost monotypic vegetation in the vicinity of lakes and rivers was thus taller (tens of centimetres) and both vegetatively and reproductively more productive.

A recent attempt at interpretation taking into account taphonomic biases (Edwards 1990) found some evidence for regional differences in vegetation, but whether or not this resulted from failure of the rhyniophytoids to migrate further inland or that they had, and were subsequently ousted by *Zosterophyllum*, cannot be determined. The second explanation (i.e. competitive replacement) is far more likely for the late Gedinnian–early Siegenian radiation of zosterophylls. In the previous section we have emphasized those innovations which appear to increase the vigour and fitness of the plants in terms of greater height-prolonged vegetative growth and increased ground cover, more sophisticated controls over dispersal, and increased reproductive capacity. Whether or not these plants were able to colonize a wider range of environments, perhaps drier or less stable, remains conjectural. Clearly, they would still be limited to damp places by the constraints of the homosporous pteridophyte life cycle, although it had been suggested that extensive clonal growth would reduce dependence on sexual reproduction. In addition, *Sciadophyton*, a probable gametophyte, and possibly that of *Zosterophyllum*, may have possessed homoiohydric characters.

There remains the possibility that, because the facies in which the older rhyniophytoids occur are not so frequent from the late Gedinnian onwards, such plants existed but were not preserved. Detailed quantitative analysis of dispersed spore assemblages considered in conjunction with the rapidly increasing data base on *in situ* spores may eventually resolve the problem. Current megafossil evidence indicates that a few rhyniophytes survived the *Zosterophyllum* competition and these possessed derived characters in branching pattern (e.g. in Emsian *Renalia*), and/or dehiscence mechanisms (e.g. *Renalia* and *Uskiella*).

Vegetational change in the mid-Siegenian–Emsian

Both lack of space and familiarity with the raw data preclude extensive treatment here. The best Siegenian and early Emsian records come from mainland Europe (e.g. Gerrienne 1983; Schweitzer 1983; Stockmans 1940) while the most complete record throughout the Emsian derives from the north-eastern seaboard of North America (see e.g. Gensel and Andrews 1984). However, detailed numerical analysis is hampered because some taxa (e.g. *Baragwanathia abitibiensis* from the Sextant Formation of northern Ontario, Richardson and McGregor 1986, and *Koniora andrychoviensis*, Zdebska 1982) cannot be precisely located within the Emsian. As yet there is little well-founded evidence for provincialism; differences probably reflect the patchy nature of the record. Thus, for example, research on early Emsian assemblages from Belgium (Gerrienne 1983) demonstrated the presence, for the first time in Europe, of a number of North American taxa.

Analysis of the taxonomic composition of late Siegenian/early Emsian assemblages is complicated by the high percentage of taxa that lack sufficient diagnostic anatomical characters for identification and/or present new combinations that cannot readily be accommodated in existing classifications (e.g. *Estinnophyton*, *Sartilmania*), some of which may be advanced rhyniophytes or even transitional between trimerophytes and 'ferns' or progymnosperms (Gensel 1984). Such taxa are included under 'others' in Figs 16.4 and 16.6.

Whereas this analysis has strongly supported Knoll's (1986*a*) hypothesis that rhyniophytes were competitively replaced by zosterophylls, his contention that zosterophylls and lycophytes were replaced as principal contributors to infloral diversity by trimerophytes receives only limited quantitative support at the species level. Trimerophytes (first recorded in Laurussia as *Dawsonites* sp. (Croft and Lang 1942) early in the Siegenian) began to diversify at the beginning of the Emsian, when zosterophylls reached maximum species diversity. The greater number of zosterophyll genera reflects the fact that they are based on a variety of combinations of vegetative and sporangial characteristics (see discussion in Edwards *et al.* 1989). In contrast, proliferation of trimerophytes largely results from the increase in the number of species of *Psilophyton*, where gross morphology and lateral branches with trusses of elongate sporangia are conservative generic characteristics.

There is thus no clear-cut picture of one group rising to dominance and replacing another. What is evident, as Knoll (1986*a*) noted, is that successively younger taxa tend to be larger in all dimensions and are far more complex morphologically (although still based primarily on axial branching systems) and that taxa must have had increased capacity for harvesting light, and absorbing and transporting water and nutrients (Knoll 1986*a*, p. 139) and also, particularly in the case of trimerophytes, increased capacity for spore reproduction.

Towards the end of the Lower Devonian zosterophylls existed alongside the trimerophytes but their decline in numbers was associated with the appearance of taxa usually found in the Middle Devonian (e.g. *Calamophyton*, *Leclercqia*) and others of less certain affinity which show innovation in reproductive (*Chaleuria*, Andrews *et al.* 1974) and vegetative (*Gothanophyton*, Remy and Hass 1986*b*) characters.

Schweitzer (1983) presented a detailed ecological setting for Rhenish Lower Devonian land plants with monotypic swards or 'meadows' growing at various levels in the littoral zone including intertidal and salt marsh. The most extensive review of vegetation towards the end of the Devonian centres on Gaspé and Maine assemblages (Andrews *et al* 1977; Gensel and Andrews 1984; Kasper *et al.* 1988). Based on sedimentological studies and the predominant occurrence of monotypic assemblages, these authors reconstruct a coastal low-lying marshland colonized by dense monospecific stands of tall trimerophytes interspersed with mats of zosterophylls and (pre)lycophytes with more pronounced rhizomatous growth and occasional upright shoots. Distribution of taxa was perhaps controlled by small-scale topograhical features associated with differences in soil moisture.

Apologia

It is undeniable that, in the quantitative analysis of raw data leading to attempts at elucidation of evolutionary patterns and processes in the fossil record, palaeobotany lags well behind invertebrate palaeontology, although recent analyses by North American workers (e.g. Niklas *et al.* 1980, 1985; Knoll *et al.* 1984; Knoll 1986*b*) are beginning to rectify the balance. Detailed assessments of diversification of early land plants as reported here flounder due to incompleteness of the terrestrial record and the fragmentary and unproductive nature of the megafossils themselves. Our perhaps excessive caution in handling the raw data has been matched or even negated by our lack of caution in our overly optimistic and imaginative attempts to ascribe adaptive significance to all morphological variation, especially since our knowledge of the nature and diversity of the physical, chemical, and biotic environment which these early land plants inhabited is so rudimentary.

To overemphasize the limitations of the data base on early land plants and their environment, as done yet again here, may seem defeatist, but we are slightly encouraged in this approach by the demonstration of similar difficulties encountered in explaining character diversity in extant vegetation.

Acknowledgements

We are grateful to Drs M. Fairon-Demaret, U. Fanning, P. Gerrienne, P. Kenrick and J. B. Richardson for allowing us to use unpublished data and for invaluable advice.

References

Andrews, H. N., Gensel, P. G. and Forbes, W. H. (1974). An apparently heterosporous plant from the Middle Devonian of New Brunswick. *Palaeontology*, **17**, 387–408.

Andrews, H. N., Kasper, A. E., Forbes, W. H., Gensel, P. G. and Chaloner, W. G. (1977). Early Devonian flora of the Trout Valley Formation of Northern Maine. *Review of Palaeobotany and Palynology*, **23**, 255–85.

Banks, H. P. (1968). The early history of land plants. In *Evolution and environment* (ed. E. T. Drake), pp. 73–107. Yale University Press, New Haven, Connecticut.

Beerbower, R. (1985). Early development of continental ecosystems. In *Geological factors and the evolution of plants* (ed. B. H. Tiffney), pp. 47–91. Yale University Press, New Haven, Connecticut.

Burdon, J. J. (1980) Intra-specific diversity in a natural population of *Trifolium repens*. *Journal of Ecology*, **68**, 717–36.

Chaloner, W. G. (1976). The evolution and adaptive features in fossil exines. In *The evolutionary significance of the exine*, Linnean Society Symposium Series, no. 1 (ed. I. K. Ferguson and J. Muller), pp. 1–14. Academic Press, London.

Crane, P. R. (1989). Patterns of evolution and extinction in vascular plants. In *Evolution and the fossil record* (ed. K. C. Allen and D. E. G. Briggs), pp. 153–87. Belhaven Press, London.

Croft, W. N. and Lang, W. H. (1942). The Lower Devonian flora of the Senni Beds of Monmouthshire and Breconshire. *Philosophical Transactions of the Royal Society of London*, **B231**, 131–64.

DiMichele, W. A. and Phillips, T. L. (1985). Arborescent lycopod reproduction and palaeoecology in a coal-swamp environment of late Middle Pennsylvanian age (Herrin Coal, Illinois, USA). *Review of Palaeobotany and Palynology*, **44**, 1–26.

Edwards, D. (1970). Further observations on the Lower Devonian plant,

Gosslingia breconensis Heard. *Philosophical Transactions of the Royal Society of London*, **B231**, 131–64.

Edwards, D. (1975). Some observations on the fertile parts of *Zosterophyllum myretonianum* Penhallow from the Lower Old Red Sandstone of Scotland. *Transactions of the Royal Society of Edinburgh*, **69**, 251–65.

Edwards, D. (1990). Constraints on Silurian and early Devonian phytogeographic analysis based on megafossils. In *Palaeozoic palaeogeography and biogeography* (ed. W.S. McKerrow and C.R. Scotese), pp. 233–42. Geological Society Memoir no. 12. Geological Society, London.

Edwards, D. and Edwards, D.S. (1986). A reconsideration of the Rhyniophytina. In *Systematic and taxonomic approaches in palaeobotany*, Systematics Association Special Volume, no. 31 (ed. R.A. Spicer and B.A. Thomas), pp. 199–220. Clarendon Press, Oxford.

Edwards, D. and Fanning, U. (1985). Evolution and environment in the late Silurian–early Devonian: the rise of the pteridophytes. *Philosophical Transactions of the Royal Society of London*, **B309**, 147–65.

Edwards, D., Fanning, U. and Richardson, J. B. (1986). Stomata and sterome in early land plants. *Nature*, **323**, 438–40.

Edwards, D., Kenrick, P. and Carluccio, L.M. (1989). A reconsideration of *Psilophyton princeps* (Croft and Lang, 1942), a zosterophyll widespread in the Lower Old Red Sandstone of South Wales. *Botanical Journal of the Linnean Society*, **100**, 293–318.

Edwards, D., Fanning, U. and Richardson, J.B. (in press). *In situ* spores in early land plants from Wales and the Welsh Borderland: a review. *Proceedings of the Geological Association*.

Edwards, D.S. (1980). Evidence for the sporophytic status of the Lower Devonian plant *Rhynia gwynne-vaughanii* Kidston and Lang. *Review of Palaeobotany and Palynology*, **29**, 177–88.

Edwards, D.S. (1986). *Aglaophyton major*, a non-vascular land-plant from the Devonian Rhynie Chert. *Botanical Journal of the Linnean Society*, **93**, 173–204.

Fanning, U., Richardson, J.B. and Edwards, D. (1988). Cryptic evolution in an early land plant. *Evolutionary trends in plants*, **2**, 13–24.

Fanning, U., Edwards, D. and Richardson, J.B. (1990). Further evidence for diversity in late Silurian land vegetation. *Journal of the Geological Society of London*, **147**, 725–8.

Fanning, U., Edwards, D. and Richardson, J.B. (in press). A new rhyniophytoid from the late Silurian of the Welsh Borderland. *Neues Jahrbuch für Geologie und Paläontologie*.

Garratt, M.J., Tims, J.D., Richards, R.B., Chambers, T.C. and Douglas, J.G. (1984). The appearance of *Baragwanathia* (Lycophytina) in the Silurian. *Botanical Journal of the Linnean Society*, **89**, 355–8.

Gensel, P.G. (1976). *Renalia hueberi*, a new plant from the Lower Devonian of Gaspé. *Review of Palaeobotany and Palynology*, **22**, 19–37.

Gensel, P.G. (1984). A new Lower Devonian plant and the early evolution of leaves. *Nature*, **309**, 785–7.

Gensel, P.G. and Andrews, H.N. (1984). *Plant life in the Devonian*. Praeger Publishers, New York.

Gerrienne, P. (1983). Les plantes emsiennes de Marchin (Vallée du Hoyoux, Belgique). *Annales de la Société Géologique de Belgique*, **106**, 19–35.

Gerrienne, P. (1988). Early Devonian plant remains from Marchin (North of Dinant Synclinorium, Belgium), I. *Zosterophyllum deciduum* sp.nov. *Review of Palaeobotany and Palynology*, **55**, 317–35.

Gray, J. (1985*a*). The microfossil record of early land plants: advances in understanding of early terrestrialization, 1970–1984. *Philosophical Transactions of the Royal Society of London*, **B309**, 167–95.

Gray, J. (1985*b*). Ordovician Silurian land plants: the interdependence of ecology and evolution. *Special Papers in Palaeontology*, **32**, 281–94.

Grime, J. P. and Hodgson, J. G. (1987). Botanical contributions to contemporary ecological theory. *New Phytologist*, **106**, (Suppl.), 283–95.

Harper, J. L. (1982). After description. In *The plant community as a working mechanism*, Special Publication Series of the British Ecological Society, no. 1, (ed. E. I. Newman), pp. 11–25. Blackwell Scientific Publishers, Oxford.

Holland, C. H. (1985). Synchronology. *Philosophical Transactions of the Royal Society of London*, **B309**, 11–27.

Kasper, A. E., Gensel, P. G., Forbes, W. H. and Andrews, H. N. (1988). Plant paleontology in the State of Maine—a review. *Maine Geological Survey. Studies in Maine Geology*, **1**, 109–28.

Kenrick, P. and Edwards, D. (1986). A new zosterophyll from the Lower Devanian of Wales. *Botanical Journal of the Linnean Society*, **92**, 269–83.

Kenrick, P. and Edwards, D. (1988). A new zosterophyll from a recently discovered exposure of the Lower Devonian Senni Beds in Dyfed, Wales. *Botanical Journal of the Linnaean Society*, **98**, 97–115.

Kevan, P. G., Chaloner, W. G. and Savile, D. B. O. (1975). Interrelationships of early terrestrial arthropods and plants. *Palaeontology*, **18**, 391–417.

Knoll, A. H. (1986*a*). Patterns of change in plant communities through geological time. In *Community ecology* (ed. J. Diamond and T. Case), pp. 126–48. Harper and Row, New York.

Knoll, A. H. (1986*b*). Patterns of extinction in the fossil record of vascular plants. In *Extinctions* (ed. M. Nitecki), pp. 21–68. University of Chicago Press, Chicago.

Knoll, A. H. Niklas, K. J., Gensel, P. G. and Tiffney, B. H. (1984). Character diversification and patterns of evolution in early vascular plants. *Paleobiology*, **10**, 34–47.

McGregor, D.C. and Camfield, M. (1976). Upper Silurian? to Middle Devonian spores of the Moose River Basin, Ontario. *Bulletin of the Geological Survey of Canada*, **263**, 1–63.

McKerrow, W.S., Lambert, K. St. J. and Cocks, L.R.M. (1985). The Ordovician, Silurian and Devonian periods. In *The Phanerozoic timescale* (ed. N.J. Snelling). Special publications of the Geological Society of London, **21**, 73–80.

Niklas, K.J. and Banks, H. B. (1990). A reevaluation of the Zosterophyllophytina with comments on the origin of the lycopods. *American Journal of Botany*, **77**, 274–83.

Niklas, K.J., Tiffney, B. H. and Knoll, A. H. (1980). Apparent changes in the diversity of fossil plants. A preliminary assessment. In *Evolutionary botany*,

Vol. 12, (ed. M.K. Hecht, W.C. Steere, and B. Wallace), pp. 1-89. Plenum Press, New York.
Niklas, K.J. Tiffney, B.H. and Knoll, A.H. (1985). Patterns in vascular land plant diversification: an analysis at the species level. In *Phanerozoic diversity patterns: profiles in macroevolution* (ed. J.W. Valentine), pp. 97-128. Princeton University Press, Princeton, New Jersey.
Pirozynski, K.A. and Malloch, D.W. (1975). The origin of land plants: a matter of mycotrophism. *Biosystems*, **6**, 153-64.
Rayner, R.J. (1983). New observations on *Sawdonia ornata* from Scotland. *Transactions of the Royal Society of Edinburgh: Earth Sciences*, **74**, 79-93.
Remy, W. (1982). Lower Devonian gametophytes: relation to the phylogeny of the land plants. *Science*, **215**, 1625-7.
Remy, W. and Hass, H. (1986*a*). Das Ur-Landpflanzen Konzept-unter bensonderer Berücksichtigung der Organization AltDevonischer Gametophyten. *Argumenta Palaeobotanica*, **7**, 173-214.
Remy, W. and Hass, H. (1986*b*). *Gothanophyton zimmermannii* nov. gen., nov. spec., eine Pflanze mit komplexem Stelär-Körper aus dem Emsian. *Argumenta Palaeobotanica*, **7**, 9-69.
Richardson, J.B. and Edwards, D. (1989). Sporomorphs and plant megafossils. In *A global standard for the Silurian system*, National Museum of Wales Geological Series, no. 9. (ed. C.H. Holland and M.G. Bassett), pp. 216-26. National Museum of Wales, Cardiff.
Richardson, J.B. and McGregor, D.C. (1986). Silurian and Devonian spore zones of the Old Red Sandstone Continent and adjacent regions. *Bulletin of the Geological Survey of Canada*, **364**, 1-79.
Schweitzer, H.J. (1983). Die Unterdevonflora des Rheinlandes. *Palaeontographica* Abt, **B189**, 1-138.
Selden, P.A. and Edwards, D. (1989). Colonization of the land. In *Evolution and the fossil record* (ed. K.C. Allen and D.E.G. Briggs), pp. 122-52. Belhaven Press, London.
Sepkoski, J.J. (1979). A kinetic model of Phanerozoic taxonomic diversity. II. Early Phanerozoic families and multiple equilibria. *Paleobiology*, **5**, 222-51.
Sepkoski, J.J. (1981). A factor analytic description of the Phanerozoic marine fossil record. *Paleobiology*, **7**, 36-53.
Shute, C.H. and Edwards, D. (1989). A new rhyniopsid with novel sporangium organization from the Lower Devonian of South Wales. *Botanical Journal of the Linnean Society*, **100**, 111-37.
Stebbins, G.L. (1970). Adaptive radiation of reproductive characteristics in angiosperms I. Pollination mechanisms. *Annual Review of Ecology and Systematics*, **1**, 307-26.
Stockmans, F. (1940). Végétaux éodéonians de la Belgique. *Mémoires Institut Royal des Sciences Naturelles de Belgique*, **93**, 1-88.
Streel, M., Higgs, K., Loboziak, S., Riegel, W. and Steemans, P. (1987). Spore stratigraphy and correlations woth faunas and floras in the type marine Devonian of the Ardenne-Rhenish regions. *Review of Palaeobotany and Palynology*, **50**, 211-29.
Strother, P.R. (1988). New species of *Nematothallus* from the Silurian Bloomsburg Formation of Pennsylvania. *Journal of Paleontology*, **62**, 967-82.

Taylor, T. N. (1988). The origin of land plants some answers, more questions. *Taxon*, **37**, 805–33.

Tiffney, B. H. and Niklas, K. J. (1985). Clonal growth in land plants—a paleobotanical perspective. In *Population biology and evolution* (ed. J. B. C. Jackson, L. W. Buss, and R. E. Cook), pp. 35–66. Yale University Press, New Haven, Connecticut.

Walton, J. (1964). On the morphology of *Zosterophyllum* and some early land plants. *Phytomorphology*, **14**, 155–60.

Whittaker, R. H. (1975). Design and stability of plant communities. In *Unifying concepts in ecology* (ed. W. W. van Dobben and R. H. Lowe-McConnell), pp. 169–81. Junk, The Hague.

Zdebska, D. (1982). A new zosterophyll from the Lower Devonian of Poland. *Palaeontology*, **25**, 247–63.

17. Angiosperm radiation and patterns of Cretaceous palynological diversity

PETER R. CRANE
SCOTT LIDGARD
Department of Geology, Field Museum of Natural History, Chicago, USA

Abstract

Cladistic analyses indicate that angiosperms, Bennettitales, Gnetales, and *Pentoxylon* together form a monophyletic group, and imply that angiosperms may have diverged from related taxa as early as the Triassic. However, quantitative evaluations of systematic diversity in Cretaceous macrofloras and palynofloras show that angiosperms did not radiate to floristic prominence until the Aptian–Albian. Analyses of 1125 Cretaceous palynofloras distributed between palaeolatitudes 80°N and 20°S indicate that the major increase in angiosperm diversity occurs first at low palaeolatitudes (20°N–20°S) but expands rapidly to higher palaeolatitudes in the Northern Hemisphere during the mid- to Late Cretaceous. By the Cenomanian angiosperm pollen 'species' are significant in palynofloras from all palaeolatitudes sampled, and most of this angiosperm diversity is accounted for by triaperturate, or triaperturate-derived, pollen diagnostic of one angiosperm subclade (non-magnoliid dicotyledons). Most non-angiosperm groups either decline, or remain unchanged, in their importance within successive Cretaceous palynofloras. However, ephedroid pollen, characteristic of Gnetales, shows a marked increase in diversity at low palaeolatitudes in parallel with the initial angiosperm radiation. Considered in the context of previous ideas on the causal basis of angiosperm diversification these coincident patterns suggest that the mid-Cretaceous evolution of both groups may have been influenced by similar biological and/or environmental factors.

Major Evolutionary Radiations (ed. P. D. Taylor and G. P. Larwood), Systematics Association Special Volume No. 42, pp. 377–407, Clarendon Press, Oxford, 1990.

Introduction

The radiation (diversification) of angiosperms is often considered among the most significant evolutionary events in the history of life with profound consequences for the subsequent development of terrestrial ecosystems. However, despite various hypotheses offered to account for angiosperm diversity and vegetational dominance (Stebbins 1974, 1981; Raven 1977; Regal 1977; Burger 1981; Doyle *et al.* 1982; Knoll 1984; Doyle and Donoghue 1986; Wing and Tiffney 1987; Bond 1989), surprisingly little attention has focused on surveying and quantifying available palaeobotanical evidence of the ecological changes that occurred during the critical mid-Cretaceous phase of angiosperm diversification. Modifications in the floristic composition of terrestrial vegetation are an important component of such changes, and causal explanations of angiosperm diversification need to be evaluated in the context of large-scale temporal and geographic trends in the diversity of other plant groups.

Previous analyses of Cretaceous floristic change have either focused on local stratigraphic sequences (Doyle and Hickey 1976; Hickey and Doyle 1977) or have synthesized data predominantly from middle and high palaeolatitude areas (Dorf 1955; Cousminer 1961; Niklas *et al.* 1980, 1983, 1985; Knoll 1986; Crane 1987*a*; Lidgard and Crane 1988, 1990, and others; but see Muller 1984). In this chapter we contribute to a broader geographical perspective using palynological data to assess trends in floristic diversity across a range of palaeolatitudes (80°N–20°S). We briefly review previous work on the origin and early diversification of angiosperms, and then summarize large-scale patterns of floristic change during the Cretaceous using trend surfaces fitted to within-palynoflora diversity data reported in the literature. The results substantiate latitudinal gradients in floristic composition previously recognized as palaeofloristic provinces (e.g. Brenner 1976; Herngreen and Chlonova 1981) and also show that different plant groups exhibit qualitatively different temporal and palaeolatitudinal trends in diversity that have important implications for ecological interpretations of Cretaceous vegetational change.

Angiosperm origin

The most significant conclusion to emerge from recent cladistic analyses of seed plant phylogeny is that angiosperms and Gnetales, along with the fossil groups Bennettitales and *Pentoxylon*, together form a well-supported monophyletic taxon (Fig. 17.1A; anthophytes *sensu* Doyle and Donoghue 1986; Crane 1985). Although the precise pattern of relationships among these groups is unresolved, in part because of inadequate knowledge of the extinct taxa, they are separated from other seed plants by a range of vegetative and reproductive features. Recognition of the anthophyte clade, combined with the presence of both Bennettitales and Gnetales in the Triassic (Crane 1985, 1988; Doyle and Donoghue 1986) implies that the angiosperm lineage may have diverged from related 'gymnosperms' significantly earlier than has often been supposed (Crane 1985; Doyle and Donoghue 1986; see also Crane *et al.* 1989*a*).

Despite the possibility of a pre-Cretaceous origin of angiosperms, direct evidence of early Mesozoic flowering plants has not been recognized, and the relationships of potentially relevant Triassic and Jurassic fossils (e.g. Cornet 1986, 1989*a, b*) are currently too poorly understood for them to be accepted as early representatives of the angiosperm clade (see also Crane 1987*b*). The earliest unequivocal angiosperm pollen is reported from the Hauterivian (Brenner 1984; Hughes and McDougall 1986), and triaperturate pollen diagnostic of non-magnoliid dicotyledons is first encountered at around the Barremian–Aptian boundary (Brenner 1976; Doyle *et al.* 1982). Through the mid-Cretaceous (Aptian to Turonian) angiosperm pollen, leaves, and reproductive structures increase in both abundance and diversity (Brenner 1963; Doyle and Hickey 1976; Hickey and Doyle 1977; Crane 1987*a*; Friis *et al.* 1987; Lidgard and Crane 1988, 1990). By the late Albian–early Cenomanian there are several fossil angiosperms that are well understood and clearly related to extant families such as Chloranthaceae (Crane *et al.* 1989*b*; Pedersen *et al.*, in press), Lauraceae (Drinnan *et al.*, in press *a*), Magnoliaceae (Dilcher and Crane 1984), Buxaceae (Drinnan *et al.*, in press *b*), and Platanaceae (Friis *et al.* 1988). Based on their position within the angiosperm clade (Fig. 17.1B) such extinct plants presumably possessed the full complement of angiosperm synapomorphies. Irrespective of whether pre-Cretaceous angiosperms are eventually recognized, these orderly and synchronous temporal patterns clearly document that the major diversification of angiosperms, of essentially modern form, began during the mid-Cretaceous (Doyle 1969; Muller 1970, 1984; Wolfe *et al.* 1975; Doyle and Hickey 1976; Lidgard and Crane 1988, 1990; Crane 1989).

Angiosperm diversification and Cretaceous floristic change

1. *Potomac Group sequence*

One of the classic mid-Cretaceous sequences in which the angiosperm diversification has been carefully documented, based on both leaves and

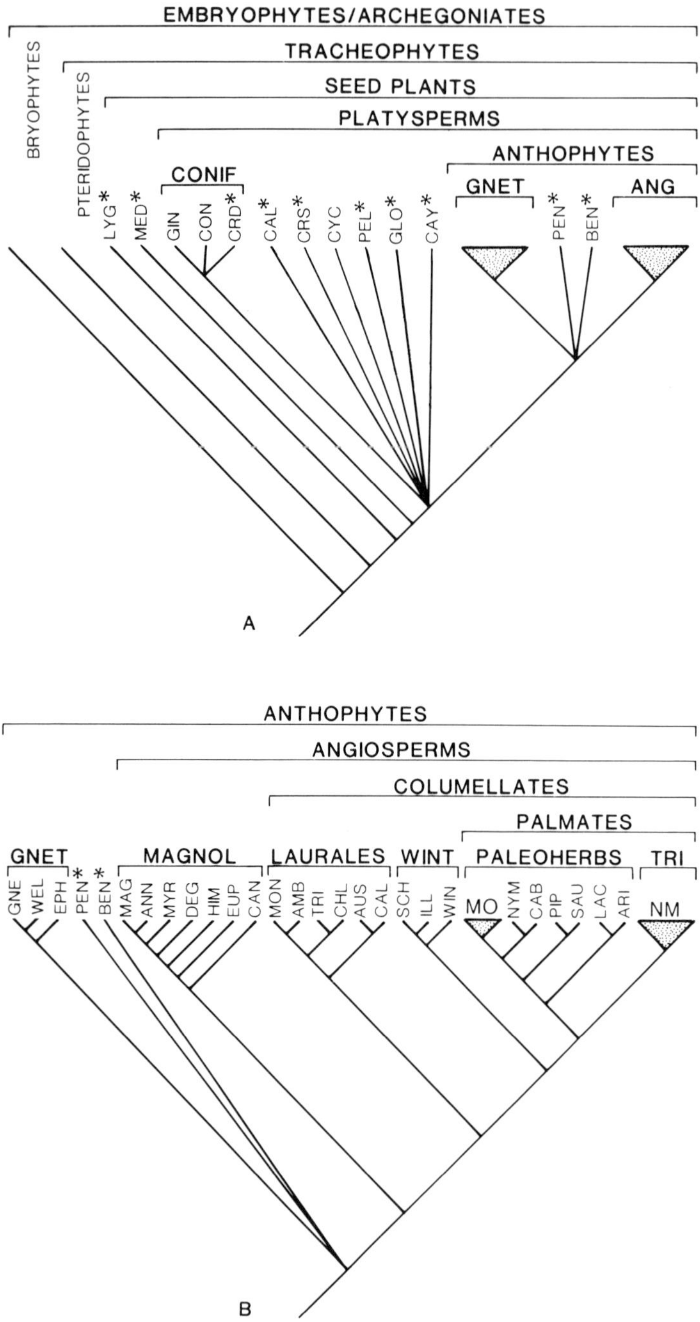
EMBRYOPHYTES/ARCHEGONIATES
TRACHEOPHYTES
SEED PLANTS
PLATYSPERMS
ANTHOPHYTES
BRYOPHYTES
PTERIDOPHYTES
CONIF
GNET
ANG
LYG*
MED*
GIN
CON
CRD*
CAL*
CRS*
CYC
PEL*
GLO*
CAY*
PEN*
BEN*
A
ANTHOPHYTES
ANGIOSPERMS
COLUMELLATES
PALMATES
GNET
MAGNOL
LAURALES
WINT
PALEOHERBS
TRI
GNE
WEL
EPH
PEN*
BEN*
MAG
ANN
MYR
DEG
HIM
EUP
CAN
MON
AMB
TRI
CHL
AUS
CAL
SCH
ILL
WIN
MO
NYM
CAB
PIP
SAU
LAC
ARI
NM
B

pollen, is the Potomac Group of eastern North America (Brenner 1963; Doyle and Hickey 1976; Hickey and Doyle 1977). In Zone I of this sequence (Aptian–lower Albian) palynological analyses (Doyle and Robbins 1977) have recognized 13 angiosperm pollen 'species', while six 'species' of angiosperm leaves have been recognized in coeval

---

Fig. 17.1. Phylogenetic relationships of land plants (extinct taxa marked by asterisks); see Crane (1985), Doyle and Donoghue (1986), and Donoghue and Doyle (1989*b*) for details of cladistic analyses and characters. A, Phylogenetic relationships among embryophytes based on strict consensus of the cladistic analyses of Crane (1985) and Doyle and Donoghue (1986) but also incorporating re-evaluation of the position of cycads (Crane 1988). 'Bryophytes' and 'pteridophytes' are paraphyletic groups of non-tracheophyte embryophytes and non-seed plant tracheophytes, respectively. Note the anthophyte clade comprising angiosperms, Gnetales, and the two extinct taxa Bennettitales and *Pentoxylon*. Abbreviations of taxa: LYG, *Lyginopteris*, representative of 'lyginopterid seed ferns' which are probably a paraphyletic, relatively plesiomorpic, group of early seed plants; MED, Medullosales; GIN, *Ginkgo*; CON, conifers; CRD, cordaites; CAL, *Callistophyton*; CRS, corystosperms; CYC, Cycadales; PEL, peltasperms; GLO, glossopterids; CAY, *Caytonia*; GNET, Gnetales; PEN, *Pentoxylon*; BEN, Bennettitales; ANGIO, angiosperms; CONIF, coniferophytes. LYG, MED, CAL, CRS, PEL, GLO, and CAY have traditionally been grouped together as 'seed ferns' (Pteridospermales). B, Phylogenetic relationships among anthophytes based on the representative most parsimonious cladogram of angiosperms provided by Donoghue and Doyle (1989*a*, fig. 2). Abbreviations of taxa: GNE, *Gnetum*; WEL, *Welwitschia*; EPH, *Ephedra*; PEN, *Pentoxylon*; BEN, Bennettitales; MAG, Magnoliaceae; ANN, Annonaceae; MYR, Myristicaceae; DEG, Degeneriaceae; HIM, Himantandraceae; EUP, Eupomatiaceae; CAN, Canellaceae; MON, Monimiaceae *sensu lato*, Gomortegaceae, Hernandiaceae, Lauraceae; AMB, Amborellaceae; TRI, Trimeniaceae; CHL, Chloranthaceae; AUS, Austrobaileyaceae; CAL, Calycanthaceae; SCH, Schizandraceae; ILL, Illiciaceae; WIN, Winteraceae; MO, monocotyledons; NYM, Nymphaeaceae; CAB, Cambombaceae; PIP, Piperaceae; SAU, Saururaceae; LAC, Lactoridaceae; ARI, Aristolochiaceae; NM, non-magnoliid dicotyledons (Nelumbonaceae, Papaverales, Ranunculales plus subclasses Asteridae, Caryophyllidae, Dilleniidae, Hamamelidae (e.g. Buxaceae, Platanaceae), Rosidae *sensu* Cronquist 1981): GNET, Gnetales; MAGNOL, Magnoliales *sensu* Donoghue and Doyle 1989*b*; WINT, Winterales; TRI, triaperturates (dicotyledons with triaperturate or triaperturate-derived pollen); palmates includes angiosperms with palmately veined leaves or derivative forms; columellates includes angiosperms with a columellate or columellate-derived exine structure. Note that taxa with monosulcate or monosulcate-derived pollen (monocotyledons and magnoliid dicotyledons) constitute a grade within angiosperms as a whole (cf. non-magnoliid dicotyledons).

macrofloras (Fig. 17.2; Berry 1911). In Zone IIB (middle–upper Albian) the same studies recognized 32 and 19 species of angiosperm pollen and leaves, respectively (Fig. 17.2), including a greater variety of structural types indicative of taxa that are derived relative to those in Zone I (Doyle and Hickey 1976; Hickey and Doyle 1977). Associated changes in the diversity of other components of Potomac Group floras have received less consideration. In the macrofloras there is a pronounced decrease in the diversity of free-sporing plants and non-conifer gymnosperms (Fig. 17.2A). Neither of these trends is reflected clearly in the palynofloras (Fig. 17.2B).

At the local scale of the Potomac Group, increasing angiosperm diversity is accompanied by an increase in the variety of depositional environments in which angiosperm macrofossils occur, perhaps indicating ecological expansion and also suggesting ecological displace-

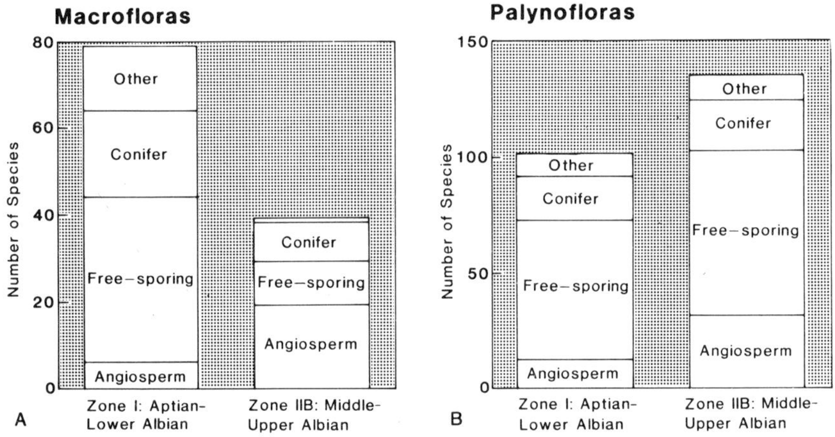

Fig. 17.2. Floristic change in the Potomac Group of eastern North America between the Aptian–lower Albian (Zone I) and the middle–upper Albian (Zone IIB). Other = other (non-conifer) 'gymnosperms'. A, Diversities of leaf macrofossil 'species' are based on lists in Berry (1911, pp. 90–5). Localities and age assignments are as in Hickey and Doyle (1977, fig. 3) but excluding Drewrys Bluff and Vinegar Hill (Zone I) and West Brothers and Red Point (Zone IIB) for which macrofossil lists are not available. Recent collections from Drewrys Bluff, West Brothers, and Red Point support the trends illustrated here (Crane, unpublished). Other 'gymnosperms' in Zone II not labelled. B, Diversities of palynomorph 'species' are based on pollen and spore taxa reported by Brenner (1963), with the exception of angiosperms. Angiosperm data is based on a re-analysis of Potomac Group angiosperm pollen 'species' by Doyle and Robbins (1977, plates 1–4). Note the difference in scale relative to Fig. 17.2A.

ment of the pre-existing flora (Hickey and Doyle 1977). More extensive sampling could provide further resolution of local vegetational changes but detailed taphonomic analyses are needed to assess the extent to which such patterns may be biased by sedimentological parameters. In addition, such analyses alone cannot resolve whether or not Potomac Group patterns accurately describe vegetational changes occurring over a wider geographic area. A complementary approach to sampling, aimed at providing much broader (through stratigraphically and sedimentologically less precise) regional coverage, is necessary to assess the general significance of ecological interpretations based on geographically restricted sequences.

2. *Regional comparisons of macrofloras and palynofloras*

Patterns of per cent within-flora diversity compiled from 91 macrofloras and 860 palynofloras, occurring principally in lowland areas between 25° and 65° palaeolatitude in the Northern Hemisphere (Fig. 17.3;

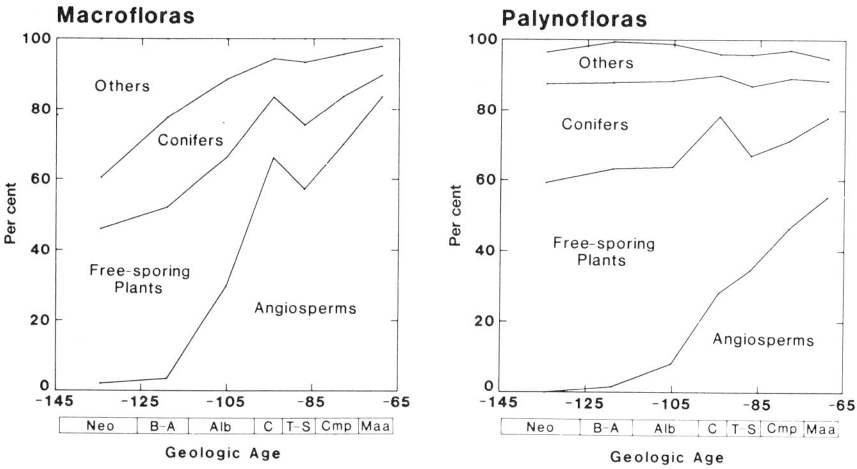

Fig. 17.3. Temporal trends in mean within-flora percentage 'species' diversity for macrofloras and palynofloras calculated for seven successive age intervals through the Cretaceous, based on Lidgard and Crane (1990, figs 1–3, see also fig. 4 and Appendices I and II). Other = other (non-conifer) 'gymnosperms'. Linear trends such as those presented here incorporate a variety of simplifying assumptions (Crane and Lidgard, in prep.). The choice of intervals represents the most successful of several attempts to maintain uniform numbers of floras per interval, and minimize the overlap of age estimates for individual floras across interval boundaries (see discussion in Crane and Lidgard, in prep.). For Figs 17.3–17.5: NEO, Neocomian; B–A, Barremian–Aptian; ALB, Albian; C, Cenomanian; T–S, Turonian–Santonian; CMP, Campanian; MAA, Maastrichtian.

Lidgard and Crane 1990), broadly confirm that the floristic trends seen in the Potomac Group also apply at a regional scale (compare Figs 17.2 and 17.3). Most significantly, while there is much variation among floras (Fig. 17.4), the relative contribution of angiosperms to both macrofloras and palynofloras increases dramatically through the mid- and Late Cretaceous (Fig. 17.3). The rate and magnitude of this increase in percentage values is less pronounced in palynological data, probably because of factors that differentially affect the representation and recognition of angiosperm leaves and pollen grains in the palaeobotanical record (Prentice 1988; Spicer 1989; Lidgard and Crane 1990 and references therein). Representational effects include

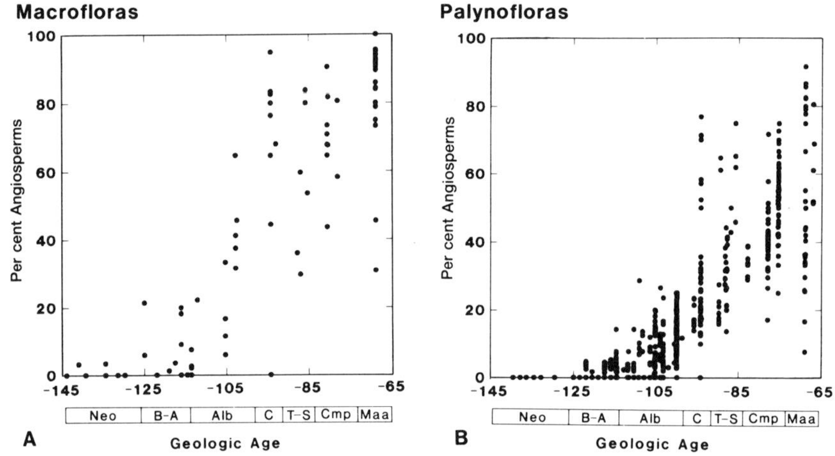

Fig. 17.4. Percentage contribution of angiosperm 'species' to Cretaceous leaf macrofloras (A) and palynofloras (B) from middle to upper middle palaeo-latitudes (25–65° N) of Laurasia and Northern Gondwana. Floras are plotted at the midpoints of their possible age range, using the time-scale of Harland *et al.* (1982). Each point thus represents the percentage of angiosperms among all 'species' occurring in an individual time-averaged assemblage (flora), as reported in the primary literature (sources listed in Lidgard and Crane 1990; see also Bambach 1977 for discussion of this approach). Individual floras have a minimum of 10 'species' in order to reduce variance due to extremely small sample sizes, while at the same time including as many floristic samples as possible. Within this data, angiosperms are questionably reported from four pre-Aptian macrofloras: Kootenay, Canada, one 'species' based on fragmentary putatively monocotyledonous leaf/stem material (Bell 1956, p. 138); Wealden, E. Germany, one 'species' assigned to *Pandanus* (Berry 1911, p. 113); Horsetown, USA, six 'species' of which Berry (1911, p. 124) questions the assignment of three to angiosperms, and comments on the poor preservation of two others; Upper Knoxville, USA, two 'species', neither of which shows unequivocal angiosperm characteristics (Berry 1911, p. 123).

the possibility that angiosperms were particularly diverse near environments suitable for the preservation of macrofossils, or are underrepresented in palynofloras because of the greater range of habitats sampled. Similarly, low pollen production in entomophilous taxa, or poor preservability of the pollen of certain Cretaceous plants (e.g. Lauraceae) may also result in palynological underrepresentation. Recognition effects include the possibility that intraspecific variability among fossil leaves may have been mistakenly reported as multiple species, or that small or featureless mid-Cretaceous angiosperm pollen species may have been overlooked. Significantly, where angiosperm pollen has been studied in detail (e.g. Potomac Group; Fig. 17.2B) the discrepancy between macrofossil and palynomorph diversity, as a proportion of the total flora recovered, is less pronounced or even reversed (Lidgard and Crane 1990).

Among other groups of plants, 'pteridophytes' and 'bryophytes' (free-sporing plants) show a clear decline in within-flora percentage diversity through the mid-Cretaceous in parallel with the angiosperm diversification (Fig. 17.3). Conifers show little change in their contribution to the diversity of macrofloras or palynofloras (Fig. 17.3). Other (non-conifer) gymnosperms show a clear decline in their relative macroflora diversity (Fig. 17.3) but show little change in their contribution to palynofloras (Fig. 17.3). A similar discrepancy is also seen in the Potomac Group data where a decline in the relative diversity of other gymnosperms based on leaves (Fig. 17.2A) is not apparent in palynofloras (Fig. 17.2B). To a large extent this may reflect the different systematic resolution provided by leaves and pollen. While the leaves of several gymnosperm groups (e.g. Bennettitales, cycads, *Ginkgo*, and Caytoniales, Corystospermales) are easily distinguished, their pollen morphology is relatively uniform (Frederiksen 1980; Lidgard and Crane 1990). Within the limits of our environmental and geographical sampling, the similarity of macroflora and palynoflora results indicates that the overall patterns of floristic change are robust and accurately reflect major modifications in the composition of Cretaceous vegetation. For Cretaceous angiosperms, conifers, and free-sporing plants, the macrofossil and palynological records provide very similar indications of temporal trends in diversity despite substantially different recognition and representation biases inherent in the records of leaves and palynomorphs (Frederiksen 1980; DiMichele and Wing 1988; Lidgard and Crane 1990). This conclusion provides a more secure basis for interpreting palynological diversity data in areas where macrofloras are sparse (e.g. high and low palaeolatitudes), and thus an opportunity to assess latitudinal variation in patterns of Cretaceous floristic change using pollen/spore assemblages.

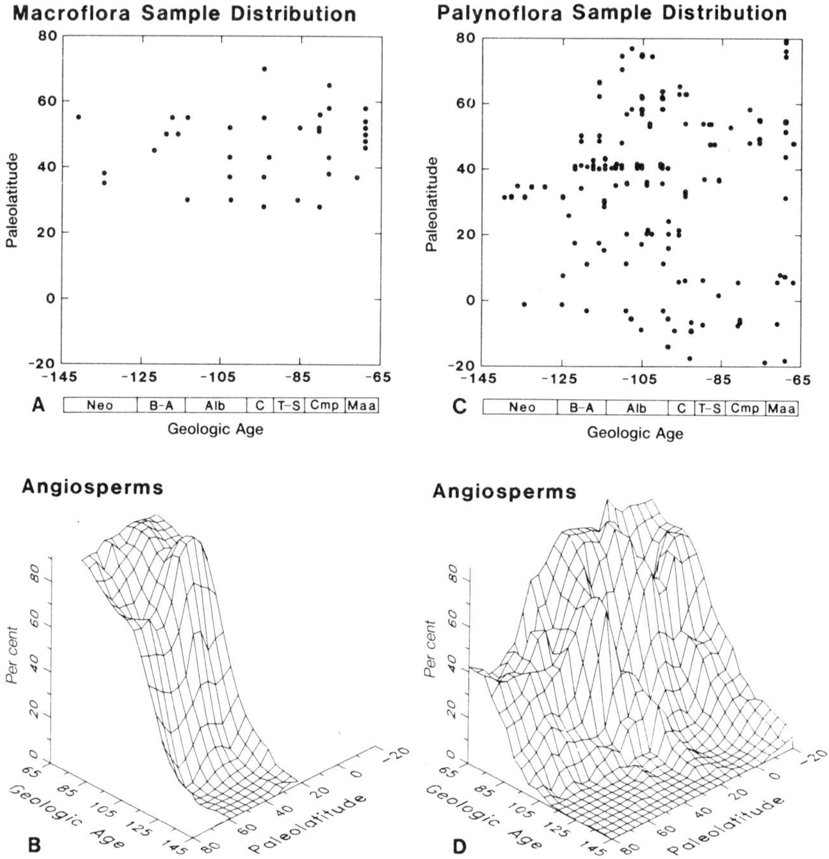

Fig. 17.5. Temporal and latitudinal trends in the relative diversity of angiosperm 'species' in Cretaceous palynofloras. A, Distribution of sampled Cretaceous macrofloras in North America with respect to palaeolatitude and geologic age (Ma) (sources listed in Upchurch and Wolfe 1990; see also Crane 1987*a*; Lidgard and Crane 1990; Crane and Lidgard, in prep.). Forty-three macrofloras are included, but diversity values from floras with the same combination of palaeolatitude and geologic age were averaged, resulting in 35 unique sample points. B, Temporal and palaeolatitudinal trends in relative diversity of angiosperms based on leaf 'species' in Cretaceous macrofloras shown in A. Percentage estimates for successive evenly spaced points on gridded surfaces were calculated from unevenly spaced percentage values derived as in Fig. 17.4. Surfaces were fitted using a moving average method (kriging: SURFER 1987) that is highly sensitive to local changes in 'species' percentages relative to the latitudinal and temporal distribution of samples (Davis 1986; Macdonald and Waters 1987). Values for single floras deviate from the fitted surface (cf. Fig. 17.4), but the surface nonetheless provides an

3. *Temporal and latitudinal trends in palynofloras*

Relative diversity data compiled from 43 North American macrofloras (Crane 1987*a*; Upchurch and Wolfe 1990) clearly show a marked increase in angiosperm diversity through the mid-Cretaceous. These floras span approximately 45° of palaeolatitude (Fig. 17.5A), and a trend surface fitted to the within-palynoflora angiosperm percentage values provides an indication of two weak tendencies in these data (Fig. 17.5B). First, angiosperms become floristically prominent slightly earlier at lower palaeolatitudes and, second, by the end of the Cretaceous angiosperms appear to comprise a greater proportion of the species in macrofloras from lower latitude areas.

The weak patterns recognized among macrofloras (Fig. 17.5B) are seen more clearly in palynological data from North and South America, western Europe, and West Africa, which spans a greater range of palaeolatitudes (80°N–20°S, Fig. 17.5C). Angiosperms first become floristically significant at low palaeolatitudes (20°N–20°S) but their rise to prominence is delayed at higher palaeolatitudes (Fig. 17.5D). Many mid-Cretaceous palynofloras from high palaeolatitudes contain no, or very few, angiosperm pollen grains long after they have become a major floristic component of palynofloras from equatorial regions (e.g. compare Singh 1964 with Doyle *et al.* 1982). Also, by the end of the Cretaceous, angiosperms are floristically most significant at low palaeolatitudes where they typically comprise more than 80 per cent of

---

accurate 'best fit' of the overall trends. A limitation of kriging and other surface-fitting methods is the potential for error or uncertainty in estimating percentage values for gridded points where sampling is sparse, but for all results presented in this chapter surfaces fitted by least squares locally weighted moving average methods and low-order polynomial trend surface methods (Wilkinson 1988) corroborate the trends shown here. C, Distribution of sampled Cretaceous palynofloras with respect to palaeolatitude and geologic age (Ma) based on the literature sources cited in the Appendix. Palaeolatitude and palaeolongitude coordinates were obtained using the computerized projections of the Paleogeographic Atlas Project at the University of Chicago. 1125 palynofloras are included, but diversity values for floras with the same combination of palaeolatitude, palaeolongitude, and geological age were averaged, resulting in 252 unique sample points. While sampling remains sparse at low and high palaeolatitudes during the Neocomian, sample distribution is greatly increased through the critical mid-Cretaceous interval relative to macrofloras. D, Temporal and palaeolatitudinal trends in relative diversity of angiosperms based on pollen 'species' in Cretaceous palynofloras shown in C. Surface fitted by kriging as in Fig. 17.5B. Note the generally earlier rise to prominence of angiosperms in macrofloras (compare Fig. 17.4).

the palynomorph species present. At high palaeolatitudes, they generally account for only 40 per cent of the species present (Fig. 17.5D).

Two major components can be distinguished within the palynological data on angiosperm diversification. First, monocotyledons and dicotyledons of the subclass Magnoliidae *sensu* Cronquist (1981, but excluding Nelumbonaceae, Papaverales, Ranunculales) produce pollen with a single distal aperture (monosulcate) or obvious derivative forms (Figs 17.1B, 17.6A,B). This apertural configuration is similar to that of many other seed plants (e.g. cycads) but the pollen of most angiosperms (except Magnoliales, Fig. 17.1B) can be distinguished by the presence of columellate pollen wall architecture. Second, non-magnoliid dicotyledons (subclasses Hamamelidae, Dilleniidae, Asteridae, Caryophyllidae, Rosidae, and the Nelumbonaceae, Papaverales, Ranunculales, *sensu* Cronquist 1981) produce pollen with three equatorially-arranged apertures (triaperturate) or obvious derivative forms (Figs 17.1B, 17.6C,D). In contrast to monosulcate pollen, which is characteristic of a *grade* of relatively plesiomorphic angiosperms, the triaperturate condition defines a monophyletic group that includes 72 per cent of extant angiosperm species (Fig. 17.1B). Polyporate or other kinds of angiosperm pollen that could potentially be derived from either monosulcate or triaperturate forms are consistently a relatively minor component of diversity in Cretaceous palynofloras (Fig. 17.7C, D).

Except for the earliest pre-Aptian (pre-tricolpate) phase of angiosperm evolution most of the angiosperm pollen species recognized in palynofloras are triaperturate (non-magnoliid) forms (Fig. 17.7B). Thus, the diversification of a particular angiosperm subclade (non-magnoliid dicotyledons) is responsible for the major temporal and palaeolatitudinal trends in the within-palynoflora diversity of angiosperms as a whole (Figs 17.5D, 17.7). Comparisons of mid-Cretaceous macrofossils and pollen with extant taxa show that the early non-magnoliid dicotyledons included taxa closely related to extant Trochodendrales (Crane 1989), possibly Myrothamnaceae (Doyle 1969, personal communication), Buxaceae (Drinnan *et al.*, in press *b*), Platanaceae (Friis *et al.*, 1988), and a variety of Rosiidae (e.g. Basinger and Dilcher 1984; Crane, Friis, and Pedersen, in progress).

Monosulcate and monosulcate-derived forms, indicative of a grade of angiosperms comprising monocotyledons and magnoliid dicotyledons, consistently account for a smaller proportion of within-palynoflora diversity than the triaperturate pollen of non-magnoliid dicotyledons (Fig. 17.7A, B). During the mid-Cretaceous these taxa included forms closely related to monocotyledons (Doyle 1973; Walker and Walker 1986) as well as extant Magnoliidae such as Chloranthaceae (Crane *et al.*

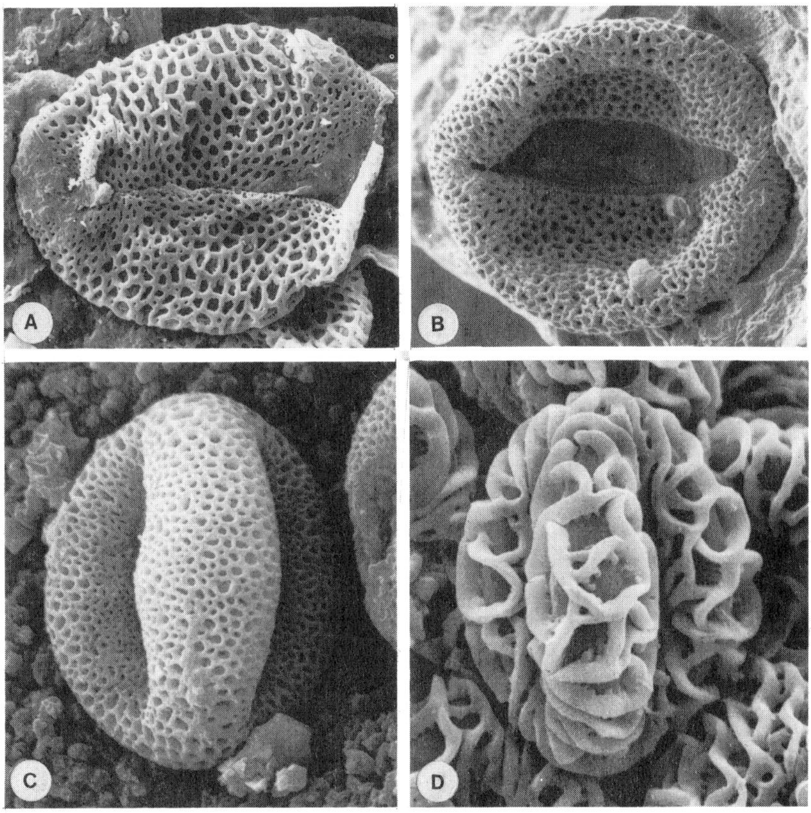

Fig. 17.6. Scanning electron micrographs of monosulcate and triaperturate mid-Cretaceous angiosperm pollen from the Potomac Group of eastern North America. A, *Liliacidites* sp., monosulcate pollen probably produced by extinct monocotyledons (see Walker and Walker 1984); distal view showing single aperture, West Brothers loc., × 1800. B, *Clavatipollenites* sp., monosulcate pollen probably produced by plants closely related to extant Chloranthaceae (see Fig. 17.1B: Pedersen *et al.* in press); distal view showing single aperture, Mauldin Mtn loc., × 1800. C, *Tricolpites* sp. cf. *minutus*, tricolpate pollen produced by an extinct taxon of Platanaceae (see Friis *et al.* 1988); equatorial view showing two of the three longitudinal apertures, Mauldin Mtn loc., × 4500. D, cf. *Retitricolpites paraneus* Norris, tricolpate grain probably produced by an extinct taxon of 'lower' Hamamelidae (Drinnan *et al.*, in press *b*); equatorial view showing two of the three longitudinal apertures, Mauldin Mtn loc., × 3000.

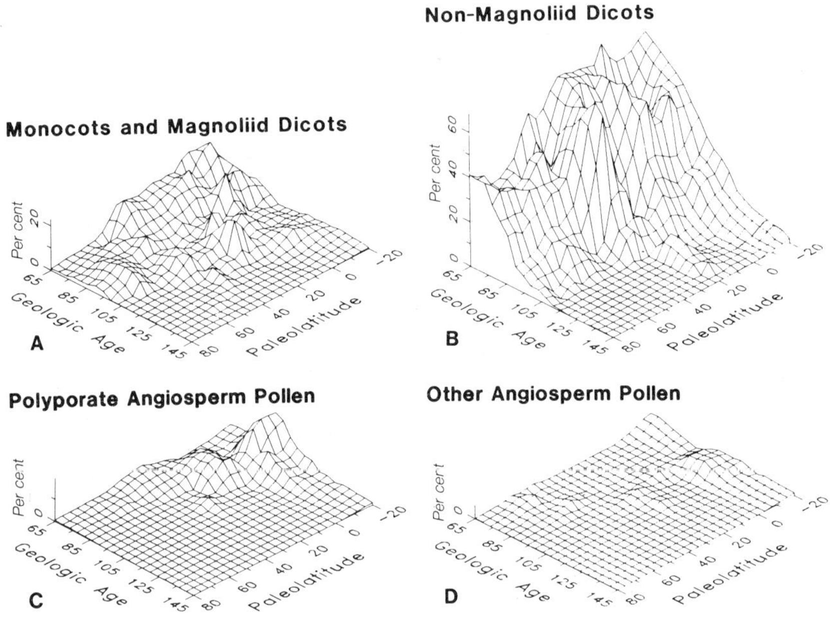

Fig. 17.7. Temporal and latitudinal trends in the relative diversity of 'species' of four major kinds of angiosperm pollen in Cretaceous palynofloras. Surfaces fitted by kriging as in Fig. 17.5B; palynoflora data as for 17.5C. A, Within-palynoflora percentage diversity for 'species' of monosulcate and monosulcate-derived pollen (indicative of monocotyledons and magnoliid dicotyledons—see text). Note that monosulcate or monosulcate derived pollen appears early in the fossil record particularly at low palaeolatitudes but does not become a major floristic component until the latest Cretaceous among low to middle palaeo-latitude palynofloras. B, Within-palynoflora percentage diversity for 'species' of triaperturate and triaperturate-derived pollen (diagnostic of non-magnoliid dicotyledons—see text). Note that triaperturate pollen accounts for the majority of angiosperm 'species' diversity and clearly shows a latitudinally diachronous pattern of increasing floristic dominance through the Cretaceous. C, Within-palynoflora percentage diversity for 'species' of polyporate pollen (representing diverse angiosperms taxa). D, Within-palynoflora percentage diversity for other angiosperm pollen 'species' (representing diverse angio-sperm taxa). Note that pollen of uncertain relationship within angiosperms is a consistently minor component of most Cretaceous palynofloras.

1989*b*; Pedersen *et al.*, in press), Lauraceae (Drinnan *et al.*, in press *a*), Magnoliaceae (Dilcher and Crane 1984), Winteraceae (Walker *et al.* 1983), and possibly Myristicaceae (Walker and Walker 1984). Systematic resolution in our data is currently insufficient to separate reliably the pollen of monocotyledons and magnoliid dicotyledons and thus test the hypothesis, based on extant taxa and fossils, that the initial diversification of monocotyledons was centred in middle palaeolatitudes of the Northern Hemisphere (Walker and Walker 1986). In the Late Cretaceous monosulcate or monosulcate-derived pollen is most conspicuous in Campanian–Maastrichtian palynofloras from mid- to low palaeolatitude areas (Fig. 17.7A) and this appears to reflect increasing diversity of palms (Arecaceae) (see Muller 1984).

Among other plant groups (Figs 17.8–17.10) the most pronounced diversity trend is the diversification of distinctive polyplicate pollen grains ('ephedroids') characteristic of Gnetales, in mid-Cretaceous palynofloras from low palaeolatitudes (Fig. 17.8A; Crane and Lidgard 1989). Simple fossil ephedroid pollen (e.g. *Ephedripites*) closely resembles the distinctive polyplicate grains of extant *Ephedra* and *Welwitschia* (Fig. 17.1B), has similar pollen wall ultrastructure to the extant taxa (Trevisan 1980), and has been recorded in close association with gnetalean macrofossils from at least two mid-Cretaceous localities (Krassilov 1986; Crane and Upchurch 1987). In Barremian–Santonian palynofloras from low palaeolatitudes such typical ephedroid pollen types intergrade with a variety of more complex forms with polar auriculae (e.g. *Steevsipollenites*), elater-like extensions of the exinal folds (e.g. *Elateroplicites*, *Elaterosporites*), or club-like projections of the pollen wall (e.g. *Galeacornea*). This structural variety substantially exceeds that among pollen of extant Gnetales and reaches its mid-Cretaceous peak in regions as widely separated as Peru (Brenner 1968), Ecuador (Hunter, personal communication), Brazil (Herngreen 1973, 1975), West Africa (Doyle *et al.* 1982), South China (Yu Jingxian 1981), and Papua-New Guinea (Lowe, personal communication). The initial increase in diversity parallels closely the similar low-palaeolatitude increase in the diversity of angiosperm pollen (compare Figs 17.5D and 17.8A). However, while angiosperm diversity continues to expand through the Late Cretaceous, and becomes increasingly significant at middle and high palaeolatitudes (Fig. 17.5D), the diversity of ephedroid pollen peaks at around the Cenomanian, subsequently declines, and never becomes a significant component in palynofloras from outside the low-palaeolatitude Northern Gondwana palaeofloristic province (Fig. 17.8A; Brenner 1976; Herngreen and Chlonova 1981).

Latitudinal and temporal trends in the within-palynoflora diversity of conifer pollen are much less pronounced than those for ephedroid or

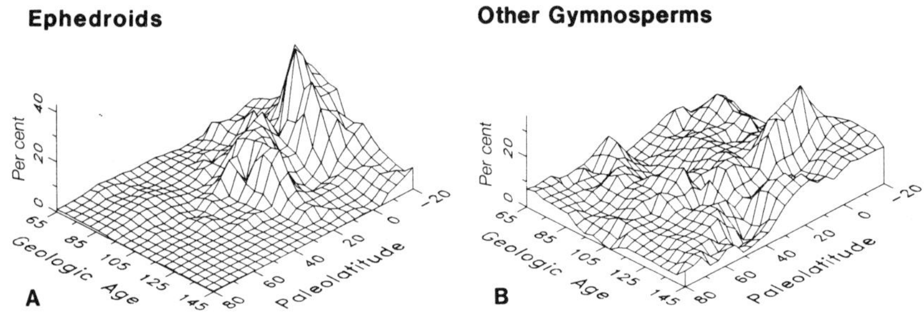

Fig. 17.8. Temporal and latitudinal trends in the relative diversity of ephedroid and other (non-conifer) 'gymnosperm' pollen 'species' in Cretaceous palynofloras. Surfaces fitted by kriging as in Fig. 17.5B; palynoflora data as for 17.5C. A, Trends in the relative diversity of ephedroid pollen 'species' indicate a striking pattern of increase coincident with angiosperm radiation at low palaeolatitudes during the mid-Cretaceous (compare Fig. 17.5D; Crane and Lidgard 1989). However, unlike angiosperms, ephedroids never become important floristic components at middle and high palaeolatitudes, and decline precipitously in relative diversity among latest Cretaceous palynofloras. B, Percentage diversity levels for other (non-conifer) 'gymnosperm' groups decline through the Cretaceous, and are generally higher among palynofloras from low palaeolatitudes than among those from high palaeolatitudes.

angiosperm grains, and only a slight decline in diversity occurs between the Early and Late Cretaceous (Fig. 17.9A). Throughout the Cretaceous the highest percentage diversity of conifer pollen is maintained in mid- to high-palaeolatitude areas, and this is particularly clear for saccate grains that account for the majority of Cretaceous conifer pollen species (Fig. 17.9B). Note, however, that among conifers with non-saccate pollen the diversity of extinct Cheirolepidiaceae may be seriously underestimated in data from middle and low palaeolatitudes. Studies of *in situ* pollen show that several different cheirolepidiaceous taxa produced *Classopollis* pollen that cannot be distinguished in dispersed palynofloras (Watson 1988).

Non-conifer, non-ephedroid gymnosperm pollen (Fig. 17.1A) (treated here as other gymnosperms) generally comprises less than 20 per cent of species in Cretaceous palynofloras (Fig. 17.8B) and is most diverse in Early Cretaceous low-palaeolatitude areas. Subsequently, this heterogeneous pollen group declines in importance, and typically accounts for less than 10 per cent of the species recorded in latest Cretaceous palynofloras.

Spores of free-sporing plants (bryophytes and pteridophytes, Fig. 17.1A) are a major component of Cretaceous palynofloras in all

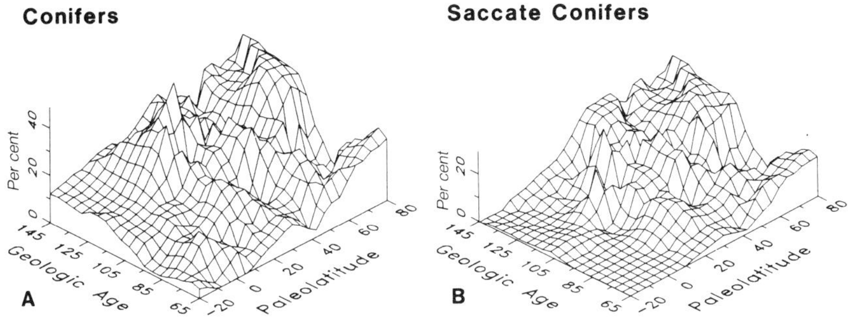

Fig. 17.9. Temporal and latitudinal trends in the relative diversity of conifer pollen 'species' in Cretaceous palynofloras. Surfaces fitted by kriging as in Fig. 17.5B but note that *axes showing palaeolatitude and geologic age are reversed relative to Figs 17.5, 17.7, and 17.8* to show complementary trends. Palynoflora data as for Fig. 17.5C. A, Conifers show only a slight decline in relative diversity through the Cretaceous, but are consistently more important elements in palynofloras from mid- to high palaeolatitudes than in palynofloras from low palaeolatitudes. B, Relative diversity patterns of saccate conifer pollen 'species' indicate that this group is the principal component of the overall conifer diversity trends at middle and high palaeolatitudes throughout the Cretaceous.

areas but tend to be most diverse in assemblages from mid- to high palaeolatitudes (Fig. 17.10A). At all latitudes, but particularly between 20°N and 20°S, the within-palynoflora diversity of spores exhibits a marked mid-Cretaceous decline. A significant component is the reduced importance of ribbed or striate spores, most of which were probably produced by schizaeaceous ferns (Fig. 17.10B). In mid- and high-palaeolatitude floras this is paralleled by a decline of putative gleicheniaceous spores (*Gleicheniidites* or similar genera, Fig. 17.10C). These results are consistent with the general conclusion that characteristic Mesozoic fern families such as Dipteridaceae, Gleicheniaceae, Marattiaceae, Matoniaceae, Osmundaceae, and Schizaeaceae were largely replaced during the Late Cretaceous and Early Tertiary by more derived filicalean ferns (Crane 1987*a*).

Discussion

The palynological results presented here document a major rise in angiosperm diversity, that was initially most pronounced at low palaeolatitudes, but that expanded rapidly to higher palaeolatitudes in the Northern Hemisphere during the mid- to Late Cretaceous. While

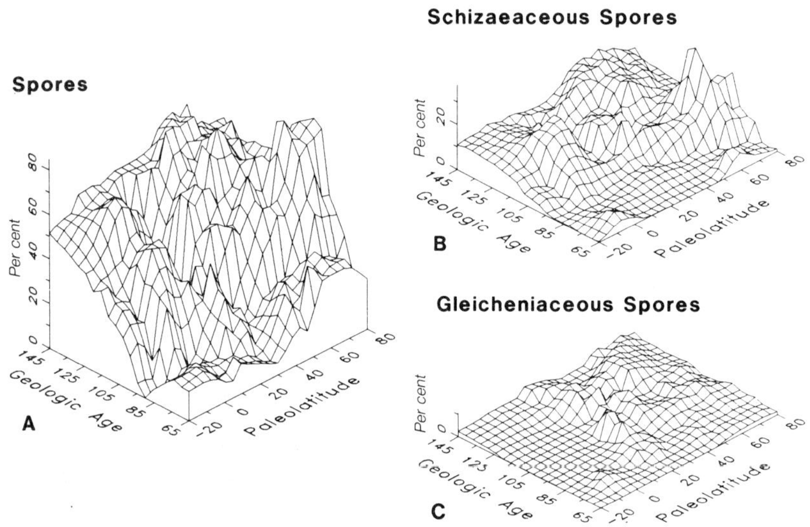

Fig. 17.10. Temporal and latitudinal trends in the relative diversity of spore 'species' in Cretaceous palynofloras. Surfaces fitted by kriging as in Fig. 17.5B, but note that *axes showing palaeolatitude and geologic age are reversed relative to Figs 17.5, 17.7, and 17.8* to show complementary trends. Palynoflora data as for Fig. 17.5C. A, Trends in relative diversity indicate that free-sporing plants, the dominant floristic components of the Early Cretaceous at all palaeolatitudes, decline dramatically in importance among mid- and Late Cretaceous palynofloras, and the decline is registered initially at low palaeolatitudes. B, Relative diversity patterns of spore 'species' characteristic of schizaeaceous ferns are an important component of the overall spore trends, and also show a latitudinally diachronous pattern. C, Gleicheniaceous ferns show limited representation as spore 'species' at middle to high palaeolatitudes during the Early Cretaceous but subsequently decline in their relative importance.

this general pattern seems clear (Fig. 17.7), the underlying causes and attendant changes in the diversity of other groups (Figs 17.8–17.10) are uncertain. Many of the causal explanations offered have not distinguished clearly between increasing angiosperm diversity and increasing angiosperm abundance, and a variety of angiosperm features have also been interpreted as 'key innovations' without adequately considering the occurrence of these characters in other groups and angiosperm subclades. In the following paragraphs we briefly review some of these causal explanations in the context of latitudinal and temporal trends in palynological diversity.

It has been suggested that the presence of specialized water-conducting cells (vessels) in angiosperm wood conferred increased tolerance to high evapo-transpiration rates, particularly in regionally (and probably seasonally) arid mid-Cretaceous equatorial areas. These may also have permitted the development of laminar, reticulate-veined foliage that would have been competitively advantageous with respect to the perhaps more limited light-gathering capabilities of contemporary gymnosperms (Doyle *et al.* 1982). The effects of such vegetative differences may have been particularly pronounced during seedling growth (Bond 1989).

The origin of flowers and entomophily may have provided new possibilities for reproductive isolation, and allowed flowering plants to overcome some of the limitations of anemophily that is widespread in many other seed plants (Raven 1977; Regal 1977; Burger 1981). Wind pollination is typically correlated with seasonal environments, high pollen production, and the maintenance of relatively dense population structure (Whitehead 1969, 1983) and thus may limit diversity via constraints on isolating mechanisms (Regal 1977; Burger 1981; Doyle *et al.* 1982). In contrast, insect pollination permits effective outcrossing at lower population densities and in a potentially greater range of environments (Burger 1981; Doyle *et al.* 1982). Entomophily may therefore not only have accelerated speciation rate by facilitating reproductive isolation, but may also have contributed to the maintenance of high diversity levels. More scattered distributions, perhaps favoured by new modes of dispersal associated with the origin of the carpel (Regal 1977; Burger 1981), may have reduced the effects of species-specific pests and increased possibilities for geographical isolation (Doyle *et al.* 1982; Doyle and Donoghue 1986). Ovule enclosure perhaps also facilitated the development of incompatibility systems and could have intensified pollen (male gamete) competition leading to the production of competitively superior sporophytes (Mulcahy 1979).

Several authors have suggested that some or all of the morphological characteristics underlying angiosperm reproductive biology, such as reduction, simplification, and aggregation of sporophylls to form a flower, enclosure of the ovules, truncation of the gametophyte phase of the life cycle, and extreme vegetative flexibility, may have been secondary effects of selection for precocious reproduction (progenesis) and a weedy life history in open, perhaps seasonally arid environments (Takhtajan 1969; Stebbins 1974, 1981). Similar selective pressures may also have been involved in the origin of the characteristic angiosperm leaf (Hickey and Doyle 1977) and the progenetic expression of metaxylem characteristics (e.g. scalariform pitting) in angiosperm secondary xylem (Doyle 1978; Doyle and Donoghue 1986). Disturbance by herbivorous dinosaurs could have further favoured taxa with *r*-selected traits (Wing and Tiffney 1987). Under these scenarios, secondary morphological effects of progenesis would have had incidental, but profound, consequences for angiosperm history via enhanced ecological capabilities, and effects on both speciation and extinction rates.

Different authors have emphasized different elements in this plexus of ideas. One extreme argues that the 'success' of angiosperms was largely due to 'basically superior' competitive abilities (Knoll 1984), while another argues that angiosperm success 'is not due to any competitively advantageous feature(s) common to all angiosperms, but rather to the wide array of adaptive types [generated] as a consequence of high speciation rates' (Doyle and Donoghue 1986, pp. 391–2). Complex interactions between factors make this issue difficult to resolve and underline the problems involved in assessing the significance of different characters in population and species-level processes, particularly over long periods of time that also encompass changing biotic, climatic, and edaphic conditions. However, patterns of floristic change provide part of the empirical background against which such contrasting causal hypotheses must be judged.

Of particular interest in connection with the initial angiosperm diversification at low palaeolatitudes is the dramatic coincident increase in the diversity of ephedroid pollen (Fig. 17.8A), which contrasts strikingly with the virtually stable, or generally declining diversity of conifers, free-sporing plants and other (non-conifer, non-ephedroid) gymnosperms (Figs 17.8B, 17.9, 17.10). Furthermore, many of the features interpreted as key innovations in angiosperms also occur in certain groups of gymnosperms. Cladistic analyses show that angiosperms and Gnetales (as well as anthophytes in general) are united by a range of features including aggregation of the sporophylls into flower-like reproductive structures (implying insect pollination), scalariform pitting in the secondary xylem, once pinnate microsporophylls, granular exine structure, loss of sacci on pollen grains, and reduction of megaspore membrane (also in *Caytonia*) (Crane 1985; Doyle and Donoghue 1986). Similar characteristics have also originated independently in different anthophyte lineages, for example, extreme reduction of the microgametophyte, tetrasporic megagametophyte development, simple stamens, laminar reticulate-veined leaves, vessels, and cellular embryogeny both in angiosperms and within Gnetales (Crane 1985; Doyle and Donoghue 1986). These character distributions suggest that 'factors favoring progenesis were operating on the anthophytes since their origin but were continued and intensified independently in angiosperms and Gnetales' (Doyle and Donoghue 1986, p. 388). Such biological similarities, combined with parallel patterns of increasing diversity, imply that the early radiation of both taxa may have been influenced by similar biotic and/or environmental factors.

This idea receives further support from evidence that angiosperms and Gnetales were associated ecologically during the mid-Cretaceous. At the Drewrys Bluff locality in eastern North America (Potomac Group, Zone I, Aptian–lower Albian) gnetalean macrofossils occur with ferns and herbaceous angiosperms in a fossil assemblage that may reflect early colonizing vegetation of disturbed flood-plain environments (Crane and Upchurch 1987). Similarly, in the mid-Cretaceous of West Africa and eastern South America the abundance of ephedroid and angiosperm pollen is often positively correlated, while the abundance of both groups tends to be negatively correlated with high abundance of *Classopollis* (Doyle *et al.* 1982; Crane and Lidgard, unpublished). More detailed analyses of angiosperm–gnetalean co-occurrence are needed to assess the extent to which the broad association of both groups in low-latitude areas is also manifested in ecological association at a local population level.

The recognition of latitudinal, temporal, and ecological parallels between the early radiation of angiosperms and Gnetales has important implications. First, it suggests that ideas on the causal basis of the initial mid-Cretaceous angiosperm radiation might focus most appropriately on characters expressed in both groups (e.g. insect pollination, vessels, reticulate-veined leaves) rather than on features that occur only in angiosperms (carpel, stigmatic pollen germination). Second, it raises the possibility that changes in the environment (e.g. Arthur *et al.* 1985; Reyment and Bengston 1986; Leckie 1989), or in other elements in the biota (Friis *et al.* 1987), may have been more influential in effecting mid-Cretaceous floristic trends than has been previously recognized. Early Cretaceous angiosperms and Gnetales may have possessed similar biological attributes that predisposed similar responses to environmental change or, alternatively, some innovation may have arisen and spread more or less simultaneously in the two different lineages in response to similar biotic or environmental selective pressures.

Equally significant, however, are the very different patterns of angiosperm and gnetalean diversity during the Late Cretaceous. If common factors stimulated parallel radiations, they must have become relatively less influential in post-Cenomanian time when gnetalean diversity declined and angiosperm diversity continued to increase. This dissimilarity, contrasted with the similarity of the angiosperm and gnetalean patterns during the mid-Cretaceous, emphasizes the difficulty of generalizing about the long-term macroevolutionary consequences of particular traits. Cladistic analyses can clarify the level at which such features define groups, and can facilitate direct comparisons of sister taxa. However, such analyses do not provide guidance on which characters, in the large sequence of nested synapomorphies generated by diversification, are likely to be of particular long-term evolutionary significance. In attempting to account for the vast number of extant angiosperm species, should the focus be on angiosperm synapomorphies, the synapomorphies of non-magnoliid dicotyledons, or on diagnostic characters at some other more, or less, inclusive level? In addition, sequential character evolution within a clade may itself modify ecological parameters and thus disrupt the specific conditions under which particular features contribute to enhanced *relative* fitness, or elevated *relative* diversification rates. Thus, while it is possible that the divergent Late Cretaceous angiosperm and gnetalean patterns may have been induced by environmental change (cf. Cenomanian–Turonian extinctions in the marine realm, Sepkoski 1986), it is likely that they partially reflect biotic change, including the origin of new subclades within angiosperms.

Conclusions

While interpretation of palynological diversity data is complicated by problems associated with the representation and recognition of different taxa in pollen–spore assemblages, these disadvantages are offset by a level of ecological and geographic sampling exceeding that currently available for macrofloras (Batten 1984; Lidgard and Crane 1990). The temporal and palaeolatitudinal trends in within-palynoflora diversity recognized here need to be tested with data from other areas, particularly from the Southern Hemisphere, but the patterns documented to date provide a preliminary floristic framework within which to evaluate hypotheses of Cretaceous vegetational change. Furthermore, although these results are not geographically comprehensive, they provide an improved basis for assessing the significance of floristic trends observed in local stratigraphic sequences.

Measures of within-palynoflora diversity show clearly that angiosperms diversified first in equatorial areas, but then expanded rapidly through the mid-Cretaceous, to become significant in palynofloras from a broad range of palaeolatitudes in the Northern Hemisphere. The major component of this palynological diversification is a marked radiation of triaperturate pollen diagnostic of non-magnoliid dicotyledons. Increasing angiosperm diversity is accompanied by stable, or declining, patterns of within-palynoflora diversity for most other plant groups. While some of these patterns may reflect competitive interactions with angiosperms, the Gnetales (as measured by increasing diversity of ephedroid pollen) show a remarkable radiation through the mid-Cretaceous at low palaeolatitudes. The latitudinal, temporal, and ecological similarities between the mid-Cretaceous diversification of angiosperms and Gnetales suggest that the initial radiation of both groups may have been influenced by similar biological and/or environmental factors. Further information on mid-Cretaceous fossil plants from low palaeolatitudes, integrated with information from sedimentology and co-occurring macrofossils, will be needed to develop an empirical palaeoecological background against which to evaluate adaptational and macroevolutionary hypotheses of angiosperm diversification.

Acknowledgements

We thank M. E. Collinson, A. N. Drinnan, E. M. Friis, J. J. Sepkoski, Jr, S. L. Wing, and particularly J. A. Doyle for helpful discussion. We gratefully acknowledge B. Ballard, M. Cotton, A. Fernald, C. A. Hult, J. Kissinger, and C. Nuñez for assistance in compiling and editing data; A. Lottes, A. M. Ziegler, and the University of Chicago Paleogeographic Atlas Project for providing palaeolatitudinal assignments; E. Zeiger for typing the manuscript; and C. Richardson for preparing the figures. We also thank P. D. Taylor and G. P. Larwood for the opportunity to participate in this symposium. This work was supported in part by NSF research grants BSR 8314592 and BSR 8708460 (P.R.C.). Acknowledgement is made to the Donors of the Petroleum Research Fund, administered by the American Chemical Society for partial support of this research (S.L.).

References

Arthur, M. A., Dean, W. E. and Schlanger, S. O. (1985). Variations in the global carbon cycle during the Cretaceous related to climate, volcanism, and changes in atmospheric CO_2. In *The carbon cycle and atmospheric CO_2: natural variations Archean to present*, Geophysical Monographs, no. 32 (ed. E. T. Sundquist and W. S. Broecker), pp. 504–29. American Geophysical Union, Washington, DC.

Bambach, R. K. (1977). Species richness in marine benthic habitats through the Phanerozoic. *Paleobiology*, **3**, 152–67.

Basinger, J. F. and Dilcher, D. L. (1984). Ancient bisexual flowers. *Science*, **224**, 511–13.

Batten, D. J. (1984). Palynology, climate and the development of Late Cretaceous floral provinces in the Northern Hemisphere; a review. In *Fossils and climate* (ed. P. J. Brenchley), pp. 127–64. John Wiley, London.

Bell, W. A. (1956). Lower Cretaceous floras of Western Canada. *Geological Survey of Canada, Memoirs*, **285**, 1–331.

Berry, E. W. (1911). The Lower Cretaceous floras of the world, correlation of the Potomac Formations, Pteridophyta, Cycadophytae, Gymnospermae,

Monocotyledonae, Dicotyledonae. In *Upper Cretaceous* (ed. W.B. Clark), pp. 99–165, 214–596. Maryland Geological Survey, Baltimore.

Bond, W.J. (1989). The tortoise and the hare: ecology of angiosperm dominance and gymnosperm persistence. *Biological Journal of the Linnean Society*, **36**, 227–49.

Brenner, G.J. (1963). The spores and pollen of the Potomac Group of Maryland. *Department of Geology, Mines and Water Resources, State of Maryland, Bulletin*, **27**, 1–215.

Brenner, G.J. (1968). Middle Cretaceous spores and pollen from northeastern Peru. *Pollen et Spores*, **10**, 341–83.

Brenner, G.J. (1976). Middle Cretaceous floral provinces and early migrations of angiosperms. In *Origin and early evolution of angiosperms* (ed. C.B. Beck), pp. 23–47. Columbia University Press, New York.

Brenner, G.J. (1984). Late Hauterivian angiosperm pollen from the Helez Formation, Israel. *Sixth International Palynology Conference, Calgary, Abstracts*, **6**, 15.

Burger, W.C. (1981). Why are there so many kinds of flowering plants? *Bioscience*, **31**, 572, 577–81.

Cornet, B. (1986). The reproductive structures and leaf venation of a Late Triassic angiosperm, *Sanmiguelia lewisii*. *Evolutionary Theory*, **7**, 231–309.

Cornet, B. (1989*a*). The reproductive morphology and biology of *Sanmiguelia lewisii*, and its bearing on angiosperm evolution in the Late Triassic. *Evolutionary Trends in Plants*, **3**, 25–51.

Cornet, B. (1989*b*). Late Triassic angiosperm-like pollen from the Richmond Rift Basin of Virginia, U.S.A. *Palaeontographica Abt. B*, **213**, 37–87.

Cousminer, H.L. (1961). Palynology, paleofloras and paleoenvironments. *Micropaleontology*, **7**, 365–8.

Crane, P.R. (1985). Phylogenetic analysis of seed plants and the origin of angiosperms. *Annals of the Missouri Botanical Garden*, **72**, 716–93.

Crane, P.R. (1987*a*). Vegetational consequences of the angiosperm diversification. In *The origins of angiosperms and their biological consequences* (ed. E.M. Friis, W.G. Chaloner, and P.R. Crane), pp. 107–40. Cambridge University Press, Cambridge.

Crane, P.R. (1987*b*). Review of Cornet, B., The leaf venation and reproductive structures of a Late Triassic angiosperm, *Sanmiguelia lewisii*. *Taxon*, **36**, 778–9.

Crane, P.R. (1988). Major clades and relationships in the 'higher' gymnosperms. In *Origin and evolution of gymnosperms* (ed. C.B. Beck), pp. 218–72. Columbia University Press, New York.

Crane, P.R. (1989). Paleobotanical evidence on the radiation of non-magnoliid dicotyledons. *Plant Systematics and Evolution*, **162**, 165–91.

Crane, P.R. and Lidgard, S. (1989). Angiosperm diversification and paleolatitudinal gradients in Cretaceous floristic diversity. *Science*, **246**, 675–8.

Crane, P.R. and Lidgard, S. (in prep.). Angiosperm diversity and large scale floristic change through the Cretaceous: analyses of macrofossil data from the Northern Hemisphere.

Crane, P. R. and Upchurch, G. R. (1987). *Drewria potomacensis* gen. et sp. nov., an early Cretaceous member of the Gnetales from the Potomac Group of Virginia. *American Journal of Botany*, **74**, 1722–36.

Crane, P.R., Donoghue, M.J., Doyle, J.A. and Friis, E.M. (1989*a*). Angiosperm origins. *Nature*, **342**, 131.

Crane, P.R., Friis, E.M. and Pedersen, K.R. (1989*b*). Reproductive structure and function in Cretaceous Chloranthaceae. *Plant Systematics and Evolution*, **165**, 211–26.

Cronquist, A. (1981). *An integrated system of classification of flowering plants*. Columbia University Press, New York.

Davis, J.C. (1986). *Statistics and data analysis in geology*. Wiley, New York.

Dilcher, D. L. and Crane, P. R. (1984). *Archaeanthus*: an early angiosperm from the Cenomanian of the western interior of North America. *Annals of the Missouri Botanical Garden*, **71**, 351–83.

DiMichele, W. A. and Wing, S. L. (ed.) (1988). Methods and applications of plant paleoecology. *Paleontological Society Special Publication*, **3**, 1–171.

Donoghue, M.J. and Doyle, J. A. (1989*a*). Phylogenetic studies of seed plants and angiosperms based on morphological characters. In *The hierarchy of life* (ed. B. Fernholm, K. Bremer, and H. Jornvall), pp. 181–93. Elsevier Excerpta Medica, Amsterdam.

Donoghue, M.J. and Doyle, J.A. (1989*b*). Phylogenetic analysis of angiosperms and the relationships of 'Hamamelidae'. In *Evolution, systematics and fossil history of the Hamamelidae* (ed. P.R. Crane and S. Blackmore), pp. 17–45. Clarendon Press, Oxford.

Dorf, E. (1955). Plants and the geologic time scale. *Geological Society of America Special Paper*, **62**, 575–92.

Doyle, J. A. (1969). Cretaceous angiosperm pollen of the Atlantic Coastal Plain and its evolutionary significance. *Journal of the Arnold Arboretum*, **50**, 1–35.

Doyle, J.A. (1973). The monocotyledons: their evolution and comparative biology. V. Fossil evidence on the early evolution of the monocotyledons. *Quarterly Review of Biology*, **48**, 399–413.

Doyle, J.A. (1978). Origin of angiosperms. *Annual Review of Ecology and Systematics*, **9**, 365–92.

Doyle, J. A. and Donoghue, M.J. (1986). Seed plant phylogeny and the origin of angiosperms: an experimental cladistic approach. *Botanical Review*, **52**, 321–431.

Doyle, J.A. and Hickey, L.J. (1976). Pollen and leaves from the mid-Cretaceous Potomac Group and their bearing on early angiosperm evolution. In *Origin and early evolution of angiosperms* (ed. C.B. Beck), pp. 139–206. Columbia University Press, New York.

Doyle, J.A. and Robbins, E.I. (1977). Angiosperm pollen zonation of the continental Cretaceous of the Atlantic Coastal Plain and its application to deep wells in the Salisbury Embayment. *Palynology*, **1**, 43–78.

Doyle, J. A., Jardiné, S. and Doerenkamp, A. (1982). *Afropollis*, a new genus of early angiosperm pollen, with notes on the Cretaceous palynostratigraphy and paleoenvironments of northern Gondwana. *Bulletin des Centres de Recherches Exploration-Production Elf-Aquitaine*, **6**, 39–117.

Drinnan, A. N., Crane, P. R., Friis, E. M. and Pedersen, K. R. (in press *a*) Lauraceous flowers from the mid-Cretaceous of eastern North America. *Botanical Gazette.*

Drinnan, A. N., Crane, P. R., Friis, E. M. and Pedersen, K. R. (in press *b*). Angiosperm flowers and tricolpate pollen of buxaceous affinity from the mid-Cretaceous of eastern North America. *American Journal of Botany.*

Frederiksen, N. O. (1980). Significance of monosulcate pollen abundance in Mesozoic sediments. *Lethaia*, **13**, 1–20.

Friis, E. M., Chaloner, W. G. and Crane, P. R. (ed.). (1987). *The origins of angiosperms and their biological consequences*. Cambridge University Press, Cambridge.

Friis, E. M., Crane, P. R. and Pedersen, K. R. (1988). Reproductive structures of Cretaceous Platanaceae. *Biologiske Skrifter Danske Videnskabernes Selskab*, **31**, 1–55.

Harland, W. B., Cox, A. V., Llewellyn, P. G., Pickton, C. A. G., Smith, A. G. and Walters, R. (1982). *A geologic time scale.* Cambridge University Press, Cambridge.

Herngreen, G. F. W. (1973). Palynology of the Albian–Cenomanian strata of Borehole 1-QS-1-MA, State of Maranhão, Brazil. *Pollen et Spores*, **15**, 515–55.

Herngreen, G. F. W. (1975). Palynology of Middle and Upper Cretaceous strata in Brazil. *Mededelingen Rijks Geologische Dienst. N. S.*, **26**, 39–91.

Herngreen, G. F. W. and Chlonova, A. F. (1981). Cretaceous microfloral provinces. *Pollen et Spores*, **23**, 442–555.

Hickey, L. J. and Doyle, J. A. (1977). Early Cretaceous fossil evidence for angiosperm evolution. *Botanical Review*, **43**, 3–104.

Hughes, N. F. and McDougall, A. B. (1986). Records of angiospermid pollen entry into the English Early Cretaceous succession. *Review of Palaeobotany and Palynology*, **50**, 255–72.

Knoll, A. H. (1984). Patterns of extinction in the fossil record of vascular plants. In *Extinctions* (ed. M. H. Nitecki), pp. 21–68. University of Chicago Press, Chicago.

Knoll, A. H. (1986). Patterns of change in plant communities through geologic time. In *Community ecology* (ed. J. Diamond and T. J. Case), pp. 126–41. Harper and Row, New York.

Krassilov, V. A. (1986). New floral structure from the Lower Cretaceous of Lake Baikal area. *Review of Palaeobotany and Palynology*, **47**, 9–16.

Leckie, R. M. (1989). A paleoceanographic model for the early evolutionary history of planktonic foraminifera. *Palaeogeography, Palaeoclimatology, Palaeoecology*, **73**, 107–38.

Lidgard, S. and Crane, P. R. (1988). Quantitative analyses of the early angiosperm radiation. *Nature*, **331**, 344–6.

Lidgard, S. and Crane, P. R. (1990). Angiosperm diversification and Cretaceous floristic trends: a comparison of palynofloras and leaf macrofloras. *Paleobiology*, **16**, 73–93.

MacDonald, G. M. and Waters, N. M. (1987). An evaluation of automated and mapping algorithms for the analysis of Quaternary pollen data. *Review*

of Palaeobotany and Palynology, **51**, 289–307.
Mulcahy, D. L. (1979). The rise of angiosperms: a genecological factor. *Science,* **206**, 20–3.
Muller, J. (1970). Palynological evidence on early differentiation of angiosperms. *Biological Reviews of the Cambridge Philosophical Society*, **45**, 417–50.
Muller, J. (1984). Significance of fossil pollen for angiosperm history. *Annals of the Missouri Botanical Garden*, **71**, 419–33.
Niklas, K. J., Tiffney, B. H. and Knoll, A. H. (1980). Apparent changes in the diversity of fossil plants: a preliminary assessment. In *Evolutionary Biology*, Vol. 12 (ed. M. K. Hecht, W. C. Steere, and B. Wallace), pp. 1–89. Plenum Press, New York.
Niklas, K. J., Tiffney, B. H. and Knoll, A. H. (1983). Patterns in vascular land plant diversification. *Nature*, **303**, 614–16.
Niklas, K. J., Tiffney, B. H. and Knoll, A. H. (1985). Patterns in vascular land plant diversification: a factor analysis at the species level. In *Phanerozoic diversity patterns: profiles in macroevolution* (ed. J. W. Valentine), pp. 97–128. Princeton University Press, Princeton, New Jersey.
Pedersen, K. R., Crane, P. R., Drinnan, A. N. and Friis, E. M. (in press). Fruits from the mid-Cretaceous of eastern North America with pollen grains of the *Clavatipollenites*-type. *Grana*.
Prentice, I. C. (1988). Records of vegetation in time and space: the principles of pollen analysis. In *Vegetation history* (ed. B. Huntley and T. Webb III), pp. 17–42. Kluwer Academic Publishers, Boston.
Raven, P. H. (1977). A suggestion concerning the Cretaceous rise to dominance of the angiosperms. *Evolution*, **31**, 451–2.
Regal, P. J. (1977). Ecology and evolution of flowering plant dominance. *Science*, **196**, 622–9.
Reyment, R. A. and Bengtson, P. (1986). *Events of the mid-Cretaceous*. Pergamon Press, Oxford.
Sepkoski, J. J. (1986). Phanerozoic overview of mass extinction. In *Patterns and processes in the history of life* (ed. D. M. Raup and D. Jablonski), pp. 277–94. Springer-Verlag, Berlin.
Singh, C. (1964). Microflora of the Lower Cretaceous Mannville Group, east-central Alberta. *Research Council of Alberta Bulletin*, **15**, 1–238.
Spicer, R. A. (1989). The formation and interpretation of plant fossil assemblages. *Advances in Botanical Research*, **16**, 95–191.
Stebbins, G. L. (1974). *Flowering plants: evolution above the species level.* Harvard University Press, Cambridge, Massachusetts.
Stebbins, G. L. (1981). Why are there so many species of flowering plants? *Bioscience*, **31**, 573–7.
SURFER (1987). Golden Software. Golden, Colorado.
Takhtajan, A. (1969). *Flowering plants: origin and dispersal.* Oliver and Boyd, Edinburgh.
Trevisan, L. (1980). Ultrastructural notes and considerations on *Eucommiidites* and *Monosulcites* pollen grains from Lower Cretaceous sediments of southern Tuscany (Italy). *Pollen et Spores*, **22**, 85–132.

Upchurch, G. R. and Wolfe, J. A. (1990). Cretaceous vegetation of the western interior and adjacent regions of North America. *Geological Association of Canada Special Papers*.
Walker, J. W. and Walker, A. G. (1984). Ultrastructure of Lower Cretaceous angiosperm pollen and the origin and early evolution of flowering plants. *Annals of the Missouri Botanical Garden*, **71**, 464–521.
Walker, J. W. and Walker, A. G. (1986). Ultrastructure of Early Cretaceous angiosperm pollen and its evolutionary implications. In *Pollen and spores: form and function* (ed. S. Blackmore and I. K. Ferguson), pp. 203–17. Academic Press, London.
Walker, J. W., Brenner, G. J. and Walker, A. G. (1983). Winteraceous pollen in the Lower Cretaceous of Israel: early evidence of a magnolialean angiosperm family. *Science*, **220**, 1273–5.
Watson, J. (1988). The Cheirolepidiaceae. In *Origin and evolution of gymnosperms* (ed. C. B. Beck), pp. 382–447. Columbia University Press, New York.
Whitehead, D. R. (1969). Wind pollination in the angiosperms: evolutionary and environmental considerations. *Evolution*, **23**, 28–35.
Whitehead, D. R. (1983). Wind pollination: some ecological and evolutionary perspectives. In *Pollination biology* (ed. L. Real), pp. 97–108. Academic Press, New York.
Wilkinson, L. (1988). Sygraph. Systat Inc., Evanston, Illinois.
Wing, S. L. and Tiffney, B. H. (1987). The reciprocal interaction of angiosperm evolution and tetrapod herbivory. *Review of Palaeobotany and Palynology*, **50**, 179–210.
Wolfe, J. A., Doyle, J. A. and Page, V. M. (1975). The bases of angiosperm phylogeny: paleobotany. *Annals of the Missouri Botanical Garden*, **62**, 801–24.
Yu Jingxian (1981). Late Cretaceous sporo-pollen assemblages of Sache District, Xinjiang. *Acta Geologica Sinica*, **55**, 93–102.

Appendix

Palynoflora bibliography

Aboul-Ela, N. M. and Mabrouk, B. (1978). Palynological studies on some Nubia Sandstone samples in Ed-Dueim area, Sudan. *Revista Española de Micropaleontologia*, **10**(3), 395–406.
Aboul-Ela, N. M. (1978). The palynological characteristics of the Upper Cretaceous sediments in Abu Tartar Plateau, Kharga Oasis, Egypt. *Revista Española de Micropaleontologia*, **10**(3), 421–42.
Anan-Yorke, R. and Stelck, C. R. (1978). Microfloras from Upper Albian *Neogastroplites* Zone, Sikanni Chief River, Northeastern British Columbia. *Geological Association of Canada Special Paper*, **18**, 473–85.
Anderson, R. Y. (1960). Cretaceous–Tertiary palynology, eastern side of San Juan Basin, New Mexico. *New Mexico Bureau of Mines and Mineral Resources, Institute of Mining and Technology, Memoir*, **6**, 1–58.
Barss, M. S., Buzak, J. P. and Williams, G. L. (1979). Palynological zonation

and correlation of sixty-seven wells, eastern Canada. *Geological Survey of Canada Paper*, **78-24,** 1-118.

Batten, D.J. (1973). Use of palynologic assemblage-types in Wealden correlation. *Palaeontology*, **16,** 1-40.

Batten, D.J. and Li, W.B. (1987). Aspects of palynomorph distribution, floral provinces and climate during the Cretaceous. *Geologisches Jahrbuch*, **A96,** 219-37.

Bond, T.A. (1972). A Lower Cretaceous (Aptian-Albian) palynological assemblage from the DeQueen Formation, Pike County, Arkansas. *Pollen et Spores*, **14,** 173-86.

Brenner, G.J. (1963). The spores and pollen of the Potomac Group of Maryland. *Department of Geology, Mines and Water Resources, State of Maryland, Bulletin*, **27,** 1-215.

Brenner, G.J. (1968). Middle Cretaceous spores and pollen from northeastern Peru. *Pollen et Spores*, **10,** 341-83.

Brenner, G.J. (1974). Palynostratigraphy of the Lower Cretaceous Gevar'Am and Talme Yafe Formations in the Gevar'Am 2 Well (Southern Coastal Plain, Israel). *Israel Geological Survey Bulletin*, **59,** 1-27.

Brideaux, W.W. and McIntyre, D.J. (1975). Miospores and microplankton from Aptian-Albian rocks along Horton River, District of MacKenzie. *Geological Survey of Canada Bulletin*, **252,** 1-85.

Couper, R.A. (1958). British Mesozoic microspores and pollen grains. A systematic and stratigraphic study. *Palaeontographica Abt. B*, **103,** 75-179.

Delcourt, A. and Sprumont, G. (1959). Spores, grains de pollen, Hystrichospheres et Peridiniens dans le Wealdien de Feron-Glageon. *Annales de la Société Géologique du Nord*, **79,** 29-64.

Groot, J.J., Penny, J.S. and Groot, C.R. (1961). Plant microfossils and age of the Raritan, Tuscaloosa and Magothy Formations of the eastern United States. *Palaeontographica Abt. B*, **108,** 121-40.

Habib, D. (1969). Middle Cretaceous palynomorphs in a deep-sea core from the Seismic Reflector Horizon A outcrop area. *Micropaleontology*, **15,** 85-101.

Habib, D. (1970). Middle Cretaceous palynomorph assemblages from clays near the Horizon Beta deep-sea outcrop. *Micropaleontology*, **16,** 345-79.

Hall, J.W. and Norton, N.J. (1967). Palynological evidence of floristic change across the Cretaceous-Tertiary boundary in eastern Montana (U.S.A.). *Palaeogeography, Palaeoclimatology, Palaeoecology*, **3,** 121-31.

Herngreen, G.F.W. (1971). Palynology of a Wealden Section (Lower Cretaceous) in the 'Carrière de Longueville', the Boulannais (France). *Review of Palaeobotany and Palynology*, **12,** 271-302.

Herngreen, G.F.W. (1973). Palynology of Albian-Cenomanian strata of Borehole 1-QS-1-MA, State of Maranhão, Brazil. *Pollen et Spores*, **15,** 515-55.

Herngreen, G.F.W. (1974). Middle Cretaceous palynomorphs from northeastern Brazil. *Sciences Géologiques, Bulletin*, **27,** 101-16.

Herngreen, G.F.W. (1975). Palynology of Middle and Upper Cretaceous strata in Brazil. *Mededelingen Rijks Geologische Dienst. N.S.*, **26** (3), 39-91.

Hochuli, P. and Kelts, K. (1980). Palynology of Middle Cretaceous Black Clay Facies from the Deep Sea Drilling Project Sites 417 and 418 of the western North Atlantic. *Initial Reports of the Deep Sea Drilling Project* 51, 52, 53, Part 2, pp. 897–935. U.S. Government Printing Office, Washington, DC.
Hopkins, W.S., Jr and Sweet, A.R. (1976). Miospores and megaspores from the Lower Cretaceous Mattagami Formation of Ontario. *Geological Survey of Canada Bulletin*, **256**, 55–71.
Jardiné, S. and Magloire, L. (1965). Palynologie et stratigraphie du Crétace des Bassins du Senegal et de Côte d'Ivoire. *Mémoires du Bureau de Recherches Géologiques et Minières*, No. 32. Colloque International de Micropaléontologie (Dakar, 6–11 Mai 1963), pp. 187–245.
Jarzen, D.M. (1982). Palynology of Dinosaur Provincial Park (Campanian), Alberta. *National Museums of Canada Syllogeus Series*, **38**, 1–69.
Kemp, E.M. (1970). Aptian and Albian miospores from Southern England. *Palaeontographica Abt. B*, **131**, 73–143.
Kotova, I.Z. (1978). Spores and pollen from Cretaceous deposits of the eastern North Atlantic Ocean, Deep Sea Drilling Project, Leg 41, Sites 367 and 370. *Initial Reports of the Deep Sea Drilling Project*, supplement to volumes XXXVIII, XXXIX, XL, and XLI, pp. 841–81. U.S. Government Printing Office, Washington, DC.
Médus, J. (1987). Analyse quantitative des palynoflores du Campanien de Sedano, Espagne. *Review of Palaeobotany and Palynology*, **51**, 309–26.
Miki, A. (1977). Late Cretaceous pollen and spore floras of northern Japan: composition and interpretation. *Journal of the Faculty of Science, Hokkaido University*, (IV), **17**, 399–436.
Norris, G. (1967). Spores and pollen from the Lower Colorado Group (Albian–?Cenomanian) of Central Alberta. *Palaeontographica Abt. B*, **120**, 72–115.
Oltz, D.F., Jr (1969). Numerical analyses of palynological data from Cretaceous and Early Tertiary sediments in east-central Montana. *Palaeontographica Abt. B*, **128**, 90–166.
Playford, G. (1971). Palynology of Lower Cretaceous (Swan River) strata of Saskatchewan and Manitoba. *Palaeontology*, **14**, 533–65.
Pocock, S.A.J. (1962). Microfloral analysis and age determination of strata at the Jurassic-Cretaceous boundary in the western Canada plains. *Palaeontographica Abt. B*, **111**, 1–95.
Renéville, P. de and Raynaud, J.-F. (1981). Palynologie du stratotype du Barrémien. *Bulletin des Centres de Recherches Exploration–Production Elf-Aquitaine*, **5**, 1–171.
Reyre, Y. (1966). Palynologie du Crétaçé Moyen du Sahare Tunisien. *Revue de Micropaléontologie*, **9**, 3–18.
Reyre, Y. (1973). Palynologie du Mésozoique Saharien. *Mémoires du Muséum National d'Histoire Naturelle, Series C. Sciences de la Terre*, **27**, 1–284.
Rouse, G.E. and Srivastava, S.K. (1972). Palynological zonation of Cretaceous and Early Tertiary Rocks of the Bonnet Plume Formation, northeastern Yukon, Canada. *Canadian Journal of Earth Sciences*, **9**, 1163–79.
Schrank, E. and Perch-Nielsen, K. (1985). Late Cretaceous palynostrati-

graphy in Egypt with comments on Maastrichtian and Early Tertiary calcareous nannofossils. *Newsletters on Stratigraphy*, **15**, 81-99.

Singh, C. (1964). Microflora of the Lower Cretaceous Mannville Group, east-central Alberta. *Research Council of Alberta, Bulletin*, **15**, 1-238.

Singh, C. (1971). Lower Cretaceous microfloras of the Peace River area, northwestern Alberta. *Research Council of Alberta, Bulletin*, **28**, 1-542.

Singh, C. (1983). Cenomanian microfloras of the Peace River area, northwestern Alberta. *Research Council of Alberta, Bulletin*, **44**, 1-193.

Soliman, H.A. and Sultan, I. (1976). Spores et pollen des Grès de Baharia, Désert Ouest, Égypte. *Revue de Micropaléontologie*, **19**, 108-11.

Srivastava, S.K. (1967). Palynology of Late Cretaceous mammal beds, Scollard, Alberta, (Canada). *Palaeogeography, Palaeoclimatology, Palaeoecology*, **3**, 133-50.

Stanley, E.A. (1965). Upper Cretaceous and Paleocene plant microfossils and Paleocene dinoflagellates and hystrichosphaerids from northwestern South Dakota. *Bulletins of American Paleontology*, **49**, 179-384.

Stone, J.F. (1973). Palynology of the Almond Formation (Upper Cretaceous), Rock Springs Uplift, Wyoming. *Bulletins of American Paleontology*, **64**, 1-135.

Sultan, I.Z. (1985). Palynological studies in the Nubia Sandstone Formation, east of Aswan, southern Egypt. *Neues Jahrbuch für Geologie und Paläontologie Monatshefte*, **10**, 605-17.

Sultan, I.Z. (1986). Palynostratigraphy of Lower Cretaceous sediments in the Nile Delta Region, Egypt. *Revista Española de Micropaleontologia*, **18**, 55-70.

Tschudy, B.D. (1973). Palynology of the Upper Campanian (Cretaceous) Judith River Formation, north-central Montana. *United States Geological Survey, Professional Paper*, **770**, 1-42.

Wilson, M.A. (1978). Palynology of three sections across the Uppermost Cretaceous/Paleocene boundary in the Yukon Territory and District of MacKenzie, Canada. *Palaeontographica Abt. B*, 166, 99-183.

Wingate, F.H. (1980). Plant microfossils from the Denton Shale Member of the Bokchito Formation (Lower Cretaceous, Albian) in southern Oklahoma. *Oklahoma Geological Survey Bulletin*, **130**, 1-93.

18. The causes of the diversification of life

M.J. BENTON
Department of Geology, University of Bristol, Bristol, UK

Abstract

Diversification on the large scale has been explained in many different ways. Equilibrium and non-equilibrium models of various kinds have been proposed, but both kinds suffer from unsupported assumptions. More specific explanations of diversification fall into four main categories: (1) the patterns are artefacts of a biased fossil record; (2) the patterns are real and caused mainly by physical perturbation; (3) the patterns are real and caused mainly by biotic factors; or (4) the patterns are real but stochastic. At least 14 specific explanations have been given, but most of these fail to explain long-term global patterns. Three main candidate models for large-scale diversification are tested with data from the fossil record of tetrapods, and the strongest seems to be expansion of adaptive space, followed by subdivision of niches and specialization, and then by provinciality caused by topographic and climatic change. These tests illustrate some of the complexities of disentangling the different explanations for diversification, but they point to ways in which such tests may be quantified.

Introduction

Life has diversified massively during the course of geological time. It should be possible to establish the broad pattern of the global diversification of life since its origin some 3.7 billion years ago, and to determine the causes of that diversification. Further, if all organisms are related to each other in a great phylogenetic tree that derives from a simple common ancestor, the causes of diversification will describe major phylogenetic processes affecting a single clade.

Major Evolutionary Radiations (ed. P.D. Taylor and G.P. Larwood), Systematics Association Special Volume No. 42, pp. 409–430, Clarendon Press, Oxford, 1990.

So far, palaeobiologists have approached this problem by a piecemeal approach, plotting the patterns of diversification of relatively well preserved segments of the fossil record. An early atempt by Valentine (1969) showed the diversification of benthic marine invertebrates over the past 600 Ma (Fig. 18.1(a)). An initial low level of 50 or so families in the Early Cambrian was followed by a rapid rise to 250 or so in the Late Ordovician, a sudden drop at the end of the Permian, and a continuing climb ever since. Valentine (1969) argued that this pattern was essentially real, and that it reflected changes in the physical environment, such as continental drift. Valentine and Moores (1972) argued, for example, that the great extinction at the end of the Permian was caused by the coalescence of continents, and continental shelf seas, into a single supercontinent, Pangaea. The consequent loss of shelf habitats supposedly led to widespread extinction.

Raup (1972) challenged Valentine's model, and suggested that much of the pattern of diversification in the Phanerozoic was produced by systematic sources of error. For example, he argued, the volume of sedimentary rock preserved, and its area of exposure today, diminish the further back one goes in time. Hence, early Palaeozoic floras and faunas are less well represented than are more recent ones, and their global diversity appears depressed. There are also human factors involved. It is a well known fact that more palaeontologists devote their efforts to describing Mesozoic and Cainozoic fossils than Palaeozoic ones. 'Palaeontologist Interest Units' can be assigned to each geological period, and these show a systematic bias against earlier faunas which

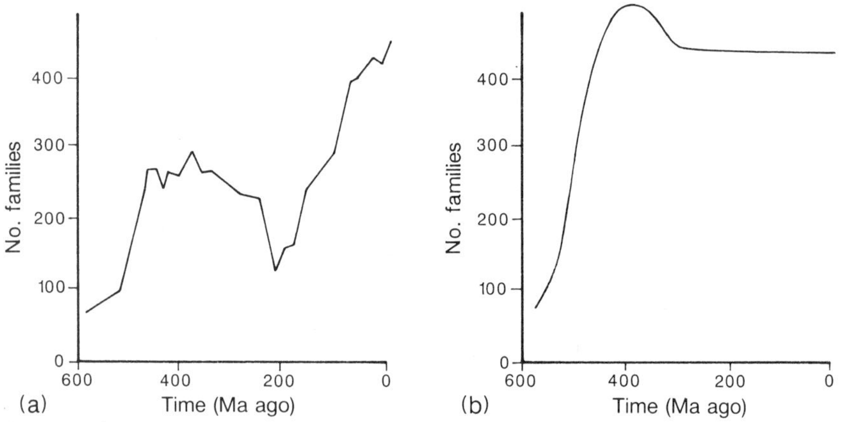

Fig. 18.1. Two views of the pattern of diversification of marine invertebrates during the Phanerozoic (the past 600 Ma): (a) the empirical model (after Valentine 1969); and (b) the bias simulation model, in which estimated systematic biases have been removed (after Raup 1972).

will thus appear less diverse (Sheehan 1977). Raup (1972) presented a 'bias simulation model' (Fig. 18.1(b)) which was meant to represent the true pattern of diversification of marine benthic organisms through time, after the various systematic sources of error had been accounted for. This model shows a rapid rise to a peak in the Cambrian and Ordovician, followed by an equilibrium level.

Gould *et al.* (1977) and Sepkoski (1978) supported Raup's (1972) model, but later studies have tended to confirm Valentine's (1969) empirical model (Fig. 18.1(a)). For example, Sepkoski's (1981, 1984) and Sepkoski *et al.*'s (1981) detailed studies of the diversification of marine animals have produced essentially the same pattern (see below), as have the compilations by Niklas *et al.* (1983) on the species diversity of vascular land plants, and Benton (1985*a*) on the family diversity of non-marine tetrapods (Fig. 18.2(a)–(c)).

The aims of this chapter are to look at the broad patterns of the

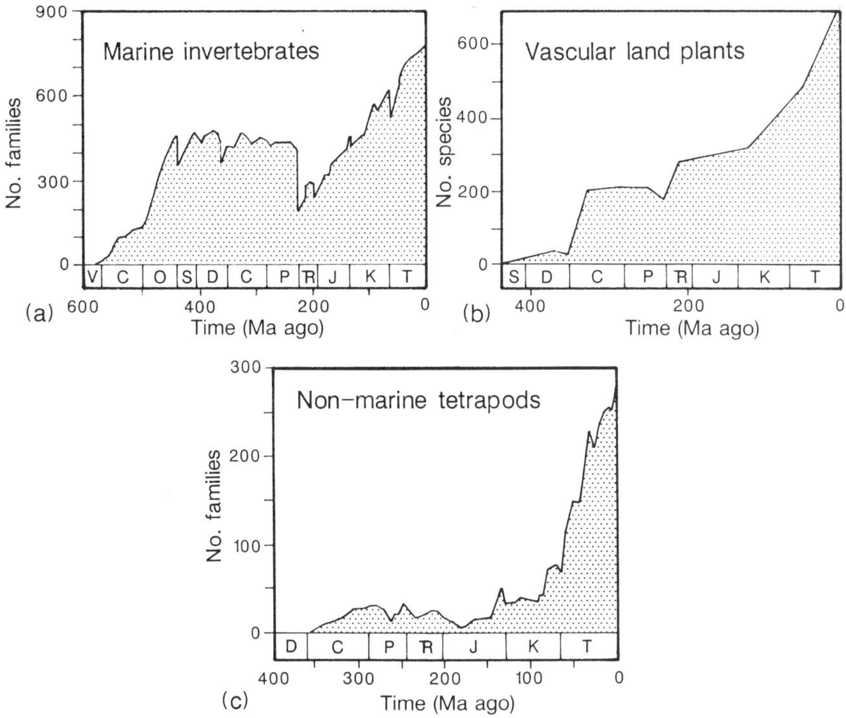

Fig. 18.2. The diversification of three groups of multicellular organisms during the Phanerozoic: (a) marine invertebrates (after Erwin *et al.* 1987); (b) vascular land plants (after Niklas *et al.* 1980); and (c) non-marine tetrapods (after Benton 1985a). Standard stratigraphic abbreviations are used for the geological periods.

diversification of life, to review the numerous available explanations for diversification, and to apply tests of these postulated causes to some examples of the diversification of tetrapods. First, a comparison of the two broad ways of viewing diversification, by means of equilibrium and non-equilibrium models, will be given.

Equilibrium and non-equilibrium models

1. *Equilibrium models*

Many palaeobiologists assumed that the appropriate approach to understanding global diversification would be within an equilibrium framework. The models were developed as extensions of logistic models from the ecological literature, of which Macarthur and Wilson's (1967) theory of island biogeography is an example. All of the parameters were scaled up from the ecological scale to the evolutionary: the island

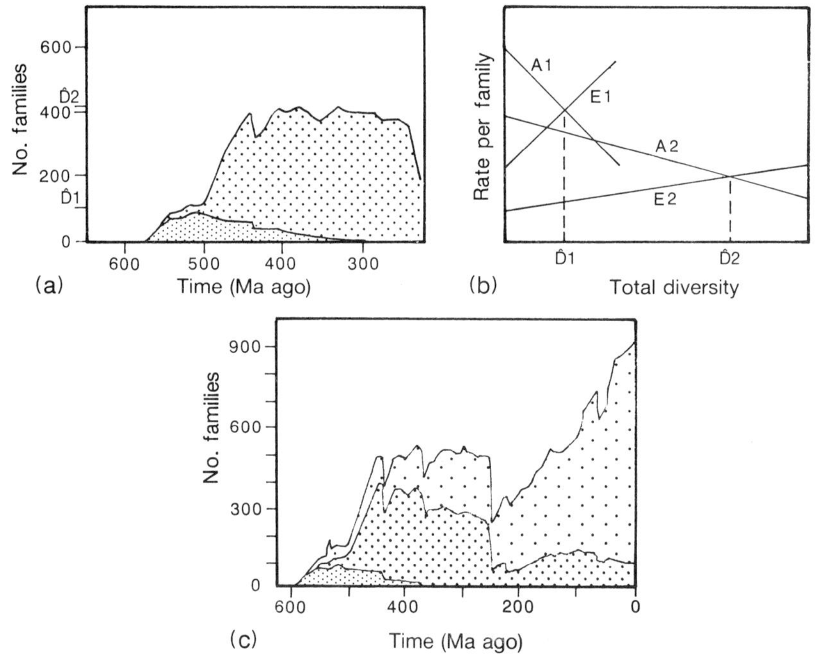

Fig. 18.3. The kinetic equilibrium model for the Phanerozoic diversification of marine metazoans: (a) two faunal phases for the Palaeozoic part of the pattern (after Sepkoski 1978); (b) the equilibrium model in which faunal phase (1, 2) has a characteristic rate of origin (A1, A2), extinction (E1, E2), and equilibrium diversity ($\hat{D}1$, $\hat{D}2$) (after Sepkoski 1978); (c) three faunal phases for the whole Phanerozoic record (after Sepkoski 1984). (See text for discussion).

became the world (or world shallow seas); rates of immigration and local extinction became rates of global taxon origination and extinction; and the equilibrium level was for families or species globally, rather than for species on an island. Sepkoski (1978, 1979, 1981) applied these models to the record of marine animal ordinal and family diversification. He determined first that the Palaeozoic faunas fell into two major 'evolutionary faunas', the first in the Cambrian of under 100 families, and the second in the Ordovician of about 350 families (Fig. 18.3(a)). The model (Fig. 18.3(b)) seeks to explain how each 'evolutionary fauna' had different overall rates of origination and extinction, and hence different equilibrium levels. The model was later extended (Sepkoski 1984) to include a third 'evolutionary fauna' which rose to importance after the end-Permian extinction event. The third 'fauna' has already risen past 1800 families, and it has not yet apparently reached its equilibrium level (Fig. 18.3(c)).

Hoffman (1981, 1983, 1985, 1989), Cracraft (1985), and others, have criticized these kinds of equilibrium models, partly because of problems in extrapolating from the ecological level to the global level. For example, local species immigration rates are very different from global speciation or family origination rates: species splitting in phylogeny cannot simply be equated with the local addition of existing species to an otherwise unchanged fauna. A further criticism is that, in island biogeography theory, local extinction rate increases with increasing species diversity, whereas the opposite appears to have been the case, at least for families, during the global diversification of marine animals (Raup and Sepkoski 1982; Van Valen 1984).

2. *Non-equilibrium models*

Various kinds of non-equilibrium models for total global diversification have been developed recently. For example, Kitchell and Carr (1985) argued that their modelled ecological system never reaches an equilibrium level because of evolutionary innovations and mass extinctions. Their model matches the empirical data, but they retained an assumption that rates of origination and extinction are diversity-dependent.

This diversity-dependent assumption has been questioned by Hoffman (1981, 1985), Walker and Valentine (1984), and Cracraft (1985). These authors argue that a diversity-dependent model for speciation assumes that there is a fixed number of available niches, and that new species generally arise by displacing pre-existing species. This idea that there is an equilibrium level of species diversity, a level of saturation with species, has also been criticized by Gould and Calloway (1980), Rieppel (1984), and Benton (1987*a*), who note that species, and higher taxa, can invade empty adaptive zones, and increase overall species numbers on a global scale.

Hoffman (1986, 1989) and Hoffman and Fenster (1986) have developed a non-equilibrium, non-diversity-dependent 'neutral model' for total global diversification. Here, both origination and extinction rates are assumed to be independent of total diversity, and of each other, and each is modelled as a random walk. Computer simulations of this 'neutral', or null, model produced graphs that appeared to match the empirical patterns just as well as the equilibrium model of Sepkoski (1984) or the non-equilibrium diversity-dependent model of Kitchell and Carr (1985), but the models were not truly null (see below).

3. *Resolution?*

It is hard at present to choose between the equilibrium and the non-equilibrium approaches to modelling global biotic diversification. There are clear parallels in this debate with the controversy over equilibrium theory and diversity dependence in community ecology (e.g. Harvey *et al.* 1983; Strong *et al.* 1984; Diamond and Case 1986; Benton 1987*a*). The criticisms of the equilibrium approach, especially those concerning its theoretical basis, require answering. Sounder justification is needed in support of equilibrium models of global diversification than a mere scaling-up of the concepts of community ecology (Webb 1969; Rosenzweig 1975).

The non-equilibrium models are also problematical. Is there an ultimate, but unattainable, equilibrium number of species on a global scale (Kitchell and Carr 1985)? Is it possible to develop an entirely stochastic null model, as Hoffman (1986) has attempted? It may be that, as in modern ecological examples, most so-called 'null models' still include some biological factors or biases (Harvey *et al.* 1983). In any case, if the data fit a null model, the deterministic models are not thereby disproved.

The causes of global biotic diversification

1. *Suggested explanations*

Many explanations have been advanced for the patterns of apparent diversity increase observed in the fossil records of most groups of organisms (Figs 18.1, 18.2). These suggestions are listed below under four main headings that include proposals that the patterns are: (1) artefactual; (2) real and caused primarily by physical factors; (3) real and caused primarily by biotic factors; and (4) real, but neutral in cause. The proposed explanations are summarized briefly, and key references are given, as appropriate. They are then discussed in more detail, and tests are proposed.

Artefact The pattern of increase is an artefact of the patchy quality of the fossil record and the way in which we study it (Raup 1972, 1976*a, b*).

1. The volume of unmetamorphosed sedimentary rock increases towards the present.

2. The area of exposure of such fossiliferous rock increases towards the present.

3. Palaeontologists devote more attention to younger faunas and floras and hence name more species, genera, and families.

4. The 'pull of the recent', in which our knowledge of modern species enhances our knowledge of close relatives of the recent geological past.

Real and caused by physical factors

5. Increased endemicity (provinciality), as a result-of global changes in topography or climate, such as the break-up of Pangaea, changing patterns of ocean circulation, and global climatic gradients (Valentice and Moores 1972; Valentine *et al.* 1978; Valentine 1980; Benton 1985*b*).

6. Fragmentation of physical habitats, by changes in topography or climatic zones, for example, and associated fragmentation of faunas.

7. Continuous change in the physical environment maintains a level of biotic turnover and diversification—'lithospheric complexity' (geographic and climatic barriers to migration, global climatic gradients, topography) is in constant flux and stimulates constant diversification (Cracraft 1985).

Real and caused by biological factors

8. Increased endemicity, and increased speciation rate, as a result of genetic, physiological, or behavioural changes that reduced gene flow.

9. Reduction in the probability of extinction by:
   (a) optimization of fitness (Raup and Sepkoski 1982);
   (b) reduction in levels of diffuse competition (Van Valen 1984);

(c) increase in the species: family ratio through time (Flessa and Jablonski 1985);
(d) increase in mean extinction resistance as the set of families increases in mean age (Boyajian 1986).

10. Increase in the overall adaptive space occupied by a clade.

11. Subdivision of niches, so that later forms occupy narrower niches and have more specialized adaptations than their forebears.

12. The 'progressive' quality of lineages which leads to increased diversity by:
    (a) an internal motor of improvement;
    (b) escalation (Vermeij 1987), or a Red Queen model (Van Valen 1973), in which all lineages specialize in order to remain in evolutionary equilibrium;
    (c) 'stepping up' of the recovery time and scope following mass extinctions—surviving clades of each event are able to radiate faster and further than those surviving earlier events;
    (d) large-scale competition between 'evolutionary faunas' and attainment of ever-higher intrinsic equilibrium levels after each major replacement (Sepkoski 1984).

Real, but neutral in cause

13. 'Cladistic inevitability'—a clade must have one ancestor by definition and, if the clade survives for long, its diversity is more likely to increase than to remain constant at one (it cannot decrease, or the clade would have disappeared and would not be the subject of a study of diversification): a geometric argument.

14. Null, neutral, or stochastic, model—the increase in biotic diversity is the sum of untold numbers of microevolutionary processes acting on all species, and has nothing to do with grand unifying processes (Hoffman 1986, 1989).

2. *Critique*

I shall survey the 14 proposed explanations for the increase in the diversity of life in order, and select those for testing that seem most likely to have played a significant part.

The suggestion (Raup 1972) that much of the pattern is an artefact of the incompleteness of the fossil record, and of our biased methods of palaeontological study (points 1–4), has been widely discounted. The sources of error are not denied, but they do not seem to be the main determinants of the pattern established from the fossil record. Valentine (1973) argued that many of the sources of error identified by Raup (1972) might indeed apply at the species level, but that these biases would be greatly reduced when genera or families were used, the normal practical analytical level. In other words, one species record, out of the dozens possible, is enough to establish the presence of a family in rocks of a particular age. Analyses at these higher taxic levels are robust enough to withstand a rather poor-quality fossil record without being seriously distorted.

Two further specific tests were applied to the data of Valentine (1969) and Raup (1972) in order to determine who was correct. Sepkoski *et al.* (1981) compared global diversity curves based on several sets of data that are, to some extent, independent of each other:

(1) marine trace fossil species (burrows, tracks, and trails);

(2) invertebrate species per million years, based on sampling from the *Zoological Record*;

(3) species richness in well-preserved marine benthic communities;

(4) generic diversity from compilations in the *Treatise on invertebrate paleontology*;

(5) family diversity of marine Metazoa.

These data sets all yielded very similar patterns and confirmed the general correctness of Valentine's (1969) model. All five graphs showed low levels at the start of the Cambrian, a higher 'plateau' during the rest of the Palaeozoic, a sharp drop in the Late Permian, and a rising curve thereafter. Signor (1982) came to a similar conclusion by a different route. He made estimates of the various sources of error noted by Raup (1972) and others, and removed them as far as possible, by computer modelling, from the data. He found that the empirical model still held up after correction, and that this was true for orders, families, genera, and species.

Of the abiotic factors, increasing endemicity caused by continental drift, changes in patterns of ocean circulation, or global climatic change (point 5) are possible factors and they are testable (see below), but habitat fragmentation on a smaller scale (point 6), although testable, is unlikely to be a major factor in overall global biotic diversification. It is hard to envisage any form of habitat fragmentation that would have acted in a single direction in a relatively constant way through Phanerozoic time, and that would have affected the ecologically diverse tetrapods in similar ways. Cracraft's (1985) suggestion of the constant flux in the physical environment as a stimulus for biotic diversification (point 7) deliberately minimizes the role of biological interactions, and it is hard to see how it could be quantified and tested.

Of the biotic factors, increased endemicity caused by genetic or physiological change (point 8) would probably be impossible to test in fossil taxa, and it suffers from the same criticism as point 6: although it could be a factor that would produce local increases in diversity, it is hard to see how such a phenomenon could produce sustained unidirectional change on a global and long-term scale.

Reduced probability of extinction of families (point 9) is an empirical observation from the fossil record of marine animals (Raup and Sepkoski 1982; Van Valen 1984, 1985). Boyajian's (1986) model in which reduced extinction rates are tied to a net increase through time of extinction-resistant families, and Flessa and Jablonski's (1985) explanation in terms of increase in the species: family ratio through time have to be considered before the other two, rather vague, biological explanations given by Raup and Sepkoski (1982) and Van Valen (1984). Both of these imply a progressive and competitive element to the history of life, which has been disputed (see below). However, a declining probability of extinction is not true of all fossil records, since Benton (1985*a*) found a *rising* trend in total extinction rates and a constant probability of extinction for non-marine tetrapods, and Niklas *et al.* (1983) reported a complex pattern of changing extinction rates for vascular land plants: declining within particular major plant groups, but increasing from group to group through time. The suggestion that the probability of extinction has declined through time must be further tested for all groups of plants and animals, and the significance of Flessa and Jablonski's (1985) and Boyajian's (1986) models must be assessed. The explanation of increased biotic diversification by a reduced probability of extinction will not be considered further here for the reasons noted, and because it is not necessarily in itself an explanation, but merely another way of saying the same thing. It does not explain, for example, the magnitude of biotic diversification, nor the appearance of new body plans and new modes of life.

Increases in overall adaptive space occupied (point 10) and subdivision of niches (point 11) are possible explanations of increasing diversity, and they will be further discussed and tested below.

The 'progressive' models for diversification (point 12) all suffer from at least some of the problems of the 'Whig interpretation of history' (Schopf 1979)—they can only be identified as progressive *after* the event, and the explanations of improvement given by evolutionists may have no relation to what was really going on at the time—many of the models tread perilously close to teleology. Gould (1985), Benton (1987*a*), and others have argued that there is very little evidence for progress in a law-like or broad way in the history of life, and that the concept of progress does not sit comfortably in modern evolutionary theory. Four models of a progressive quality of lineages have been noted above (point 12, a–d), of which (a) is pure teleology and would not now be accepted by evolutionists. Progress by escalation, or by the Red Queen model (b) both suffer from the problems of scale: it has yet to be demonstrated how either model could act in a global long-term way that would set it above the vicissitudes of large environmental changes and extinction events. As Vermeij (1989) notes, biotic factors may be significant at low taxic levels, but physical factors predominate in affecting the biosphere globally. Explanations (c) and (d), improvements in recovery rates after extinction events and large-scale competition, are both assumptions rather than independently supported theories. They take the form 'group A radiated faster than group B and replaced it, so group A must have been competitively superior'. They are *post hoc* explanations, they involve beliefs that certain clades are overwhelmingly better than others, and that interclade competition is possible globally and over millions of years, and they suffer from all of the other problems of such scenarios outlined by Benton (1987*a*): confusion of pattern and mechanism; oversimplification; lack of proof; and incorrent scaling of concepts. Indeed, Erwin *et al.* (1987) gave evidence that might contradict explanation (c); the early Palaeozoic radiation of marine metazoans went further in terms of morphological innovation and per-taxon rate of family appearances than did the Mesozoic radiation. I suggest that the 'progressive' explanations (point 12) are either teleological or statements of faith, and they are untestable and will not be considered further.

The final group of models, the neutral or null models, have a certain attraction for palaeobiologists who believe that the antidote to competitive macroevolution is stochastic macroevolution. Cladistic inevitability (point 13) must take a part in all patterns of diversification, and stochastic branching patterns can apparently produce 'clades' that match the empirical pattern (Gould *et al.* 1977). However, Stanley *et al.* (1981) argued that the model was incorrectly scaled and that a truly null model of stochastic species splitting did not match reality. Proper account had to be taken of times of varying origination and extinction rates, and of the intervention of particular biotic and physical factors. Random splitting may explain parts of a pattern of diversification, but by no means the bulk of it. It is unclear also whether such a model can explain the overall diversification of the biosphere, and the appearance of major new body plans.

The second neutral explanation (point 14), that of Hoffman (1986, 1989) and Hoffman and Fenster (1986), suffers from similar problems. These authors had to maintain the rate of origination higher than the rate of extinction permanently in order to produced a realistic curve. A straightforward random walk model with no such constraint would be as likely to decline as to increase. In addition, they inserted and removed high numbers of clades at times equivalent to the Cambro-Ordovician radiation and the Late Permian extinction.

This leaves three explanations of diversification for testing: increasing endemicity (point 5); increase in adaptive space occupied (point 10); and subdivision of niches (point 11). These are all testable, and some preliminary analyses (Benton and Blacker, manuscript) suggest that all three factors have played a role in the diversification of the tetrapods — especially the last two.

Tests of postulated causes of diversification

1. *The data and methods*

The data used in this study were based on the fossil record of marine and non-marine tetrapods. This group was chosen since it is a well-established clade, much of its phylogeny is relatively well worked out by cladistic analysis (e.g. Benton 1988*b*), it is known over a long span of time (the past 400 Ma, since the Late Devonian), it is ecologically diverse, and ecologies of extinct and living forms can be discriminated into many categories (Benton 1988*a*, 1989*a, b*). The data on family names and stratigraphic durations are given by Benton (1987*b*, 1988*a*), and these were updated for the purposes of the present study. Data on basic ecological divisions, body size, and geographic distribution were added to the data base for this study, and a second parallel data base on well-preserved faunas (fossil Lagerstätten) was also constructed (Benton and Blacker, manuscript).

2. *The role of physical change and endemicity*

It was expected that the break-up of Pangaea after the Triassic would have led to increased endemicity of terrestrial tetrapod families at least. Earlier, Benton (1985*b*) established this in a preliminary study of Mesozoic tetrapods: whereas Triassic families had nearly world-wide distributions, those that arose in the Cretaceous were generally restricted to a single modern continental area. The present study, based on a larger and updated data set, did not support this finding (Fig. 18.4). All the major changes in levels of continental endemicity

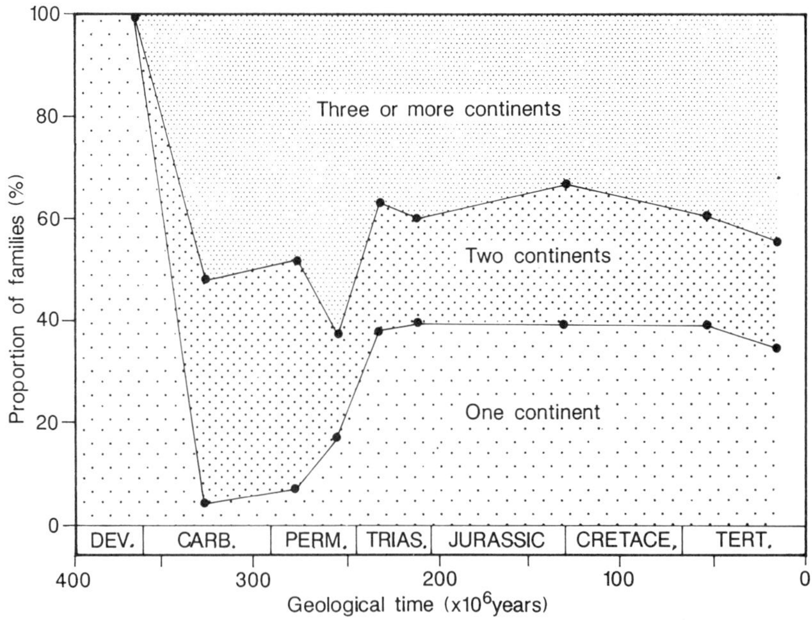

Fig. 18.4. Variations in the ultimate geographic distribution of non-marine tetrapod families that arose during each time interval. Family data were recorded by stratigraphic stage, and were then grouped into broad time intervals in order to provide large enough samples throughout ($n = 44$–360; mean = 146 families per time interval).

took place in the Carboniferous and Permian when tetrapod families became more restricted in their geographic distribution, possibly as a result of increasing north–south climatic differentiation. From the Triassic to the present, about 35 per cent of families spread to become nearly global in distribution (present in three or more continents), about 25 per cent spread to two continents, and about 40 per cent were endemic to one continent for their entire existence.

These results are no doubt affected by a number of factors that may require further consideration before a definitive test can be made. First, the left-hand side of the graph, in the Devonian and Carboniferous in particular, is based on relatively small numbers of families (1–30 families present per stratigraphic stage). In addition, this older fossil record may give poorer distributional data than later fossil records, and may underestimate the numbers of continents ultimately reached by each family. In addition, the ultimate distribution of a family may depend on its stratigraphic duration, and durations that are underestimated either by gaps in the fossil record, or by the cut-off line of the present day, may underestimate the final number of continents that members of a particular family will reach. Of course, the effect of the 'pull of the recent', noted above, will also distort records of familial distributions in more recent parts of the Cainozoic in the opposite direction, by expanding their distributions relative to families known only as fossils.

Valentine and co-workers (1978; Valentine 1980) developed a model of marine bioprovinces through time that depended upon continental drift, patterns of ocean circulation, and global climatic gradients. They argued that the diversification of marine life was controlled by the numbers of available marine bioprovinces which were determined by purely physical factors of the environment. Bakker (1977) provided a comparable model for diversification and extinction of terrestrial vertebrates.

3. *The role of increases in adaptive space*

Increases in the breadth of the adaptive space occupied by a clade would seem to be an obvious candidate for an explanation of diversification. In the case of the tetrapods, the earliest representatives, the two or three species of ichthyostegids and acanthostegids of the Late Devonian (Famennian) were semi-aquatic piscivores that lived in and close to fresh waters. During the Carboniferous and Permian, many lineages of tetrapods became more fully terrestrial in habits, and various gliding and flying forms appeared in the Permian and Triassic. Fully marine forms rose in the Permian (mesosaurs), Triassic (ichthyosaurs, nothosaurs, placodonts), Jurassic (crocodilians, plesiosaurs), Cretaceous (mosasaurs), Eocene (whales), and Oligocene (seals). Fliers became more diverse after the evolution of birds in the Jurassic and bats in the Eocene. These qualitative observations translate into quantitative data (Fig. 18.5(a)). The proportion of families remaining in the ancestral freshwater habitat declined from 100 per cent to a steady 10–15 per cent after the Late Permian. Tetrapods expanded first into terrestrial habitats in the Carboniferous and Permian, into flying, arboreal, and marine habitats in the Permian, and into burrowing habitats in the Jurassic. Occupancy of all of these additional zones of adaptive space has varied in proportion through time, and also in the precise nature of the occupants. Once conquered, however, it seems that these additional adaptive zones have been reconquered successively

---

or the majority of, species within a family (full details in Benton and Blacker, manuscript). Families were grouped into broader time units than stages, as in Fig. 18.4.

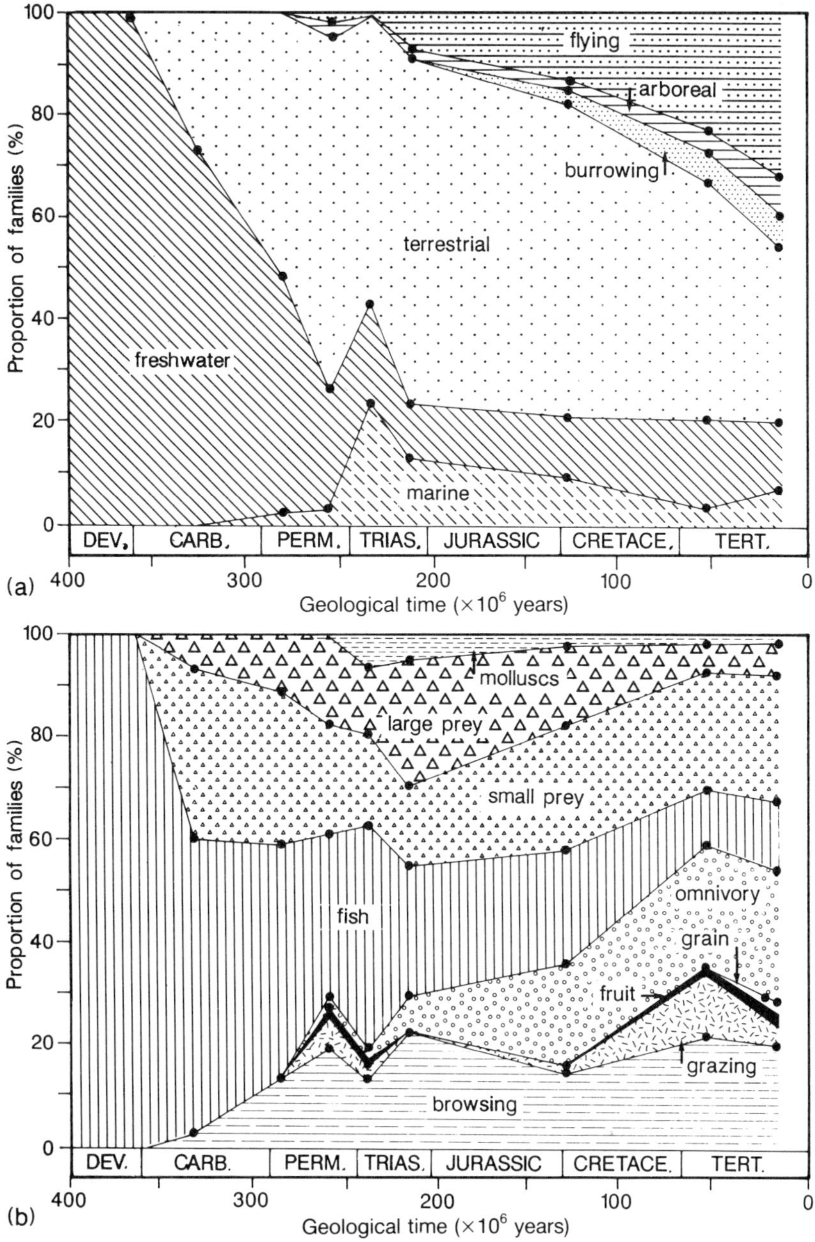

Fig. 18.5. Proportions of (a) broad habitat types and (b) diets of terrestrial and maritime tetrapod families through time. The habitats and diets were determined from the primary literature, and they represent the activities of all,

and independently, and a tetrapod presence has been maintained for all but relatively short intervals following major extinction events.

The expansion of adaptive space can also be documented in terms of diets (Fig. 18.5(b)). Fish-eating has fallen from 100 per cent dominance of tetrapod diets in the Late Devonian and Early Carboniferous to about 15 per cent in the Tertiary. Diets broadened to include insectivory and carnivory in the Carboniferous, broadly adapted browsing herbivory in the Permian, omnivory after the Late Permian, and ever-more specialized herbivorous or carnivorous modes after that (e.g. grazing, fruit, grain, molluscs).

These analyses suffer from some of the problems noted above for the study of endemicity, and especially from the fact that the earlier fossil record had yielded a relatively low number of tetrapod families. In addition, it is probably harder to determine the diets and habitat preferences of early tetrapods that have no close living analogues than those of Tertiary tetrapods. However, many of the usual problems of the fossil record are avoided in a study of this kind since the palaeobiologist needs only one or two well-preserved taxa in an extinct family in order to study their ecology and biology, and there are exquisitely preserved Devonian and Carboniferous tetrapods, just as there are excellent complete skeletons in the Tertiary.

Expansions of adaptive space have been proposed as a cause of the diversification of marine invertebrates (e.g. Bambach 1977, 1985; Ausich and Bottjer 1985) and of vascular land plants (e.g. Niklas *et al.* 1980). In the first case, marine invertebrates adopted entirely new life styles after the Cambrian, such as pelagic carnivory (cephalopods, vertebrates), benthic mobile suspension feeding (bivalves), carnivory (cephalopods, eurypterids, crabs, starfishes), deep infaunal deposit-feeding (bivalves), and shallow passive infaunal suspension feeding (bivalves, rostroconchs). New habitats conquered in the Mesozoic included pelagic herbivory (vertebrates), shallow infaunal passive deposit feeding and carnivory (bivalves), deep infaunal suspension feeding (bivalves, polychaetes, crabs), and deep infaunal active carnivory (polychaetes). Ausich and Bottjer (1985) focus on the expansion in adaptive space occupied by benthic marine invertebrates in terms of increases in tiering, that is, the vertical division of space above and below the sediment–water interface. The invention of invertebrate ‘trees’, such as long-stemmed crinoids, allowed these organisms, and a host of others that lived upon them, to occupy and feed in water up to 1 m from the sediment level from Silurian times onwards. In the other direction, the ability to burrow allowed invertebrates to occupy progressively deeper levels in the sediment, from maximum depths of 60 mm in the early Palaeozoic, to 120 mm in the Early

Devonian (bivalves), and 1 m from the Early Permian onwards (bivalves, arthropods). Similarly, in a comparison of the early Palaeozoic and the Mesozoic radiations of marine metazoans, Erwin *et al.* (1987) argued that filling of empty adaptive space was the main force behind diversification.

For vascular land plants, major jumps in diversity are linked to the evolution of heterospory and the seed habit, the origin of trees, the herbaceous habit in angiosperms, and other shifts into new adaptive space (Niklas *et al.* 1980).

4. *The role of subdivision of niches*

The story of the evolution of many clades includes as much specialization, or niche subdivision, as it does expansion of adaptive space. For example, the early tetrapods had generalized dentitions, and it is assumed that many of them could have eaten a mixed diet of fish, insects, worms, small tetrapods, and even fruit and other nutritive plant parts. There is little evidence for highly specialized diets and other habits as are seen today in many groups. This argument is far too weak, however, without some firmer quantitative evidence.

An attempt has been made to test the idea of niche subdivision by comparing typical faunas through time, in a test analogous to that applied by Bambach (1977, 1985) to marine faunas. It is valid to compare an individual exceptionally-preserved fauna (Lagerstätte) from the Carboniferous with one from the Eocene, say, since levels of preservation appear to be equivalent. In other words, Lagerstätten of all ages can show preservation of tiny animals, soft parts such as skin, scales, and hair, and wholly soft-bodied organisms. A measure of the subdivision of niches can be obtained by counting the numbers of species in comparable faunas, that is, groups of tetrapods found in a single deposit in a single locality or small sedimentary basin. A preliminary survey of 100 such faunas, spanning the past 350 Ma in Europe and North America (Benton and Blacker, manuscript) shows a marked increase in mean faunal diversities from 18 in the Carboniferous to 35–51 in the Miocene and Pliocene (Fig. 18.6). The maximum figures rose from 25 in the Carboniferous to 80–100 in the Late Cretaceous and Tertiary.

This test requires refinement since a part of the increase in mean and maximum faunal diversities is almost certainly the result of the conquest of new adaptive space. Tertiary faunas typically include arboreal mammals and birds, for example. However, a preliminary survey on the present data set of the relative influences of niche subdivision and adaptive space expansion suggests that most of the increase in faunal diversity in caused by the former factor.

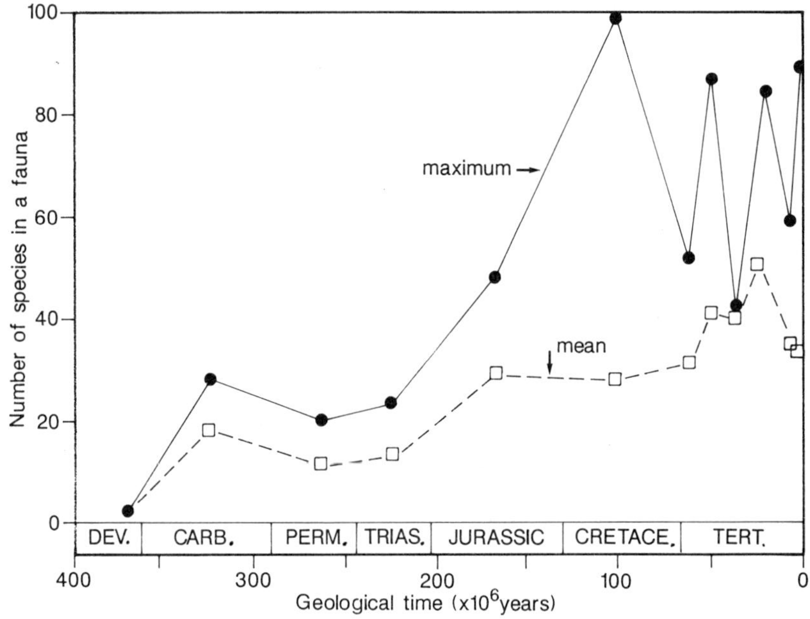

Fig. 18.6. Maximum and minimum mean numbers of tetrapod species in well-preserved terrestrial faunas. This graph is based on a sample of 100 faunas (details in Benton and Blacker, manuscript), with 1–20 faunas sampled per stratigraphic period (or epoch in the Cainozoic). Each 'fauna' is the species list of a single quarry or restricted sedimentary basin. Note the overall rise, with fluctuations in the maximum curve probably largely the result of variations in preservation quality.

Conclusion

What are the relative roles of major physical change, adaptive space expansion, and specialization in producing diversification? Cracraft (1985) has argued that most macroevolutionary patterns can be tied to changes in the physical environment. However, the present study suggests that certain long-term physical changes that might have been expected to produce major effects (e.g. continental drift, climatic differentiation) are not adequate agents of diversification, for tetrapods at least. Doubt has also been cast on the role of continental drift and sea-level change in the history of marine invertebrates. Amalgamations of continents, such as occurred during the Permian, may indeed produce reductions in the circumcontinental shelf seas and global marine regressions (Valentine and Moores 1972), but the area of shallow water around oceanic islands increases. Nearly 90 per cent of marine families today are represented around oceanic islands, and these would survive a major regression (Jablonski and Flessa 1986). The global effects of continental drift and marine regression on terrestrial faunas, argued eloquently by Bakker (1977), are even harder to confirm.

The relative roles of increased adaptive space and increased specialization in producing diversification are hard to determine, but the former may have had the main influence in tetrapod radiations. Bambach (1977, 1985) also found evidence that marine invertebrates have increased their diversity mainly by expansions of utilized adaptive space. Sepkoski (1988) has tested the relationship between increases in within-community, or alpha, diversity and between-region, or gamma, diversity. He found that the increases in alpha diversity detected by Bambach (1977, 1985) in offshore-shelf marine communities of 50–70 per cent during the Ordovician were insufficient to account for the total global increases in generic diversity of 300 per cent. The majority of the increase was probably explained by increases in between-community, or beta, diversity, in other words, by increases in the numbers of broad habitat types occupied, a conclusion that appears to correspond to the findings from the present study of terrestrial tetrapods.

Certain aspects of diversification patterns may of course be 'one-off' phenomena, dependent on specific historical circumstances, such as the break-up of Pangaea, latitudinal climatic diversification, the diversification of fishes (hence providing new adaptive space for predators), and other environmental changes. However, biological and ecological factors, such as increases in the overall adaptive space occupied by a clade, and subdivision of niches, seem to lie behind the long-term patterns of net increasing diversity. There is certainly no evidence for an inevitable motor of change that drives diversity ever upwards.

Acknowledgements

I thank Anne Blacker for compiling much of the data used in this study, and the Leverhulme Trust for funding this part of the work. Mrs Pat Baldero drafted the diagrams. I thank Richard Fortey, David Jablonski, Philip Signor, and Paul Taylor for useful critical comments.

References

Ausich, W.I. and Bottjer, D.J. (1985). Phanerozoic tiering in suspension-feeding communities on soft substrata: implications for diversity. In *Phanerozoic diversity patterns; profiles in macroevolution* (ed. J.W. Valentine), pp. 255–74. Princeton University Press, Princeton, New Jersey.

Bakker, R.T. (1977). Tetrapod mass extinctions — a model of the regulation of speciation rates and immigration by cycles of topographic diversity. In *Patterns of evolution, as illustrated by the fossil record* (ed. A. Hallam), pp. 439–68. Elsevier, Amsterdam.

Bambach, R.K. (1977). Species richness in marine benthic habitats through the Phanerozoic. *Paleobiology* **3,** 152–67.

Bambach, R.K. (1985). Classes and adaptive variety: the ecology of diversification in marine faunas through the Phanerozoic. In *Phanerozoic diversity patterns; profiles of macroevolution* (ed. J.W. Valentine), pp. 191–253. Princeton University Press, Princeton, New Jersey.

Benton, M.J. (1985*a*). Mass extinctions among non-marine tetrapods. *Nature, London*, **321,** 857–61.

Benton, M.J. (1985*b*). Patterns in the diversification of Mesozoic non-marine tetrapods and problems in historical diversity analysis. *Special Papers in Palaeontology*, **33,** 185–202.

Benton, M.J. (1987*a*). Progress and competition in macroevolution. *Biological Reviews*, **62,** 305–38.

Benton, M.J. (1987*b*). Mass extinctions among families of non-marine tetrapods: the data. *Mémoires de la Société géologique de la France*, **150,** 21–32.

Benton, M.J. (1988*a*). Mass extinctions and the fossil record of reptiles: paraphyly, patchiness and periodicity. In *Extinction and survival in the fossil record*, Systematics Association Special Volume, no. 34 (ed. G.P. Larwood), pp. 269–94. Clarendon Press, Oxford.

Benton, M.J. (ed.) (1988*b*). *The phylogeny and classification of the tetrapods*, Systematics Association Special Volume, no. **35A,B**. Clarendon Press, Oxford.

Benton, M.J. (1989*a*). Patterns of evolution and extinction in vertebrates. In *Evolution and the fossil record* (ed. K.C. Allen and D.E.G. Briggs), pp. 218–41. Belhaven Press, London.

Benton. M.J. (1989*b*). Mass extinctions among tetrapods and the quality of the fossil record. *Philosophical Transactions of the Royal Society*, **B325,** 369–86.

Boyajian, G.F. (1986). Phanerozoic trends in background extinction: consequence of an aging fauna. *Geology*, **14,** 955–8.

Cracraft, J. (1985). Biological diversification and its causes. *Annals of the Missouri Botanical Garden*, **72,** 794–822.

Diamond, J.M. and Case, T.J. (1986). *Community ecology*. Harper and Row, New York.

Erwin, D.H., Valentine, J.W. and Sepkoski, J.J., Jr (1987). A comparative study of diversification events: the early Paleozoic versus the Mesozoic. *Evolution*, **41,** 1177–86.

Flessa, K.J. and Jablonski, D. (1985). Declining Phanerozoic background extinction rates: effect of taxonomic structure? *Nature, London*, **313,** 216–18.

Gould, S.J. (1985). The paradox of the first tier: an agenda for paleobiology. *Paleobiology*, **11,** 2–12.

Gould, S.J. and Calloway, C.B. (1980). Clams and brachiopods — ships that pass in the night. *Paleobiology* **6,** 383–96.

Gould, S.J., Raup, D.M., Sepkoski, J.J. Jr and Simberloff, D.S. (1977). The shape of evolution: a comparison of real and random clades. *Paleobiology*, **3,** 23–40.

Harvey, P.H., Colwell, R.K., Silvertown, J.W. and May, R.M. (1983).

Null models in ecology. *Annual Reviews in Ecology and Systematics*, **14**, 189-211.

Hoffman, A. (1981). Stochastic versus deterministic approach to paleontology: the question of scaling or metaphysics? *Neues Jahrbuch für Geologie und Paläontologie, Abhandlungen* **162**, 80-96.

Hoffman, A. (1983). Paleobiology at the crossroads: a critique of some modern paleobiological research programs. In *Dimensions of Darwinism* (ed. M. Grene), pp. 241-71. Cambridge University Press.

Hoffman, A. (1985). Island biogeography and paleobiology: in search for evolutionary equilibria. *Biological Reviews*, **60**, 455-71.

Hoffman, A. (1986). Neutral model of Phanerozoic diversification: implications for macroevolution. *Neues Jahrbuch für Geologie und Paläontologie, Abhandlungen*, **172**, 219-44.

Hoffman, A. (1989). *Arguments on evolution: a paleontologist's perspective*. Oxford University Press, New York.

Hoffman, A. and Fenster, E.J. (1986). Randomness and diversification in the Phanerozoic: a simulation. *Palaeontology*, **29**, 655-63.

Jablonski, D. and Flessa, K.W. (1986). The taxonomic structure of shallow-water marine faunas: implications for Phanerozoic extinctions. *Malacologia*, **27**, 43-66.

Kitchell, J.A. and Carr, T.R. (1985). Nonequilibrium model of diversification: faunal turnover dynamics. In *Phanerozoic diversity patterns; profiles in macroevolution* (ed. J.W. Valentine), pp. 277-310. Princeton University Press, Princeton, New Jersey.

MacArthur, R.H. and Wilson, E.O. (1967). *The theory of island biogeography*. Princeton University Press, Princeton, New Jersey.

Niklas, K.J., Tiffney, B.H. and Knoll, A.H. (1980). Apparent changes in the diversity of fossil plants. *Evolutionary Biology*, **12**, 1-89.

Niklas, K.J., Tiffney, B.H. and Knoll, A.H. (1983). Patterns in vascular land plant diversification. *Nature, London*, **303**, 614-16.

Raup, D.M. (1972). Taxonomic diversity during the Phanerozoic. *Science*, **177**, 1065-71.

Raup, D.M. (1976*a*). Species diversity in the Phanerozoic: a tabulation. *Paleobiology*, **2**, 279-88.

Raup, D.M. (1976*b*). Species diversity in the Phanerozoic: an interpretation. *Paleobiology*, **2**, 289-97.

Raup, D.M. and Sepkoski, J.J. Jr (1982). Mass extinctions in the marine fossil record. *Science*, **215**, 1501-3.

Rieppel, O. (1984). The problem of extinction. *Zeitschrift für zoologisch Systematik und Evolutionsforschung*, **22**, 81-5.

Rosenzweig, M.L. (1975). On continental steady states of species diversity. In *Ecology and evolution of communities* (ed. M.L. Cody and J.M. Diamond), pp. 121-40. Belknap Press, Cambridge, Massachusetts.

Schopf, T.J.M. (1979). Evolving paleontological views on deterministic and stochastic approaches. *Paleobiology*, **5**, 337-52.

Sepkoski, J.J. Jr (1978). A kinetic model of Phanerozoic taxonomic diversity. I. Analysis of marine orders. *Paleobiology*, **4**, 223-51.

Sepkoski, J.J. Jr (1979). A kinetic model of Phanerozoic taxanomic diversity. II. Early Phanerozoic families and multiple equilibria. *Paleobiology*, **5**, 222–51.

Sepkoski, J.J. Jr (1981). A factor analytic description of the Phanerozoic marine fossil record. *Paleobiology*, **7**, 36–53.

Sepkoski, J.J. Jr (1984). A kinetic model of Phanerozoic taxonomic diversity. III. Post-Paleozoic families and mass extinctions. *Paleobiology*, **10**, 246–67.

Sepkoski, J.J. Jr (1988). Alpha, beta, and gamma: where does all the diversity go? *Paleobiology*, **14**, 221–34.

Sepkoski, J.J. Jr, Bambach, R.K., Raup, D.M. and Valentine, J.W. (1981). Phanerozoic marine diversity and the fossil record. *Nature, London*, **293**, 435–7.

Sheehan, P.M. (1977). Species diversity in the Phanerozoic, a reflection of labor by systematists? *Paleobiology*, **3**, 325–8.

Signor, P.W. III (1982). Species richness in the Phanerozoic: compensating for sampling bias. *Geology*, **10**, 625–8.

Stanley, S.M., Signor, P.W., Lidgard, S. and Karr, A.F. (1981). Natural clades differ from 'random' clades: simulations and analyses. *Paleobiology*, **7**, 115–27.

Strong, D.R., Simberloff, D., Abele, L.G. and Thistle, A.B. (ed.) (1984). *Ecological communities: conceptual issues and the evidence*. Princeton University Press, Princeton, New Jersey.

Valentine, J.W. (1969). Patterns of taxonomic and ecological structure of the shelf benthos during Phanerozoic time. *Palaeontology*, **12**, 684–709.

Valentine, J.W. (1973). Phanerozoic taxonomic diversity: a test of alternate models. *Science*, **180**, 1078–9.

Valentine, J.W. (1980). Determinants of diversity in higher taxonomic categories. *Paleobiology*, **6**, 444–50.

Valentine, J.W. and Moores, E.M. (1972). Global tectonics and the fossil record. *Journal of Geology*, **80**, 167–84.

Valentine, J.W., Foin, T.C. and Peart, D. (1978). A provincial model of Phanerozoic marine diversity. *Paleobiology*, **4**, 55–66.

Van Valen, L.M. (1973). A new evolutionary law. *Evolutionary Theory*, **1**, 1–30.

Van Valen, L.M. (1984). A resetting of Phanerozoic community evolution. *Nature, London*, **307**, 50–2.

Van Valen, L.M. (1985). How constant is extinction? *Evolutionary Theory*, **7**, 93–106.

Vermeij, G.J. (1987). *Evolution and escalation*. Princeton University Press, Princeton, New Jersey.

Vermeij, G.J. (1989). Evolution in the long run. *Paleobiology*, **15**, 199–203.

Walker, T.D. and Valentine, J.W. (1984). Equilibrium models of evolutionary species diversity and the number of empty niches. *American Naturalist*, **124**, 887–99.

Webb, S.D. (1969). Extinction-origination equilibria in late Cenozoic land mammals of North America. *Evolution*, **23**, 688–70.

Systematics Association Publications

1. Bibliography of key works for the identification of the British fauna and flora *3rd edition* (1967)
   Edited by G. J. Kerrich, R. D. Meikle and N. Tebble
2. Function and taxonomic importance (1959)
   Edited by A. J. Cain
3. The species concept in palaeontology (1956)
   Edited by P. C. Sylvester-Bradley
4. Taxonomy and geography (1962)
   Edited by D. Nichols
5. Speciation in the sea (1963)
   Edited by J. P. Harding and N. Tebble
6. Phenetic and phylogenetic classification (1964)
   Edited by V. H. Heywood and J. McNeill
7. Aspects of Tethyan biogeography (1967)
   Edited by C. G. Adams and D. V. Ager
8. The soil ecosystem (1969)
   Edited by H. Sheals
9. Organisms and continents through time (1973)†
   Edited by N. F. Hughes

Published by the Association (out of print)

Systematics Association Special Volumes

1. The new systematic (1940)
   Edited by Julian Huxley (Reprinted 1971)
2. Chematoxonomy and serotaxonomy (1968)*
   Edited by J. G. Hawkes
3. Data processing in biology and geology (1971)*
   Edited by J. L. Cutbill
   Out of print
4. Scanning electron microscopy (1971)*
   Edited by V. H. Heywood
   Out of print

5. Taxonomy and ecology (1973)*
Edited by V. H. Heywood
6. The changing flora and fauna of Britain (1974)*
Edited by D. L. Hawksworth
Out of print
7. Biological identification with computers (1975)*
Edited by R. J. Pankhurst
8. Lichenology: progress and problems (1976)*
Edited by D. H. Brown, D. L. Hawksworth, and R. H. Bailey
9. Key works to the fauna and flora of the British Isles and northwestern Europe, *4th Edition* (1978)*
Edited by G. J. Kerrich, D. L. Hawksworth, and R. W. Sims
10. Modern approaches to the taxonomy of red and brown algae (1978)*
Edited by D. E. G. Irvine and J. H. Price
11. Biology and systematics of colonial organisms (1979)*
Edited by G. Larwood and B. R. Rosen
12. The origin of major invertebrate groups (1979)*
Edited by M. R. House
13. Advances in bryozoology (1979)*
Edited by G. P. Larwood and M. B. Abbot
14. Bryophyte systematics (1979)*
Edited by G. C. S. Clarke and J. G. Duckett
15. The terrestrial environment and the origin of land vertebrates (1980)*
Edited by A. L. Panchen
16. Chemosystematics: principles and practice (1980)*
Edited by F. A. Bisby, J. G. Vaughan, and C. A. Wright
17. The shore environment: methods and ecosystems (2 Volumes) (1980)*
Edited by J. H. Price, D. E. G. Irvine and W. F. Farnham
18. The Ammonoidea (1981)*
Edited by M. R. House and J. R. Senior
19. Biosystematics of social insects (1981)*
Edited by P. E. Howse and J.-L. Clément
20. Genome evolution (1982)*
Edited by G. A. Dover and R. B. Flavell
21. Problems of phylogenetic reconstruction (1982)*
Edited by K. A. Joysey and A. E. Friday
22. Concepts in nematode systematics (1983)*
Edited by A. R. Stone, H. M. Platt, and L. F. Khalil
23. Evolution, time and space: the emergence of the biosphere (1983)*
Edited by R. W. Sims, J. H. Price, and P. E. S. Whalley
24. Protein polymorphism: adaptive and taxonomic significance (1983)*
Edited by G. S. Oxford and D. Rollinson
25. Current concepts in plant taxonomy (1983)*
Edited by V. H. Heywood and D. M. Moore
26. Databases in systematics (1984)*
Edited by R. Allkin and F. A. Bisby
27. Systematics of the green algae (1984)*
Edited by D. E. G. Irvine and D. M. John

28. The origins and relationships of lower invertebrates (1985)‡
Edited by S. Conway Morris, J. D. George, R. Gibson, and H. M. Platt
29. Infraspecific classification of wild and cultivated plants (1986)‡
Edited by B. T. Styles
30. Biomineralization in lower plants and animals (1986)‡
Edited by B. S. C. Leadbeater and R. Riding
31. Systematic and taxonomic approaches in palaeobotany (1986)‡
Edited by R. A. Spicer and B. A. Thomas
32. Coevolution and systematics (1986)‡
Edited by A. R. Stone and D. L. Hawksworth
33. Key works in the fauna and flora of the British Isles and north-western Europe, *5th edition* (1987)‡
Edited by R. W. Sims, P. Freeman, and D. L. Hawksworth
34. Extinction and survival in the fossil record (1987)‡
Edited by G. P. Larwood
35A. The phylogeny and classification of the tetrapods. Volume 1: Amphibians, reptiles, birds (1988)‡
Edited by M. J. Benton
35B. The phylogeny and classification of the tetrapods. Volume 2: Mammals (1988)‡
Edited by M. J. Benton
36. Prospects in systematics (1988)‡
Edited by D. L. Hawksworth
37. Biosystematics of haematophagous insects (1988)‡
Edited by M. W. Service
38. The chromophyte algae: problems and perspectives (1980)‡
Edited by J. C. Green, B. S. C. Leadbeater, and W. Diver
39. Electrophoretic studies of agricultural pests (1989)‡
Edited by H. D. Loxdale and J. den Hollander
40A. Evolution, systematics, and fossil history of the Hamamelidae. Volume 1: Introduction and 'lower' Hamamelidae (1989)‡
Edited by P. R. Crane and S. Blackmore
40B. Evolution, systematics, and fossil history of the Hamamelidae. Volume 2: 'Higher' Hamamelidae (1989)‡
Edited by P. R. Crane and S. Blackmore
41. Scanning electron microscopy in taxonomy and functional morphology (1990)
Edited by D. Claugher
42. Major evolutionary radictions (1990)
Edited by P. D. Taylor and G. P. Larwood

* Published by Academic Press for the Systematics Association
†Published by the Palaeontological Association in conjunction with the Systematics Association
‡Published by the Oxford University Press for the Systematics Association.

Subject and stratigraphical index